SOME USEFUL NUMBERS IN ASTRONOMY

Speed of light:	c	$= 2.998 \times 10^8$ m/s
Astronomical unit:	AU	$= 1.499 \times 10^{11}$ m
Light year:	LY	$= 9.46 \times 10^{15}$ m
		$= 6.32 \times 10^4$ AU
Acceleration of gravity:	a	$= 9.8$ m/s^2
Absolute temp. for 0°C:	T	$= 273$ K
Mass of Earth:	m	$= 6.0 \times 10^{24}$ kg
Diameter of Earth:	d	$= 1.3 \times 10^7$ m
Age of Earth:	t	$= 4.5 \times 10^9$ yr
Mass of Sun:	M	$= 2.0 \times 10^{30}$ kg
Diameter of Sun:	D	$= 1.4 \times 10^9$ m
Luminosity of Sun:	L	$= 4.0 \times 10^{26}$ watts
Hubble constant:	H	$= 25$ km/s per 10^6 LY
Age of "empty" universe:	$1/H$	$= 13 \times 10^9$ yr
Critical density of universe:	ρ	$= 10^{-29}$ g/cm^3

ABELL'S

EXPLORATION

OF THE

UNIVERSE

SEVENTH EDITION

David Morrison
**Chief, Space Science Division
NASA Ames Research Center**

Sidney Wolff
**Director
National Optical Astronomy Observatories**

Andrew Fraknoi
**Chair, Astronomy Department
Foothill College**

Saunders Golden Sunburst Series
SAUNDERS COLLEGE PUBLISHING
HARCOURT BRACE JOVANOVICH COLLEGE PUBLISHERS
Philadelphia   Fort Worth   Chicago   San Francisco
Montreal   Toronto   London   Sydney   Tokyo

Text Typeface: Bembo and Helvetica
Compositor: General Graphic Services, Inc.
Publisher: John Vondeling
Developmental Editor: Jennifer Bortel
Managing Editor: Carol Field
Project Editor: Anne Gibby
Copy Editor: York Production Services
Manager of Art and Design: Carol Bleistine
Text Designer: CIRCA 86
Text Artwork: J & R Studio
Layout Artist: Carmen DiBartolomeo/CIRCA 86
Director of EDP: Tim Frelick
Sr. Production Manager: Charlene Squibb
Marketing Manager: Marjorie Waldron

About the cover: Hubble Space Telescope Johnson Space Center.

Printed in the United States of America

EXPLORATION OF THE UNIVERSE, Seventh Edition

ISBN 0-03-001034-9

Library of Congress Catalog Card Number: 94-38828

2345   032   987654321

PREFACE FOR THE STUDENT

In college textbooks, there is a long tradition that the preface of the book is written to be read by the instructor and the rest of the book by the student. The preface for each new edition, for example, has a listing of all the sections of the book that have been improved. We are sure that you as a student are hoping to use only the one edition of the book that you are presently reading, so you probably don't care very much how this edition differs from the last one. Still, many students wind up reading the preface (it does come first) and wondering why it doesn't say much to them.

So, we have decided to begin with a preface to our student readers. It's not a preface about the subject matter of astronomy, because that is introduced in the Prologue. Instead, we want to tell you a little about the book and give you some hints for the effective study of astronomy. (Your professor will probably have other, more specific suggestions for doing well in your class.)

Astronomy, the study of the universe beyond the confines of our planet, is one of the most exciting and rapidly changing branches of science. Even scientists from other fields, when their colleagues aren't listening, often confess to having a lifelong interest in astronomy, although they wound up doing biology, chemistry, or engineering. There are fewer than 10,000 professional astronomers in the entire world, but astronomy has a much larger group of *amateur astronomers* whose hobby is observing the sky, who spend many evenings with a telescope under the stars, and who occasionally make a discovery, such as a new comet or exploding star. And, many people in all walks of life are fascinated just to read about such bizarre objects as black holes and quasars that astronomers are uncovering. Others are intrigued by the scientific search for planets or life in other star systems. And quite a few people like to follow the results of space exploration, such as the recent repair of the Hubble Space Telescope by the Shuttle astronauts or the Galileo mission to probe the giant planet Jupiter. Hearing about some astronomical event in the news media may be what first sparked your interest in taking this course.

But some of the things that make astronomy interesting can also make it a challenge for the beginner. The universe, as you probably know, is a big place, and full of objects and processes that do not necessarily have familiar counterparts on Earth. Like a visitor to a new country, it may take you a while to feel familiar with the territory or the local customs. Astronomy has its own specialized vocabulary, with which you will need to get acquainted. And the pace of discovery in astronomy never lets up, making the task of keeping up with it all a nontrivial enterprise.

To assist students taking their first college-level course in astronomy, we have built a number of special features into this book, and we invite you to make use of them:

All technical terms are printed in **boldface** type the first time they are used and clearly defined in the text; the definitions are summarized in Appendix 2 (the glossary) in alphabetical order so that you can refer to them at any time.

We have taken care to provide the background information from other sciences that you will need to appreciate astronomy fully. Chapter 7, for example, provides an extensive introduction to the nature of light and atoms and should be helpful if you have not had this in a previous course.

The book begins with a historical summary of astronomy, then surveys the universe, starting at home and finishing with the properties of the entire cosmos. Don't worry if your instructor (like many who use this book) doesn't assign all the chapters or doesn't assign the chapters in order. Throughout the book we have put "directional signs" leading you to earlier material you will need to know before tackling the current section.

Topics that are more mathematical or technical and sections that are not essential to following the main thread of the discussion are set off with colored bars and set in a different (and slightly smaller) typeface. Your instructor will tell you which of these sections you are expected to study and which can be omitted.

We have used tables throughout the book to bring together numerical data for your convenience. For example, there are boxes summarizing the important properties of each planet in the solar system, including the Earth. For students who want to see more of the data that astronomers use, there are extensive appendices at the back of the book, giving you the latest information on many aspects of astronomy.

Each chapter concludes with a summary of the essential points in the chapter and thought questions and problems to help you review and apply what you have learned. Problems marked with an asterisk are more challenging.

We conclude this Preface with a few suggestions for studying astronomy, which come from good teachers and good students from around the country:

First and foremost, the best advice we can give you is to be sure to leave enough time in your schedule to study the material in this class *regularly*. It sounds obvious, but many college students include too many activities in their schedules and lives and find themselves falling behind in some classes. No matter what anyone tells you, it is not easy to catch up with a subject like astronomy by trying to do everything just before an exam. Try to put aside some part of each day or every other day when you can have uninterrupted time for reading and studying astronomy.

Try to read each assignment in the book twice, once before it is discussed in class and once afterward. Take notes or use an outliner to highlight ideas that you may want to review later. (Also, take some time to coordinate the notes from your reading with the notes you take in class.) Many students today arrive in college without good note-taking habits. If you are not as good a note-taker as you would like to be, get some assistance. Many colleges and universities have a student learning center that offers short courses, workbooks, or videos on developing good study habits. Take time to find out what your school offers.

Form a small astronomy study group with people in your class. Get together as often as you can and discuss the topics that may be giving group members some trouble. Make up sample exam questions, and be sure everyone in the group can answer them confidently. (If you have always studied alone, you may at first resist this idea, but don't be too hasty in your judgment. Several studies have shown that a group whose members participate equally can be a very effective way of studying subjects with a lot of new information, such as a foreign language, law, or astronomy.)

Before each exam, make a concise outline of all the main ideas in class discussion and your reading. You can compare your outline with those of other students as a check on your own studying.

If you find a particular topic in the text or class especially difficult or interesting, don't hesitate to make use of the resources in your library for additional study. Appendix 1 lists some of the best introductory magazines and books in astronomy that students may find useful.

Don't be unnecessarily hard on yourself! If you are new to astronomy, many of the ideas and terms in this book will be new to you. As with any new language, it may take a while for you to be a good conversationalist in astronomy. Practice as much as you can, but also realize that it is natural to be occasionally overwhelmed by the vastness of the universe and the variety of things that are going on in it.

We hope you will enjoy reading this text as much as we enjoyed writing it. We are always glad to hear from students who have used the text, and we invite you to send us your reactions to the book and suggestions for how we can improve future editions. You can send your comments to: Andrew Fraknoi, Astronomy Department, Foothill College, 12345 El Monte Rd., Los Altos Hills, CA 94022, USA. We will not send you the answers to the chapter problems or do your homework for you, but all other thoughts are welcome.

David Morrison, Sidney Wolff, and Andrew Fraknoi
September 1994

PREFACE FOR THE INSTRUCTOR

This is the 7th edition of *Exploration of the Universe* since the late George Abell introduced the text in 1964. Our objectives remain the same: We want to present a comprehensive, up-to-date summary of the facts of astronomy, but that by itself is not enough. We also want to provide an introduction to the rational exploration of nature and give the student a clear view of the history and character of science and scientific thinking.

At the same time that astronomers are making remarkable strides in understanding the universe, many students in introductory classes seem less and less prepared for taking college-level science. While there are exceptions, students (especially those not majoring in science or engineering) seem to have increasingly limited backgrounds in science and little preparation for the sort of problem solving and skeptical thinking a science course will demand of them. Although we may rage against this trend, its reality cannot be denied, and many instructors have had to make provisions in their astronomy courses to deal with the resulting problems. In the current revision we have tried to assist in these efforts by paying even greater attention to pedagogical concerns, without reducing the authoritative coverage of astronomical topics for which the book has received praise over the years.

NEW FEATURES OF THIS EDITION

With this edition we add a new co-author, Andrew Fraknoi, who has taught introductory astronomy classes for 22 years and has developed a wide range of educational materials for the Astronomical Society of the Pacific. As a result, we have revised a number of the features of the book to help students learn astronomy more effectively:

A new Preface for the Student suggests ways to develop better study habits in astronomy.

An expanded Prologue introduces the study and ideas of astronomy more thoroughly for students whose background in science may not be strong.

New chapter opener pages highlight important ideas for each chapter and tie chapters together more effectively.

Several parts of the text have been re-edited for clarity and teaching effectiveness.

Captions for figures throughout the book have been expanded and clarified.

A new Epilogue ties together the grand threads that were developed throughout the book.

Astronomy itself has changed dramatically since 1964, and the new edition of the text is designed to keep pace with the rapid increase in information and the many new ideas that are the result. This new edition features updated and rewritten sections on asteroids, comets, and collisions in the solar system, the galactic center, galaxy clusters and the large-scale structure of the universe, and cosmology, among others. New images from ground-based and space instruments are used to highlight and clarify the points made in the text. There is coverage of such recent developments as the repair of the Hubble Space Telescope, the potential discovery of a small galaxy closer than the Magellanic Clouds, and the collision of Comet Shoemaker-Levy 9 with Jupiter.

ANCILLARIES

A complete teaching package supports this text:

The **Instructor's Manual** has been significantly expanded and rewritten and now comes in two parts. Part One includes answers to end of chapter questions, while Part Two is an extensive resource guide, listing a wide range of teaching ideas and written and audio-visual materials for both the beginning and the veteran college teacher. (For example, you will find additional reading suggestions for many topics in astronomy, drawn from recent books and magazines, as well as suggestions for interesting term paper topics.)

A **Test Bank** by Harry Robinson of Bryant College contains true-false, multiple-choice, and fill-in-the-blank questions for tests and quizzes. A **Computerized Test Bank** is available in Macintosh and IBM (both DOS and Windows) format.

The **Saunders Astronomy Transparency Collection** is a comprehensive set of 205 color transparency acetates of conceptually based artwork for use with an overhead projector. Over half are enlarged reproductions of figures from the textbook. The others are supplemental illustrations that complement the text figures. A detailed guide, arranged by topic, accompanies the collection. This collection is also available in a **35-mm slide** format.

We have added the **Saunders Astronomy Transparency Collection Supplement,** a set of 25 new images, many from the Hubble Space Telescope and major observatories.

Also new is the **Saunders Astronomy Videodisc,** Volume I: The Solar System. The two-sided CAV disc contains one hour of up-to-date still images of the solar system (including images contained in the Transparency Collection), animations, and live video and is free to adopters of the textbook. A laserdisc player bar code guide linking the images to the text is also available.

LectureActive™ Software helps instructors customize lectures by giving them quick, efficient access to the video clip and still frame data on the Saunders Astronomy Videodisc. LectureActive is available in Windows and Macintosh formats and is free to adopters.

The Saunders Astronomy Video Tape is a 93-minute VHS tape with five NASA/JPL movies, including a segment on images obtained from the Magellan Venus Orbiter.

A selection of **NOVA Videotapes** is available to qualified adopters. See your sales representative for details.

Other features of the text that adopters have praised have remained the same in this edition:

Full-color diagrams are used as teaching tools, not merely as a cosmetic device.

Peripheral or more technical material is printed in a different typeface and set off by colored bars. These sections can be omitted without disrupting the flow of the chapter.

We have avoided mathematics beyond algebra but have used algebra when it helps to illuminate or clarify a concept. Sections with anything but the most elementary mathematics are set in a different typeface and can be included or omitted at the discretion of the instructor.

We have tried to limit technical jargon where it is not necessary. At the same time, we do not hesitate to introduce the terminology astronomers use regularly, and we have carefully defined all terms that may be new to students.

We portray astronomy as a human endeavor and have included descriptions and photographs of some of the key men and women who have created our science over the years.

We welcome comments and suggestions about the text and the ancillaries from adopters and potential adopters (and even from graduate students, who might occasionally look back to this text when they are preparing for their prelim exams.) Address your cards and letters to: Andrew Fraknoi, Astronomy Dept., Foothill College, 12345 El Monte Rd., Los Altos Hills, CA 94022 or e-mail: FRAKNOI@ADMIN.FHDA.EDU.

We would like to thank the many colleagues who have provided information and critiques for this and all the previous editions. We are grateful to Don Davis and Don Dixon for permission to reproduce their fine paintings, and to David Malin for his assistance and his superb astronomical photographs. Others who have helped with images include Ray Villard at the Space Telescope Science Institute, Jurri Van der Woude at the Jet Propulsion Laboratory, and Scott Hildreth and Sally Stephens at the Astronomical Society of the Pacific. We very much appreciate the assistance of John Vondeling, Jennifer Bortel, Charlene Squibb, Sue Westmoreland, and other members of the staff at Saunders College Publishing, for graciously accommodating the needs of three authors with busy schedules and strong opinions.

Janet Morrison read much of the text and offered many useful editorial suggestions. Andrew Fraknoi would like to thank Dennis Schatz of the Pacific Science Center, Alan Friedman of the New York Hall of Science, and Cary Sneider of the Lawrence Hall of Science for helpful discussions.

We dedicate the 7th edition to Alex Fraknoi, who came into the universe during the same period this new edition was born. May a future edition of this text still be helping students with their exploration of the universe by the time he gets to college.

David Morrison, Sidney Wolff, and Andrew Fraknoi
September 1994

CONTENTS

ABOUT THE AUTHORS

David Morrison received his Ph.D. from Harvard University. He was at the University of Hawaii from 1969 to 1988, where his positions included Professor of Astronomy, Chair of the Astronomy Graduate Program, Director of the Infrared Telescope Facility at Mauna Kea Observatory, and University Vice-Chancellor for Research and Graduate Education. Dr. Morrison currently heads the space science program at the NASA Ames Research Center. His primary research interests are in planetary science. Dr. Morrison is the author of more than 100 professional articles and of several books, including *The Planetary System, Cosmic Catastrophes, Exploring Planetary Worlds,* and two other astronomy texts from Saunders. He has served as President of the Astronomical Society of the Pacific, Chair of the Astronomy Section of the American Association for the Advancement of Science, and President of the Planetary Commission of the International Astronomical Union. A celestial object, asteroid 2410 Morrison, is named for him.

Sidney C. Wolff received her Ph.D. from the University of California at Berkeley, and then joined the Institute for Astronomy at the University of Hawaii. During the seventeen years Dr. Wolff spent in Hawaii, the Institute for Astronomy developed Mauna Kea into the world's premier international observatory. Dr. Wolff became Associate Director of the Institute for Astronomy in 1976 and Acting Director in 1983. She earned international recognition for her research, particularly on stellar atmospheres—the evolution, formation, and composition of stars. In 1984, she was named Director of the Kitt Peak National Observatory, and in 1987 became Director of the National Optical Astronomy Observatories. She is the first woman to head a major observatory in the United States. As Director of NOAO, she oversees a staff of 460 and facilities used by nearly 1000 visiting scientists annually. Recently, Dr. Wolff has also been acting as Director of the Gemini Project, which is an international program to build two state-of-the-art 8-m telescopes. Wolff has served as President of the Astronomical Society of the Pacific and is the second woman to be elected President of the American Astronomical Society. She is also a member of the Board of Trustees of Carleton College, a liberal arts school that excels in science education. Wolff is the author of more than 70 professional articles and a book, *The A-Type Stars: Problems and Perspectives.*

Andrew Fraknoi is the Chair of the Astronomy Department at Foothill College near San Francisco and an Educational Consultant for the Astronomical Society of the Pacific (where he directs Project ASTRO, a program to bring astronomers into fourth- through ninth-grade classrooms). From 1978 to 1992 he was Executive Director of the Society, as well as Editor of *Mercury* Magazine and the *Universe in the Classroom* Newsletter. He has taught astronomy and physics at San Francisco State University, Cañada College, and the University of California Extension Division. He is author of *The Universe in the Classroom*, co-author of *Effective Astronomy Teaching and Student Reasoning Ability*, and scientific editor of *The Planets* and *The Universe*, two collections of science and science fiction. In the past 22 years he has given over 400 public lectures on astronomical topics. For five years he was the lead author of a nationally syndicated newspaper column on astronomy, and he appears regularly on radio and television explaining astronomical developments. He has received the Annenberg Foundation Prize of the American Astronomical Society and the Klumpke-Roberts Prize of the Astronomical Society of the Pacific for his contributions to the public understanding of astronomy. Asteroid 4859 was named Asteroid Fraknoi in 1992 in recognition of his work in astronomy education.

Frontispiece: Jupiter bearing the scars of impacts by the multiple components of Comet Shoemaker-Levy 9, which crashed into the planet in July 1994, each releasing millions of megatons of energy. Each dark spot is made up of dust released in an impact and deposited in the jovian stratosphere. (NASA/STScI)

PROLOGUE AND BRIEF
TOUR OF THE UNIVERSE

There is nothing like astronomy to pull the
stuff out of man.

His stupid dreams and red-rooster importance:
let him count the star-swirls.
From *Star-Swirls* by Robinson Jeffers

We invite you to join us in a grand journey of
exploration. The territory we shall explore encompasses
vast regions of space and time, but at the end of our
journey, like all smart travelers, we will have learned at
least as much about ourselves as about the worlds we
have seen. Along the way, we shall see that the emergence of our species on the small planet we call Earth is
intimately connected with events and processes in the
larger universe. The fact that you, our gentle reader, are
here today is as much the result of the actions of stars
billions of years ago as the actions of your parents
decades ago.

We propose to explore more than just the objects
that populate our universe; we shall pay equal attention
to the process by which we come to understand the
cosmos and the tools we need to increase that understanding. After all, the exploration of the universe must
today still be carried out mostly from Earth. Although
we have built four robot spacecraft that have passed
beyond the orbit of the most distant known planet,
humanity itself still inhabits only one world and must
glean information about the rest of the universe mostly
from the messages that universe is kind enough to send

us. Since the *stars* are the fundamental building blocks of
the universe, decoding the message of starlight has been
the great challenge and triumph of modern astronomy.
By the time you have finished reading this book, you
will be conversant with much of that message.

THE NATURE OF ASTRONOMY

Strictly speaking, astronomy can be defined as the study
of the objects that lie beyond the atmosphere of our
own planet and the processes by which they interact
with one another. But, as we shall see, it is much more
than that. It is also humanity's attempt to chronicle the
history of the universe, from the instant of its birth in
the cataclysmic explosion we call the big bang to the
present moment in which you read this sentence. The
full understanding of that history is not the province of
astronomers alone; for example, we need the help of
physicists to understand the principles of motion in
space, the help of chemists to learn more about the way
the elements in our bodies were forged, the help of
geologists to get to know the forces that govern the
surfaces of planets, and the help of biologists to help us
piece together how ancient molecules—some forming
in space before the Earth ever formed—could eventually evolve into our readers.

Putting this history together is an audacious undertaking—especially by a creature of modest stature, liv-

ing on a rocky ball circling a nondescript star in the suburbs of the Milky Way—and we are far from finished. On the one hand, the thickness of this book is a testament to how much we have already learned. On the other hand, much of science is a "progress report," constantly changing as new techniques and instruments allow us to probe the universe more deeply.

By the way, have you thought how much of the power of the expression, "on the one hand, on the other hand," derives from our having two hands? Eight-handed creatures out there would find it an unimpressive figure of speech. Throughout this book we try to point out ways that our thinking may be "Earth chauvinist"—determined by local conditions and not universal necessities. For example, we will want to be careful not to be "light chauvinists"—to rely only on the kind of light our human eyes are able to see. The universe actually sends out a far wider range of light, much of which our eyes filter out in their attempt to make sense of the terrestrial scene. But this invisible light can tell us a great deal about what is happening in the universe, much of which we simply cannot learn from the more familiar visible rays. Modern instruments have given us the power to have "x-ray vision," "radio vision," and many other forms of vision that were not part of our original sensory equipment.

In considering the history of the universe, we will see again and again that the cosmos *evolves*: it changes in profound ways over long periods of time. Although certain religious groups and sensational media have tried to create the impression that the concept of evolution is controversial, in science this is simply not true. Few theories are better established and have more solid evidence behind them than the idea that the universe is not the same today as it was long ago. You will see, for example, that the universe could not have produced the readers of this book during the first generation of stars after the big bang. The ingredients of an astronomy student simply did not exist in sufficient numbers in those early days; the universe still had to make the carbon, the calcium, the oxygen, and the iron needed to make life as we know it. Today, many billions of years later, the universe has evolved to be a more hospitable place for life. Indeed, tracing the evolutionary processes that continue to shape the universe is one of the most important and satisfying parts of modern astronomy.

THE NATURE OF SCIENCE

Science, unlike religion or philosophy, accepts nothing on faith. The ultimate judge in science is always the experiment or observation—what nature itself reveals. Science is not merely a body of knowledge, as is often taught in our schools. Much more, science is a *method*, by which we attempt to understand nature and how it behaves. This method begins with many observations over a period of time. From the trends in the observa-

tions, scientists may come up with a model of the particular phenomenon we want to understand. Such models are always approximations of nature itself and are subject to further testing.

To take a concrete astronomical example, ancient astronomers constructed a *model* (partly from observations, partly from philosophical beliefs) in which the Earth was the center of the universe, and everything moved around it. At first, all observations of the Sun, Moon, and planets could be fitted to this model, but eventually, better observations required the model to add circle upon circle to the movements of the planets to keep the Earth at the center. As the centuries went by, and improved instruments were developed for keeping track of celestial objects, the model could no longer explain all the observed facts. A new model, with the Sun at the center, fit the experimental evidence better and, after a period of philosophical struggle, became accepted as our view of the universe.

When they are first proposed, new models or ideas are sometimes called *hypotheses*. Many students today think that there are no new hypotheses in a science like astronomy—that everything important has already been learned. Nothing could be farther from the truth. Throughout this book, you will find discussions of recent and occasionally still controversial hypotheses in astronomy—the significance of huge chunks from space hitting the Earth for the development of life on our planet, the existence of vast quantities of invisible "dark matter" that could make up the bulk of the universe, the presence of a strange "black hole" at the center of the Milky Way. All such hypotheses are built on difficult observations done at the forefront of our technology, and all require further testing before we fully incorporate them into our standard astronomical models.

Over the centuries, scientists have extracted from countless observations certain fundamental principles that are called *scientific laws*. These are, in a sense, the rules of the game nature plays. One remarkable discovery about nature that underlies everything you will read in this book is that the laws we discover on Earth seem to have universal applicability. The rules that govern the behavior of gravity on Earth, for example, are the same rules that determine the motion of two stars in a system so far away that they are invisible to the unaided eye. Note that without the existence of universal laws, we could not do much astronomy. If each pocket of the universe had not only different objects, but also completely different rules, we would have little chance of interpreting what happened in that neighborhood. But the consistency of the laws of nature gives us enormous power to understand distant objects without traveling to them and relearning the local laws. (In the same way, if every state of the U.S. or province of Canada had completely different laws, it would be very difficult to carry out commerce or even understand the behavior of people in different regions. But a consistent set of laws allow us to apply what we learn or practice in one state to any other state.)

Today, for a variety of reasons, there are people who claim that the laws of the natural world (as discovered by science) have sometimes been, or could under the right circumstances be, suspended. This is an enticing fantasy, but, despite many attempts, not a shred of evidence exists to support such a belief. Although it would be nice if we could suspend the law of gravity and float freely through effort of will alone, in real life such experiments generally result in broken bones.

This is not to say that new experiments and observations do not lead to more sophisticated models, models that could even include new phenomena and new rules. The theory of relativity (which we shall study in Chapters 8 and 31) is a perfect example of such a transformation, which took place not long ago. But wishing isn't going to bring such new models into existence; only the patient process of observing nature ever more finely can reap such rewards.

One important problem about describing scientific models has to do with the limitations of language. When, especially in beginning books like this one, we try to describe complex phenomena in everyday terms, the words themselves may not be adequate to the job. For example, you may have heard the structure of the atom likened to a miniature solar system. Although some aspects of our modern model of the atom do remind us of planetary orbits, many other aspects are fundamentally different. (But thinking of the words literally, many a youngster has wondered whether electrons might not be little planets, with little electron students studying electron astronomy books.)

This is why, often to the annoyance of those who are not mathematically inclined, scientists often prefer to describe their theories using equations rather than words. In this book, designed to introduce the field of astronomy, we have used mainly words to discuss what scientists have learned, and have not used math beyond basic algebra. But if this course piques your interest and you go on in science, more and more of your studies will involve the language of mathematics.

NUMBERS IN ASTRONOMY

You may have heard a television reporter refer to a large number (like the national debt) as astronomical. In astronomy, we do have to deal with distances on a scale you may never have thought about before and numbers larger than any you may have encountered. Most students take a while to navigate among the millions and billions that astronomers tend to throw about in their everyday discourse.

In this book we adopt two approaches to make dealing with astronomical numbers a little bit easier. First, we use a system for writing large and small numbers called *powers-of-ten* notation (or sometimes *scientific nota-*tion). This does away with the huge number of zeros that can be so time-consuming and discouraging for students first approaching science.

In this system, if you want to write a figure like $490,000 (which is definitely *not* the starting salary for an astronomer, but might be for a national television star), you write 4.9×10^5. The superscript number after 10, called an *exponent*, keeps track of the number of times you have to multiply 10 together to get the number you want. In our example, 10 is multiplied by itself 5 times, and $10 \times 10 \times 10 \times 10 \times 10 = 100,000$. Multiply $100,000$ by 4.9 and you get our astronomical starting salary. Another way to remember the basics of this notation is to note that 5 is the number of places you have to move the decimal point to the right to convert 4.9 to 490,000.

Small numbers are written with negative exponents. Three millionths (0.000003) is expressed as 3.0×10^{-6}. One reason this notation is so popular among scientists (trust us, it is, even if you at first don't like it) is that it makes arithmetic a lot easier. To multiply two numbers in scientific notation, you need only add their exponents: thus, $10^3 \times 10^9 = 10^{12}$. To divide numbers, you can just subtract exponents. (For more on this system, see Appendix 3.)

The second way we try to make numbers simple is to use a consistent set of units—the *international metric system*. Unlike the British system, where it takes a completely ridiculous number like 5280 feet to equal a mile, metric units are related by powers of ten: a kilometer, for example, equals a thousand meters. The metric system, which has been adopted by every major country in the world except the U.S., is summarized in Appendix 4, and should be part of your vocabulary if you want to face the future unafraid.

Light Years

To give you a chance to practice using scientific notation, and to set the scene for the tour of the universe in the next section, let us define a common unit astronomers use to describe distances in the universe. A light year (LY) is the distance that light travels in one year. Because light always travels at the same speed, and because its speed turns out to be the fastest possible speed in the universe, it makes a good standard for keeping track of distances. Some students complain about this name for a unit of distance—light years seem to imply that we are measuring time, not distance. But the same use of language is common in everyday life, for example, when we tell a friend to meet us at a movie theater that's 20 minutes away.

So, how many kilometers are there in a light year? (First of all, in case you are not yet a metric system fan, we should tell you that a kilometer is about 0.6 mile.) Light travels at the amazing pace of 3×10^8 meters per second (m/s). Since there are a thousand (10^3) meters in a kilometer, the speed becomes 3×10^5 km/s. Think

FIGURE P.1 Part of the remnant of an exploded star in the southern constellation of Vela. Such exploding stars were crucial to the development of life in the universe. The diagonal line is the path of an artificial satellite that happened to cross the field while one of the exposures for this image was being made. (Photo by David Malin; copyright Anglo-Australian Telescope Board)

away—more than 40,000 billion km. This is why astronomers are skeptical that UFOs are extraterrestrial spacecraft that are coming here across these vast distances, picking up two drunken fishermen or loggers, and then going right home. It seems like such a small reward for such a large investment.

Consequences of Light Travel Time

There is another reason why the speed of light is such a natural unit of distance for astronomers. Information about the universe comes to us almost exclusively via radiation (of which light is one example), and all such radiation travels at the speed of light—that is, one light year every year. This sets a limit as to how quickly we can learn about events in the universe. If a star is 100 LY away, the light we see tonight left that star 100 years ago and is just arriving in our neighborhood. The soonest we can learn about any changes in that star—its blowing up, for example—is 100 years after the events occurred at the star. For a star 500 LY away, the radiation we detect tonight left 500 years ago and is carrying 500-year-old news.

Some students, accustomed to CNN and other news media known for "instant world coverage," at first find this frustrating. "You mean, when I see that star up there," they ask, "I won't know what's really happening there *now* for another 500 years?" But that's not really the right way to think about the situation. For astronomers, *now* is when the light reaches us here on Earth. There is simply no way for us to know anything about that star (or other object) until the radiation from it makes its way to us; despite the fondest dreams of science fiction writers, instant communication through the universe is not possible.

And what at first may seem a great frustration is actually a tremendous boon in disguise. If astronomers really want to piece together what happened in the universe since its beginnings, they need to find evidence about each epoch of the past. Where can we find evidence today about cosmic events that happened billions of years ago? The delay in the arrival of light provides such evidence automatically. The further out in space we look, the longer the light has taken to get here, and the longer ago it left its place of origin. By looking billions of light years into space, astronomers are seeing billions of years into the past. In this way, we can actually reconstruct the history of the cosmos and get a sense of how it has evolved with time.

This is one reason why astronomers seek to build telescopes that can collect more and more of the faint light (and other radiation) the universe sends us. The more light we collect, the fainter the objects we can make out. On average, fainter objects are farther away, and can thus tell us about periods of time even deeper in the past. New instruments, such as the Hubble Space Telescope and Hawaii's Keck Telescope (Chapters 9

about that—light covers 300,000 kilometers every second. In 1 second, it can travel seven times around the circumference of the Earth; a commercial airplane, in contrast, would take about two days to go around once, not counting time to refuel.

Now that we know how far light goes in a second, we can calculate how far it goes in a year. There are 60 (6×10^1) seconds in each minute, and 6×10^1 minutes in every hour. Thus light covers 3×10^5 km/s $\times$ 3.6×10^3 s/hr $= 1.08 \times 10^9$ km/hr. There are 2.4×10^1 hours in a day and 365.24 (3.65×10^2) days in a year. Multiplying the product of those two numbers, 8.77×10^3 hr/yr by 1.08×10^9 km/hr gives 9.46×10^{12} km in a light year. That's almost 10 trillion kilometers that light covers in a year. A string 1 LY long could fit around the circumference of the Earth 236 million times!

You might think that such a long unit would more than reach to the nearest star. But the stars are far more remote than our imaginations (or episodes of *Star Trek*) might lead us to believe. Even the nearest star is 4.3 LY

FIGURE P.2 The Hubble Space Telescope, repaired in December 1993, is an example of the new generation of astronomical instruments in space. (NASA)

and 10), are giving astronomers unprecedented views of deep space and deep time.

But the delay of light has another interesting philosophical consequence. If you take a snapshot of your family or a group of friends, you can write a date on the back and correctly claim that the picture records all of you as you were on that date. But if you take a snapshot of a section of the night sky, the picture represents a more complicated record. On that single image, you may have recorded a star 150 LY away, shown as it looked 150 years ago. There, too, is cluster of stars about 2,000 LY away, shown as it was some 2,000 years in the past. (One of its stars may in the meantime have exploded, but the news of the explosion is still making its way toward us.) And faintly visible on the image is a galaxy of stars 100 million LY away, whose light left for its journey to Earth long before the human species evolved on the surface of our planet. The only date you can fairly write on the back of that photo is when the light from each of these sources arrived on the Earth—a date of only local interest. There is no "universal now" in the cosmos, only a "local now" for each observer.

A TOUR OF THE UNIVERSE

Let us now take a brief, introductory tour of the universe as astronomers understand it today, just to get acquainted with the sorts of objects and distances we will be looking at throughout the text. We begin at home, with the Earth, a nearly spherical planet about 13,000 kilometers in diameter (Figure P.3). A space traveler entering our planetary system would easily distinguish the Earth by the large amount of liquid water that covers some ⅔ of its crust. If the traveler had equipment to receive radio or television signals or came close enough to see the lights of our cities at night, she would soon find signs that this watery planet had intelli-

FIGURE P.3   The Earth is a planet, as becomes evident when we view it from a perspective in space. (NASA)

FIGURE P.5   The largest planet in our solar system is Jupiter. We could fit almost 11 Earths side by side into its equator, and it contains as much mass as all the other planets combined. This image, which also shows two of its large satellites, was taken in February 1979 with the Voyager 1 spacecraft. (JPL/NASA)

gent life (of course, depending on what television signal the traveler tuned to, the conclusion might be modified to "semi-intelligent" life).

Our nearest astronomical neighbor is the Earth's satellite, commonly called the Moon. Figure P.4 shows the Earth and the Moon to scale on the same diagram. The Moon's distance from Earth is about 30 times the Earth's diameter, or about 384,000 km, and it takes about a month for the Moon to revolve around the Earth. The Moon's diameter is 3476 km, about ¼ the size of the Earth. (Many textbooks, including ours, are forced to show diagrams of a Moon much closer to the Earth when illustrating concepts like eclipses. We do this because the publisher doesn't think you'll buy a book whose dimensions are big enough that we could show everything properly to scale; the publisher is probably right! But keep Figure P.4 in mind as you continue your exploration of the Earth-Moon system in the course.)

Light (or radio) waves take 1.3 seconds to travel between the Earth and the Moon. If you've seen videos of the Apollo flights to the Moon, you may recall that

there was a delay of about 3 seconds between the asking of a question by Mission Control and a reply by the astronauts. The reason is not that the astronauts were thinking slowly, but that it took the radio waves about 3 seconds to make the round trip.

The Earth revolves about our star, the Sun, which is about 150 million km away—about 400 times as far as the Moon. We call the average Earth–Sun distance an astronomical unit (AU). Yes, it is a slightly Earth-chauvinist term, but then it's only Earth people who need to use these terms (at least for now). Light takes a little more than 8 minutes to travel 1 AU, which means our latest news from the Sun is always 8 minutes old. It takes the Earth one year (3×10^7 seconds) to go around the Sun at that distance; to make it around, we must travel at approximately 110,000 km/hr. Since gravity holds us firmly to the Earth and there is no resistance to the Earth's motion in the vacuum of space,

Earth

Moon

FIGURE P.4   The Earth and Moon, drawn to scale.

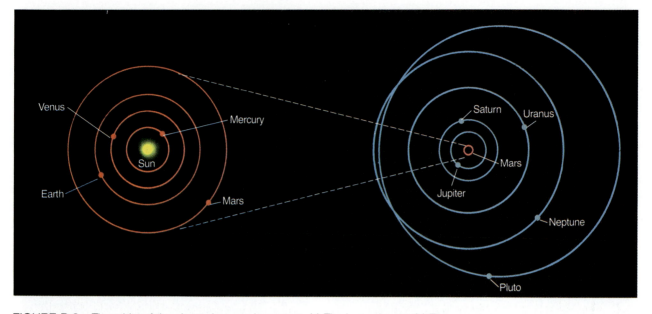

FIGURE P.6   The orbits of the planets in our solar system. (a) The inner planets. (b) The outer planets (note the change of scale.)

we participate in this breakneck journey without being particularly aware of it day by day.

The diameter of the Sun is about 1.5 million km; as we shall see in later chapters, our Earth could fit comfortably inside one of the minor eruptions that occur on the surface of our star. If the Sun were the size of a basketball, the Earth would be a small apple seed some 30 meters from the ball. By a happy coincidence, the Sun is about 400 times the diameter of the Moon; recall that it was also 400 times farther away than the Moon. This means that the two are about the same size as seen from the Earth, and, when they line up just right, the Moon can exactly cover or *eclipse* the Sun. Because the Moon is moving (very) slowly away from the Earth, this was not true in the distant past, and will not be true in the distant future. From this point of view, humanity picked a good time to get started on our planet.

The Earth is only one of nine planets that we have discovered revolving around the Sun. These nine planets, along with their satellites, and swarms of smaller bodies, make up the solar system, what we might call the family of the Sun. The definition of a planet is that it must be a body of some significant size, orbiting a star, and not making its own light. (If a large body produces its own light, as well as other radiation, then it is called a star.) We see the nearby planets in our skies only because they reflect the light of the Sun. If they were much further away, the tiny amount of light that they manage to reflect would not be visible to us. For this reason, astronomers have never seen planets around other stars; the stars—as we have seen—are just too far away for the reflected light from small planets to be detectable with even the largest telescopes we have today. We suspect (and have some indirect evidence) that such planets should be there, but we don't have the technology to find them directly.

Jupiter, the largest planet in the solar system, is about 143,000 km in diameter, 11 times the size of the Earth. Its distance from the Sun is five times the Earth's, or 5 AU. On the scale in which the Sun is a basketball, Jupiter would be about the size of a grape, and would be located about 150 meters from the basketball. That's quite a separation; if you'll forgive the mixed sport metaphor, 150 meters is about 1½ football fields in length! The orbits of the planets are shown schematically in Figure P.6.

Usually, the most distant planet is little Pluto, but until the year 1999, Pluto's strange orbit actually carries it inside the orbit of Neptune. Still, if we take the average of Pluto's eccentric orbit, it is about 40 AU or 5.9 billion km from the Sun. On our basketball scale, Pluto would be a grain of sand about 1 km from the ball.

The Sun is our local star, and all the other stars are also suns: enormous balls of glowing gas, generating vast amounts of energy by nuclear reactions deep within—very much as nuclear bombs did on Earth. (Note that we put the last phrase in a hopeful past tense.) The other stars look faint only because they are so far away. Continuing our basketball analogy, Proxima Centauri, the nearest other star (4.3 LY away), would be almost 7,000 km from the basketball.

When we look up at the star-studded country sky on a clear night, all the stars we can see with the unaided eye turn out to be part of a single collection of stars we call the Milky Way Galaxy, or just simply our Galaxy. (When astronomers are referring to the Milky Way, they capitalize Galaxy; when we talk about other galaxies of stars, we put the word in lower case.) The Sun is one of hundreds of billions of stars that compose the Galaxy; its extent, as we shall see, staggers the human imagination.

Let's make a rough scale drawing showing the stars

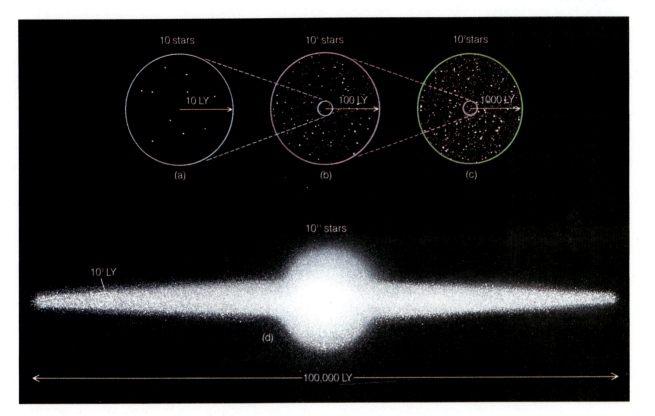

FIGURE P.7 The distribution of stars around the Sun within (a) 10 LY, (b) 100 LY, (c) 1000 LY, and (d) the Galaxy (with its disk seen edge-on).

within 10 LY of the Sun (Figure P.7). The small circle labeled (*a*) represents a sphere 10 LY in radius centered on the Sun. We find roughly 10 stars in this sphere. Now we change scale: the circle labeled (*b*) represents a sphere 100 LY in radius. Note that all of (*a*) is now just a small circle in the center. Sphere (*b*) contains about 10,000 (10^4) stars, far too many to make the job of counting them pleasant or to make the job of naming them reasonable. And yet in going out to a distance of 100 LY, we have traversed only a minuscule part of the extent of the Milky Way Galaxy.

Now let's draw a circle of radius 1000 LY (*c*), in which, once again, our previous circle is just a small center. Within the 1000 LY sphere, we would find some 10 million (10^7) stars. In the bottom half of Figure P.7, we change scale and examine the entire Galaxy, a wheel-shaped system whose diameter is 100,000 LY across (seen edge-on in our figure). If we could move outside the Galaxy and look down on the wheel (or disk) of the Milky Way from above, it would probably resemble the galaxy in Figure P.8, its spiral structure outlined by the blue light of hot adolescent stars.

The Sun is located about 30,000 LY from the center of the Galaxy, a location with nothing much to distinguish it. From within the Milky Way Galaxy, we cannot see through it to its far rim (at least not with ordinary light) because interstellar space (the space between the stars) is not completely empty. It contains a sparse distribution of gas (mostly the simplest element, hydrogen) intermixed with tiny solid particles that we

call interstellar dust. This gas and dust collect into enormous clouds in many places in the Galaxy and become the raw material for future generations of stars (and probably planets.)

Typically, the interstellar material is so extremely sparse that the space between the stars is a far, far better

FIGURE P.8 This galaxy of billions of stars, called by its catalogue number M83, about 10 million LY away, is thought to be similar to our own Milky Way Galaxy. Here we are seeing the giant wheel-shaped system face on, as if we were looking down on the disk of stars. (Cerro Tololo Interamerican Observatory, NOAO)

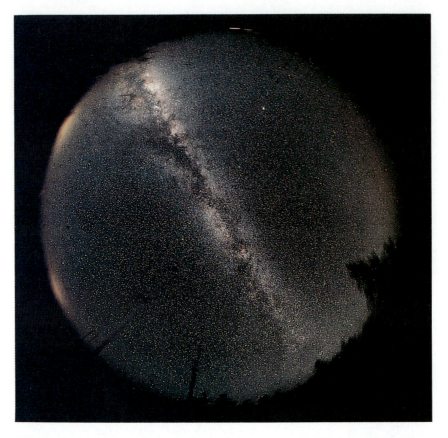

FIGURE P.9   Because we are inside the Galaxy, we see its disk in cross-section flung across the sky like a great white avenue of stars. This full sky view, taken with a special lens from the summit of Mount Graham in southern Arizona, shows the Milky Way with its myriad stars and dark rifts of dust. The outer circle is the horizon, and you can see, in addition to fir trees and test towers, the lights of several Arizona cities around it. (Roger Angel, Stewart Observatory/University of Arizona)

vacuum than anything we can produce in terrestrial laboratories. Yet, the dust in space, building up over thousands of light years, can obscure the light of more distant stars. Like the distant buildings that disappear from our view on a smoggy day in Los Angeles, the more distant regions of the Milky Way cannot be seen behind the layers of interstellar smog. Luckily, astronomers have found that stars and raw material "shine" with various forms of invisible radiation that do penetrate the smog, and so we have been able to build up a pretty good map of the Galaxy in recent years.

Not all stars live by themselves, as the Sun does. Many are born in double or triple systems, with two or three stars revolving about each other (and even larger numbers of partners are possible). Because the stars influence each other in such close systems, multiple stars allow us to measure characteristics of stars that we cannot discern from observing single stars. In a number of places, enough stars have formed together that we recognize them as **star clusters.** Some of the largest of the more than a thousand star clusters that astronomers have catalogued can contain hundreds of thousands of stars, and take up volumes of space hundreds of light years across (Figure P.10).

Because stars live a long time (compared to the people who like to watch them), you often hear them referred to as eternal. But, in fact, no star can last forever. Since the business of stars is making energy, and since energy production requires some sort of fuel to be used up, eventually all stars will run out of fuel. (This news should not make you run out to stock up on thermal underwear—our Sun still has at least 4 or 5 billion years to go.) But ultimately, the Sun and all stars *will* die, and it is in their death throes that some of the most intriguing and important processes of the universe stand revealed. For example, we shall see that many of the atoms that now make up our bodies were once inside a number of stars that exploded at the end of their lives, recycling their material back into the reservoir of the Galaxy. In this sense, all of us are literally made of star dust.

THE UNIVERSE ON THE LARGE SCALE

In a very rough sense, you could think of the solar system as your house or apartment and the Galaxy as your town, made up of many houses and buildings. In

FIGURE P.10   In a cloud of cosmic raw material called the Rosette Nebula, a cluster of bright hot stars can be seen forming in the upper right. (Image by David Malin; copyright Anglo-Australian Telescope Board.)

the 20th century, astronomers have been able to show that just as our planet is made up of many, many towns, so the universe is made up of enormous numbers of galaxies. (We define the universe to be everything that exists that is accessible to our observations.) Galaxies stretch as far in space as our telescopes can see, many billions of them within the reach of modern instruments. When they were first discovered, some astronomers called galaxies "island universes," and the term is an apt one: they do appear like islands of stars in the vast dark seas of intergalactic space.

The nearest galaxy, just discovered in 1993, is a small one that lies 75,000 LY from the Sun in the direction of the constellation Sagittarius, where the smog in our own galaxy makes it especially difficult to discern. (A *constellation* is one of the 88 sections into which astronomers divide the sky, each named after a prominent star pattern within it.) The existence of this Sagittarius dwarf galaxy is, in fact, still controversial and will require other observations before all astronomers believe it. Beyond, about 160,000 LY away, lie a pair of small galaxies, first recorded by Ferdinand Magellan's crew as he sailed around the world, and thus called the Magellanic Clouds. All three of these small galaxies are "satellites" of the Milky Way, interacting with it through the force of gravity. Ultimately, all three may even be swallowed by our much larger Galaxy.

The nearest large galaxy is a spiral quite similar to our own, located in the constellation of Andromeda and thus often called the Andromeda Galaxy. (It is also known by one of its catalogue numbers, M31. Given the number of galaxies, no one in his right mind would suggest giving all of them proper names. In fact, astronomers today try to denote most astronomical objects by

FIGURE P.11   A cluster of galaxies in the constellation of Hercules, almost 500 million LY away. (Palomar Observatory, Caltech)

numbers that define their location in the sky.) M31 is about 2 million LY away and is, with the Milky Way, part of a small cluster of over 30 galaxies that we call the Local Group.

At distances of about 10 to 15 million LY are other small groups, and then at about 50 million LY is a more impressive system with thousands of member galaxies called the Virgo Cluster. We have discovered that galaxies are found mostly in clusters, and some of the clusters themselves also form into larger groups we call superclusters. Our Local Group and the Virgo Cluster are part of such a supercluster that stretches over a diameter of at least 60 million LY. We are just beginning to explore the structure of the universe at these enormous scales and are already finding some surprising results (Chapter 34.)

At even greater distances, where many ordinary galaxies are too dim to see, we find the quasars. These are the brilliant centers of fainter galaxies, glowing with the light of some extraordinarily energetic process (perhaps a giant black hole swallowing whole neighborhoods of raw material; we'll describe these bizarre objects in Chapter 31). Whatever they are, their brilliance makes the quasars the most distant beacons we can see in the oceans of space; they allow us to probe the universe ten billion or more LY away, and thus ten billion or more years in the past.

With the quasars we see a substantial way back to the big bang explosion that marks the beginning of time. Beyond the quasars, we can detect only the feeble glow of the explosion itself, filling the universe and thus coming to us from all directions in space. The discovery of this "afterglow of creation" was one of the most exciting events in 20th-century science, and its ramifications are still being explored.

For now, we should just remember that such ideas are far easier to state than to discover. Measurements of the properties of galaxies and quasars in remote locations require large telescopes, sophisticated light-amplifying devices, and painstaking labor. At observatories around the world, astronomers and their students are at work on such questions as the large-scale structure of the universe every clear night, mostly observing one star or one galaxy at a time, fitting their results into the tapestry of our understanding.

THE UNIVERSE OF THE VERY SMALL

The foregoing discussion should impress on you that the universe is extraordinarily large and extraordinarily empty. The universe on average is 10,000 times more empty than our Galaxy. Yet, as we have seen, even the Galaxy is mostly empty space. The air we breathe has about 10^{19} atoms in each cubic centimeter—and we think of air as pretty empty stuff. In the interstellar gas of the Galaxy, there is about *one* atom in every cubic centimeter. Intergalactic space is so sparsely filled, that

to find one atom, on average, we must search through a cubic *meter* of space! Most of the universe is fantastically empty; places that are dense, such as the bodies of our readers, are tremendously rare.

Yet even the familiar solids, such as this book, are mostly space. If we could take such a solid apart, piece by piece, we would eventually reach the molecules of which it is formed. Molecules are the smallest particles into which matter can be divided while still retaining its chemical properties. A molecule of water (H_2O), for example, consists of two hydrogen atoms and an oxygen atom, bonded together. Molecules are built up of atoms, which are the smallest particles of an element that can still be identified as that element. For example, an atom of gold is the smallest piece you can have of gold (although one atom won't impress your sweetheart very much!) Nearly 100 different kinds of atoms (elements) exist in nature, but most of them are rare, and only a handful account for more than 99 percent of everything with which we come in contact. The most abundant elements in the cosmos today are listed in Table P.1.

All atoms consist of a central, positively charged nucleus, surrounded by negatively charged electrons. The bulk of the matter in each atom is in the nucleus, which consists of a certain number of positive protons and a roughly equal number of electrically neutral neutrons, all tightly bound together in a very small space. What defines each element is the number of protons in its atoms; thus, any atom with 6 protons in its nucleus is called carbon, any with 50 protons is called tin, and any with 69 protons is called ytterbium. (Ytterbium, as you can probably guess, is not big on the cosmic "hit parade" of elements, but we like its name. For a list of the elements, see Appendix 18.) Generally, in a place like Earth, an atom has as many electrons as protons. Because protons and electrons have charges that are equal in magnitude but opposite in sign, they cancel each other out, and the atom is electrically neutral. Inside stars, on the other hand, where conditions are extremely hot, electrons can leave their atoms (a process

TABLE P.1	THE COSMICALLY ABUNDANT ELEMENTS	
Element	Symbol	Number of Atoms per Million Hydrogen Atoms
Hydrogen	H	1,000,000
Helium	He	68,000
Carbon	C	420
Nitrogen	N	87
Oxygen	O	690
Neon	Ne	98
Magnesium	Mg	40
Silicon	Si	38
Sulfur	S	19
Iron	Fe	34

called ionization), and the atom would then have a net positive charge.

The distance from an atomic nucleus to its electrons is typically 100,000 times the size of the nucleus itself. This is why we say that even solid matter is mostly space. The typical atom is far emptier than the solar system out to Pluto (the distance from the Earth to the Sun, for example, is only 100 times the size of the Sun.) Here is another reason why atoms are not like miniature solar systems.

Remarkably, physicists have discovered that everything that happens in the universe, from the smallest atoms to the largest superclusters of galaxies, can be explained through the action of only four forces: gravity, a force called electromagnetism (which combines the actions of electricity and magnetism), and two forces that act at the nuclear level. We will get to know these forces in much more detail throughout the book, but the fact that there are four forces (and not a million, or just one) has puzzled physicist and astronomers for many years and has led to a quest for a unified picture of nature. We will return to this quest in the last chapter of the book.

A CONCLUSION AND A BEGINNING

If you are typical of students who are new to astronomy, you have probably reached the end of our tour with mixed emotions. On the one hand, you may be fascinated by some of the new ideas you've read about and may be eager to learn more. On the other hand (presuming you still have only two hands), you may be feeling somewhat overwhelmed by the number of topics we have covered and the number of new words and ideas we have introduced. Learning astronomy is a little like learning a new language: At first it seems there are so many new expressions that you'll never learn them all, but with some practice, you soon develop a facility with them.

You may also feel a bit small and insignificant at the end of our tour, dwarfed by the cosmic scales of distance and time. Such a feeling of insignificance is not a bad thing from time to time—sometimes we wish more of our politicians and film stars felt it. And just before a difficult exam, or when you've just ended a treasured relationship, it can certainly help to see your problems in a cosmic perspective. But there is another way to look at what we've learned at the end of our tour.

Let us consider the history of the universe from the big bang to today and compress it, for easy reference, into a single year. (We have borrowed this idea from Carl Sagan's Pulitzer Prize–winning book, *The Dragons of Eden,* published in 1977 by Random House.) On this scale, the big bang happened at the first moment of January 1, the solar system formed around September 9, and the oldest rocks we can date on Earth go back to the beginning of October. Where in this cosmic year would the origin of human beings fall? The answer turns out to be the evening of December 31! The invention of the alphabet (a development dear to the hearts of textbook authors) doesn't occur until the 50th second of 11:59 P.M. on December 31. And the beginnings of modern astronomy are a mere fraction of a second before the New Year. Seen in a cosmic context, the amount of time we have had to study the stars is very small, and our success at piecing together as much of the story as we have is remarkable.

Certainly, our attempts to understand the universe are not complete. Most likely, our grandchildren's grandchildren will find some of what is in this book a bit primitive. But as you read our current progress report on the exploration of the universe, take a few minutes every once in a while just to savor how much we have already learned.

We begin *our* journey of exploration by examining the beginnings of astronomy itself—humanity's first attempts to understand its place in the cosmos and the patterns and cycles of the sky. While many of these early ideas are remarkable for their insight and clarity of observation, they were limited by humanity's lack of tools—the telescopes and other instruments that have made modern astronomy possible. As a result, our remote ancestors could not yet understand the vast scales of time and space that undergird our modern view of the universe (and are a main theme throughout this book). As we shall see in the next few chapters, coming to appreciate these great vistas of space and time was a slow task that was to take many centuries.

In the process, humanity would lose its comfortable central place in the cosmos but gain a much more profound understanding of our connections with the stars. This understanding is far more exciting than anything the ancient *astrologers* described in this chapter (or their modern counterparts) came up with—and it has the added advantage, as Carl Sagan has said, of being true.

As the Earth rotates on its axis, the stars appear to turn around the north and south celestial poles. In this long-exposure photograph, Australian photographer David Malin captured about 10.5 hours of the turning of the sky around the south celestial pole. Each star makes a circular trail as the hours pass. The motions visible in the night sky—which, before the advent of electricity and television, were a much more intimate part of people's lives—fascinated and challenged our ancestors. The dome at bottom left houses the largest light-gathering telescope in Australia, the Anglo-Australian telescope.

(Copyright Anglo-Australian Telescope Board)

EARLY ASTRONOMY: MYTH AND SCIENCE

Claudius Ptolemaeus (Ptolemy) (second century) was one of the great astronomers of antiquity. Based on the work of many earlier thinkers, he devised a system of cosmology, with the Earth at its center, that described the motions of the planets so satisfactorily that there was no substantial change until the time of Copernicus, 13 centuries later. (Burdy Library, photograph by Owen Gingerich)

Speculation about the nature of the universe must date from prehistoric times. In many ancient civilizations the regularity of celestial motions was recognized, and attempts were made to predict celestial events. The invention of a calendar required some knowledge of astronomy, since the basic units of the calendar are the day of the Earth's spin, the month (originally, the cycle of the Moon's phases), and the year of seasons.

The Babylonians, Assyrians, and pre-Christian Egyptian astronomers knew the approximate length of the year. By a few centuries before Christ (B.C.), the Egyptians had adopted a calendar based on a 365-day year. Of particular significance to them was the date when the bright star Sirius could first be seen in the dawn sky, rising just before the Sun. This predawn rising of Sirius coincided fairly well with the average time of the annual flooding of the Nile, which gave the astronomer-priests the ability to predict when this economically important event could be expected to occur.

The Chinese had a working calendar and had determined the length of the year several centuries before Christ. They recorded comets, meteors, fallen meteorites, and sunspots visible to the unaided eye. Later the Chinese also kept records of "guest stars," stars that are normally too faint to be seen but suddenly flare up to become visible for a few weeks or months. The most significant of the Chinese observations of such outbursts was that of the exploding star of A.D. 1054 in the constellation of Taurus.

There is evidence of ancient astronomical knowledge in many parts of the world. The Mayan Indians in Central America developed a sophisticated calendar and made astronomical observations a thousand years ago. The Polynesians learned to navigate by the stars over distances of hundreds of kilometers of open ocean. Monuments of astronomical significance were built by Bronze Age people in northwestern Europe, especially in the British Isles. The best-preserved of these structures is Stonehenge (Figure 1.1) in southwest England, a monument in which some of the stones are aligned with the directions of the Sun at its rising and setting at critical times of the year (such as the beginnings of summer and winter).

All of these ancient peoples depended on some knowledge of the sky for their survival. Astronomy was born of the necessity to keep track of time and the seasons. Their knowledge was based on direct observations of the appearance and motion of the heavens—concepts much more familiar in those days before the development of cities, artificial lighting, and air pollution.

1.1 THE HEAVENS ABOVE

Our senses suggest to us that the Earth is the center of the universe—the hub around which the heavens turn. This **geocentric** view was held almost universally until

FIGURE 1.1   Stonehenge, a megalithic monument, possibly an observatory or calendar-keeping device, located on Salisbury Plain of England. (David Morrison)

the European Renaissance. It is simple, logical, and seemingly self-evident. Further, the geocentric perspective reinforces philosophical and religious systems that teach the unique role of humans as the central focus of the cosmos. However, the geocentric view happens to be wrong. One of the great themes of intellectual history—and this book—is the overthrow of the geocentric perspective and the steps by which we have re-evaluated the place of our world in the cosmic order.

(a) The Celestial Sphere

Our study of astronomy begins with a view of the heavens above us, a view identical to the perspective available before the invention of the telescope. If we look up on a clear night, we get the impression that the sky is a great hollow spherical shell with the Earth at the center. The early Greeks regarded the sky as just such a **celestial sphere** (Figure 1.2). Some even thought of it as an actual sphere of crystalline material, with the stars embedded in it like tiny jewels. At any one time we see only a hemisphere overhead, but with the smallest effort of imagination we can envision the remaining hemisphere, that part of the sky that lies below the horizon.

The Sun, Moon, and stars rise and set as the Earth turns within this imaginary sphere. In ancient times, of course, people did not realize that the Earth was a planet, spinning about its axis. Their concept of the Earth was restricted to the apparently flat world that they could see with their own eyes. It is easy to re-create this viewpoint just by finding a dark, quiet spot from which to look at the stars above. If we watch the sky for several hours, we see that the celestial sphere appears gradually to turn around us. Stars rise and set, moving

across the vault of heaven in the course of the night. The ancients, unaware of the Earth's rotation, imagined that the celestial sphere rotated about an axis that passed through the Earth. As it turned, the celestial sphere carried the stars up in the east, across the sky, and down in the west.

As the Earth rotates, our **horizon,** that line in the distance at which the ground seems to dip out of sight, providing a demarcation between Earth and sky, follows along with us. (The horizon may at times be hidden from view by mountains, trees, buildings, or, in large cities, smog.) As our horizon tips down in the direction that the Earth's rotation carries us, stars hitherto hidden beyond it appear to rise above it. In the opposite direction the horizon tips up, and stars hitherto visible appear to set behind it. Analogously, as we round a curve in a mountain road, new scenery comes into view while old scenery disappears behind us. The point directly overhead is called the **zenith,** and the zenith and horizon together define a coordinate system (the *altitude-azimuth system*; see Appendix 6) to describe the apparent positions of objects in the sky.

As the celestial sphere rotates, all of the objects in the sky maintain their positions with respect to each other. A grouping of stars like the Big Dipper has the same shape wherever we see it in the sky, although its apparent orientation with respect to objects on Earth shifts during the night. Even celestial objects that we know have their own motions, such as planets, seem fixed relative to the stars over the period of a single night. Only the meteors—"shooting stars" that flash into view for a few seconds—move appreciably with respect to the celestial sphere, and they are very close, located within the atmosphere of the Earth.

FIGURE 1.2   Time exposure showing trails left by stars as a consequence of the apparent rotation of the celestial sphere. The bright short trail near the center was made by Polaris (the North Star), which is about 1° away from the north celestial pole. The telescopes of the National Optical Astronomy Observatories on Kitt Peak near Tucson, Arizona, are seen in the foreground. (National Optical Astronomy Observatories)

The pole or point about which the celestial sphere appears to pivot lies along an extension of the line through the Earth's North and South Poles. As the Earth rotates about its polar axis, the sky appears to turn in the opposite direction about those **north** and **south celestial poles** (Figure 1.3). Halfway between them,

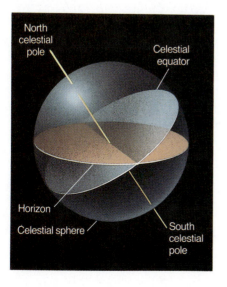

FIGURE 1.3   The celestial sphere, showing celestial poles, celestial equator, and horizon.

and separating the sky into its northern and southern halves, is the **celestial equator**—just like the Earth's equator, which separates the Northern and Southern Hemispheres of our planet. The celestial poles and celestial equator define another coordinate system that is useful in astronomy, especially for describing the positions of the stars.

Since the turning of the sky is just a reflection of the turning of the Earth, observers at different points on our planet would see it quite differently. An observer at the North Pole would see the north celestial pole directly overhead (at the zenith). The stars would all appear to circle about the sky parallel to the horizon, none rising or setting. An observer at the Earth's equator, on the other hand, would see the celestial poles at the north and south points on the horizon. As the sky apparently turned about these points, all the stars would rise straight up in the east and set straight down in the west. For an observer at an arbitrary place in the Northern Hemisphere (for example, in Greece), the north celestial pole would appear at a point between the zenith and the north point on the horizon, equal in elevation angle to the latitude of the observer (as explained in Figure 1.4). The stars that were not always above the horizon would rise at an oblique angle in the east, arc across the sky, and set obliquely in the west.

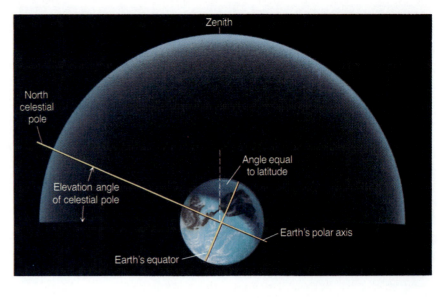

FIGURE 1.4   At the North Pole (latitude 90°), the celestial pole is overhead (altitude 90°). At the equator (latitude 0°), the pole is on the horizon (altitude 0°). In general, the altitude of the celestial pole is equal to the latitude of the observer. Here we see the dome of the sky from the perspective of an observer at roughly 30° north latitude.

(b)   Rising and Setting of the Sun

We have described the appearance of the night sky. The situation during the day is similar, except that the brilliance of the Sun renders the stars and planets invisible. (The Moon is still easily seen in the daylight, however.) We can think of the Sun as being located at a position on the hypothetical celestial sphere. When the Sun rises—that is, when the rotation of the Earth carries the Sun above the horizon—sunlight scattered about by the molecules of the atmosphere produces the blue sky that hides the stars that are also above the horizon.

For thousands of years, astronomers have been aware that the Sun gradually changes its position, moving each day about 1° to the east relative to the stars. (A degree [°] is an angle equal to 1/360 of a full circle. Astronomers also use smaller units of angular measure: An arcminute is 1/60 of a degree, and an arcsecond is 1/60 of an arcminute.) The Sun's apparent path around the celestial sphere, which reflects the revolution of the Earth about the Sun, is called the **ecliptic.** Each day the Sun rises about 4 min later with respect to the stars; the Earth must make just a bit more than one complete rotation (with respect to the stars) to bring the Sun up again.

As we look at the Sun from different places in our orbit, we see it projected against different stars in the background, or we would, at least, if we could see the stars in the daytime. In practice, we must deduce what stars lie behind and beyond the Sun by observing the stars visible in the opposite direction at night. After a year, when the Earth has completed one trip around the Sun, the Sun will appear to have completed one circuit of the sky along the ecliptic. We have a similar experience if we walk around a campfire at night; we see the flames appear successively in front of each person seated about the fire.

It was noted by the ancients that the ecliptic does not lie in a plane perpendicular to the line between the celestial poles, but is inclined at an angle of about 23° to that plane. This angle is called the **obliquity** of the ecliptic and was measured surprisingly accurately by several ancient observers. The obliquity of the ecliptic is responsible for the seasons, as we describe later.

(c)   Fixed and Wandering Stars

The Sun is not the only object that moves among the stars. The Moon and each of the five planets visible to the unaided eye—Mercury, Venus, Mars, Jupiter, and Saturn—also slowly change their positions from day to day. The Moon, being the Earth's nearest celestial neighbor, has the fastest apparent motion; it completes a trip around the sky in about one month. During a single day, of course, the Moon and planets all rise and set, as does the Sun. But, like the Sun, they have independent motions among the stars, superimposed on the daily rotation of the celestial sphere.

The Greeks of 2000 years ago distinguished between what they called the fixed stars, the stars that maintain fixed patterns among themselves throughout many generations, and the wandering stars or planets. The word planet means "wanderer" in Greek. Today, we do not regard the Sun and Moon as planets, but the ancients sometimes applied the term to all seven of the moving objects in the sky. Much of ancient astronomy was devoted to observing and predicting their motions. In most Romance languages the planets still give the names for the seven days of the week, although modern English retains only Sunday (Sun), Monday (Moon), and Saturday (Saturn).

The individual paths of the Moon and planets in the sky all lie close to the ecliptic, although not exactly on it. The reason is that the paths of the planets about the Sun, and of the Moon about the Earth, are all in nearly the same plane, as if they were marbles rolling about on the top of a table. The planets and Moon are always found in the sky within a belt 18° wide centered on the ecliptic, called the **zodiac.** The apparent motions of the planets in the sky result from a combination of their actual motions and the motion of the Earth about the Sun, and consequently they are, as we shall see, somewhat complex.

(d) Constellations

The backdrop for the motions of the "wanderers" in the sky is the canopy of fixed stars. Ancient Greeks, Chinese, and Egyptians divided the sky into **constellations,** apparent patterns of stars (Figure 1.5). Modern astronomers still make use of these constellations to denote approximate locations in the sky, much as geographers use political areas to denote the locations of places on the Earth. The modern boundaries between the constellations are imaginary lines in the sky running north–south and east–west, so that each point in the sky falls in one of 88 constellations or sky sectors.

Many of the constellations are of Greek origin and bear names that are Latin translations of those given them by the Greeks. Ten of the 12 constellations along the zodiac are named for animals, which is how the zodiac was named (*zoo* being the Latin root for "animal"). Today, the layperson is often puzzled because the constellations seldom resemble the people or animals for which they were named. In all likelihood, the Greeks themselves did not name groupings of stars because they resembled actual people or objects, but rather named sections of the sky in honor of the characters in their mythology. They then fitted the configurations of stars to the animals and people as best they could.

FIGURE 1.5 The winter constellation of Orion, the hunter, as illustrated in the 17th-century atlas by Hevelius. (J.M. Pasachoff and the Chapin Library)

1.2 EARLY GREEK ASTRONOMY

The science of astronomy as we know it originated with the Greeks in the first millennium B.C. They went beyond the simple observations discussed above to try to develop theories or models for the nature of the heavenly objects and their motions. Greek philosophers also sought to understand the place of the Earth and of humanity in the cosmic scheme. However, most of the ancient Greek thinkers were not scientists in the modern sense. They were often more interested in solving abstract mathematical problems or debating the fine points of logic and rhetoric than in making original observations. Yet, in that Greek reservoir of ideas and inspiration, many observations were carried out, with the result that science in general and astronomy in particular were raised to a level unsurpassed until the 16th century.

(a) The Ionian School of Greek Astronomers

The earliest Greek scientists were the Ionians, who lived in what we now call Turkey. Pythagoras (who died about 497 B.C.) was originally an Ionian, but he later founded a school of his own in southern Italy. He pictured a series of concentric spheres, in which each of the seven moving objects—the planets, the Sun, and the Moon—was carried by a separate sphere from the one that carried the stars, so that the motions of the planets resulted from independent rotations of the different spheres about the Earth. Pythagoras, who was also a mystic, thought that these motions gave rise to harmonious sounds, the music of the spheres, which only the most gifted ear could hear.

Pythagoras thought that the Earth, Moon, and other heavenly bodies were spherical. It is doubtful that he had a sound reason for this belief, but it may have stemmed from the realization that the Moon shines only by reflected sunlight and that the Moon's sphericity is indicated by the curved shape of the demarcation line between its illuminated and dark portions. If he reasoned that the Moon is round, the sphericity of the Earth might have seemed to follow by analogy.

(b) Phases of the Moon

Aristotle (384–322 B.C.), most famous of the Greek philosophers and one of the few who can be thought of as a scientist in the modern sense, wrote encyclopedic treatises on nearly every field. Aristotle's writings tell us that such phenomena as phases of the Moon and eclipses were understood at least as early as the fourth century B.C.

The Moon's changing shape during the month results from the fact that it is not itself luminous but shines by reflected sunlight. Because of its sphericity, only half of the Moon is illuminated—the half turned toward the Sun. The apparent shape of the Moon in our sky then depends on how much of its sunlit hemisphere is turned to our view.

Even in Aristotle's time it was known that the Sun is more distant than the Moon. This was evident from the fact that the Moon occasionally passes between the Earth and Sun and temporarily hides the Sun from view (a **solar eclipse**). Thus, when the Moon is in the same general direction from Earth as the Sun (position *A* in Figure 1.6), its sunlit side is turned away from the Earth. Because its other side—the side turned toward us—is dark and invisible, we do not see the Moon in that position. The phase of the Moon is then *new*. (Perhaps it would seem more reasonable to call it "no moon" instead of "new moon," for we do not see any Moon at all.)

A few days after the new phase, the Moon reaches position *B* in Figure 1.6, and from the Earth we see a small part of its sunlit hemisphere. The illuminated crescent increases in size on successive days as the Moon moves farther around the sky away from the direction of the Sun. After about a week, the Moon is one-quarter of the way around the sky from the Sun (position *C*) and is at the *first quarter* phase. The line from the

Earth to the Moon is at right angles to the line from the Earth to the Sun, and half of the Moon's daylight side is visible. Note that at first quarter (a quarter of the way around the sky) the illumination corresponds to a "half moon."

During the week after the first quarter phase, we see more and more of the Moon's illuminated hemisphere, until the Moon (at *E* in Figure 1.6) and the Sun are opposite each other in the sky, and we have *full* phase. (The traditional terminology is confusing, since the name of the phase now refers to the illumination [full], while at first quarter it referred to the fraction of the way around the orbit the Moon had traveled [1/4]. To be consistent, the "first quarter" should have been "first half." But astronomical terminology is not always consistent, alas.) During the next two weeks the Moon goes through the same phases in reverse order—through third (or last) quarter and back to new.

If you find difficulty in picturing the phases of the Moon from this verbal account, try a simple experiment. Stand about 6 ft in front of a bright electric light outdoors at night and hold up a small round object, such as a tennis ball or an orange. Move the object around your head, and you can simulate the phases of the Moon during the month. But the best way to become fully

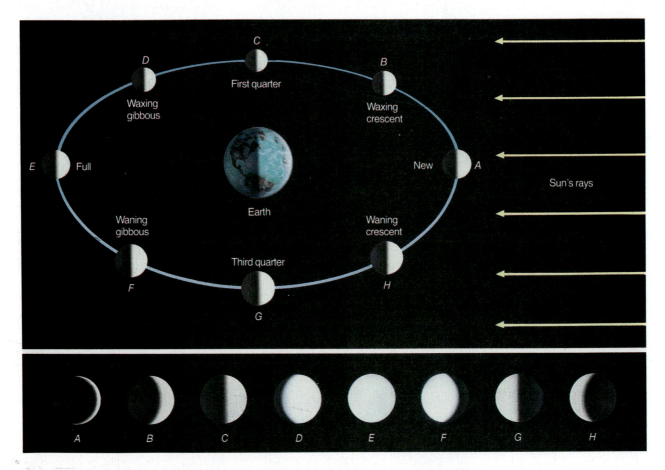

FIGURE 1.6   Phases of the Moon for a month. In the upper part, we look from a perspective in space. Now imagine yourself standing on the Earth, facing the Moon in each situation. You would observe the phases shown in the strip below.

acquainted with these lunar phases is to watch the real Moon in the sky. Observe its shape, its direction from the Sun, and its time of setting and rising.

(c) Shape of the Earth

Another important topic discussed by Aristotle was the shape of the Earth, and he cited several arguments for the Earth's sphericity. A **lunar eclipse** occurs when the Moon passes through the shadow of the Earth. Aristotle noted that during a lunar eclipse, as the Moon enters or emerges from the Earth's shadow, the shape of the shadow seen on the Moon is always round (Figure 1.7). Only a spherical object always produces a round

shadow. As a second argument, Aristotle explained that northbound travelers observe the stars near the north celestial pole to be higher in the sky than they are observed to be at home, and different stars pass through the zenith. Conversely, when one travels to more southern latitudes, the stars near the north celestial pole are seen lower in the sky, and some stars that are never above the horizon at home are seen to rise and move across the southern sky. The only possible explanation is that the travelers' horizons had tipped to the north or south, respectively, which indicates that they must have moved over a curved surface of the Earth.

Aristotle pointed out that the apparent daily motion of the sky can be explained by a hypothesis or model of the rotation of either the celestial sphere or the Earth.

9:04 P.M.

11:32 P.M.

9:38 P.M.

12:04 A.M.

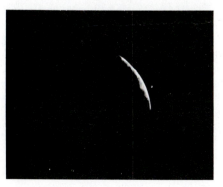

FIGURE 1.7 A lunar eclipse, with the Moon moving into the Earth's shadow. Note the curved shape of the shadow—evidence for a spherical Earth that has been recognized since antiquity. (Yerkes Observatory)

10:01 P.M.

12:23 A.M.

However, he rejected the latter (correct) explanation. He also considered the possibility that the Earth revolved about the Sun rather than the Sun about the Earth. He discarded this hypothesis in the light of an argument that has been used many times since. Aristotle explained that if the Earth moved about the Sun we would be observing the stars from successively different places along our orbit, and their apparent directions in the sky would then change continually during the year.

Any apparent shift in the direction of an object as a result of motion of the observer is called **parallax.** The annual shifting in the apparent directions of the stars that results from the Earth's orbital motion is called stellar parallax. For the nearer stars it is observable with modern telescopes, but it is impossible to measure with the unaided eye because of the great distances of even the nearest stars. As late as the 16th century, failure of astronomers to observe an annual stellar parallax was used as an argument for a stationary Earth (Section 21.1).

1.3   LATER GREEK ASTRONOMY

The early Greeks were aware of, and to some extent understood, the phenomena of the sky. Additional progress was made in the centuries following Aristotle, especially by the school of astronomers centered in the Egyptian city of Alexandria.

(a)   Aristarchus of Samos

Aristarchus of Samos (ca. 310–230 B.C.) is credited with the first arguments that the Earth revolves about the Sun—the *heliocentric* hypothesis, as opposed to the geocentric ideas of his predecessors. We know of his heliocentric ideas, however, only from the writings of others. Only one manuscript of Aristarchus survives, entitled "On the Sizes and Distances of the Sun and Moon." But this document alone is remarkable and deserves some discussion.

Aristarchus opened his treatise with several postulates, the "givens" that are needed to proceed with a geometrical proof. The most essential of these are (1) that the Moon receives its light from the Sun, (2) that it appears half full when the angle in the sky between the Moon and Sun is 3° less than a right angle (that is, 87°), and (3) that the diameter of the Earth's shadow at the Moon's distance is twice the size of the Moon. (He also implicitly assumed that the Moon's orbit about the Earth is a perfect circle.) From these assumptions, Aristarchus, using the rules of Euclidean geometry, determined that (1) the distance of the Sun is about 19 times the distance of the Moon and (2) the ratio of the Sun's diameter to that of the Earth is approximately 7. (The correct values for these numbers are 400 and 109, respectively.)

A reading of his account suggests that to Aristarchus the entire exercise was no more than an interesting geometry problem. There is no mention at all of where the numbers used in his postulates came from. Perhaps he made crude estimates; perhaps they were someone else's estimates; we do not know. But there is no suggestion that Aristarchus himself actually made any careful observations or measurements. To him, it was, we repeat, an exercise in geometry. On the other hand, the basic ideas were ingenious and beautiful in their simplicity. Moreover, these ideas were applied later by other astronomers—especially by Hipparchus—to attempt an *accurate* determination of the size and distance of the Moon.

It is interesting to see how the method of Aristarchus can work. The following, we emphasize, is not the procedure or reasoning of Aristarchus (whose geometry is actually rather tedious) but is a description of how we would be able to derive these astronomical dimensions, given Aristarchus' assumptions.

The Moon appears exactly half full (first and last quarters) when the terminator—the line dividing the light and dark halves—is a perfectly straight line as viewed from the Earth. But the Moon is spherical, and the terminator, being a line upon its surface, must be curved. Thus, the only way it can appear straight is for us to view it exactly edge on. That is, the plane of the terminator must contain the line of sight from the Earth to the center of the Moon. When that is true, the line from the Moon to the Earth must be at right angles to the line from the Moon to the Sun. In Figure 1.8 these right angles are *EMS* and *EM'S*. Now we see that because the Sun is not infinitely far away, by assumption, the points *M'*, *E*, and *M* do not lie along a straight line. Hence the Moon, moving at a uniform rate, should require a shorter time to go from *M'* to *M* than from *M* to *M'*. We could use the difference between these intervals from third quarter to first quarter and from first quarter to third quarter to determine the angle *M'EM*. For example, if the period from *M* to *M'* were, say, twice that from *M'* to *M*, the angle *M'EM* would be a third of a circle, or 120°, and the angle *SEM* would be 60°. We have no idea how Aristarchus arrived at the figure 87°. Even with our modern equipment of the late 20th century, we could not observe the instants of quarter moon with sufficient accuracy to determine the *ES/EM* ratio meaningfully, because of the Sun's great distance.

However determined, the angle *M'EM* can be constructed inside a circle representing the Moon's orbit; the lines *MS* and *M'S*, drawn tangent to the circle at *M* and *M'*, intersect at *S*, thus determining the position of the Sun and hence its distance in terms of the size of the Moon's orbit.

To find the relative sizes of the Sun and Moon, we use the information that the Earth's shadow at the Moon's distance is twice the size of the Moon (the correct ratio is about 8 to 3). It is well known that the Sun and Moon appear to be the same angular size in the sky, each about 1/2° in diameter. If, as Aristarchus had determined, the Sun is 19 times as distant as the Moon, it must also be 19 times as big to appear the same size. Aristarchus grossly overestimated the angular sizes of the Sun and Moon to be about 2° each. (Perhaps the error was intentional to emphasize the geometry rather than reality.) With such data we could find the relative sizes of the

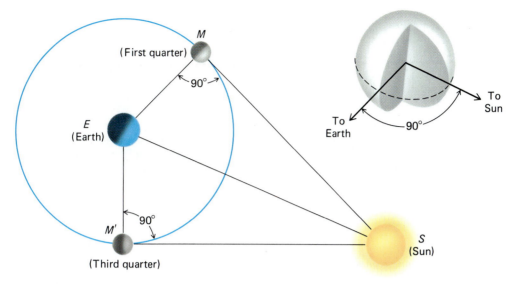

FIGURE 1.8   Aristarchus' method of determining the relative distances of the Sun and Moon from the Earth.

Earth, the Moon, and the Sun by geometrical construction.

We illustrate the geometrical principles of the construction in Figure 1.9. First, at the center of the Earth, we draw two lines that intersect at an angle of 1/2°. During a lunar eclipse the Sun and Moon are opposite in the sky; thus in the Moon's direction the 1/2° angle can be considered as representing the angular diameter of the Sun, and in the Sun's direction the angular diameter of the Moon. The Sun and Moon can now be drawn in, and at arbitrary distances from Earth, as long as the Sun's distance is 19 times the Moon's. Now at the Moon, the diameter of the Earth's shadow, *AA'*, can be constructed at twice the size of the Moon. Because the rays of sunlight, in which the Earth casts its shadow, travel in straight lines, the lines *AB* and *A'B'*, drawn tangent to the Sun at *B* and *B'*, must also be tangent to the Earth. Thus, finally, the sphere of the Earth can be drawn in to proper scale. We have now constructed a scale drawing of

the Earth, Moon, and Sun. We need only measure with a ruler to obtain their relative sizes.

Perhaps it was his finding that the Sun was seven times the Earth's diameter that led Aristarchus to the conclusion that the Sun, not the Earth, was at the center of the universe. At any rate, he is the first person of whom we have knowledge who professed a belief that the Earth goes about the Sun. He also postulated that the stars must be extremely distant to account for the fact that their parallaxes could not be observed.

(b)   Measurement of the Earth by Eratosthenes

Aristarchus had derived the dimensions of the Sun and Moon, but only in terms of the size of the Earth. The

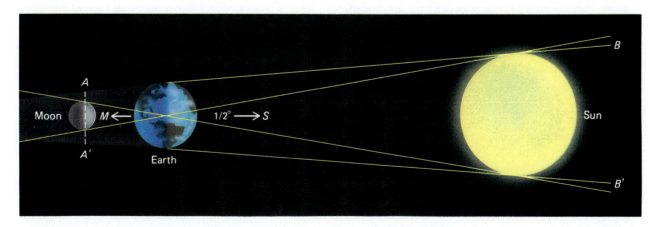

FIGURE 1.9   The principle by which Aristarchus could determine the relative sizes of the Sun, Moon, and Earth.

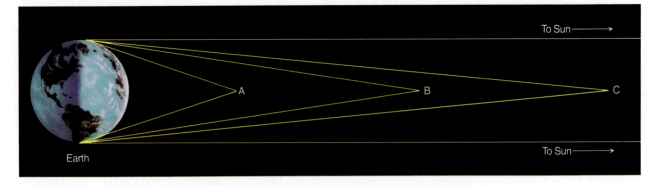

FIGURE 1.10   The more distant an object, the more nearly parallel are the rays of light coming from it.

latter was not accurately known to him. The first fairly accurate determination of the Earth's diameter was made by Eratosthenes (276–195 B.C.), an astronomer who lived in Alexandria in Egypt.

To appreciate Eratosthenes' technique for measuring the Earth, we must understand that the Sun is so distant from the Earth compared with its size that the Sun's rays approach all parts of the Earth along essentially parallel lines. To see why, imagine a light source near the Earth, say at position *A* in Figure 1.10. Its rays strike different parts of the Earth along diverging paths. From a light source at *B*, or at *C*, still farther away, the angle between rays that strike extreme parts of the Earth is smaller. The more distant the source, the smaller the angle between the rays. For a source infinitely distant, the rays travel along parallel paths. The Sun is not, of course, infinitely far away, but light rays striking the Earth from a point on the Sun diverge from each other by too small an angle to be observed with the unaided eye. As a consequence, if people all over the Earth who could see the Sun were to point at it, their fingers would all be pointing in the same direction—they would all be parallel to each other.

Eratosthenes noticed that at Syene, Egypt (now modern Aswan), on the first day of summer, sunlight struck the bottom of a vertical well at noon, which indicated that Syene was on a direct line from the center of the Earth to the Sun. At the corresponding time and date in Alexandria, 5000 stadia north of Syene (the stadium was a Greek unit of length), he observed that the Sun was not directly overhead but slightly south of the zenith, so that its rays made an angle with the vertical equal to 1/50 of a circle (about 7°). Yet the Sun's rays striking the two cities are parallel to each other. Therefore (Figure 1.11), Alexandria must be 1/50 of the way around the round Earth's circumference from Syene, and the Earth's circumference must be 50 × 5000, or 250,000, stadia.

It is not possible to evaluate precisely the accuracy of Eratosthenes' solution because there is doubt about which of the various kinds of Greek stadia he used. If it

was the common Olympic stadium, his result was about 20 percent too large. According to another interpretation, he used a stadium equal to about 1/6 km, in which case his figure was within 1 percent of the correct value of 40,000 km. The diameter of the Earth is found from the circumference, by dividing the latter by π = 3.14.

(c)   Hipparchus and the Zenith of Greek Astronomy

The greatest astronomer of pre-Christian antiquity was Hipparchus, who was born in Nicaea in Bithynia. The dates of his life are not accurately known, but he carried out his work at Rhodes, and possibly also at Alexandria, in the period from 160 to 127 B.C. Many of the phenomena Hipparchus detected are quite subtle, and the measurements he made—all without optical aid—were remarkable for the time.

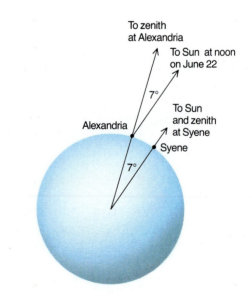

FIGURE 1.11   Eratosthenes' method of determining the size of the Earth.

Hipparchus erected an observatory on the island of Rhodes and built instruments with which he measured as accurately as possible the directions of objects in the sky. He compiled a star catalogue of about 850 entries. He designated for each star its celestial coordinates, that is, quantities analogous to latitude and longitude that specify its position (direction) in the sky. He also divided the stars according to their apparent brightness into six categories, or **magnitudes,** and specified the magnitude of each star. A modification of his magnitude system is still in use by astronomers today (Chapter 22).

In the course of his observations of the stars, and in comparing his data with older observations, Hipparchus made one of his most remarkable discoveries: The position in the sky of the north celestial pole had altered over the previous 150 years. Hipparchus correctly deduced that the axis about which the celestial sphere appears to rotate continually changes its position. The explanation for this phenomenon is that the direction of the Earth's rotational axis is slowly changing because of the gravitational influence of the Moon and the Sun, much as a top's axis describes a conical path as the Earth's gravitation tries to tumble the top over. This variation in the orientation of the Earth's axis, called **precession,** requires about 26,000 years to complete one cycle.

Hipparchus, refining the technique first applied by Aristarchus, also obtained a good estimate of the Moon's size and distance. He used the correct value of 1/2° for the angular diameters of the Sun and Moon and also the correct value of 8:3 for the ratio of the diameter of the Earth's shadow to the diameter of the Moon, as derived from observations of lunar eclipses. He concluded the Moon's distance to be 59 times the Earth's radius; the correct number is 60.

(d) Motions of the Sun and Moon

Hipparchus' study of the motion of the Sun deserves special mention. We now know that the Earth's true orbit around the Sun is not a circle but an ellipse. The Earth's distance from the Sun and its orbital speed both vary slightly. Thus the apparent rate of motion of the Sun relative to the fixed stars also varies during the year. But the ancient Greeks assumed that all celestial motion was circular, and thus the varying speed of the Sun was difficult for them to understand.

Eudoxus of Cnidas (about 400 B.C.) had accounted for the Sun's motion approximately by representing it with a series of rotating spheres pivoted one on the other. Later the mathematician Apollonius of Perga suggested that the motions of all the heavenly bodies could be represented equally well by a combination of uniform circular motions. By uniform circular motion is meant a motion at a uniform speed about the circum-

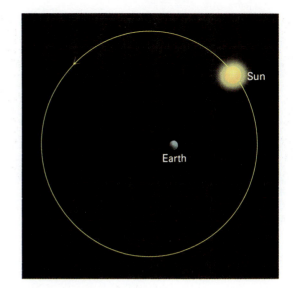

FIGURE 1.12 The eccentric.

ference of a circle. Because the circle is the simplest geometrical figure and uniform motion seemed the most natural kind, Hipparchus, following the suggestion of Apollonius, attempted to find a combination of uniform circular motions that would account for the Sun's apparently irregular behavior.

The plan he adopted was to represent the Sun's orbit by an **eccentric,** which is a circle with the Earth slightly off center (Figure 1.12). The scheme was successful because the true orbit of the Earth is very close to a circle with the Sun just off center. One effect of the Sun's variable speed on the ecliptic is to produce an inequality in the lengths of the seasons. Hipparchus remeasured the small differences between the seasons' durations and found that the Earth and Sun were nearest each other in early December, which was correct at that time. (The date has changed over the thousands of years and now occurs in early January.)

Hipparchus pointed out that he could also have represented the Sun's apparent motion by presuming it to move on the circumference of a kind of circle called an **epicycle,** whose center, in turn, revolves about the Earth in a circle called a **deferent** (Figure 1.13). However, he considered the eccentric a simpler and thus preferable system.

The Moon's motion is more complicated, and Hipparchus was not quite so successful in finding a geometrical scheme to describe it. According to the model he adopted, the Moon moved in a circle about a point near the Earth (an eccentric), but the center of the eccentric also revolved slowly about the Earth. The apparent motions of the planets are even more complicated than that of the Moon, and Hipparchus was unable to fit the planets into a logical scheme of circular orbits as he had for the Sun and Moon.

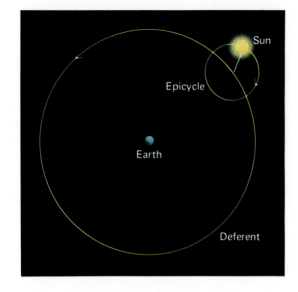

FIGURE 1.13   The deferent and epicycle.

1.4   PTOLEMY AND HIS HERITAGE

(a)   Ptolemy of Alexandria

The last great Greek astronomer of antiquity was Claudius Ptolemy (or Ptolemaeus), who lived in Alexandria around A.D. 140, when Egypt was part of the Roman empire. He compiled a series of 13 volumes on astronomy, usually known by its later Arabic title as the *Almagest.* Not all of the *Almagest* deals with Ptolemy's own work, for it includes a compilation of the astronomical achievements of the past, principally those of Hipparchus. In fact, it is our main source of information about Greco-Roman astronomy.

One of Ptolemy's accomplishments was a new measurement of the distance to the Moon. The method he used, the principle of which is illustrated in Figure 1.14, makes use of the Moon's parallax, discussed by Hipparchus in connection with solar eclipses. Suppose we could observe the Moon directly overhead. We would have to be, then, at position *A* on the Earth, on a line between the center of the Earth *E*, and the center of the Moon *M*. Suppose that at the same time someone else at position *B* were to observe the angle *ZBM* between the Moon's direction and the point directly over his head, *Z*. The angle *MBE* would then be determined in the triangle *MBE* (it is 180° minus angle *ZBM*). The distance from *A* to *B* determines the angle *BEM*. For example, if *A* is 1/12 of the way around the Earth from *B*, the angle *BEM* is 30°. The side *BE* is, of course, the radius of the Earth. We therefore know two angles and an included side of the triangle *MBE*. It is now possible to determine, either by trigonometry or by geometrical construction, the distance *EM* between the centers of the Earth and Moon.

In practice, we do not need another observer at *B*, for the rotation of the Earth will carry us over there in a few hours anyway, and we can observe the angle *ZBM* then. We shall have to correct, however, for the motion of the Moon in its orbit during the interval between our two observations; the Moon's motion being known, the correction is a detail easily accomplished. Using the principle described, Ptolemy determined the Moon's distance to be 59 times the radius of the Earth—very nearly the correct value.

(b)   Ptolemy's Scheme of Cosmology

Ptolemy's most important original contribution was a geometrical representation of the solar system that predicted the motions of the planets with considerable accuracy. The study of the structure and motions of the universe is called **cosmology.** In those days, the known universe was limited to the planets, and the central problem of cosmology was to understand and predict their complex apparent motions. Hipparchus, having determined by observation that earlier theories of planetary motion did not fit their actual behavior, and not having enough data on hand to solve the problem himself, instead amassed observational material for pos-

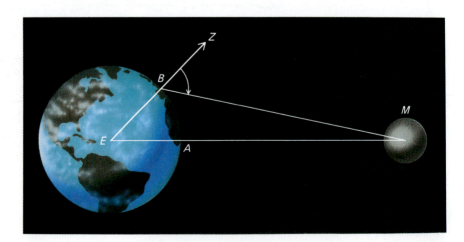

FIGURE 1.14   Ptolemy's method of finding the distance to the Moon.

terity to use. Ptolemy supplemented the material with observations of his own and with it produced a cosmological model that endured until the time of Copernicus more than a thousand years later.

The complicating factor in the analysis of the planetary motions is that their apparent wanderings in the sky result from the combination of their own motions and the Earth's orbital revolution. Notice, in Figure 1.15, the orbit of the Earth and the orbit of a hypothetical planet farther from the Sun than the Earth. The Earth travels around the Sun in the same direction as the planet and in nearly the same plane, but the Earth has a higher orbital speed. Consequently, it periodically overtakes the planet, like a faster race car on the inside track. The apparent directions of the planet, seen from the Earth, are shown at successive intervals of time along lines $AA'A''$, $BB'B''$, and so on. In the right side of the figure, we see the resultant apparent path of the planet among the stars. From positions B to D, the planet appears to drift backward, to the west in the sky, even though it is actually moving to the east. Similarly, a slowly moving car appears to drift backward with respect to the distant scenery when we pass it in a faster-moving car. As the Earth rounds its orbit toward position E, the planet again takes up its usual eastward motion in the sky.

The temporary westward motion of a planet as the Earth swings between it and the Sun is called **retrograde motion.** Retrograde motion is easy to understand in the context of a moving Earth. But Ptolemy required a different explanation for retrograde

motion, since he assumed that the planet was revolving about a stationary Earth.

Ptolemy solved the problem by having a planet P (Figure 1.16) revolve in an epicyclic orbit about C. The center of the epicycle C in turn revolved in the deferent about the Earth. When the planet is at position x, it is moving in its epicyclic orbit in the same direction as the point C moves about the Earth, and the planet appears to be moving eastward. When the planet is at y, however, its epicyclic motion is in the direction opposite to the motion of C. By choosing the right combination of speeds and distances, Ptolemy succeeded in having the planet moving westward at the right speed at y and for the correct interval of time. However, because the planets, as well as the Earth, travel about the Sun in elliptical orbits, their actual behavior cannot be represented accurately by so simple a scheme of uniform circular motions. Consequently, Ptolemy made the deferent an eccentric, centered not on the Earth, but slightly away from the Earth at A. Furthermore, he had the center of the epicycle, C, move at a uniform angular rate, not around A, or E, but at point B, called the *equant*, on the opposite side of A from the Earth.

It is a tribute to the genius of Ptolemy as a mathematician that he was able to conceive such a complex model to account successfully for the observations. His hypothesis, with some modifications, was accepted as absolute authority until the 17th century. In the *Almagest*, however, Ptolemy made no claim that his cosmological model described reality. He intended his

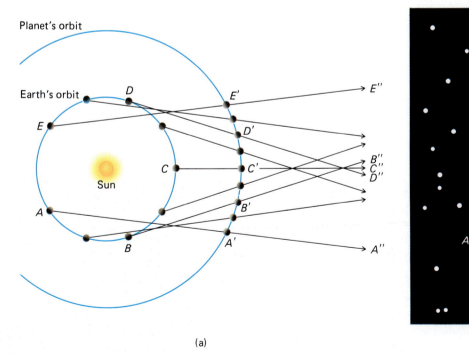

(a)   (b)

FIGURE 1.15 Retrograde motion of a planet external to the Earth's orbit. (a) Actual positions of the planet and the Earth. (b) The apparent path of the planet as seen from the moving Earth, against the background of stars.

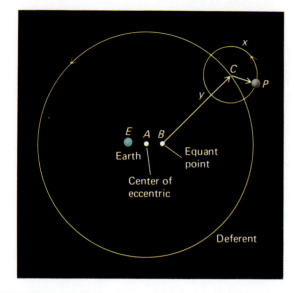

FIGURE 1.16 Ptolemy's cosmological system based on geocentric motion; the diagram shows the deferent, epicycle, eccentric, and equant.

scheme rather as a mathematical representation to predict the positions of the planets at any time—a kind of conceptual computer program.

(c) Ptolemy's Scheme of Astrology

In addition to his work in astronomy, Ptolemy also developed many of the concepts of the mystical religion called **astrology.** The basic idea of astrology is that events on Earth are influenced and perhaps controlled by the positions and motions of the planets. Underlying this idea is the concept that the celestial bodies exert some mysterious and powerful force upon us.

Modern research has shown that all matter in the universe is composed of atoms—and the same kinds of atoms. Thus our space probes on Mars and Venus, or our telescopic spectra of the light from the most remote quasars, indicate that these objects are made of the same stuff that makes up our own bodies. The ancients did not know this, of course. They assumed that the luminous orbs in the sky are made of "heavenly" substances and not of the "earthly" elements we find on Earth. In fact, the realization that celestial worlds are actually worlds and not ethereal substance is relatively recent in the history of science.

In the geocentric view, the planets were close to the Earth and revolved about it. Presumably, they existed for some purpose associated with the Earth and its human population. Small wonder, then, that the ancients regarded the planets (including the Sun and Moon) as having special significance. Often the planets came to be associated with the gods of ancient mythologies; in some cases, they were themselves thought of as gods. Even in the comparatively sophisticated Greece of antiquity, the planets had the names of gods and were credited with having the same powers and influences as

the gods whose names they bore. Astrology was a logical product of this magic-based philosophy. What is surprising is that this ancient belief system should still be popular in the world today.

Astrology began, we think, in the valley of the Euphrates and Tigris rivers about 3000 years ago. The Mesopotamians and the Babylonians, believing that the planets and their motions influenced the fortunes of kings and nations, practiced what we call mundane astrology. When the Babylonian culture was absorbed by the Greeks, their astrology gradually influenced the entire Western world and eventually spread to the Orient as well. By the second century B.C. the Greeks had democratized astrology by developing the tradition that the planets influenced the life of every individual. In particular, they believed that the configuration of the planets at the moment of a person's birth affected his or her personality and fortune. This form of astrology, known as natal astrology, reached its acme with Ptolemy. His *Tetrabiblos*, a treatise on astrology, remains the bible on the subject even today.

The key to natal astrology is the *horoscope*, a chart that shows the positions of the planets in the sky at the moment of an individual's birth. The planets are located in the sky with respect to the fixed stars on the celestial sphere by specifying their positions in the zodiac. For the purposes of astrology, the zodiac is divided into 12 sectors, called signs, each 30° long. In addition, the constantly turning celestial sphere, with its stars and the planets, must have its orientation specified with respect to the Earth at the time and place of the subject's birth. For this purpose the sky is divided into 12 regions, called houses, that are fixed with respect to the horizon. Each day the turning sky carries the planets and signs through all of the houses. An example of a horoscope (author George Abell's) is shown in Figure 1.17.

There are more or less standardized rules for the interpretation of the horoscope, many or most of which (at least in Western schools of astrology) are derived from the *Tetrabiblos*. Each sign, each house, and each planet, the last supposedly acting as a center of force, is associated with particular matters. However, the interpretation of a horoscope is a very complicated business, and whereas the rules may be standardized, how each rule is to be weighed and applied is a matter of judgment—and "art."

(d) Astrology Today

Astrologers today use the same basic principles written down by Ptolemy nearly 2000 years ago. They cast horoscopes (a process much simplified by the development of appropriate computer programs) and suggest interpretations. A popular modern variant of natal astrology is Sun-sign astrology, which uses only one element of the horoscope, the sign occupied by the Sun at the time of a person's birth. Although even professional astrologers do not place much trust in such a limited scheme, which tries to fit everyone into just 12 groups, Sun-sign astrology is the mainstay of newspaper astrology columns and party games. Apparently, many people take it quite seriously; a recent poll showed that more than half of the teenagers in the U.S. said

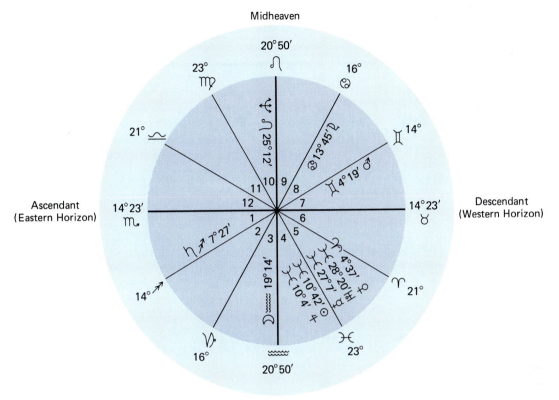

FIGURE 1.17 Natal horoscope of George O. Abell, who was born in Los Angeles, California, on March 1, 1927, at 10:50 P.M., PST. The 12 pie-shaped sectors represent the 12 houses, and the outer circular zone represents the zodiac. The boundaries between the houses (cusps) intersect the ecliptic in the zodiacal signs indicated by their symbols in the outer circular zone. The number beside each sign symbol is the angular distance of the cusp from the beginning of that sign. The position of each planet is shown in the house it occupied at the instant of birth. Beside the symbol for the planet is the symbol of the zodiacal sign it was also in at that time and the angular distance of the planet from the beginning of that sign. The places where the horizon intersects the zodiac are shown, as well as the highest point of the ecliptic in the sky (midheaven) and its lowest point below the horizon (astrological nadir).

they "believed in astrology."

Today, with our knowledge of the nature of the planets as physical bodies, composed as they are of rocks and fluids, it is hard to imagine that the directions of these planets in the sky at the moment of one's birth could have anything to do with one's personality or future. The gravitational influence of the Moon and Sun on tides is unquestionable, but tides produced on a person by a book held in his or her hand are millions of times stronger than those produced by all the planets combined. Jupiter (and to a lesser extent, the other planets) emits radio waves, but their detection requires large radio telescopes. The feeble radio signals from a small 1000-watt transmitter 100 miles away reach us with a strength millions of times greater than the radio waves from Jupiter and can be picked up by a pocket radio. Moreover, the distances of the planets from the Earth vary greatly, and any gravitational and radiation effects would vary as the inverse square of their distances—factors ignored by astrology.

Some astrologers argue that there are unknown forces exerted by the planets that depend on their configurations with respect to each other and with respect to arbitrary coordinate systems—forces for which there is not a whit of solid evidence. Are astronauts on the Moon similarly affected by the same kind of force exerted by the Earth? Or is the Earth alone subject to these unknown laws of nature?

Another curious aspect of astrology is its exclusive emphasis on the configurations of the planets at *birth*. What about the forces that might influence us at conception? Isn't our genetic makeup, determined by the particular joining of a sperm with an egg, much more important for determining our personality than the circumstances of our birth? Why don't these mysterious planetary forces act to influence us at conception?*

---

* For more on questions about astrology and some of the tests of its efficacy, see *Sky & Telescope* magazine, August 1989, p. 146.

In the most orthodox astrology, one's entire life (and death) is predetermined by the natal horoscope. If a man dies in an auto accident at the age of 63 because someone else ran a stoplight, are we supposed to assume that all of the complicated chain of events that led to the circumstances of his being in that accident were blueprinted by the planets at the instant of his birth, but that all would have been different if he had been born 2 hr later? Most of us find this assumption so incredible that we would need the most overwhelming evidence of its validity before taking it seriously. In the tens of centuries of astrology, no such evidence has been presented.

One could argue, on the other hand, that astrology works only statistically and that astrological influences are important only as tendencies, everything else being equal. In that case the reality of astrological effects could be tested only statistically. During the past few years, a number of statistical tests of the predictive power of natal astrology have been carried out. The simplest of these examine Sun-sign astrology to determine whether some signs are more likely than others to be associated with such objective measures of success as winning Olympic medals, earning high corporate salaries, achieving elective office, or attaining high military rank. You could make such a test yourself using, for example, the birth dates of all members of Congress or of all members of the U.S. Olympic Team. But more sophisticated studies have also been done, involving horoscopes calculated for thousands of individuals. The results of all of these studies are the same: There is no evidence that natal astrology has any predictive power, even in a statistical sense.

In retrospect, we can understand the belief in astrology on the part of ancient peoples, who thought the heavenly bodies to be made of celestial material different from the elements that compose the Earth and to be placed in the sky by their gods for the benefit of humanity. In the light of modern knowledge, the astrological claims seem ludicrous. Virtually all scientists reject astrology as an unfounded superstition. Yet it continues to appeal to the popular fancy. The hope of predicting the future by magical or mystical means, and perhaps of transferring one's responsibilities and the blame for one's failures and misfortunes to an omnipotent power, continues to be a strong attraction.

One fact remains: The practice of astrology in ancient times required knowledge of the motions of the planets in order to construct horoscopes for past or future events. The quest to find a mechanism for charting the planets, joined with a natural curiosity about nature, stimulated centuries of observations and calculations, leading, as we shall see, to our modern science and technology.

S U M M A R Y

1.1 The direct evidence of our senses supports a **geocentric** perspective, with the **celestial sphere** pivoting on the **celestial poles** and rotating about a stationary Earth. We see only half of this sphere at one time, limited by the **horizon;** the point directly overhead is the **zenith.** The Sun's annual path on the celestial sphere is the **ecliptic,** a line that runs through the center of the **zodiac,** the 18°-wide strip of sky within which are found the Moon and planets. The fixed stars are organized into 88 **constellations.**

1.2 Greeks such as Aristotle recognized that the Earth and Moon were spheres and understood the phases of the Moon and the nature of lunar and solar **eclipses.** They knew that the Sun was more distant than the Moon, but they were not sure either was as large as the Earth. Because of his inability to detect stellar **parallax,** Aristotle rejected the idea that the Earth moved.

1.3 Aristarchus postulated that the Earth circled the Sun, and Eratosthenes measured the size of the Earth with surprising precision. Hipparchus carried out many astronomical observations, made a star catalogue, defined the system of stellar **magnitudes,** and discovered **precession** from the apparent shift of the position of the north celestial pole. He also developed an elaborate model for the motions of the Sun and Moon based on uniform motion along an **eccentric,** which is a circle offset from the center of the Earth.

1.4 Ptolemy of Alexandria summarized classical astronomy in his *Almagest* and **astrology** in his *Tetrabiblos.* His geocentric **cosmology** explained complex planetary motions, including **retrograde motion,** with high accuracy. This model, based on combinations of uniform circular motion using the **epicycle** and **deferent,** was accepted as authority for more than a thousand years.

E X E R C I S E S

THOUGHT QUESTIONS

1. Where on Earth are all stars above the horizon at one time or another? Where is only half the sky ever above the horizon?

2. Show with a simple diagram how the lower parts of a ship disappear first as it sails away from you on a spherical Earth. Use the same diagram to show why lookouts on

old sailing ships could see farther from the masthead than from the deck. Would there be any advantage to posting lookouts on the mast if the Earth were flat? (Note that these nautical arguments for a spherical Earth were quite familiar to Columbus and other mariners of his time.)

3. About what time of day or night does the Moon rise when it is full? When it is new?

4. Why can an eclipse of the Moon never occur on the day following a solar eclipse?

5. As seen by a terrestrial observer, which (if any) of the following can never appear in the opposite direction in the sky from the Sun? In the same direction? At an angle of 90° from the Sun?
   a. Mars
   b. a star
   c. the Sun
   d. Earth
   e. Jupiter
   f. the Moon
   g. Venus
   h. Mercury

6. Why would Eratosthenes' method not have worked if the Earth were flat, like a pancake?

7. From our modern perspective, what are the strongest arguments you could use to convince a skeptical neighbor
   a. that the Earth is not flat and
   b. that the Earth is not stationary but goes around the Sun?

Problems

8. The north celestial pole appears at an altitude above the horizon that is equal to the observer's latitude. Identify Polaris, the North Star, which lies very close to the north celestial pole. Measure its altitude. (This can be done with a protractor. Alternatively, you can use your fist, which, extended at arm's length, spans a distance approximately equal to 10°.) Compare this estimate with your latitude. The next time you travel several hundred miles north or south, determine the altitude of Polaris again. Can you detect a difference? (This experiment cannot easily be performed in the Southern Hemisphere because Polaris itself is, of course, not visible, and there is no bright star near the south celestial pole.)

9. The Moon moves with respect to the background stars. Go outside at night and note the position of the Moon relative to nearby stars. Repeat the observation a few hours later. How far has the Moon moved relative to the stars? (For reference, the diameter of the Moon is about 1/2°.) Based on your estimate of the Moon's motion, how long will it take for the Moon to return to the position relative to the stars in which you first observed it?

10. Look up the names of the days of the week in French, Italian, and Spanish, and compare them with the names of the planets.

11. The Earth's diameter is about 3.7 times the diameter of the Moon. What is the angular diameter of the Earth as seen by an observer on the Moon?

12. Suppose Eratosthenes had found that at Alexandria at noon on the first day of summer the line to the Sun makes an angle of 30° with the vertical. What then would he have found for the Earth's circumference?

13. Suppose Eratosthenes' results for the Earth's circumference were quite accurate. If the diameter of the Earth is 12,740 km, evaluate the length of his stadium in kilometers.

14. You are on a strange planet. You note that the stars do not rise or set, but that they circle around parallel to the horizon. Then you travel over the surface of the planet in one direction for 10,000 km, and at that new place you find that the stars rise straight up from the horizon in the east and set straight down in the west. What is the circumference of the planet?
    Answer: 40,000 km

15. Is retrograde motion observed for a planet, like Venus, which has an orbit inside that of the Earth? Explain.

★16. According to the geocentric theory of the orbits of the planets, Venus moves on a circle that always lies between the Earth and the Sun. What phases would Venus go through? How does this prediction differ from that of the heliocentric theory?

A remarkable photograph showing both the Earth and the Moon, taken December 16, 1992, with the Galileo spacecraft. Galileo was on its way toward Jupiter and looked back at our home system from a distance of 6.2 million km. The contrast in this photo was adjusted to make the dimmer Moon somewhat brighter and thus easier to see. Project scientists named the spacecraft after the 17th-century astronomer who first observed the moons of Jupiter with a telescope.

(JPL/NASA photo)

On February 16, 1600, in the Roman Field of Flowers, the Italian monk and astronomical author Giordano Bruno was burned at the stake by the Inquisition. He was put to death in part because in his writings he advocated the Copernican system, in which the Earth was not the center of the universe, and because he taught that there were many worlds like the Earth in the vastness of space.★

Although it came too late to save Bruno's life, the new century was to bring a wind of change that would sweep away the Earth-centered view that had dominated European thinking for centuries and usher in a new method of discovering truth about the natural world. From the ashes of that fire in 1600, through the actions of the pioneering astronomers described in this chapter, would come the birth of the scientific method and the beginnings of our modern perspective on the cosmos.

With the work of Copernicus, Kepler, Tycho, and Galileo, the Earth became just another planet revolving about the Sun, and the long quest to understand humanity's true place in the universe could finally get under way. We begin our exploration of that quest in this chapter, but we will continue to follow the story throughout the book.

★ The authors of this text certainly hope that any of our readers who may be unhappy with us will advocate less drastic measures.

Nicolaus Copernicus (1473–1543), cleric and scientist, played a leading role in the emergence of modern science. Although he could not prove that the Earth revolves about the Sun, he presented such compelling arguments for this idea that he turned the tide of cosmological thought.

COPERNICUS AND THE HELIOCENTRIC HYPOTHESIS

In the millennium following Ptolemy, the most significant astronomical investigations were made by the Hindus and Arabs. The Hindus invented our system of numbers with place counting by tens. The Arabs brought the Hindu system of numbers to Europe and developed trigonometry. They also had access to many of the writings of the Greek astronomers, and their culture provided continuity between the Greco-Roman world and the development of modern astronomy in the Renaissance.

Astronomy made no major advances in medieval Europe, where the prevailing philosophy was acceptance of the dogma of authority. Medieval cosmology combined the crystalline spheres of Pythagoras (as perpetuated by Aristotle) with the epicycles of Ptolemy. Astrology was widely practiced, however, and an interest in the motions of the planets was thus kept alive. Then came the Renaissance; in science the rebirth was begun by Nicolaus Copernicus. His theory that the Earth revolves about the Sun was (pun intended) a revolution in cosmological thought, laying the foundations upon which Galileo and Kepler so effectively built in the following century.

2.1 COPERNICUS

Nicolaus Copernicus (in Polish, Mikolaj Kopernik, 1473–1543) was born in Torun on the Vistula in mod-

ern-day Poland. His training was in law and medicine, but Copernicus' main interest was astronomy and mathematics. His great contribution to science was a critical reappraisal of the existing theories of planetary motion and the development of a new model of the solar system. His unorthodox idea that the Sun is at the center of the solar system is called the **heliocentric** (Sun-centered) model of cosmology.

His ideas were presented in detail in his book *De Revolutionibus*, published in the year of his death (Figure 2.1). In it, Copernicus set forth certain postulates from which he derived his system of planetary motions. His postulates included the traditional assumptions that the universe is spherical and that the motions of the heavenly bodies must be made up of combinations of uniform circular motions. Yet, he evidently found something orderly and pleasing in the new idea of the heliocentric system, and his defense of it was elegant and persuasive. His model, although not widely accepted until more than a century after his death, was ultimately of immense influence.

(a) Planetary Motions According to Copernicus

A person moving at a uniform speed is not necessarily aware of this motion. We have all seen an adjacent train, car, or ship appear to change position, only to discover that it is we who are moving. Copernicus argued that

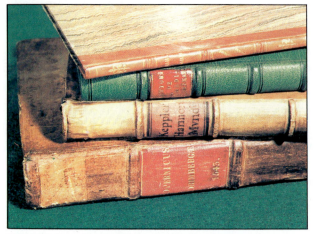

FIGURE 2.1 First editions of some books of great historical interest, including *De Revolutionibus*. (Crawford Collection, Royal Observatory Edinburgh)

the apparent annual motion of the Sun about the Earth could be represented equally well by a motion of the Earth about the Sun. He also reasoned that the apparent rotation of the celestial sphere could be accounted for by assuming that the Earth rotates about a fixed axis while the celestial sphere is stationary.

The most important hypothesis Copernicus adopted in *De Revolutionibus* was that the Earth was but one of six (then known) planets that revolve about the Sun. Given this, he was able to work out the correct general picture of the solar system. He placed the planets, starting nearest the Sun, in the correct order of Mercury, Venus, Earth, Mars, Jupiter, and Saturn (Figure 2.2). Further, he deduced that the nearer a planet is to the Sun, the greater is its orbital speed. The retrograde motions of the planets were easily understood without the necessity for epicycles (Section 1.4), simply as a consequence of the moving Earth. Also, Copernicus worked out the correct approximate scale of the solar system. However, because Copernicus retained the Aristotelian concept of uniform circular motion in his heliocentric theory, he had trouble explaining the details of planetary motions that we now know are a result of their noncircular orbits. As a consequence, Copernicus had to resort to a number of epicycles in order to achieve the accuracy he required in predicting the positions of the planets in the sky.

The heliocentric model of Copernicus did not prove that the Earth revolves about the Sun. In fact, with some adjustments, the old Ptolemaic system could have accounted as well for the motions of the planets in the sky. But the Ptolemaic cosmology was clumsy and lacked the beauty and symmetry of its successor. Copernicus made the Earth an astronomical body, which brought a kind of unity to the universe. In addition, his new cosmology had the revolutionary implication that the Earth was a small and perhaps not so important place. We were no longer the central element in the cosmic order!

In Copernicus' time, few people would have supposed that ways existed to prove whether the heliocentric or the older geocentric system was correct. A long philosophical tradition, going back to the Greeks and defended by the Catholic Church, held that pure human thought combined with divine revelation represented the path to truth. Nature and the evidence of our senses were suspect. In this environment, there was little motivation to carry out observations or experiments to try to distinguish between competing theories for the solar system (or anything else).

It should not surprise us, therefore, that the heliocentric idea was debated for more than half a century without any tests being applied to determine its validity. (In fact, Ptolemy's geocentric system was still taught at Harvard University in the first years after its founding in 1636.) Contrast this with the situation today, when scientists rush out to try to test each new theory. In the three centuries since Copernicus, we have learned that nature is orderly and consistent, and that experiments and observations can therefore be used as a check on theory. And, in addition, we have developed sophisticated technologies to aid in our probing of the natural world.

When a new hypothesis or theory is proposed in science, it should first be checked for consistency with what is already known. Copernicus' heliocentric idea passed this test, for it allowed planetary positions to be calculated at least as well as the Ptolemaic theory. The next step is to see what predictions made by the new theory differ from those of competing ideas. In the case of the heliocentric theory, one example is that if Venus

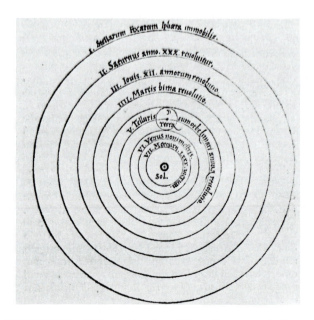

FIGURE 2.2 Heliocentric plan of the solar system in the first edition of Copernicus' *De Revolutionibus*. (Crawford Collection, Royal Observatory Edinburgh)

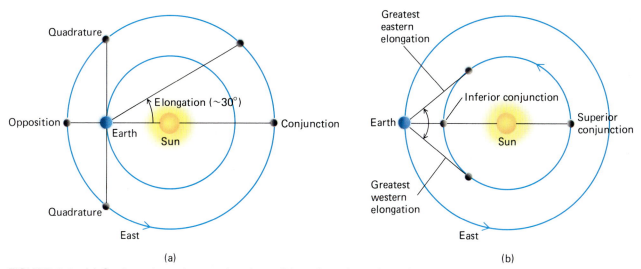

FIGURE 2.3 (a) Configurations of a superior planet; (b) configurations of an inferior planet.

circles the Sun, it should go through the full range of phases just as the Moon does, whereas if it and the Sun circle the Earth, it should not. In 1610, Galileo used his new telescope to discover that Venus does go through the full range of phases. But before the telescope, no one could have imagined testing this prediction.

Today, the predictions of competing theories are rarely as simple as the example of the phases of Venus. Since most predictions in science are *quantitative*, involving numerical values that can be calculated from mathematical equations, the checks usually require accurate measurements of the phenomena being studied. Often it is only through increasing the precision of the observations that a distinction between two competing theories can be made. For example, ordinary gravitational theory and Einstein's theory of relativity make nearly identical predictions about everyday experience; these theories diverge only under extreme conditions, such as the surface of a collapsed star.

(b) The Scale of the Solar System

Using his new heliocentric model, Copernicus worked out the scale of the solar system. To understand how, it will be helpful to define a few terms that describe the positions of planets in their orbits. These are illustrated in Figure 2.3.

A **superior planet** is any planet whose orbit is larger than that of the Earth, that is, a planet that is farther from the Sun than the Earth is (e.g., Mars, Jupiter, and Saturn). An **inferior planet** is a planet closer to the Sun than the Earth is (e.g., Venus and Mercury).

Every now and then, the Earth passes between a superior planet and the Sun. Then that planet appears in the opposite direction in the sky from the Sun. At such time, the planet rises at sunset, is above the horizon all night long, and sets at sunrise. We look one way to see the Sun and in the opposite direction to see the planet. The planet is then said to be in **opposition.** On other occasions, a superior planet is on the other side of the Sun from the Earth. It is then in the same direction from the Earth as the Sun is and of course is not visible. At such time, the planet is said to be in **conjunction.**

Between these extremes (but not halfway between), a superior planet may appear 90° away from the Sun in the sky, so that a line from the Earth to the Sun makes a right angle with the line from the Earth to the planet. Then the planet is said to be at **quadrature.** At quadrature, a planet rises or sets at either noon or midnight.

The angle formed at the Earth between the Earth-planet direction and the Earth-Sun direction is called the planet's **elongation.** In other words, the elongation of a planet is its angular distance from the Sun as seen from the Earth. At conjunction, a planet has an elongation of 0°, at opposition 180°, and at quadrature 90°.

An inferior planet can never be at opposition, for its orbit lies entirely within that of the Earth. Its largest possible angular distance from the Sun, on either the east or the west side, is called its greatest eastern elongation or greatest western elongation. When an inferior planet passes between the Earth and Sun, it is in the same direction from Earth as the Sun and is said to be in inferior conjunction. When it passes on the far side of the Sun from the Earth, and is again in the same direction as the Sun, it is said to be at superior conjunction.

Copernicus was able to find the planets' distances from the Sun relative to the Earth's. For the sake of illustration, let us assume that the orbits of the planets are precisely circular, even though that assumption is an

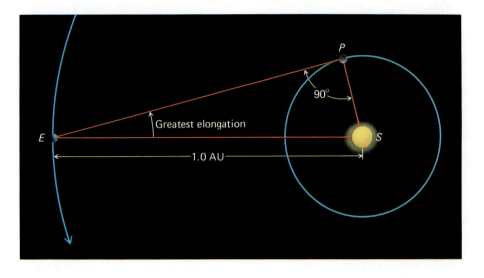

FIGURE 2.4 Determination of the distance of an inferior (inner) planet from the Sun, relative to the Earth's distance (which is called an astronomical unit, or AU).

oversimplification. The problem is particularly simple for the inferior planets. When an inferior planet is at greatest elongation (Figure 2.4), the line of sight from the Earth to the planet, EP, must be tangent to the orbit of the planet, and hence perpendicular to the line from the planet to the Sun, PS. We have, therefore, a right triangle, EPS. The angle PES is observed (it is the greatest elongation), and the side ES is the Earth's distance from the Sun. The planet's distance from the Sun, side PS, can then be found, in terms of the Earth's distance, by geometrical construction or by trigonometric calculation.

As an illustration of the procedure by which the distance of a superior planet can be found, suppose (Figure 2.5) the planet P is at opposition. We can now time the interval until the planet is next at quadrature; the planet is then at P' and the Earth at E'. With a knowledge of the revolution periods of the planet and

the Earth, we can calculate the fractions of their respective orbits that have been traversed by the two bodies. Thus the angles PSP' and ESE' can be determined, and subtraction gives the angle $P'SE'$ in the right triangle $P'SE'$. The side SE' is the Earth's distance from the Sun, so enough data are available to solve the triangle and find the planet's distance from the Sun, $P'S$ (again in terms of the Earth's distance), by construction or calculation.

The values obtained by Copernicus for the distances of the various planets from the Sun, in units of Earth's distance, are summarized in Table 2.1. Also given are the values determined by modern measurement.

TABLE 2.1	DISTANCES OF PLANETS FROM THE SUN IN AU	
Planet	Copernicus	Modern
Mercury	0.38	0.39
Venus	0.72	0.72
Earth	1.00	1.00
Mars	1.52	1.52
Jupiter	5.22	5.20
Saturn	9.18	9.54

(c) Sidereal and Synodic Periods

The time for a planet to move from one opposition (or conjunction) to the next is not the same as its true period of revolution about the Sun. Copernicus recognized the distinction between the *sidereal period* of a planet—that is, its actual period of revolution about the Sun with respect to the fixed stars—and its *synodic period*, its apparent period of revolution about the sky with respect to the Sun. The sidereal period is what we have called simply the period of revolution of a planet about the Sun. The synodic period is the time required for it to return to the same configuration, such as the

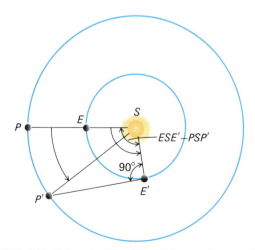

FIGURE 2.5 Determination of the distance of a superior (outer) planet from the Sun, relative to the Earth's distance.

time from opposition to opposition or from conjunction to conjunction.

Consider two planets, A and B, A moving faster in a smaller orbit (Figure 2.6). At position (1), planet A passes between B and the Sun S. Planet B is at opposition as seen from A, and A is in inferior conjunction as seen from B. When A has made one revolution about the Sun and has returned to position (1), B has, in the meantime, moved on to position (2). In fact, A does not catch up with B until both planets reach position (3). Now planet A has gained one full lap on B. Planet A has revolved in its orbit through 360° plus the angle that B has described in traveling from position (1) to position (3) in its orbit. The time required for the faster-moving planet to gain a lap on the slower-moving one is the synodic period of one with respect to the other. If B is the Earth and A an inferior planet, the synodic period of A is the time required for the inferior planet to gain a lap on the Earth; if A is the Earth and B a superior planet, the synodic period of B is the time for the Earth to gain a lap on the superior planet.

What is observed directly from the Earth is the synodic, not the sidereal, period of a planet. By reasoning along the lines outlined in the last paragraph, however, we can deduce the sidereal periods of the planets from their synodic periods. Let a planet's sidereal period be P years and its synodic period S years. In S years, the Earth, completing one revolution per year, must make S trips around the Sun. (The quantity S, of course, can be less than 1, in which case the Earth would complete less than one circuit.) The other planet, completing one revolution in P years, would make, in S years, S/P trips around the Sun.

Consider first an inferior planet. It has made one more trip around the Sun during its synodic period than has the Earth, so $(S + 1) = S/P$, which, by rearrangement of terms, can be written

$$1/P = 1 + 1/S \text{ for an inferior planet.}$$

For a superior planet, it is the Earth that gains the extra lap, and $S = (S/P + 1)$, which can be written

$$1/P = 1 - 1/S \text{ for a superior planet.}$$

As an example, consider Jupiter, whose synodic period is 1.09211 years. Since Jupiter is a superior planet, $1/P = 1 - 1/1.09211 = 1 - 0.91566$, or $1/P = 0.08434$. Thus, $P = 1/0.08434 = 11.86$ years.

2.2   TYCHO

Three years after the publication of *De Revolutionibus*, Tycho Brahe (1546–1601) was born into a family of Danish nobility. Tycho (as he is generally known) developed an early interest in astronomy and as a young man made significant astronomical observations. His work gained him the patronage of Frederick II, the Danish king, and in 1576 Tycho was able to establish a fine astronomical observatory on the Baltic island of Hveen. The chief building of the observatory was named Uraniborg. The facilities at Hveen included a library, a laboratory, living quarters, workshops, a printing press, and even a jail. There, for 20 years, Tycho and his assistants carried out the most complete and accurate pretelescopic astronomical observations yet made.

Unfortunately, Tycho was both arrogant and extravagant, and after Frederick II died, the new king, Christian IV, lost patience with the astronomer and eventually discontinued his support. Thus, in 1597 Tycho was forced to leave Denmark. He took up residence near Prague, taking with him some of his instruments and most of his records. There, as court astronomer for Emperor Rudolph II of Bohemia, Tycho Brahe spent the remaining years of his life analyzing the data accumulated over 20 years of observation. In 1600, the year before his death, he secured the assistance of a most able young mathematician, Johannes Kepler, who, like Tycho, was in exile from his native land.

(a)   Tycho's Observations

Tycho, like others of his time and before him, thought that comets were luminous vapors in the Earth's atmosphere. In 1577, however, a bright comet appeared for which he could observe no parallax. Tycho concluded that the comet was at least three times as distant as the Moon and guessed that it probably revolved around the Sun, in contradiction to earlier beliefs.

Tycho is most famous for his very accurate observations of the positions of the stars and planets. With instruments of his own design, he was able to make observations accurate to the limit of vision with the naked eye. The positions of the nine fundamental stars in his excellent star catalogue were accurate in most cases to within 1 arcminute (1/60°).

Tycho's observations included a continuous record of

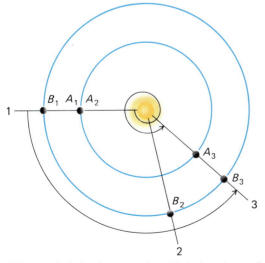

FIGURE 2.6   Relation between the sidereal and synodic periods of a planet.

the positions of the Sun, Moon, and planets. He re-evaluated nearly every astronomical constant and determined the length of the year to within 1 s. His extensive and precise observations of planetary positions enabled him to note that there were variations in the positions of the planets from those given in published tables, and he even noted regularities in the variations.

(b)   Tycho's Cosmology

Tycho rejected the Copernican heliocentric hypothesis on what seemed at the time to be very sound grounds. First, he found it difficult to reconcile a moving Earth with certain Biblical statements. The fact that he could not detect a parallax for even a single star, moreover, meant that the stars would have to be enormously distant if the Earth revolved around the Sun. Yet Tycho believed that he could measure the angular sizes of stars. The brightest of them he thought to be 2 arcminutes (1/30°) across. Now, the farther away an object is, the larger must be its true size in order that it have a given angular diameter. Tycho could not detect as much as 1 arcminute of parallax for any star, so it followed that the stars were so distant that, to have angular diameters of 2 arcminutes, their actual sizes would have to be twice the size of the entire orbit of the Earth. If they were still farther away, their diameters would have to be proportionally greater. (Later telescopic observations showed that the stars, unlike the planets, appear as luminous points; their disk-like appearance to the naked eye is illusory.)

Tycho's unhappiness with both the Ptolemaic and the Copernican models led him to suggest an original system of cosmology, although it was not worked out in full detail. To resolve the parallax problem, he envisioned a stationary Earth in the center with the Sun revolving about it. The other planets revolved about the Sun in the order Mercury, Venus, Mars, Jupiter, and Saturn (Figure 2.7). His ideas were debated among the astronomers of his time, but later deserted as the Copernican system gained favor. It is for his superb observations, and not his cosmological ideas, that we remember Tycho today.

2.3   KEPLER

Johannes Kepler (1571–1630) was born in Württemberg in southwestern Germany. He attended the University of Tübingen and studied for a theological career. There he learned the principles of the Copernican system. He became an early convert to the heliocentric hypothesis and defended it in arguments with his fellow scholars.

In 1594, because of his facility as a mathematician, he was offered a position teaching mathematics and astronomy at the secondary school at Graz, Austria. As part of his duties at Graz, he prepared almanacs that gave astronomical and astrological data. Eventually, however, the power of the Catholic church in Graz grew to the point where Kepler, a Protestant, was forced to quit his post. Accordingly, he went to Prague as an assistant to Tycho Brahe.

Tycho set Kepler to work trying to find a satisfactory theory of planetary motion—one that was compatible with the long series of observations made at Hveen. Tycho, however, was reluctant to supply Kepler with enough data to enable him to make substantial progress;

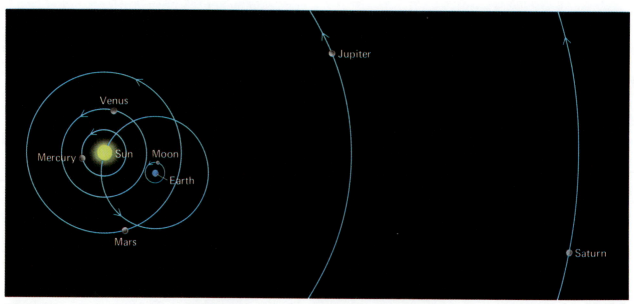

FIGURE 2.7   Tycho's model for the solar system.

perhaps Tycho was afraid of being "scooped" by the young mathematician. After Tycho's death, though, Kepler succeeded him as mathematician to Emperor Rudolph and obtained possession of the majority of Tycho's records. Their study occupied most of Kepler's time for more than 20 years.

(a) The Orbit of Mars

Kepler's most detailed study was of Mars, for which the observational data were the most extensive. He published the first results of his work in 1609 in *The New Astronomy, or Commentaries on the Motions of Mars.* He had spent several years trying without success to fit various combinations of circular motion, including eccentrics and equants, to the observed motion of Mars.

At one point he found a hypothesis that agreed with observations to within 8 arcmin (about one-quarter the diameter of the Moon), but he believed that Tycho's observations could not have been in error by even this small amount. With characteristic integrity, he discarded the hypothesis. Finally, Kepler tried to represent the orbit of Mars with an oval and soon discovered that the orbit could be fitted very well by a particular oval curve known as an ellipse.

Next to the circle, the **ellipse** is the simplest kind of closed curve. It belongs to a family of curves known as **conic sections** (Figure 2.8). A conic section is the curve of intersection between a hollow cone (whose base is presumed to extend downward indefinitely) and a plane that cuts through it. If the plane is perpendicular to the axis of the cone (or parallel to its base), the intersection is a circle. If the plane is inclined at an arbitrary angle, but still cuts completely through the surface of the cone, the resulting curve is an ellipse. If the plane is parallel to a line in the surface of the cone, it never quite cuts all the way through the cone, and the curve of intersection is open at one end. Such a curve is called a **parabola.** If the plane is inclined at an even smaller angle to the axis of the cone, an open curve results that is called a **hyperbola.** The ellipse, then, ranges from a circle at one extreme to a parabola at the other. The parabola separates the family of ellipses from the family of hyperbolas.

An interesting and important property of an ellipse is that from any point on the curve the sum of the distances to two points inside the ellipse, called the **foci** (singular: **focus**) of the ellipse, is the same. This property suggests a simple way to draw an ellipse. Tie the ends of a length of string to two tacks pushed through a sheet of paper into a drawing board, so that the string is slack. If a pencil is then pushed against the string, so that the string is held taut, and then slid against the string around the tacks (Figure 2.9), the curve that results is an ellipse. At any point where the pencil may be, the sum of the distances from the pencil to the two tacks is a constant length—the length of the string. The tacks are at the two foci of the ellipse.

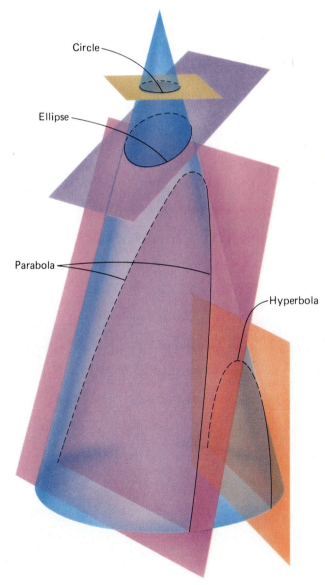

FIGURE 2.8 Conic sections.

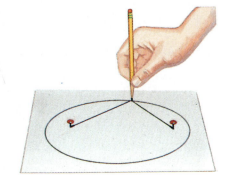

FIGURE 2.9 Drawing an ellipse.

The maximum diameter of the ellipse is called its major axis. Half the distance, that is, the distance from the center of the ellipse to one end, is the **semimajor axis.** The *size* of an ellipse is usually specified by giving the length of the semimajor axis. The *shape* of an ellipse, which depends on how close together the two foci are compared with the major axis, is specified by a quantity called the **eccentricity.** The major axis is the length of the string, and the eccentricity is the distance between the tacks divided by the length of the string.

If the foci (or tacks) coincide, the ellipse is a circle; a circle is, then, an ellipse of zero eccentricity. Ellipses of various shapes are obtained by varying the spacing of the tacks (as long as they are not farther apart than the length of the string). If one tack is removed to an infinite distance, and if enough string is available, the closed end of the resulting, infinitely long, ellipse is a parabola. A parabola has an eccentricity of 1. Figure 2.10 shows several ellipses.

Kepler found that Mars has an orbit that is an ellipse and that the Sun is at one focus (the other focus is empty). The eccentricity of the orbit of Mars is only about 0.1; the orbit, drawn to scale, would be practically indistinguishable from a circle. It is a tribute to Tycho's observations and to Kepler's perseverance that Kepler was able to determine that the orbit was an ellipse at all. Assuming that Mars was representative, he reasoned that the orbits of all planets (including the Earth) were ellipses with the Sun at one focus. This is Kepler's *first law of planetary motion.*

Kepler also needed to understand how the speed of Mars varied at different parts of its orbit. After some calculation, he found that Mars speeds up as it comes closer to the Sun and slows down as it pulls away.

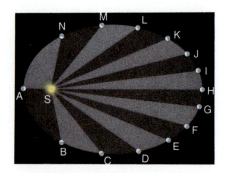

FIGURE 2.11 Kepler's law of equal areas. A planet moves most rapidly on its elliptical orbit when it is at position *A*, nearest the Sun (*S*) at one focus of the ellipse. The orbital speed of the planet varies in such a way that in equal intervals of time it moves distances *AB*, *BC*, *CD*, and so on, so that the regions swept out by the line connecting the planet to the Sun (alternating shaded and clear zones) are always the same area.

Kepler expressed this relation by imagining that the Sun and Mars are connected by a straight, elastic line. As Mars travels in its elliptical orbit around the Sun, the areas swept out in space by this imaginary line in equal intervals of time are always equal (Figure 2.11). This relation, Kepler's *second law of planetary motion,* is also called the law of equal areas.

(b) Kepler's Search for the Harmony of the Worlds

Kepler believed in an underlying harmony in nature, and he searched for numerological relations in the celestial realm. It was a great personal triumph, therefore, when he finally found a simple algebraic relation be-

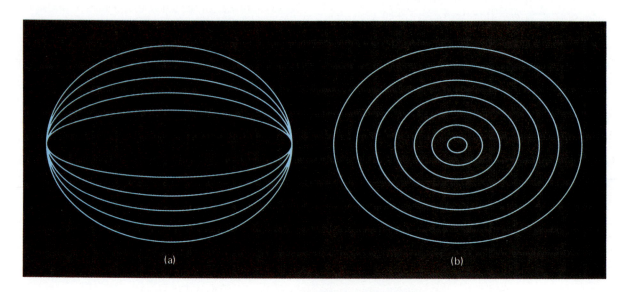

FIGURE 2.10 The ellipse. (a) A group of ellipses with the same major axis but various eccentricities. (b) Ellipses with the same eccentricity but various major axes.

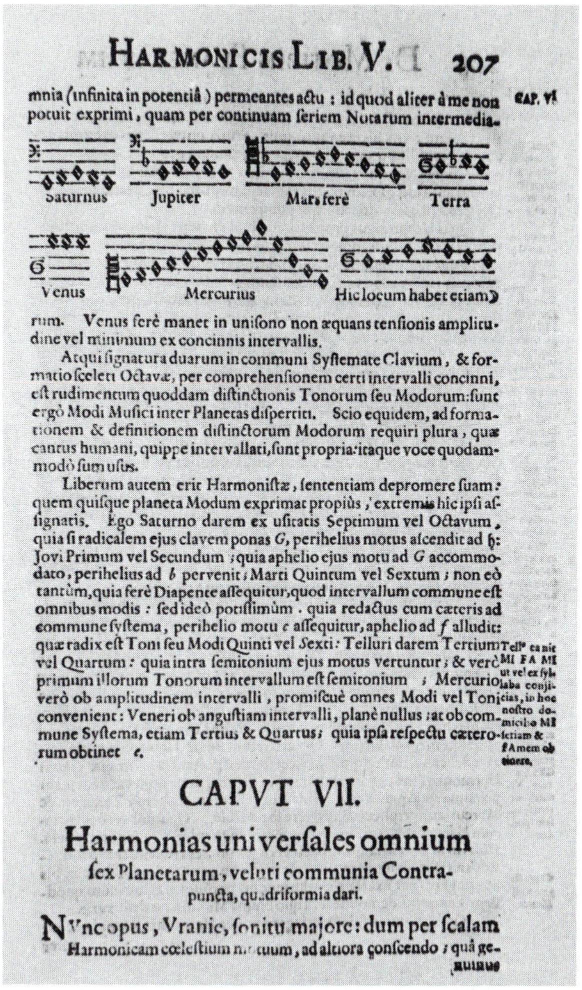

HARMONICIS LIB. V. 207

mnia (infinita in potentiâ) permeantes actu: id quod aliter à me non potuit exprimi, quam per continuam ſeriem Notarum intermedia- CAP. VI

rum. Venus ferè manet in uniſono non æquans tenſionis amplitudine vel minimum ex concinnis intervallis.

Atqui ſignatura duarum in communi Syſtemate Clavium, & formatio ſceleti Octavæ, per comprehenſionem certi intervalli concinni, eſt rudimentum quoddam diſtinctionis Tonorum ſeu Modorum: ſunt ergò Modi Muſici inter Planetas diſpertiti. Scio equidem, ad formationem & definitionem diſtinctorum Modorum requiri plura, quæ cantus humani, quippe intervallati, ſunt propria: itaque voce quodammodò ſum uſus.

Liberum autem erit Harmoniſtæ, ſententiam depromere ſuam: quem quiſque planeta Modum exprimat propiùs, extremis hîc ipſi aſſignatis. Ego Saturno darem ex uſitatis Septimum vel Octavum, quia ſi radicalem ejus clavem ponas *G*, perihelius motus aſcendit ad ♮: Jovi Primum vel Secundum; quia aphelio ejus motu ad *G* accommodato, perihelius ad *b* pervenit; Marti Quintum vel Sextum; non eò tantùm, quia ferè Diapente aſſequitur, quod intervallum commune eſt omnibus modis: ſed ideò potiſſimùm, quia redactus cum cæteris ad commune ſyſtema, perihelio motu *c* aſſequitur, aphelio ad *f* alludit: quæ radix eſt Toni ſeu Modi Quinti vel Sexti: Telluri darem Tertium vel Quartum: quia intra ſemitonium ejus motus vertuntur; & verò primum illorum Tonorum intervallum eſt ſemitonium; Mercurio verò ob amplitudinem intervalli, promiſcuè omnes Modi vel Toni convenient: Veneri ob anguſtiam intervalli, planè nullus; at ob commune Syſtema, etiam Tertius & Quartus; quia ipſa reſpectu cæterorum obtinet *e*.

Tellus canit MI FA MI ut vel ex ſyllaba conjicias, in hoc noſtro domicilio MIſeriam & FAmem obtinere.

CAPVT VII.

Harmonias univerſales omnium ſex Planetarum, veluti communia Contrapuncta, quadriformia dari.

NVnc opus, Vranie, ſonitu majore: dum per ſcalam Harmonicam cœleſtium motuum, ad altiora conſcendo; quâ genuinus

FIGURE 2.12 Detail from Kepler's *Harmony of the Worlds*. (Crawford Collection, Royal Observatory Edinburgh)

tween the sizes of the planets' orbits and their sidereal periods. Kepler published his discovery in 1619 in *Harmony of the Worlds* (Figure 2.12). The relation is Kepler's *third law of planetary motion*, also known as his harmonic law.

It is simplest to express Kepler's third law with the algebraic equation

$$P^2 = Ka^3,$$

where P represents the sidereal period of the planet, a is the semimajor axis of its orbit, and K is a numerical constant whose value depends on the kinds of units chosen to measure time and distance. It is convenient to choose for the unit of time the Earth's period—the year—and for the unit of distance the semimajor axis of the Earth's orbit —the **astronomical unit (AU).** With this choice of units, $K = 1$, and Kepler's third law can be written

$$P^2 = a^3.$$

Note that to arrive at his third law, it was not necessary for Kepler to know the actual distances of the planets from the Sun (say, in kilometers), only the distance in units of the Earth's distance, the astronomical unit. The length of the astronomical unit in kilometers was not determined accurately until later.

As an example of Kepler's third law, consider Mars. The semimajor axis (a) of Mars' orbit is 1.524 AU. The cube of 1.524 is 3.54. According to the above formula, the period of Mars, in years, should be the square root of 3.54, or 1.88 years, a result that is in agreement with observations. Table 2.2 gives for each of the six planets known to Kepler the modern values of a, P, a^3, and P^2. To the limit of accuracy of the data given, we see that Kepler's law holds exactly, except for Jupiter and Saturn, for which there are very slight discrepancies. Decades later, Newton gave an explanation for the discrepancies, but within the limit of accuracy of the observational data available in 1619, Kepler was justified in considering his formula to be exact.

Much of the rest of his book *Harmony of the Worlds* deals with Kepler's attempts to associate numerical relations in the solar system with music. He tried to derive notes of music played by the planets as they move harmoniously in their orbits. The Earth, according to Kepler, plays the notes *mi*, *fa*, *mi*, which he took to symbolize the "*miseria* (misery), *fames* (famine), *miseria*" of our planet.

TABLE 2.2 OBSERVATIONAL TEST OF KEPLER'S THIRD LAW

Planet	Semimajor Axis (AU)	Period (Years)	a^3	P^2
Mercury	0.387	0.241	0.058	0.058
Venus	0.723	0.615	0.378	0.378
Earth	1.000	1.000	1.000	1.000
Mars	1.524	1.881	3.537	3.537
Jupiter	5.203	11.862	140.8	140.7
Saturn	9.534	29.456	867.9	867.7

FIGURE 2.13 University of Padua. Galileo presented popular lectures on mathematics from this balcony. (David Morrison)

2.4 GALILEO

Galileo Galilei (1564–1642), the great Italian contemporary of Kepler, was born in Pisa. Galileo, like Copernicus, began training for a medical career, but he had little interest in the subject and later switched to mathematics. In school he incurred the wrath of his professors by refusing to accept on faith dogmatic statements based solely on the authority of great writers of the past. From his classmates he gained the nickname "Wrangler."

For financial reasons, Galileo was never able to complete his formal university training. Nevertheless, his exceptional ability as a mathematician gained him the post, in 1589, of professor of mathematics and astronomy at the University at Pisa. In 1592 he obtained a far better position at the University at Padua (Figure 2.13), where he remained until 1610, when he left to become mathematician to the Grand Duke of Tuscany. While at Padua he became famous throughout Europe as a brilliant lecturer and as a foremost scientific investigator.

(a) Galileo's Experiments in Mechanics

Galileo's greatest contributions were in the field of mechanics (the study of motion and the actions of forces on bodies). In his time, the principles of mechanics outlined by Aristotle had still not been completely discarded, and the practice of performing experiments to learn physical laws was not standard procedure. However, Galileo experimented with pendulums, with balls rolling down inclined planes, with light and mirrors, with falling bodies, and with many other objects.

Aristotle had said that heavy objects fall faster than lighter ones. Galileo argued that if a heavy and a light object were dropped together, even from a great height, both would hit the ground at practically the same time. What little difference there was could easily be accounted for by the resistance of the air. Galileo tested

these conflicting ideas by performing the experiment, demonstrating that Aristotle was wrong (Figure 2.14).

In the course of his experiments, Galileo discovered laws that invariably described the behavior of physical objects. The most far-reaching of these is the law of inertia (now known as Newton's first law of motion). The **inertia** of a body is that property of the body that resists any change of motion. It is familiar to all of us that if a body is at rest, it tends to remain at rest. Some outside influence is required to start it in motion. Rest was thus generally regarded as the natural state of matter. Galileo showed, however, that rest was no more natural than motion. If an object is slid along a rough horizontal floor, it soon comes to rest, because friction between it and the floor acts as a retarding force. However, if the floor and object are both highly polished, the body, given the same initial speed, will slide farther before coming to rest. On a smooth layer of ice, it will slide farther still. Galileo noted that the less the friction, the less the body's tendency to slow down. He reasoned that if all resisting effects could be removed, the body would continue in a steady state of motion indefinitely. In fact, he argued, not only is a force required to start an object moving from rest, but a force is also required to slow down, stop, speed up, or change the direction of a moving object.

Galileo also studied the way bodies **accelerate,** that is, change their speed, as they fall freely or roll down

FIGURE 2.14 The leaning bell-tower of the Cathedral at Pisa. While living in Pisa, Galileo carried out experiments to show that objects of different mass accelerate at the same rate. According to legend, one of his experiments consisted of dropping cannon balls of different weight from this tower. (David Morrison)

inclined planes. He found that such bodies accelerate uniformly; that is, in equal intervals of time they gain equal increments in speed. Galileo formulated these newly found laws in precise mathematical terms that enabled one to predict, in future experiments, how far and how fast bodies would move in various lengths of time. It remained for Newton to incorporate and generalize Galileo's principles into a few simple laws so fundamental that they have become the basis of a great part of our modern technology (Chapter 3).

(b) Galileo and the Heliocentric Cosmology

Sometime in the 1590s, Galileo accepted the Copernican hypothesis of the solar system. In Roman Catholic Italy, this was not a popular philosophy, for the church authorities still upheld the ideas of Aristotle and Ptolemy. It was primarily because of Galileo that in 1616 the church issued a decree stating that the Copernican doctrine was "false and absurd" and was not to be held or defended.

The prevailing notion of the time was that the celestial bodies belonged to the realm of the heavens, where all is perfect, unchanging, and incorruptible. Perpetual circular motion, being the "perfect" kind of motion, was regarded as the natural state of affairs for those heavenly bodies, just as "rest" was the natural state for objects on the Earth. Once Galileo had established the principle of inertia—that on the Earth bodies in undisturbed motion remain in motion—it was no longer necessary to ascribe any special status to the fact that the planets remain perpetually in orbit. By the same token, even the Earth could continue to move, once started. What does need to be explained is why the planets move in curved paths around the Sun rather than in straight lines. Evidently, Galileo was sufficiently imbued with Aristotelian concepts that he (like Copernicus) accepted uniform circular celestial motion without subjecting the planets to the same objective scrutiny that he applied in his terrestrial experiments.

In answer to the common objection that objects could not remain on the Earth if it were in motion, Galileo noted that if a stone is dropped from the masthead of a moving ship it does not fall behind the ship and land in the water beyond its stern, but rather lands at the foot of the mast, for the stone already has a forward inertia gained from its common motion with the ship before it is dropped. In an analogous way, objects on the Earth would not be swept off and left behind if the Earth were moving, for they share the Earth's forward motion.

(c) Galileo's Astronomical Observations

It is not certain when two pieces of curved glass were first combined to produce an instrument that enlarged distant objects, making them appear nearer. But we know that the first telescopes to attract much notice in Europe were made by the Dutch spectacle-maker Hans Lippershey in 1608. Galileo heard of the discovery in 1609, and without ever having seen an assembled telescope, he constructed one of his own with a three-power magnification; that is, it made distant objects appear three times nearer and larger. He quickly built other instruments; his best had a magnification of about 30 (Figure 2.15).

The usefulness of the telescope for terrestrial observations (including its military applications) was apparent to many people. But the idea that the telescope could also reveal new insights about the heavens was less obvious. There was a long tradition that the human eye was the best possible measure of truth, while lenses and mirrors produced distorted images and sometimes even made things look upside down. This idea is preserved today in the expression "it was done with mirrors," meaning that someone has tricked us.

Galileo himself was slow to turn his new optical toy toward the sky, and he first seems to have conducted extensive tests to convince himself that the magnified image was an accurate representation of a distant scene. Late in 1609 he began his astronomical work. While he was not the only person to try using optical aids to study

FIGURE 2.15 Telescopes used by Galileo. The longer has a wooden tube covered with paper, a focal length of 1.33 m, and an aperture of 26 mm. (Istituto e Museo di Storia della Scienza di Florenza)

the sky, Galileo applied himself to this task with his characteristic care and persistence. In 1610 he startled the world by publishing a list of his remarkable discoveries in a small book, *The Sidereal Messenger* (*Sidereus Nuncius*).

Galileo found that many stars too faint to be seen with the naked eye became visible with his telescope. In particular, he found that some nebulous blurs resolved into many stars (for example, the Praesepe cluster in Cancer) and that the Milky Way was made up of multitudes of individual stars. He found that Jupiter had four satellites or moons revolving about it with periods ranging from just under 2 days to about 17 days. (Twelve other satellites of Jupiter have been found since.) This discovery was particularly important because it showed that there could be centers of motion that in turn are in motion. It had been argued that if the Earth were in motion the Moon would be left behind, because it could hardly keep up with the rapidly moving planet. Yet here were Jupiter's satellites doing exactly that (Figure 2.16).

Another important telescopic discovery that strongly supported the Copernican view was the fact that Venus goes through phases like the Moon. In the Ptolemaic system, Venus is always closer to the Earth than is the Sun, and thus, because Venus never has more than about 45° of elongation, it would never be able to turn its fully illuminated surface to our view—it would always appear as a crescent. Galileo, however, saw that Venus went through both crescent and gibbous phases. He concluded that Venus must travel around the Sun,

FIGURE 2.16 The four Galilean satellites of Jupiter as photographed by the Voyager 1 spacecraft in 1979. (NASA/JPL)

passing at times behind and beyond it, rather than revolving directly around the Earth (Figure 2.17). Mercury also goes through all phases.

Galileo's observations revealed much about our nearest neighbor, the Moon. He saw that the Moon's surface

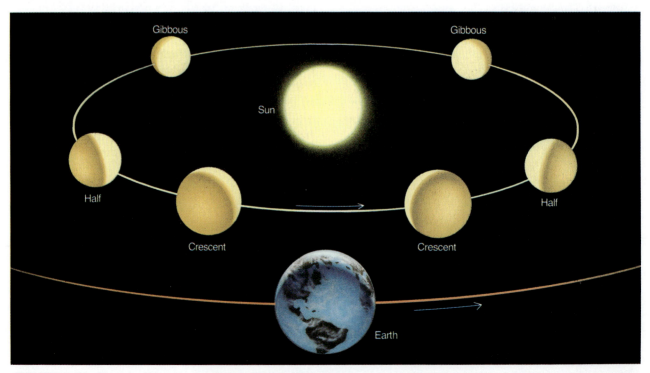

FIGURE 2.17 Phases of Venus according to the heliocentric theory.

was "uneven, rough, and full of cavities and prominences, being not unlike the face of the Earth." Galileo guessed that the prominent flat dark areas on the Moon might be water (the dark *maria*, or "seas," on the Moon were thought to be water until long after Galileo's time). These discoveries showed that the heavenly bodies, previously regarded as perfect, smooth, and incorruptible, do indeed have irregularities, as does the Earth. Further, the fact that the Moon was "not unlike the Earth" suggested that the Earth, too, could belong to the realm of celestial bodies.

One of Galileo's most disturbing observations, to his contemporaries, was of spots on the Sun, showing that this body also had "blemishes"—what we today call sunspots (Figure 2.18). **Sunspots** are now known to be large, comparatively cool areas on the Sun that appear dark because of their contrast with the brighter and hotter solar surface. Sunspots are temporary, lasting usually only a few weeks to a few months. Large sunspots actually had been observed before, with the unaided eye, but were generally regarded either as something in the Earth's atmosphere or as planets between the Earth and the Sun silhouetting themselves against the Sun's disk in the sky. In fact, some of

Galileo's critics attempted to explain the spots as inner planets orbiting about the Sun.

Galileo observed the spots to move, day by day, across the disk of the Sun. Often, after about two weeks, the same spots would reappear. Galileo explained that the spots must be either on the surface of the Sun or very close to it and that they were carried around the Sun by its own rotation. He determined the Sun's period of rotation to be a little under a month.

(d) *Dialogue on the Two Great World Systems*

As we have seen, Galileo had accumulated a great deal of evidence to support the Copernican system. By the decree of 1616 he was forbidden to "hold or defend" the odious hypothesis, but he still hoped to convert his countrymen to the heliocentric view. He prevailed upon his long-time acquaintance, Pope Urban VIII, to allow him to publish a book that explained fully all arguments for and against the Copernican system, not for the purpose of extolling it, but merely to examine it and to show those of other nationalities that Italians were not ignorant of new theories.

The book appeared in 1632 under the title *Dialogue on the Two Great World Systems* (*Dialogo dei Due Massimi Sistemi*). The *Dialogue* is written in Italian (not Latin) to reach a large audience. It is a magnificent and unanswerable argument for Copernican astronomy, in the form of a conversation among three philosophers: Salviati, the most brilliant and the one through whom Galileo generally expresses his own views; Sagredo, who is usually quick to see the truth of Salviati's arguments; and Simplicio, an Aristotelian philosopher who brings up all the usual objections to the Copernican system, which Salviati promptly shows to be absurd.

It is pointed out in the preface to the *Dialogue* that the arguments to follow are merely a mathematical fantasy and that divine knowledge assures us of the immobility of the Earth. This was thinly cloaked irony, however, and Galileo's enemies acted quickly to build a case against him. He was called before the Roman Inquisition on the charge of believing and holding doctrines that were false and contrary to Divine Scriptures. Galileo was forced to plead guilty and deny his own doctrines. His life sentence was commuted to confinement in his own home at Arcetri, near Florence, for the last ten years of his life. At the time of his inquisition, he was nearly 70.

The *Dialogue* joined Copernicus' *De Revolutionibus* on the *Index of Prohibited Books* of the Roman Catholic Church. It was removed from the *Index*, however, in 1835. In 1980, Pope John Paul II ordered a re-examination of the evidence against Galileo, which led to his exoneration and eliminated the last vestige of resistance to the Copernican revolution from the Catholic church.

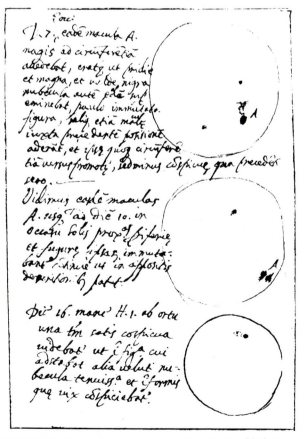

FIGURE 2.18 Galileo's drawings of sunspots. (Yerkes Observatory)

SUMMARY

2.1   The Pole Nicolaus Copernicus (1473–1543) introduced the **heliocentric cosmology** to Renaissance Europe in his book *De Revolutionibus* in 1543. Although he retained the Aristotelian idea of uniform circular motion and therefore had to include epicycles in his model, Copernicus presented a convincing case that the Earth is a planet and that the planets all circle about the Sun. He calculated the relative distances of the planets from the Sun and forever dethroned the Earth from its place at the center of the universe.

2.2   The Dane Tycho Brahe (1546–1601) was the most capable of the pretelescopic astronomical observers. Although he did not accept the heliocentric theory, his accurate observations of planetary positions provided the data base that later demonstrated the validity of the Copernican cosmology.

2.3   The German Johannes Kepler (1571–1630) succeeded Tycho as court mathematician in Prague and used Tycho's observations to derive the three fundamental laws of planetary motion that bear his name. Kepler's laws are as follows: (1) Planetary orbits are **ellipses** (described by the **semimajor axis** and **eccentricity**) with the Sun at one **focus.** (We now know that other **conic sections**—the **parabola** and **hyperbola**—are also possible orbital shapes.) (2) In equal intervals, a planet sweeps out equal areas. (3) If times are expressed in years and distances in **astronomical units,** the relationship between period and semimajor axis of an orbit is given by $P^2 = a^3$.

2.4   The Italian Galileo Galilei (1564–1642) was the father of both modern experimental physics and telescopic astronomy. His experiments in mechanics established the ideas of **inertia** and of the equal **acceleration** of falling bodies that were later incorporated in Newton's laws of motion. Galileo began telescopic observations in 1609, discovering the nature of the Milky Way, the large-scale features of the Moon, the phases of Venus, the four Galilean satellites of Jupiter, and the rotation of the Sun from observations of **sunspots.** Although accused of heresy and imprisoned by the Roman Catholic church for his support of the heliocentric cosmology, Galileo's observations and brilliant writings convinced most of his scientific contemporaries of the reality of the Copernican theory.

EXERCISES

THOUGHT QUESTIONS

1. Copernicus accepted the traditional view that celestial motions are circular and uniform. What problems did this assumption cause for him? Do you think he should have questioned this assumption, given the information available to him?

2. Which, if any, of the following can never appear at opposition? At conjunction? At quadrature?
   a. Jupiter
   b. Earth
   c. Sun
   d. Venus
   e. Saturn
   f. Mars
   g. Mercury
   h. Moon

3. a. What is the major axis of a circle?
   b. Where is the second focus of a parabola?
   c. Which conic section could have an eccentricity of 0.3?

4. Consider Kepler's third law. Carefully explain why $K = 1$ when a is measured in astronomical units and P in years.

5. A friend tells you that the Earth cannot be rotating, as the scientists claim, because if a point on the equator were being carried eastward at about 1000 mi/hr, as claimed, a baseball pitcher could throw a ball straight up in the air, and the Earth and pitcher would move to the east out from under the ball, which would land some distance behind (or to the west) of the pitcher. How might you straighten out this friend of yours?

6. Galileo's observations of the phases of Venus ruled out Ptolemy's system of cosmology. Did they also rule out Tycho Brahe's system? Why, or why not?

7. Consider three cosmological perspectives: (1) the geocentric perspective; (2) the heliocentric perspective; and (3) the modern perspective, in which the Sun is a minor star on the outskirts of one galaxy among billions. Discuss some of the cultural and philosophical implications of each point of view.

PROBLEMS

8. What would be the distance from the Sun, in astronomical units, of an inferior planet that had a greatest elongation of 30°? Assume circular orbits for the planet and the Earth.

9. Draw an ellipse by the procedure described, using a string and two tacks. Arrange the tacks so that they are separated by 1/10 the length of the string. Comment on the appearance of your ellipse. This (if you have been careful in your construction) is approximately the shape of the orbit of Mars.

10. The Earth's distance from the Sun varies from 147.2 million to 152.1 million km. What is the eccentricity of its orbit?
    Answer: 0.016

11. What is the eccentricity of the orbit of a planet whose distance from the Sun varies from 180 million to 220 million km?

12. What would be the period of a planet whose orbit has a semimajor axis of 4 AU?

13. What is the period of revolution (in years) of an asteroid with a semimajor axis of 10 AU?

14. What is the distance of an asteroid from the Sun (in AU) if it has a period of revolution of eight years?

15. What would be the distance from the Sun of a planet whose period is 45.66 days?
Answer: 0.25 AU

16. Apply Kepler's laws to the motion of Jupiter's satellites around that planet. One of the satellites has a period 5.196 times as long as another one. What would be the ratio of the semimajor axes of their orbits?
Answer: 3:1

17. Draw a picture that explains why Mercury should go through phases like the Moon, according to the heliocentric cosmology. Does Jupiter also go through phases as seen from the Earth?

★18. Comet Halley has a period of 76 years. What is its semimajor axis? If its closest point to the Sun (perihelion) is 0.6 AU, what is its farthest distance (aphelion)? (*Hint*: Remember the definition of semimajor axis.)

★19. What would be the sidereal period of an inferior planet that appeared at greatest western elongation exactly once a year?

★20. The synodic period of Saturn is 1.03513 sidereal years. What is its sidereal period?
Answer: 29.5 years

On December 25, 1968, as the Apollo 8 astronauts were returning to Earth from their historic first circumnavigation of the Moon, ground controller Mike Collins conveyed a question to the astronauts from his five-year-old son. The youngster wanted to know, he told astronaut Bill Anders, "Who was driving the spacecraft?" "That's a good question," Anders replied, "I think Isaac Newton is doing most of the driving right now."

In this chapter, we examine the grand synthesis that was Newton's great contribution to science. Before modern astronomers could begin to understand the solar system (and eventually the universe at large), they needed the basic rules through which both earthly and celestial motions could be understood. Although he built upon the observations and ideas of the scientists we discussed in the last chapter, it was Newton's genius that assembled the various elements into a single set of principles, which still serves as the backbone of physics today.

The lunar module Eagle is seen against the Moon during the flight of Apollo 11, the first mission that involved a lunar landing. The Earth is seen in the background, half in sunlight, half in darkness. The universal rules of motion that allowed the Apollo missions to travel to the Moon and that apply equally well on Earth and in lunar orbit, were formulated by the subject of this chapter, Isaac Newton.

GRAVITATION: ACTION AT A DISTANCE

Johannes Kepler (1571–1630), German mathematician and astronomer, was a contemporary of Galileo, living during the tumultuous period of the Counter-Reformation and the Thirty Years' War. His discovery of the basic quantitative laws that describe planetary motion placed the heliocentric cosmology of Copernicus on a firm mathematical basis and made possible Newton's later formulation of the laws of motion and of universal gravitation.

Kepler discovered the rules that govern planetary orbits, and Galileo discovered laws that describe the behavior of falling bodies. Later Isaac Newton unified these and other insights by showing that the force of *gravitation* that accelerates falling bodies near the Earth is the same force that keeps the Moon in its orbit around the Earth and the planets in their orbits about the Sun.

Newton's principles of mechanics and his law of gravitation are so general and powerful that they

FIGURE 3.1   Newton's birthplace in Lincolnshire, England. (G.O. Abell)

strongly influenced the way people thought about the world and their place in it. In the century following Newton's death, many people concluded that the basic rules of nature were finally known, that all that remained was to fill in minor details. This philosophy of *determinism* asserted that every event in the universe can be understood by mechanistic laws applied to the conditions immediately preceding the event. We shall see in the coming chapters that this view of the ultimate predictive power of science was too optimistic. Still, this kind of thinking, which has its origin in the astronomy of Kepler and Newton, has had a strong influence on the history of the world.

3.1   NEWTON'S PRINCIPLES OF MECHANICS

(a)   Newton's *Principia*

Isaac Newton (1643–1727) was born in Lincolnshire, England, in the year after the death of Galileo (Figure 3.1). (Newton was born on Christmas Day, 1642, according to the calendar in use at his time, but by the modern calendar his birth date was January 4, 1643.) He entered Trinity College at Cambridge in 1661 as an undergraduate student, and eight years later he was appointed Professor of Mathematics, a post that he held during most of his productive career.

As a young man in college, Newton became interested in *natural philosophy*, as science was called then. The university was closed during the plague years of 1665 and 1666, during which Newton returned to Lincolnshire. He wrote later that it was in those years that he worked out the main outline of his ideas on mechanics and gravitation. But on his later return to Cambridge, his research was mainly in mathematics and optics; it was to be nearly two decades before he turned his attention again to gravitation.

Newton's return to gravitation was almost fortuitous. Physicist Robert Hooke, architect Christopher Wren, and astronomer Edmund Halley had all come independently to some notion of the law of gravitation. They realized that there was a force of attraction between the planets and the Sun, and that this force must become weaker in proportion to the square of the distance from the Sun. None of them, however, was able to solve the mathematical problem of how a planet should move under the influence of such a force. In 1684, Halley chanced to consult Newton on the matter. He was astonished to hear that Newton had solved the problem years previously and had found that the orbit of a planet moving under such a force of gravity should be an ellipse. Although Newton was unable to find his original notes containing the mathematical proof, he was able to re-solve the problem, and a short time later he sent the demonstration of the proof to Halley.

Early the following year Newton submitted a formal paper on the subject to the Royal Society. This treatise was to become the nucleus of his great work on mechanics and gravitation, *The Mathematical Principles of Natural Philosophy*, usually known by the abbreviated form of its Latin title, *Principia* (Figure 3.2). It was published in 1686 under the imprimatur of the Royal Society of London, and supervision of its publication was in the hands of Halley. As it turned out, the Society at that time was in financial difficulties, and Halley himself covered the cost of publication from his own personal funds.

In the *Principia* Newton gives his three laws of motion:

1. Every body continues in a state of rest, or of uniform motion in a straight line, unless it is compelled to change that state by forces impressed upon it.

2. Any change of motion is proportional to the force that acts, and it is made in the direction of the straight line in which that force is acting.

3. To every action there is always an equal and opposite reaction; or, the mutual actions of two bodies upon each other are always equal and act in opposite directions.

Galileo had arrived at the first two laws, although he did not state them as precisely as Newton did. They are,

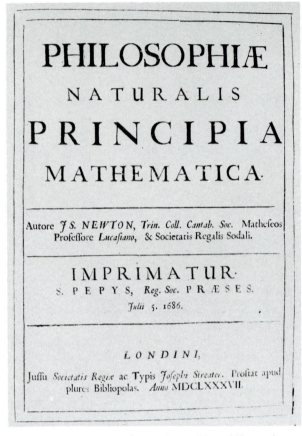

FIGURE 3.2   Title page of the first edition of Newton's *Principia*. (History of Science Collections, University of Oklahoma Libraries)

however, deeper than Galileo could have realized. To appreciate Newtonian mechanics, we must understand thoroughly the meanings of certain terms.

(b)   Units: The Metric System

One of our most basic concepts is *time*. It is said that Galileo once used the beat of his pulse as a unit of time to measure the swing of a chandelier in church. He found that the time for one swing stayed the same, even though the length of the swing died down. The story of the swinging chandelier may be apocryphal, but Galileo did discover the law that determines the period of a pendulum—that the period of its oscillation depends only on the pendulum's length and not on the amount of arc of swing. Galileo later suggested that the pendulum would be a good device to regulate a clock, thus inventing the principle of the pendulum clock.

We could choose any convenient period for a unit of time. However, the universal unit of time is now the second (s). It was originally meant to be 1/60 of a minute, which is 1/60 of an hour, which is 1/24 of the Earth's rotation period. But the Earth does not rotate quite regularly enough to serve as an accurate standard for modern measurements. In science, terms must be defined *operationally*; that is, we must supply a procedure for defining a quantity, so that someone else repeating our observation or experiment will obtain the same result. In other words, a *recipe* is better than an *artifact* for specifying fundamental units. Since 1967 the second has been defined as the duration of 9,192,631,770 periods of one of the waves emitted by a certain isotope of the cesium atom. While this may seem an esoteric definition, it actually specifies a recipe that can be repeated rather easily in the laboratory or incorporated in so-called atomic clocks, providing a universal standard for the unit of time.

Now consider the measurement of *length*. This is a quantity that must also be defined in terms of some standard "yardstick." That "yardstick" is the meter in the metric system of units, which was officially introduced in Napoleonic France in 1799. The metric system (Appendix 4) replaced earlier measures of distance based on human dimensions: the inch as the distance between knuckles on the finger or the yard as the span from the extended index finger to the nose of the king. It is in common use throughout the world today; only the U.S. among industrial nations has failed to adopt this international standard. Even in the U.S., however, metric units are used by scientists, the military, and most organizations concerned with international trade and commerce.

The meter was originally intended to be 1/10,000,000 (10^{-7}) of the distance along the Earth's surface from the equator to the pole. However, such a definition is difficult to apply. How are we to measure the length of things around us using the Earth as a yardstick? In practice, the meter was defined as the length of a metal bar kept in a vault in Paris, and all other measurements of length were referred to this standard meter.

Recently the meter has been redefined in terms of both the speed of light—the most fundamental constant of nature—and the unit of time (the second). The speed of light is 299,792,461 m/s. In other words, 1 m is the distance that light travels in a time of 1/299,792,461 s. Other units of length are derived from the meter. Thus 1 km equals 1000 m, 1 cm equals 1/100 m, and so forth. Even nonmetric units, such as the inch and the mile, are defined in terms of the meter (see Appendix 4).

Areas of surfaces are defined as square measures (square meters, or m², in the metric system). The area of a rectangular surface is its length times its width. *Volumes* of solids are cubic measures. The volume of a rectangular solid is its length times its width times its height, expressed in units of cubic meters (m^3).

Suppose we have two solids of identical shape but of different size—say, spheres 1 cm and 10 cm in radius, or a large man 2 m tall and an identical small man only 1 m tall. The surface area of one of the spheres (for example, the amount of paint needed to cover it) is proportional to the square of any of its linear dimensions, such as its radius. Thus the area of the larger sphere is 100 times that of the smaller (for 100 is the square of 10). Similarly, since 4 is the square of 2, it takes four times as much skin

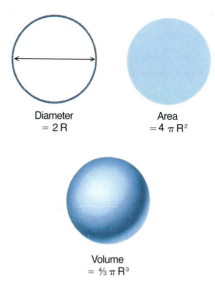

FIGURE 3.3 Diameter, surface area, and volume of a sphere.

to cover the large man as it does the small man with half the large one's linear dimensions.

The volumes of solids are proportional to the cubes of their linear dimensions. Thus the sphere 10 cm in radius can hold 1000 times as much water as the one with a radius of 1 cm. Similarly, the large man has eight times as much flesh and bones as the small one. It is convenient to remember that areas are proportional to the squares, and volumes to the cubes, of the linear dimensions of similarly shaped objects, whatever their shape may be (Figure 3.3).

(c) Speed, Velocity, and Acceleration

With ways to measure length and time we can define *speed* (distance divided by the time required to cross it). Common units of speed are meters per second (m/s), kilometers per hour (km/hr), or miles per hour (mi/hr or mph). A related concept is **velocity.** Speed and velocity are often confused, however. Velocity conveys more information than speed; it is a description of both the speed *and* the direction of motion. Velocity is an example of a **vector** (Figure 3.4), a quantity that has both size and direction. Vectors are often used in physics

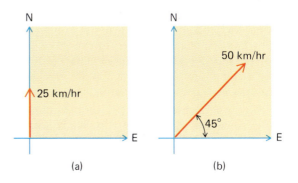

FIGURE 3.4 Two vectors, representing velocities at 25 km/hr to the north and 50 km/hr to the northeast.

and mathematics for concepts such as velocity and force, which have both a magnitude and a direction associated with them.

Any change in velocity requires *acceleration*. Acceleration, therefore, involves a *change* of speed, or of direction, or both. Starting, stopping, speeding up, slowing down, and changing direction are all accelerations. Note that this is a much broader use of the term than is common in everyday life. Most of us use acceleration to refer to an increase of speed. But slowing down, or deceleration, is also a form of acceleration, and so is a change in direction. You are accelerated in your car when you speed up or apply the brake or simply turn a corner.

Acceleration is a vector. The magnitude of acceleration is the rate at which the velocity changes, and its direction is the direction of that change. The acceleration produced by gravity at the surface of the Earth, for example, is 9.8 m/s per second (often written 9.8 m/s^2), in a direction toward the center of the Earth.

(d)   Newton's First Law—Momentum

A moving body tends to keep moving, and a stationary body tends to remain at rest. **Momentum** is a measure of this state of motion. Momentum depends on speed, for clearly a body moving at 50 km/hr has more "motion" than one moving at 10 km/hr. But momentum also depends on the amount of matter in a moving object. An automobile going at 30 km/hr certainly has more "motion" and is harder to speed up, stop, or turn than, say, a bicycle moving with the same speed.

Thus Newton defined momentum as proportional to velocity and defined the constant of proportionality as **mass.** Mass is a quantity that characterizes the total amount of material in the body and is the property that gives the body its inertia, that is, that makes it resist acceleration. We have not yet defined mass operationally, but we shall see that Newton's third law provides a means of doing so. For now it is enough to say that our intuitive idea of what is usually meant by weight (on the Earth's surface) is actually mass.

Newton's first law says that the product of a body's mass and velocity is constant if no outside force is applied to it. This means that motion is as natural as rest. With no force on it, a moving body would go in a straight line at a constant speed forever. Galileo had discovered this fact from his experiments, but Newton generalized the law and stated it more elegantly. It replaces the ancient idea that the natural path for celestial objects is the perfect circle. Instead, the natural path is a straight line. Some force is required to bend the path of a planet from a straight line and force it to orbit the Sun.

(e)   Newton's Second Law—Force

The second law of motion deals with changes in momentum. It states that if a force acts on a body, it produces a change in the momentum of the body that is in the direction of the applied force. The second law, then, defines **force.** The magnitude or strength of a force is defined as the rate at which it produces a change in the momentum of the body on which it acts. Some familiar examples of forces are the pull of the Earth, the friction of the ground or a floor on moving bodies, the impact of a bat on a baseball, the pressure exerted by air, and the thrust of a rocket engine.

Note that Newton's first law of motion is consistent with his second; when there is no force, the change in momentum is zero.

There are three ways in which the momentum of a body can be altered: Its velocity can change, or its mass, or both. Most often the mass of a body does not change when a force acts upon it—a change in momentum usually results from a change in velocity. Thus, in the vast majority of examples, the second law can be written as force = mass × acceleration, or

$$f = ma,$$

because acceleration is the rate at which velocity changes. If the acceleration occurs in the same direction as the velocity, the body speeds up. If the acceleration occurs in the opposite direction to the velocity, the body slows down. If acceleration occurs exactly at right angles to the velocity, only the direction of motion of the body, and not its speed, changes.

The acceleration of falling bodies is downward (in the direction toward which the gravitational pull of the Earth is acting). Gravity accelerates a falling body in the direction it is already moving, and so simply speeds it up. As Galileo found, this acceleration is the same for all bodies, even if their masses are different.

If a body is slid along a rough horizontal surface, it slows down uniformly in time. It is therefore accelerated in a direction opposite to its velocity (decelerated). The acceleration is produced by the force of friction between the moving body and the rough surface.

Two equal accelerations may correspond to entirely different forces if the masses involved are different. Consider the forces required to accelerate an automobile and a bicycle each to a speed of 30 km/hr in 20 s. Clearly, because of the car's greater mass, a proportionately greater force will be required to produce the necessary acceleration. Similarly, once the bodies are both moving at that speed, a far greater force is needed to stop the automobile as quickly as the bicycle.

(f)   The Third Law—Reaction

Newton's third law of motion was a new idea. It states that all forces occur as pairs of forces that are mutually equal to and opposite each other. If a force is exerted on an object, it must be exerted by something else, and the object will exert an equal and opposite force back upon that something. All forces, in other words, must be mutual forces acting between two objects or things.

If a man pushes against his car, the car pushes back against him with an equal and opposite force, but if the man has his feet firmly planted on the ground, the reaction force is transmitted through him to the Earth. Because of its enormously greater mass, the Earth accelerates far less than the car. Suppose a girl jumps off a table down to the ground. The force pulling her down is a mutual gravitational force between her and the Earth. Both she and the Earth suffer the same total change of momentum because of the influence of this mutual force. Of course, the girl does most of the moving; because of the greater mass of the Earth, it can experience the same change of momentum by accelerating only a negligible amount.

A more obvious manifestation of the mutual nature of forces between objects is familiar to all who have played baseball. The recoil of the bat shows that the ball exerts a force on the bat during the impact, just as the bat does on the ball. The momentum imparted to the bat by the ball is transmitted through the batter to the Earth, so the acceleration produced is far less than that suffered by the ball. Similarly, when a rifle is discharged, the force pushing the bullet out the muzzle is equal to that pushing backward upon the gun and marksman.

Here, in fact, is the principle of rockets—the force that discharges the exhaust gases from the rear of the rocket is accompanied by a force that shoves the rocket forward. The exhaust gases need not push against air or the Earth; a rocket operates best in a vacuum (Figure 3.5).

In all the cases considered above, a mutual force acts upon the two objects concerned. Each object always experiences the same total change of momentum, but in opposite directions. Because momentum is the product of velocity and mass, the object of lesser mass will end up with proportionately greater velocity.

We are now in a position to see how we can measure mass. Initially, some material object must be adopted as a *standard* and said to have unit mass, for example, 1 liter (l) (1000 cm³) of water, which is defined as 1 kilogram (kg) (see Appendix 4). Then the mass of any other object can be found by measuring the relative acceleration produced when the same force acts on it and the standard object.

Having found a way to measure mass, we can now express the value of a force numerically. The standard metric unit of force is the newton (N), which is the force necessary to give a mass of 1 kg an acceleration of 1 m/s². This book has a mass of about 1 kg, and you can exert a force of 1 N with a good solid shove against the book, sufficient to send it sliding across the table.

(g)   Mass, Volume, and Density

It is important not to confuse mass with volume or density. Mass is a measure of the amount of material in an object. Volume, in contrast, is a measure of the

FIGURE 3.5   The launch of the Apollo 15 rocket toward the Moon was a dramatic demonstration of Newton's third law. (NASA)

physical space occupied by a body, say in cubic centimeters or liters. In short, the volume is the "size" of an object—it has nothing to do with its mass. A penny and an inflated balloon may both have the same mass, but they have very different volumes.

The penny and balloon are also very different in **density**, which is a measure of how much mass is contained within a given volume. Specifically, density is the ratio of mass to volume:

$$\text{density} = \frac{\text{mass}}{\text{volume}}.$$

Note that often in everyday language we use "heavy" and "light" as indications of density (rather than weight), as, for instance, when we say that iron is heavy or that a puff pastry is light.

The units of density that will be used in this book are grams per cubic centimeter (g/cm³) or, equivalently, metric tons per cubic meter. Familiar materials span a considerable range in density, from gold (19 g/cm³) to artificial materials such as plastic insulating foam (less than 0.1 g/cm³) (Table 3.1). In the astronomical universe, much more remarkable densities can be found, all the way from a comet's tail (10^{-16} g/cm³) to a neutron star (10^{15} g/cm³).

To sum up, then, mass is "how much," volume is "how big," and density is "how tightly packed."

TABLE 3.1	DENSITIES OF MATERIALS
Material	Density (g/cm³)
Gold	19.3
Lead	11.4
Iron	7.9
Earth (bulk)	5.6
Rock (typical)	2.5
Water	1.0
Wood (typical)	0.8
Insulating foam	0.1
Silica gel	0.02

(h) Angular Momentum

Another useful concept is **angular momentum,** which measures the momentum of an object as it rotates or revolves about some fixed point. Just as ordinary momentum was defined as the product of two quantities—mass and speed—the angular momentum of an object is defined as the product of three quantities: mass, speed, and the distance from the fixed point around which the object turns.

If these three quantities remain constant, that is, if the motion takes place at a constant speed and at a fixed distance from the point of origin, then the angular momentum is also a constant. More generally, angular momentum is constant, or is conserved, in any rotating system in which no external forces act or in which the only forces are directed toward or away from the point of origin. An example of such a system is a planet orbiting the Sun, since the mutual gravitational forces act directly along the line joining the two objects, and there is no external force such as friction to slow the motion.

Kepler's second law is an example of the conservation of angular momentum. As each planet gets closer to the Sun in its elliptical orbit, it must move more quickly to maintain the same total orbital angular momentum.

Conservation of angular momentum applies also to a solid body spinning around its own axis, like the Earth, or to a spinning gas cloud, such as those that are the birthplaces of stars and planets. However, such objects will change their spin rate if their size or configuration is altered. Indeed, the conservation of angular momentum dictates that rate of rotation will increase if, for example, a dust cloud shrinks.

Altering the spin rate by rearranging the mass in a rotating system is exactly what is accomplished by figure skaters. Suppose a skater is spinning on the blade of her skate with arms outstretched. If she holds her body rigid, she rotates at a constant rate. However, if she pulls her arms in to her body, some parts of her mass move closer to their rotation axes and would decrease in angular momentum unless the system compensates by spinning faster. Thus, a figure skater can start a spin with her arms out and then pull them in, thereby spinning faster so that her angular momentum is conserved. She can slow down again by pushing her arms (or a free leg) out from her body.

You can perform the same experiment with an old-fashioned rotating piano stool. Purchase another copy of this book (or use another heavy book) and sit on the stool while holding both books out from your body at arm's length. Have a friend start you spinning—but not too fast, or you may not have the opportunity to finish reading even your first copy. Now pull your arms in so that the books are next to your body and you will experience a dramatic demonstration of the conservation of angular momentum.

The concept of conservation of angular momentum is important to understanding the formation of the solar system with its planets and their satellites, the death and collapse of stars, and even the formation of galaxies. We shall refer to it again in coming chapters.

3.2 UNIVERSAL GRAVITATION

The Earth exerts a force of attraction upon all objects at its surface. This is a *mutual* force; a falling apple and the Earth are pulling on each other. Newton reasoned that this force of attraction between the Earth and objects on or near its surface might extend as far as the Moon to keep the Moon in its orbit. He further speculated that there is a general force of attraction between all material bodies. If so, the attractive force between the Sun and each of the planets could provide the force necessary to keep each in its respective orbit.

Thus Newton hypothesized that there is a universal attraction between all bodies everywhere in space. Within the solar system, this gravitational force must act between the planets and the Sun. It produces an acceleration, called the **centripetal acceleration,** to pull each planet from a straight line and bend its path into an elliptical orbit, as described by Kepler's laws. Newton's great achievement was the determination of the mathematical nature of the attraction. He then went on to test the hypothesis by using it to predict new phenomena.

(a) The Mathematical Description of Gravitation

For mathematical simplification we make the assumption that planets revolve around the Sun in perfectly circular orbits. A more complicated analysis can be made to apply to the actual elliptical orbits. It can be shown from experiments that for a body of mass m to move in a circle with radius r at speed v, it must be accelerated toward the center of the circle by a force

$$F = \frac{mv^2}{r}.$$

This force is called a *centripetal force*, and it results in a *centripetal acceleration* of the body, which would otherwise move in a straight line.

Now we apply the result to a planet of mass m in orbit about the Sun. The period P of the planet, that is, the time required for the planet to go completely around the Sun, is the circumference of its orbit ($2\pi r$) divided by its speed, or

$$P = \frac{2\pi r}{v}.$$

Solving this equation for v, we find

$$v = \frac{2\pi r}{P}.$$

From Kepler's third law we know that the squares of the periods of planets are in proportion to the cubes of their distances from the Sun, the quantity we are calling r. Thus we have

$$P^2 = Ar^3,$$

where A is a constant of proportionality whose value depends on the units used to measure time and distance. Combining the last two equations, we find

$$v^2 = \frac{4\pi^2 r^2}{P^2} = \frac{4\pi^2 r^2}{Ar^3} = \frac{4\pi^2}{Ar},$$

that is,

$$v^2 \propto \frac{1}{r},$$

where the symbol $\propto$ means "proportional to."

If we substitute the above formula for v^2 into the one expressing the Sun's centripetal force on the planet, we obtain

$$\text{force} \propto \frac{m}{r^2}.$$

The centripetal force exerted on the planet by the Sun must therefore be in proportion to the planet's mass and in inverse proportion to the square of the planet's distance from the Sun. According to Newton's third law, however, the planet must exert an equal and opposite attractive force on the Sun. If the gravitational attractive force of the planet on the Sun is to be given by the same mathematical formula as that for the attractive force of the Sun on the planet, the planet's force on the Sun must be

$$\text{force} \propto \frac{M}{r^2},$$

where M is the Sun's mass. Since this is a mutual force of attraction between the Sun and planet, it must be proportional to both the mass of the Sun and the mass of the planet;

therefore, the attractive force between the two has the mathematical form

$$\text{force} \propto \frac{mM}{r^2}.$$

This is the general formula for the force of gravitation between two bodies. While we derived it for the specific case of circular orbits, Newton showed (using calculus) that the same law applies to elliptical orbits.

(b) The Law of Gravitation

For Newton's hypothesis of universal attraction to be correct, there must be an attractive force between all pairs of objects everywhere. This force is proportional to the masses of the two bodies and inversely proportional to their separation. Thus the force F between two bodies of masses m and M, and separated by a distance d, is

$$F = G\frac{mM}{d^2}.$$

Here G, the constant of proportionality in the equation, is a number called the *constant of gravitation*, whose value depends on the units of mass, distance, and force used. The actual value of G has to be determined by laboratory measurements of the attractive force between two material bodies. If metric units are used (kilograms for mass, meters for distance, and newtons for force), G has the numerical value $6.67 \times 10^{-11}\,\text{N m}^2/\text{kg}^2$. The above equation expresses Newton's law of universal gravitation.

Not only is there a force between the Sun and each planet, there is also a force between any two planets. Because of the Sun's far greater mass, the dominant force felt by any planet is that between it and the Sun. The attractive forces between the planets have relatively little influence. Similarly, there is a gravitational attraction between any two objects on Earth (for example, between two flying airplanes or between the kitchen sink and a tree outside the house), but this force is insignificant compared with the force between each of them and the very massive Earth.

Before we see how Newton tested his law of gravitation, let us investigate some of its other consequences.

(c) Acceleration and Weight

Newton hypothesized that there is a force of attraction between all pairs of bodies. The Earth is a large spherical mass that can be thought of as being composed of a large number of component parts. An object, say a person, on the surface of the Earth feels the simultaneous attractions of the many parts of the Earth pulling from many

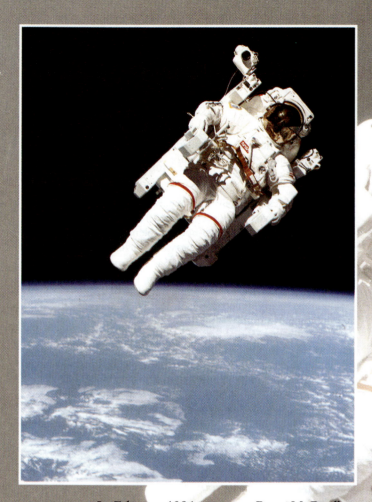

One of the great philosophical changes in humanity's outlook as a result of the work of Galileo, Kepler, and especially Newton was the realization that the same rules we learned about phenomena on Earth could be applied with equal success in the heavens. There is a universality to the workings of nature—the results of experiments in terrestrial laboratories can help us understand the orbits of the moons of Mars, the sources of energy inside stars, and the interactions of distant groups of galaxies.

With this sense of the power and efficacy of the scientific method, scientists after Newton could begin the systematic exploration of the realms beyond our Earth that is the subject of our text. And because the laws of science discovered on Earth were seen to apply not only elsewhere but elsewhen, scientists were able to use these laws to investigate the past history of celestial bodies and begin to piece together the story of how the universe has evolved to its present state.

In February 1984, astronaut Bruce McCandless, having let go of the tether to the Space Shuttle, became the first human satellite of planet Earth. Here he is shown riding the Manned Maneuvering Unit, whose nitrogen gas jets enabled him to move around in orbit, and ultimately to return to the Shuttle. As he circled our planet, he was subject to the same basic laws of motion as all other satellites and all things on Earth.

(NASA)

Sir Isaac Newton (1643–1727) had the insight to realize that the force that makes planets fall around the Sun and the force that makes apples fall to the ground are different manifestations of the same thing: gravitation. Newton's work on the laws of motion, gravitation, optics, and mathematics laid the foundation for almost all physical science up to the 20th century.

CHAPTER

4

GRAVITATION IN THE PLANETARY SYSTEM

Newton's laws of motion and the law of gravitation provide a framework within which to explore many aspects of the planetary system. With them, we can calculate the trajectories of ballistic missiles and Earth satellites, as well as the orbits of the natural members of the solar system. Aided by high-speed computers, scientists can allow for the combined gravitational attractions of many objects, permitting us to navigate interplanetary spacecraft with extraordinary precision. Newton's laws also provide an understanding for such complex phenomena as the Earth's tides. Many of these applications of Newton's ideas are discussed in this chapter.

4.1 ORBITS OF PLANETS AND SPACECRAFT

(a) Satellite Orbits

The first artificial Earth satellite was launched by the Soviet Union on October 4, 1957. Since that time, thousands of satellites have been placed into orbit around the Earth, and spacecraft have also orbited the Moon, Venus, and Mars.

Once an artificial satellite is in orbit, its behavior is no different from that of a natural satellite, such as our Moon or Deimos at Mars. If the satellite is high enough to be free of atmospheric friction, it will remain in orbit

as long as new forces aren't introduced, following Kepler's laws. Although there is no difficulty in maintaining a satellite once it is in orbit, a great deal of energy is required to lift the spacecraft off the Earth and accelerate it to orbital speed.

To illustrate how a satellite is launched, imagine a gun on top of a high mountain, firing a bullet horizontally (Figure 4.1a—adapted from a similar diagram by Newton—Figure 4.1b). Imagine, further, that the friction of the air could be removed and that all hindering objects, such as other mountains, buildings, and so on, are absent. Then the only force that acts on the bullet after it leaves the muzzle is the gravitational force between the bullet and Earth. If the bullet is fired with velocity v_a, it will continue to have that forward speed. But meanwhile the gravitational force acting upon it will accelerate it downward, so that it strikes the ground at a. However, if it is given a higher muzzle velocity v_b, its higher forward speed will carry it farther before it hits the ground, for, regardless of its forward speed, its downward gravitational acceleration is the same. Thus this faster-moving bullet will strike the ground at b.

If the bullet is given a high enough velocity, v_c, as it accelerates toward the ground, the curved surface of the Earth will cause the ground to tip out from under it, so that it remains the same distance above the ground and "falls around" the Earth in a complete circle. This is another way of saying that at a critical speed v_c the

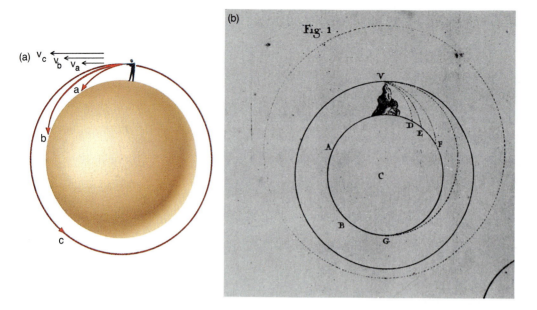

FIGURE 4.1 (a) Firing a bullet into a satellite orbit. (b) A diagram by Newton in his *De Mundi Systematic*, 1731 edition, illustrating the same concept shown in (a). (Crawford Collection, Royal Observatory Edinburgh)

gravitational force between the bullet and Earth is just sufficient to produce the centripetal acceleration needed for a circular orbit about the Earth. The speed v_c, the **circular satellite velocity** at the surface of the Earth, is about 8 km/s (or roughly 17,500 mi/hr).

Nineteenth-century novelist Jules Verne also anticipated Earth satellites. In one of his stories an enemy force was planning to bomb a city with a gigantic cannon ball. However, the cannon ball was propelled with too great a speed—in fact, the circular satellite velocity—so it passed harmlessly over the city and on into a circular orbit around the Earth.

(b) Possible Orbits

Suppose that a rocket is sent up to an altitude of a few hundred miles, then turned so that it is moving horizontally, and finally given a forward horizontal thrust. It will proceed in an orbit the size and shape of which depend critically on the exact direction and speed of the rocket at the instant of its "burnout," that is, the instant when the thrust supplied by its fuel is shut off. First, suppose that it is moving exactly horizontally, or parallel to the ground, at burnout. The possible kinds of orbits it can enter are shown in Figure 4.2.

If the rocket's burnout speed is less than the circular satellite velocity, its orbit will be an ellipse, with the center of the Earth at one focus of the ellipse. The **apogee** point of the orbit, that point that is farthest from the center of the Earth, will be the point of burnout; the **perigee** point (closest approach to the center of the Earth) will be halfway around the orbit from burnout. The corresponding farthest and closest points in an orbit about the Sun are called **aphelion** and **perihelion,** respectively (from *helios,* the Greek word for Sun).

If the burnout speed is substantially below the circular satellite velocity, most of the satellite's elliptical orbit will lie beneath the surface of the Earth (orbit A), where, of course, the satellite cannot travel. Consequently, it will traverse only a small section of its orbit before colliding with the surface of the Earth (or, more likely, burning up in the dense lower atmosphere). If the burnout speed is just slightly below the circular satellite velocity, the rocket may clear the surface (orbit B), although its orbit will probably lie too low in the atmosphere for the satellite to be long-lived.

If the burnout speed were exactly the circular satellite velocity, a circular orbit centered on the center of the Earth would result (orbit C). A slightly greater burnout speed will produce an elliptical orbit with perigee at burnout point and apogee halfway around the orbit (orbit D).

A burnout speed equal to the **velocity of escape** from the Earth's surface, also called the parabolic velocity (about 11 km/s), will put the rocket into a parabolic orbit that will just enable the vehicle to escape from the Earth into space (orbit E). A still higher burnout speed will produce a hyperbolic orbit in which the vehicle escapes the Earth with energy to spare (orbit F). The higher the burnout speed, the nearer the orbit will be to a straight line (orbit G).

(c) Application of the Energy Equation

We can apply the energy equation (Section 3.4d) to the orbit of a satellite moving about the Earth. Let us measure speed in terms of the circular satellite velocity at the Earth's surface, the masses in terms of the Earth's mass, and r and a in units of the radius of the Earth. In these units, the constant G is equal to unity, and the equation simplifies to

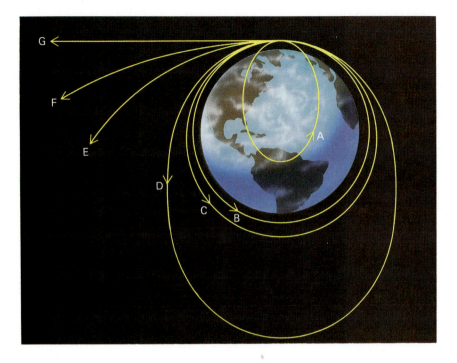

FIGURE 4.2 Various satellite orbits that result from different initial velocities parallel to the Earth's surface. *A* is an orbit that is intercepted by the solid Earth (like that of a military ballistic missile); *B* is a circular orbit; and *C, D, E,* and so on, are orbits of increasing energy but all with the same perigee at the point of injection.

$$v^2 = \frac{2}{r} - \frac{1}{a}.$$

Suppose a satellite is launched from a point near the Earth's surface (say, at an altitude of 300 km); *r* is 1.047, and *v* at that point is the burnout speed. Then the semimajor axis of the orbit, *a*, can easily be calculated if the burnout speed is known. Solving the equation above for 1/a, we have

$$\frac{1}{a} = \frac{2}{r} - v^2.$$

Negative values of *a* correspond to hyperbolic orbits.

As an example, suppose the burnout speed is 10 km/s, or about 1.263 in units of the circular satellite velocity. Then, we find for *a*

$$\frac{1}{a} = \frac{2}{1.047} - (1.263)^2 = 0.315$$

$$a = 3.17 \text{ Earth radii.}$$

Such a satellite would have an apogee distance of about 33,760 km from the center of the Earth, or about 27,381 km above the surface of the Earth.

The energy equation holds regardless of the direction the two bodies are moving with respect to each other. Note that there is no term in the equation that involves the direction in which a rocket is moving at burnout. Thus, even if the rocket were not moving parallel to the ground at burnout, the major axis of its orbit would depend only on its burnout speed (Figure 4.3). However, the eccentricity, or shape, of the orbit does depend on the direction of motion of the rocket. We see in Figure 4.3 that for a rocket launched into a satellite orbit near the surface of the Earth, unless the burnout direction is nearly parallel to the ground, the resulting orbit will be too eccentric to clear the surface of the Earth.

(d)   Launch Vehicles

Launch of an Earth satellite or an interplanetary spacecraft requires a rocket-powered launch vehicle. Only rocket engines can operate in the near-vacuum of the upper atmosphere or of near-Earth space, and only rockets have the high efficiency required to reach Earth orbit.

The simplest rocket engines are solid fuel rockets, which can be stored for long periods and require rela-

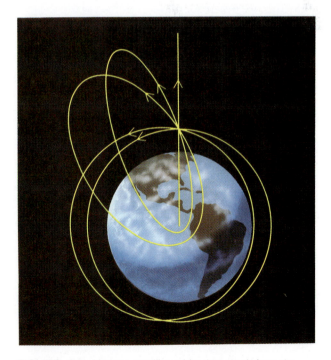

FIGURE 4.3 Various satellite orbits that result from the same initial speed but in different directions. All these orbits have the same major axis but different eccentricities.

tively little care. The 19th-century British rockets that were used in the attack on Baltimore ("the rockets' red glare"), most of the intercontinental ballistic missiles (ICBMs) of today (Minuteman, Polaris, and so forth), and the Space Shuttle booster engines are all examples of solid fuel rockets. Their main disadvantages are that they cannot be restarted in space and are less efficient than liquid fuel engines. Also, they are subject to catastrophic failures, as witnessed in the Shuttle Challenger accident in 1986.

Liquid fuel rockets were first developed in the U.S. and Germany in the 1930s and saw action in World War II in the form of the V-2 missile. The most efficient chemical rockets burn liquid hydrogen and liquid oxygen (to produce an exhaust of water vapor). Hydrogen/oxygen rockets powered the upper stages of the Saturn V Apollo vehicle used to send astronauts to the Moon, and they are used for the Centaur family of deep-space launch vehicles and for the Space Shuttle main engines (Figure 4.4). Since the hydrogen and oxygen must be stored and handled in liquid form at extremely low temperatures, other fuels are required for long-duration flights. The Shuttle's small maneuvering engines and other rockets used for interplanetary flight are powered by liquid fuels that do not require special cooling and ignite spontaneously as they are mixed in the rocket combustion chamber.

Space flight technology, so exotic a few decades ago, has spread around the world. Today there are hundreds of military and civilian launch vehicles manufactured by the governments of about a dozen nations as well as by private industry. The most powerful launch vehicles ever developed are the U.S. Saturn V (1968), with a thrust of 7.5 million lb, and the U.S.S.R. Energia (1987), with a thrust of more than 8 million lb. The U.S. discontinued the Saturn V in the early 1970s, and today the Energia is the largest and most capable space launch vehicle in operation.

(e)   Earth Satellites

Sputnik 1, the first artificial Earth satellite, had an overall weight of about 4 tons and a scientific instrumentation package of about 80 kg. The first American satellite, Explorer 1, launched January 31, 1958, was much smaller. Since then, the U.S.S.R. and the U.S. have each launched dozens of satellites each year. Thousands of satellites or fragments of satellites remain in orbit (Figure 4.5).

Most satellites are launched into low Earth orbit, since this requires the minimum launch energy. At the orbital speed of about 8 km/s, they circle the planet in about 90 min. These orbits are not stable indefinitely, since the drag generated by friction with the thin upper atmosphere eventually leads to a loss of energy and "decay" of the orbit. Upon re-entering the denser parts of the atmosphere, most satellites are burned up by atmospheric friction, although some solid parts may reach the surface. And, of course, the U.S. and Soviet shuttles and other recoverable payloads are designed to survive re-entry intact.

If a satellite is aimed for a higher orbit, it is initially injected into a somewhat eccentric orbit with apogee (highest point) near the target altitude. A second firing of the rocket engine near apogee then imparts additional energy to raise the perigee and circularize the orbit at the desired position.

For many purposes, the most desirable altitude is that at which the orbital period exactly equals the rotation period of the Earth, 24 hr. Kepler's third law tells us that this altitude above the center of the Earth must be about 40,000 km. Here, a satellite remains above the same longitude with respect to the rotating Earth beneath, and the orbit is called *geosynchronous*. If the geosynchronous orbit is also above the equator (zero inclination with respect to the Earth), the satellite remains directly above the same point at all times. It is then called *geostationary*. Geostationary orbits are ideal for most meteorological and communications applications. As we will see later, Pluto's natural satellite Charon is in a stationary orbit with respect to its planet.

FIGURE 4.4   The U.S. Space Shuttle at launch, powered by three liquid fuel engines burning liquid oxygen and liquid hydrogen, and two solid fuel boosters. (NASA)

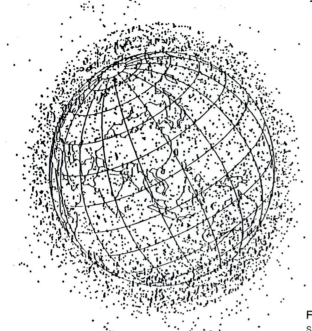

FIGURE 4.5 Plot of all known satellites and large satellite debris in Earth orbit in March 1986.

(f) Interplanetary Spacecraft

The exploration of the solar system has been carried out largely by robot spacecraft sent to the other planets (Figure 4.6). To escape Earth, these craft must achieve a velocity of more than 11 km/s, after which they coast to their targets, subject only to minor trajectory adjustments provided by small thruster rockets on board. In interplanetary flight these spacecraft follow Keplerian orbits around the Sun, modified only when they pass near one of the planets (see Section 4.1g).

Most interplanetary probes have been *flybys*, which means that they have made the relevant observations of the planets in the brief periods during which they passed near their targets. Closeup observations by flybys are generally limited to a few days or less. An extreme example is provided by the 1986 Soviet and European flybys of Comet Halley, which flashed past the small nucleus in less than a second.

While close to its target, a spacecraft is deflected into a modified orbit, either gaining or losing energy in the process. By carefully choosing the aim point in a planetary encounter, controllers have been able to redirect a flyby spacecraft to a second target. Such *gravity-assisted trajectories* were first used in 1974 to direct Mariner 10 from Venus to Mercury. Voyager 2 used a series of gravity-assisted encounters to yield successive flybys of Jupiter (1979), Saturn (1980), Uranus (1986), and Neptune (1989) (Figure 4.7).

Close flybys can be used to increase the energy of a spacecraft as well as to change its direction. This energy is acquired at the expense of the planet, but the resulting changes in the planet's orbit are far too small to be measured. The U.S. Galileo spacecraft, destined to orbit Jupiter, did not have sufficient energy at launch to reach

FIGURE 4.6 The Voyager spacecraft, an example of an interplanetary robot craft. Voyager 2, launched in 1977, has explored all four jovian planets and their ring and satellite systems. (NASA)

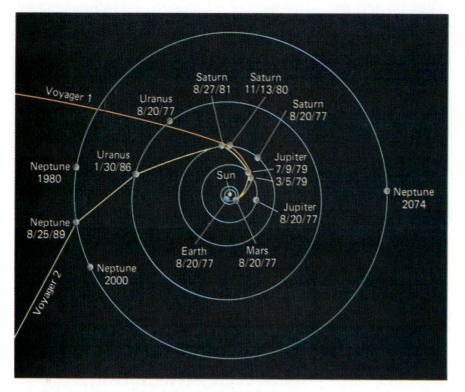

FIGURE 4.7 The flight paths of the two Voyager spacecraft through the outer solar system, taking advantage of the gravitation of each planet to adjust the trajectory toward the next target. (NASA/JPL)

the giant planet. The only way to get it to Jupiter was by close flybys, first of Venus and then twice of the Earth. The second Earth flyby, in December 1992, was at an altitude of less than 200 miles. Only then, three years after launch, was Galileo moving fast enough to reach Jupiter.

Astronomer Paul Weissman has calculated the change in the Earth's orbit from the 1992 Galileo gravity-assisted flyby. Before the flyby, the spacecraft was in an orbit about the Sun with a semimajor axis of 237,636,386 km. After the encounter, the orbital semimajor axis has increased to 472,852,194 km, corresponding to a gain of 4×10^{11} joules of energy at the expense of the Earth. Withdrawing this much energy from the Earth reduces the semimajor axis of its orbit by only 0.02 nm and decreases its period of revolution by less than 10^{-14} s—reflecting the difference in mass between spacecraft and planet.

If we wish to orbit a planet, we must slow the spacecraft with a rocket firing near its target, allowing it to be captured into an elliptical orbit. Mariner 9, in 1971, was the first spacecraft to go into orbit around another planet (Mars).

The next steps beyond a planetary orbiter are atmospheric entry probes and landers. The U.S.S.R. achieved the first successful probes of the atmosphere of Venus (in 1970), the first landers on the surface (1975), and the first instrumented balloons deployed in the venerian atmosphere (1985). The Soviets also made the first landing on Mars (in 1971), but their results were superseded by the highly successful U.S. Viking entry probes and landers of 1976. We will describe some of the results of these missions in Chapters 14 through 17.

(g) Interplanetary Trajectories

We have now learned the principles of space travel. Spacecraft, once they have left the Earth, are astronomical bodies. They obey the same laws of celestial mechanics as the planets and other natural bodies in the solar system. In other words, spacecraft travel in orbits. If the space vehicles carry auxiliary rocket engines and extra fuel, it may be possible to alter their orbits at will, but the principles remain the same.

We shall illustrate space trajectories by showing one of the many possible ways to reach each of the planets Mars and Venus. The orbits to Mars and Venus we show are those that require the expenditure of the least energy as the rocket leaves the Earth and are thus the most economical of fuel. The orbits of the successful U.S. Mariner and Pioneer Venus probes, of the Mariner and Viking Mars probes, and of the similar Soviet probes were all nearly of this type.

Suppose, for simplicity, that the orbits of Venus, Earth, and Mars are circles centered on the Sun. The least-energy orbit that will take us to Mars is an ellipse tangent to the Earth's orbit at the space vehicle's perihelion (closest approach to the Sun) and tangent to the orbit of Mars at the vehicle's aphelion (farthest point from the Sun) (Figure 4.8).

The Earth is traveling around the Sun at the right speed for a circular orbit. For us on the Earth to enter the elliptical orbit to Mars, we must achieve a speed, in the same direction as the Earth is moving, that is slightly greater than the Earth's circular velocity (which is about 30 km/s). To calculate this speed, we employ the energy equation (Section 3.4d). The major axis of the elliptical orbit we want to achieve is the sum of the radii of the orbits of the Earth and Mars. Half of this major axis is the

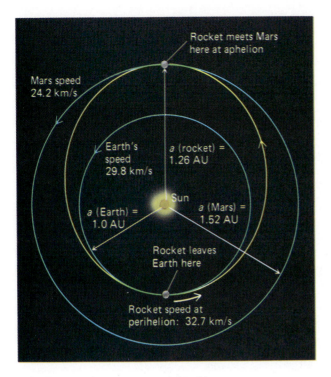

FIGURE 4.8   Least-energy orbit to Mars.

moving at the required 27 km/s and will reach the orbit of Venus along the desired elliptical orbit. The travel time to Venus, found as before from Kepler's third law, is about five months. The return from Venus to Mars (moving from a smaller orbit to a larger one) is similar to the trip from Earth to Mars already discussed.

4.2   THE MANY-BODY PROBLEM

The many-body problem is the problem of describing the motion of any body in a collection of many objects, all interacting under the influence of their mutual gravitation. In contrast to the two-body problem, there are no simple ways to describe this motion. The trajectories are not closed curves like the ellipse or the circle, and the positions cannot be calculated from simple equations like those expressing Kepler's laws. Only a large computer can handle such problems with the precision necessary, for example, to compute the path of an interplanetary probe. Such calculations are made simpler if one object dominates in mass, so that the orbits are nearly Keplerian (like the planets in the solar system). In such cases, the small departures from Keplerian motion can be treated as *perturbations* (small adjustments) of the simpler solution.

(a)   Lagrangian Points

There are some special cases of many-body problems that have interested mathematicians for centuries. One

value a. The appropriate value of a is 1.26 AU. The value r is, of course, the Earth's distance from the Sun, and $m_1 + m_2$ is the combined mass of the Sun and the spaceship (the latter is negligible). The required speed turns out to be slightly under 33 km/s. Since the Earth is already moving at 30 km/s, we need to leave the Earth with the proper speed and direction so that when we are far enough from it (where its gravitational influence on us is negligible compared with that of the Sun), we are still moving in the same direction as the Earth, with a speed relative to it of about 3 km/s.

We have now entered a trajectory that will carry us out to the orbit of Mars. The time required for the trip can be found from Kepler's third law, because our spaceship is, in effect, a planet. The period required to traverse the entire orbit is $a^{3/2}$ years if a is measured in astronomical units. The entire period of the orbit is thus $(1.26)^{3/2} = 1.41$ years. The time required to reach the aphelion point (Mars' orbit) is half of this, or about 8½ months. The trip will have to be planned very carefully so that when we reach the aphelion point of the least-energy orbit, Mars will be there at the same time.

Returning from Mars to Earth is similar to traveling from Earth to Venus and is calculated in the same way, since both involve dropping inward from the more distant planet to one closer to the Sun. The orbit to Venus is very similar to the orbit to Mars, except now it is at the aphelion point that the trajectory ellipse is tangent to the Earth's orbit, and at the perihelion point that it is tangent to the orbit of Venus (Figure 4.9). The semimajor axis of this orbit is half the sum of the radii of the orbits of the Earth and Venus, which is 0.86 AU. From the energy equation we find that the speed at the aphelion point in the orbit is about 27 km/s, about 3 km/s less (rather than more) than the Earth's speed. The space vehicle would have to leave the Earth, as before, with enough speed so that when it has left the Earth's vicinity it has a speed with respect to the Earth of 3 km/s, but in a direction opposite that of the Earth's motion. Then, relative to the Sun, the vehicle is

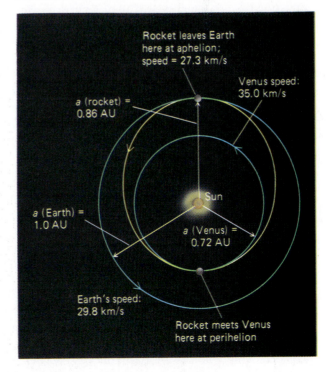

FIGURE 4.9   Least-energy orbit to Venus.

of these, involving a small object moving in the gravitational field of two large masses in circular orbits, was studied by the French mathematician Louis Lagrange (1736–1813).

Lagrange found that there are five positions in such a system where the small object, once placed, will move in a circular orbit, always maintaining a fixed orientation with respect to the two greater masses. These are known as the *Lagrangian points* associated with the two large masses (Figure 4.10). A set of Lagrangian points can be identified near each of the planets, where the two large masses are the Sun and the planet.

The Lagrangian points marked *B* in Figure 4.10 are stable, in that a small object brought near them will remain there and will not be forced away by the perturbations of other planets. (The three points marked *A* are not stable against small perturbations.) Note that these two points each mark the tip of an equilateral triangle, with the Sun and the planet at the other tips. Objects at these points follow the same orbit as the planet but lead or trail it by an angle of 60°. There are several examples in the solar system of Lagrangian orbits, including the so-called Trojan asteroids that lead and trail Jupiter (Chapter 18) and several of the small satellites of Saturn (Chapter 17). The two stable Lagrangian points of the Earth-Moon system, called L-4 and L-5, have also been suggested as suitable locations for future large space habitats.

(b)   Nonspherical Bodies

Bodies with spherical symmetry act, gravitationally, as *point masses*. That is, we can calculate their gravitational attraction for other bodies as if all of their mass were concentrated at a single point in space. In nature, however, most bodies are not exactly spherical, and the simple two-body theory does not give precise results. If the shape of a body deviates only slightly from a sphere, we usually approximate its gravitational influence by that produced by a point mass and treat the small effects of its asphericity as perturbations.

A common cause of the deformation of a star or planet from a perfect sphere is its rotation. In isolation, a nonrotating object will tend toward a perfectly spherical shape under the influence of its own gravitation, but a rotating body tends toward an **oblate spheroid,** which is flattened at the poles and bulging at the equator (like a Gouda cheese). The equatorial cross-section of an oblate spheroid is a circle, while the cross-section at the poles is an ellipse. Jupiter, for example, is noticeably flattened when seen in a telescope.

The rotational flattening of the Earth is slight but important for calculating the orbits of Earth satellites (including the Moon). The diameter of the Earth measured from pole to pole is 43 km less than the equatorial diameter. This amounts to 1 part in 298, which is referred to as the **oblateness** of the planet. The extra

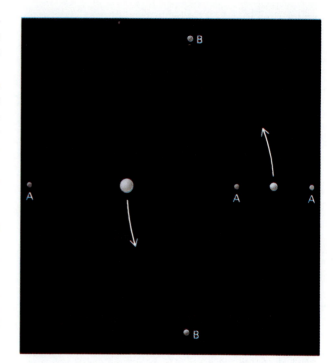

FIGURE 4.10 Lagrangian points, at each of which a body of small mass moves in a circular orbit, maintaining a fixed orientation with respect to the two larger bodies mutually revolving in circular orbits.

matter at the equator exerts an additional force of gravitational attraction on satellites, beyond what would be expected for a spherical Earth. Except for the special cases of satellites in orbits either parallel or perpendicular to the equator, this force produces easily measurable perturbations. Indeed, by accurately tracking satellites, we have mapped the gravity field of the Earth and can use this information to probe the interior structure of our planet.

4.3   DIFFERENTIAL GRAVITATIONAL FORCES

For most orbital calculations, we have seen that the gravitational effects of one body on another can be approximated by point masses. There are additional effects, however, that can distort the shape of one body in the gravitational field of another, giving rise to such phenomena as *tides* and *precession*. Tidal effects are of critical importance for understanding planetary rings (Chapter 17) and the exchange of mass between members of binary-star systems (Chapter 30).

(a)   One Body's Attraction on Two Others

A *differential gravitational force* is the difference between the gravitational forces exerted on two neighboring particles by a third, more distant, body. Both particles are, of course, attracted by the third body, around which they may be in orbit. But relative to each other, the small differential force can be quite significant. The differential force will tend to pull the two particles away from each other. If they are part of the same object, the force will distort it or perhaps even tear it apart.

As is derived below, the size of the differential gravitational force depends on the inverse cube of the distance from the third body, unlike the basic gravitational force itself, which varies with the inverse square power. Thus these forces, which give rise to tidal effects, are strongly concentrated near large bodies. If the distance decreases by half, the differential force rises by a factor of 8. We will look further at the consequences of this dependence of force on distance in Section 4.4e.

(b) Calculation of Differential Gravitational Force

The differential gravitational force can be calculated fairly easily. As an example, consider Figure 4.11a, in which three bodies are shown in a line. These are either point masses or perfectly spherical objects whose gravitational effect on external objects is the same as that produced by point masses. To the left is a large body of mass M. To the right are two bodies, each of whose masses we shall assume, for ease of calculation, to be unity—say, each has a mass of 1 kg. The first of the small bodies, body 1, is at a distance R from the large one; the other, body 2, is at a distance $R + d$.

The force of attraction between the large mass and body 1 is

$$F_1 = \frac{GM}{R^2},$$

while that between the large mass and body 2 is

$$F_2 = \frac{GM}{(R + d)^2}.$$

Note that F_2 is slightly smaller than F_1 because of the greater distance between the large mass and body 2. The difference $F_1 - F_2$ is the differential gravitational force of the large mass on the two smaller masses.

In Figure 4.11b the forces F_1 and F_2 are shown as vectors pointing toward the large mass to the left. Because the force on body 1 is greater than that on body 2, the differential force tends to separate the two bodies.

Now the center of mass of two small bodies is halfway between them. If either of the two unit masses were at that point, the attraction it would feel toward the large body, M, would be

$$F_{CM} = \frac{GM}{(R + 0.5d)^2}.$$

This force is intermediate between the force on body 1 and that on body 2. *With respect to the center of mass*, therefore, both body 1 and body 2 feel themselves pulled *outward*. If the bodies are free to move, they will separate unless their mutual gravitational attraction is great enough to hold them together. In the example described in the preceding paragraphs, the differential gravitational force ΔF was found to be

$$\Delta F = F_1 - F_2 = \frac{GM}{R^2} - \frac{GM}{(R + d)^2}.$$

Combining the two terms of ΔF, we find, with simple algebra,

$$\Delta F = GM \frac{d(2R + d)}{R^2(R + d)^2}.$$

Now let us suppose that the distance R is very much greater than the distance d. In this case, $R + d$ is so nearly equal to R that we can write

$$R + d \sim R.$$

Similarly,

$$2R + d \sim 2R.$$

With this approximation, our equation for ΔF becomes

$$\Delta F = 2GM \frac{Rd}{R^4} = 2GM \frac{d}{R^3}.$$

Now let us denote by δF the differential force corresponding to a unit separation of the two small bodies, that is, for the case where $d = 1$. Then

$$\delta F = \frac{2GM}{R^3},$$

and the total differential force is

$$\Delta F = d \times \delta F.$$

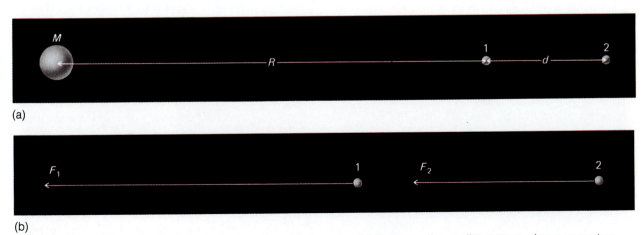

(a)

(b)

FIGURE 4.11 (a) Attraction of a large mass and two smaller ones. (b) Forces on the smaller masses, shown as vectors.

These formulas demonstrate that the differential gravitation force falls off as the cube of the distance, as described earlier.

4.4   TIDES

The effects of the differential force of the Moon's gravity on the Earth are noticed by anyone who lives near the sea, for it is this force that causes the ocean tides. While the association of high tides with the position of the Moon was noticed early in history, a satisfactory explanation of the tides awaited the development of the theory of gravitation.

(a)   Earth Tides

First, we shall consider the effects of the Moon's attraction on the solid Earth. Our planet can be regarded as being composed of a large number of particles all bound together by their mutual gravitational attraction and cohesive forces.

The gravitational forces exerted by the Moon at several arbitrarily selected places in the Earth are illustrated in Figure 4.12. These forces differ slightly from each other because of the Earth's finite size. All parts are not equally distant from the Moon, nor are they all in exactly the same direction from the Moon. If the Earth retained a perfectly spherical shape, the sum of all these forces would be equivalent to the force on a point mass, equal to the mass of the Earth, and located at the Earth's center. Such is approximately true, because the Earth is nearly spherical. It is this force acting at the center of the Earth that causes our planet to accelerate each month in an elliptical orbit about the barycenter of the Earth-Moon system.

The Earth, however, is not a point mass, and it is not perfectly rigid. Consequently, the *differential* force of the Moon's attraction on different parts of the Earth causes the Earth to distort slightly. The side of the Earth nearest the Moon is attracted toward the Moon more strongly than is the center of the Earth, which, in turn, is

attracted more strongly than is the side of the Earth opposite the Moon. Thus, the differential force tends to stretch the Earth slightly into a **prolate spheroid** with its major axis pointed toward the Moon. That is, the Earth takes on a shape like an American football, such that a cross-section perpendicular to the direction of the Moon is a circle, but the cross-section along any plane that contains the line between the centers of the Earth and Moon is an ellipse with its major axis in the Earth-Moon direction.

Figure 4.13 shows the forces (as vectors) that are acting at several points on the surface of the Earth. *In each case, the forces are shown with respect to the Earth's center.* The dashed vectors represent the forces due to

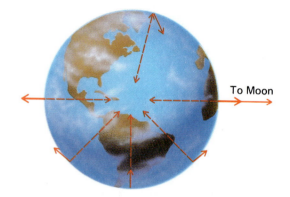

FIGURE 4.13   Gravitational and tidal forces at various places on the Earth's surface.

the Earth's gravity, that is, the weights of various parts of the Earth. The solid vectors (much exaggerated in length) represent the differential gravitational forces due to the Moon's varying force of attraction on different parts of the Earth. They are called the **tidal forces.** Those parts of the Earth closer to the Moon than the Earth's center are attracted more strongly toward the Moon than are parts of the Earth near its center. Those parts on the opposite side of the Earth are attracted less strongly than are parts at the Earth's center.

In each case, the vector representing the tidal force can be broken into two components, one in the vertical

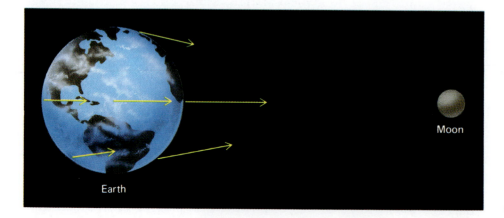

FIGURE 4.12   The Moon's differential attraction of different parts of the Earth.

direction and one in the horizontal direction, along the surface of the Earth. The effect of the vertical component of the tidal force is to change slightly the weight of the surface rocks of the Earth. The effect of the horizontal component is to attempt to cause the surface regions of the Earth to flow horizontally.

If the Earth were fluid, like water, it would distort until all the horizontal components of the tidal forces were exactly balanced by the horizontal pull of the Earth at all points throughout it. Measures have been made to investigate the actual deformation of the Earth. It is found that the solid Earth does distort, but only about one-third as much as if it were liquid, because of the high rigidity of the Earth's interior. The maximum tidal distortion of the solid Earth amounts at its greatest to only about 20 cm. Therefore, the horizonal forces are not zero.

(b) Ideal Ocean Tides

Because the Earth's solid surface does not adjust sufficiently to eliminate horizontal forces at the surface, the water on the Earth's surface is subject to horizontal forces. We shall first assume, for simplicity, that the Earth is covered uniformly by a deep ocean and investigate the nature of the tides produced in it.

The actual accelerations of the ocean waters caused by the horizontal components of the tidal force are very small. These forces, acting over a number of hours, however, produce motions of the water that result in measurable tidal bulges in the oceans. Water on the lunar side of the Earth is drawn toward the sublunar point (the point on the Earth where the Moon appears overhead). On the opposite side of the Earth, water moves in the opposite direction relative to the center of the Earth, producing a tidal bulge opposite the Moon (Figure 4.14).

It is important to understand that it is the horizontal components of the tidal forces that produce the tidal bulges in the oceans. These bulges in the oceans do not result from the Moon's lifting the water away from the

Earth. Rather, the tidal bulges result from an actual flow of water over the Earth's surface, toward the regions below and opposite the Moon, causing the water to pile up to greater depths at those places.

The tidal bulge on the side of the Earth opposite the Moon often seems mysterious to students who picture the tides as being formed by the Moon's "lifting the water away from the Earth." What actually happens, as we have seen, is that the differential gravitational force of the Moon on the Earth tends to stretch the Earth, elongating it slightly toward the Moon. The solid Earth distorts slightly, but, because of its high rigidity, not enough to reach complete equilibrium with the tidal forces. Consequently, the ocean, moving freely over the Earth's surface, flows in such a way as to increase the elongation and piles up at the two regions under and opposite the Moon.

In this section we have regarded the Earth as though its ocean waters were distributed uniformly over its surface. In this idealized picture, which is not actually realized even in the largest oceans, the tides would cause the depths of the ocean to range through only a few feet. The rotation of the Earth would carry an observer at any given place alternately into regions of deeper and shallower water. The observer, then, would see two high tides and two low tides each day.

(c) Tides Produced by the Sun

The Sun also produces tides on the Earth, although the Sun is less than half as effective a tide-raising agent as the Moon. Although the total gravitational force between the Sun and the Earth is about 180 times greater than that between the Earth and the Moon, the *differential* gravitational force is inversely proportional to the cube of the distance and so favors the Moon over the more distant Sun.

If there were no Moon, the tides produced by the Sun would be all we would experience, and the tides would be less than half as great as those we now have. The Moon's tides, therefore, dominate. On the other hand, when the Sun and Moon are lined up, that is, at new moon or full moon, the tides produced by the Sun and Moon reinforce each other and are greater than normal. These are called *spring tides*, although they have nothing to do with spring. In contrast, when the Moon is at first quarter or last quarter, the tides produced by the Sun partially cancel out the tides of the Moon, and the tides are lower than usual. These are *neap tides*. Spring and neap tides are illustrated in Figure 4.15.

(d) The Complicated Nature of Actual Tides

The "simple" theory of tides, described in the preceding paragraphs, would be sufficient if the Earth were completely surrounded by very deep oceans and if it rotated

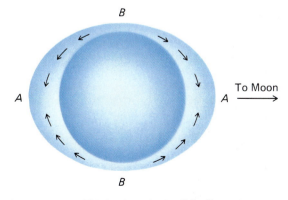

FIGURE 4.14   Tidal bulges in the "ideal" oceans.

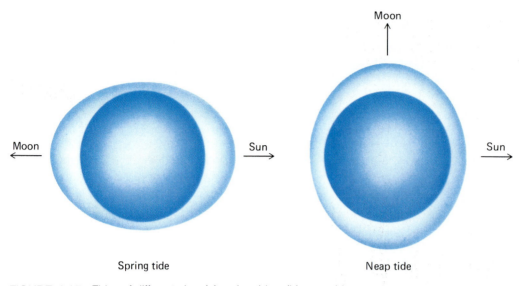

FIGURE 4.15   Tides of different size: (a) spring tides; (b) neap tides.

very slowly. However, the picture is complicated by the presence of land masses stopping the flow of water, the friction in the oceans and between oceans and the ocean floors, the rotation of the Earth, the variable depth of the ocean, winds, and so on.

Both the times and the heights of high tide vary considerably from place to place on the Earth. The Earth's rapid rotation causes the tide-raising forces within a given mass of water to vary too rapidly for the water to adjust completely to them. These forces, however, recurring periodically, set up forced oscillations in the ocean surfaces, so that the water over a large area rises and lowers in step. Consequently, the highest water does not necessarily occur when the Moon is highest in the sky (or lowest below the horizon), but rather when the oscillations of the ocean, produced by the tidal forces acting upon it, pile up the water to its greatest depth at that location.

Sometimes shallow coastal seas have such shapes and sizes that the natural frequency of oscillation of water sloshing back and forth in the sea basins is very nearly the same as that of the tidal rise and fall of the water in the adjacent ocean. Then the ocean tides can set the water in these seas into strong resonance, like wind blowing into an organ pipe. The most famous such place is the Bay of Fundy between New Brunswick and Nova Scotia. The highest tides on Earth occur at the head of the Bay of Fundy, in Minas Basin. Under favorable circumstances the tidal range here can exceed 50 ft.

(e)   Tides Elsewhere in the Solar System

Every planet produces tidal effects on its satellites, and every satellite raises tides upon its planet. One of the consequences of these forces is that nearly all of the satellites keep the same face turned toward their primaries, just as the Moon does toward the Earth. Only in this configuration does the tidal bulge remain fixed with respect to the solid body. For any other rotational rate, the satellite must turn with respect to the bulge, just as the Earth turns with respect to the tides raised by the Moon. The resulting friction has gradually slowed down the satellites until their rotation periods are the same as their periods of revolution.

The Sun also raises tides on the other planets, as well as the Earth. Since tidal effects decline as the inverse cube of the distance, they are much stronger for the inner planet, Mercury, than for any other. Mercury has, as expected, been slowed in its rotation by these solar tides, but the state it has reached is not one in which it always keeps the same face toward the Sun. Rather, Mercury has a rotation period that is just two-thirds of its period of revolution, as we will see in Chapter 13.

One of the most dramatic examples of tidal forces at work in the solar system is provided by Jupiter's volcanically active satellite Io. Io is the innermost large satellite of Jupiter, and the tidal stresses upon it by the planet are immense. As described in Chapter 17, these stresses heat Io, ultimately resulting in the remarkable volcanic eruptions discovered by the Voyager spacecraft in 1979 (Figure 4.16).

4.5   PRECESSION

The Earth, because of its rapid rotation, is not perfectly spherical but has taken on the approximate shape of an oblate spheroid; its equatorial diameter is 43 km greater than its polar diameter. As we have seen, the plane of the Earth's equator, and thus of its equatorial bulge, is inclined at about 23° to the plane of the ecliptic. The ecliptic is

FIGURE 4.16 Io, one of the Galilean satellites of Jupiter, is maintained in a constant state of volcanic activity by heating that results from tides induced on Io by Jupiter. (NASA/JPL)

inclined at 5° to the plane of the Moon's orbit. The differential gravitational forces of the Sun and Moon upon the Earth not only cause the tides but also attempt to pull the equatorial bulge of the Earth into coincidence with the ecliptic.

The latter pull is illustrated in Figure 4.17. The solid arrows are vectors that represent the attractive force of the Moon on representative parts of the Earth. The part of the Earth's equatorial bulge nearest the Moon is pulled more strongly than the part farthest from the Moon, and the Earth's center is pulled with an intermediate force. The dashed arrows show the differential forces with respect to the Earth's center. Note how they tend not only to stretch the Earth toward the Moon

but also to pull the equatorial bulge into the plane of the ecliptic. The differential force of the Sun, although less than half as effective, does the same thing. Thus, the forces of gravitational attraction of the Sun and the Moon on the Earth act in such a way as to attempt to change the direction of the Earth's axis of rotation, so that it would stand perpendicular to the orbital plane of the Earth. To understand what actually takes place, we must digress for a moment to consider what happens when a similar force acts upon a top or gyroscope.

(a) Precession of a Gyroscope

Consider the top (a simple form of gyroscope) pictured in Figure 4.18. If the top's axis is not perfectly vertical, its weight (the force of gravity between it and the Earth) tends to topple it over. The actual force that acts to change the orientation of the axis of rotation of the top is that component of the top's weight that is perpendicular to its axis. We know from watching a top spin that the axis of the top does not fall toward the horizontal. On the contrary, it moves off in a direction perpendicular to the plane defined by the axis and the force tending to change its orientation. Until the spin of the top is slowed down by friction, the axis does not change its angle of inclination to the vertical (or to the floor). Rather, it describes a conical motion (a cone about the vertical line passing through the pivot point of the top). This conical motion of the top's axis is called *precession*.

(b) Precession of the Earth

The differential gravitational force of the Sun on the Earth tends to pull the Earth's equatorial bulge into the plane of the ecliptic, and that of the Moon tends to pull the bulge into the plane of the Moon's orbit, which is nearly in the ecliptic. These forces, in other words, tend to pull the Earth's axis into a direction approximately perpendicular to the ecliptic plane. Like a top, however, the Earth's axis does not yield in the direction of these forces, but precesses. The obliquity of the ecliptic remains approximately 23°. The Earth's axis

Moon

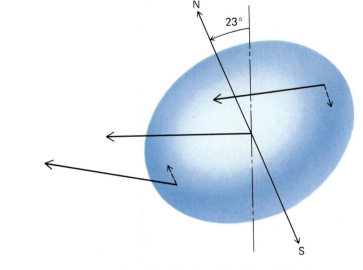

FIGURE 4.17 Differential force of the Moon on the oblate Earth tends to "erect" its axis, leading to precessional motion.

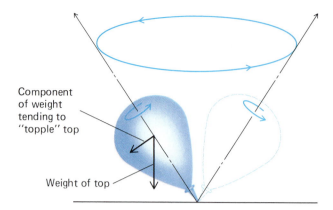

FIGURE 4.18   Precession of a top.

slides along the surface of an imaginary cone, perpendicular to the ecliptic, and with a half-angle at its apex of 23° (see Figure 4.19). The precessional motion is exceedingly slow; one complete cycle of the axis about the cone requires about 26,000 years.

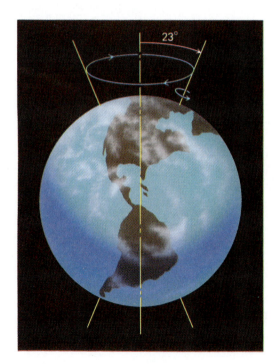

FIGURE 4.19   Precession of the Earth.

Precession does not affect positions on the surface of the Earth, but it does affect the positions of the celestial poles among the stars. In the 20th century, for example, the north celestial pole is very near the star Polaris, in the constellation of Ursa Minor. This was not always so. In the course of 26,000 years, the north celestial pole will move on the celestial sphere along an approximate circle of about 23° radius, centered on the pole of the ecliptic (where the perpendicular to the Earth's orbit intersects the celestial sphere). This motion of the pole is shown in Figure 4.20. In about 12,000 years, the celestial pole will be fairly close to the bright star Vega.

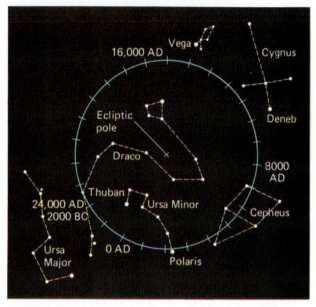

FIGURE 4.20   Precessional path of the north celestial pole among the northern stars.

As the positions of the poles change on the celestial sphere, so do the regions of the sky that are circumpolar, that is, that are perpetually above (or below) the horizon for an observer at any particular place on Earth. The Little Dipper, for example, will not always be circumpolar as seen from north temperate latitudes. Moreover, 2000 years ago, the Southern Cross was sometimes visible from parts of the continental U. S. It was by noting the very gradual changes in the positions of stars with respect to the celestial poles that Hipparchus discovered precession in the second century B.C. (Section 1.3c).

S U M M A R Y

4.1   The orbit of an artificial satellite depends on the circumstances of its launch. The **circular satellite velocity** at the Earth's surface is 8 km/s, and the **escape velocity** of 11 km/s is greater by a factor of $\sqrt{2}$. The lowest point in a satellite orbit is its **perigee,** and the highest point is its **apogee** (corresponding to **perihelion** and **aphelion** for an orbit about the Sun). A variety of liquid and solid fuel rockets are used to launch satellites and deep space probes. There are many possible

interplanetary trajectories, including those that use gravity-assisted flybys of one object to redirect the spacecraft toward its next target.

4.2   Gravitational problems that involve more than two interacting bodies are much more difficult to deal with than two-body problems. They require large computers for accurate solutions. There are five Lagrangian points associated with one body orbiting another, two of them stable. Rotation distorts spherical bodies (stars or planets) into **oblate spheroids;** the corresponding flattening is expressed by the **oblateness** of the object.

4.3   A differential gravitational force is the difference between the forces experienced by two neighboring particles due to the gravitation of a distant third body. It depends on the inverse cube of the distance from the distant body.

4.4   The twice-daily ocean tides are the result primarily of the differential gravitational force of the Moon on the material of the Earth's crust and ocean. These **tidal forces** tend to distort the Earth into a **prolate spheroid.** The horizontal component of such forces causes ocean water to flow into two tidal bulges on opposite sides of the Earth. Actual ocean tides are complicated by the additional effects of the Sun and by the shape of the coasts and ocean basins.

4.5   The precession of the Earth is also a consequence of differential gravitational forces.

E X E R C I S E S

THOUGHT QUESTIONS

1. List some of the uses of Earth satellites today and indicate how important they are to our economy and well-being.

2. As air friction causes a satellite to spiral inward closer to the Earth, its orbital speed increases. Why?

3. Earth satellites in low orbits (like the U.S. Shuttle) require 90 min for each orbit. Calculate their speed, and note that this circular satellite velocity is equal to the escape velocity divided by $\sqrt{2} = 1.414$.

4. Describe how a space vehicle must be launched if it is to fall into the Sun. Is it harder to reach the Sun or the planet Jupiter from Earth?

5. Strictly speaking, should it be a 24-hr period during which there are two high tides? If not, what should the interval be?

PROBLEMS

6. What would be the period of an artificial satellite in a circular orbit around the Earth with a radius equal to 96,000 km? (Assume that the Moon's distance and period are 384,000 km and $27\frac{1}{3}$ days, respectively.)
   Answer: About $\frac{1}{2}$ week

7. If a satellite has a nearly circular orbit at a critical distance from the Earth's center, it will have a period of revolution equal to one day and thus can appear stationary in the sky above a particular place on Earth. Calculate the radius of the orbit of such a synchronous satellite.
   Answer: About 42,400 km

8. The simplest spacecraft trajectory from one planet to another is not a straight line. Instead, it is best to follow an elliptical orbit that lies between the orbits of the two planets, touching the inner planetary orbit at the perihelion of the spacecraft orbit and the outer planetary orbit at aphelion. For such a trajectory from Earth to Jupiter, calculate the perihelion, aphelion, semimajor axis, and period of revolution of the spacecraft orbit.

9. Compute the relative tide-raising effectiveness of the Sun and the Moon. For this approximate calculation, assume that the Earth is 80 times as massive as the Moon, that the Sun is 300,000 times as massive as the Earth, and that the Sun is 400 times as distant as the Moon.
   Answer: Moon is 8/3 times as effective

★10. Verify the periods given in the text for the times required to reach the planets Venus and Mars along least-energy orbits.

★11. Show why the times at which a space vehicle can be sent to a planet on a least-energy orbit occur at intervals of a synodic period of the planet.

★12. Find the separation d between two small bodies, each of unit mass, lined up with a large body of mass M, at a distance R from the nearest of the small bodies, such that the gravitational attraction between the small bodies is just equal to the differential gravitational force between them caused by their attraction to the large body. The answer should be in terms of G, M, and R.

★13. If the precessional rate is about 50 arcsec per year, show that the complete cycle is about 26,000 years.

In this view, captured with a fisheye lens aboard the Space Shuttle on December 9, 1993, the Earth hangs above the Hubble Space Telescope as it is being repaired. The reddish continent is Australia, but its size and shape are distorted by the special lens. Because the seasons in the Southern Hemisphere are opposite ours, it was summer in Australia on this December day.

Once the bold ideas of the Renaissance had reduced the status of the Earth from the center of all creation to just one planet circling the Sun, humanity could begin to understand the Earth's true motions and relationships with celestial objects. This chapter introduces some of these ideas, which help give us our sense of time here on Earth. Ironically, in our day of artificial lights, heated and air-conditioned buildings, and instantaneous communications, fewer and fewer people pay careful attention to the cycles of time defined by the motions in the sky.

Before you start reading this chapter, take a minute and see if you can answer the following questions:

1. Why is it warmer in summer and colder in winter? What causes the seasons?

2. Supposing someone challenged you to prove that the Earth actually revolves around the Sun—how would you do it?

3. Why are there seven days in a week (and not six or eight)?

4. Where does Saturday get its name?

If you're not sure about the answer to one or more of these questions, you'll want to read Chapter 5. Once you've finished, try posing the above questions to students who are not in your astronomy class. We think you'll be surprised at how few people know the answers to such everyday questions.

Tycho Brahe (1546–1601), Danish astronomer whose extensive observations of the planets led to Kepler's laws of planetary motion. Tycho, the last major astronomer from the era before the invention of the telescope, made meticulous measurements of the lengths of the seasons, precession of the equinoxes, length of the year, and nearly every other astronomical constant known at the time. (Yerkes Observatory)

THE CELESTIAL CLOCKWORK

In the preceding chapters we were concerned with the mechanics that govern the motions of celestial bodies. Now we turn our attention to the motions of the Earth and to the relation between the Earth and the sky. It is these motions that define our measures of date and time, that give rise to the seasons, and that produce the constantly changing appearance of the sky. We begin with the rotation of the Earth about its axis.

5.1 ROTATION OF THE EARTH

(a) The Foucault Pendulum

We have seen that the apparent rotation of the celestial sphere can be accounted for either by a daily rotation of the sky around the Earth or by the rotation of the Earth itself. Since the time of Copernicus, it has been generally accepted that it is the Earth that turns, but not until the 19th century did the French physicist Jean Foucault provide a direct demonstration of this rotation. In 1851, Foucault suspended a 60-m pendulum weighing about 25 kg from the domed ceiling of the Pantheon in Paris. He started the pendulum swinging evenly by drawing it to one side with a cord and then burning the cord. The direction of swing of the pendulum was recorded on a ring of sand placed on a table beneath its point of suspension. At the end of each swing a pointed stylus attached to the bottom of the bob cut a notch in the sand

(Figure 5.1). After a few moments it became apparent that the plane of oscillation of the pendulum was slowly changing with respect to the ring of sand, and hence with respect to the Earth.

The only force acting upon the pendulum was that of gravity between it and the Earth, and this force was in a downward direction. In effect, you can think of the pendulum as if it were held by a space traveler and not attached to the Earth in any way. If the Earth were stationary, there would be no force that could cause the

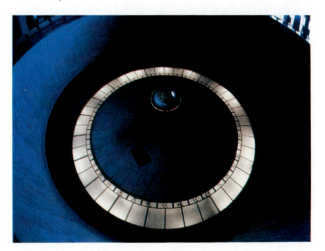

FIGURE 5.1   A Foucault pendulum at the Griffith Observatory planetarium in Los Angeles. (Griffith Observatory, courtesy of E.C. Krupp)

plane of oscillation of the pendulum to alter, and, in accord with Newton's first law, the pendulum should continue to swing in the same direction. The fact that the pendulum slowly changed its direction of swing with respect to the Earth is proof that the Earth rotates.

It is comparatively easy to visualize a Foucault pendulum experiment at the North Pole. Here we can imagine the plane of swing of the pendulum maintaining a fixed direction in space with respect to the stars, while the Earth turns under it every day. Thus, at the North (or South) Pole, a pendulum would appear to rotate its plane of oscillation once completely in 24 hr. At places other than the poles, the problem is complicated because the pendulum must always swing in a vertical plane that passes through the center of the Earth. At the equator, there would be no rotation of a Foucault pendulum at all. At intermediate latitudes we see beneath us a combination of west-east motion and a certain degree of rotation. The result is a period of rotation of the pendulum that is longer than one day. For example, at a latitude of 34° (the latitude of Los Angeles), the Foucault pendulum has a period of 43 hr.

The turning Earth also turns the support system for the pendulum, and consequently the wire and bob of the pendulum itself. However, the rotation of the wire and bob of the pendulum does not alter the direction of swing. Try the following simple experiment. Improvise a small pendulum, such as a yo-yo on its string. Swing the yo-yo to and fro, holding the end of the string in your fingers. Now twist the string in your fingers; the yo-yo will twist with the string but will not change its direction of swing.

(b)   The Coriolis Effect

The apparent rotation of the plane of oscillation of the Foucault pendulum is a demonstration of the rotation of the Earth underneath a freely moving body. Any such apparent deflection in the motion of a body, resulting from the Earth's rotation, is called the **coriolis effect.** The moving body need not be the bob of a pendulum. Any object moving freely over the surface of the Earth appears to be deflected to the right in the Northern Hemisphere (to the left in the Southern Hemisphere) because of the rotation of the Earth beneath it.

As an example of the coriolis effect, consider a projectile fired to the north from the equator. The projectile starts its northward trip with an eastward velocity that it shares with the turning Earth just before it is fired (Figure 5.2); at the equator this eastward velocity is about 1700 km/hr.

There is no westward force on the projectile to slow it down, so it continues to move eastward after being fired. Proceeding northward over the curved surface of the Earth, however, it comes closer to the axis of the Earth's rotation. To conserve its angular momentum,

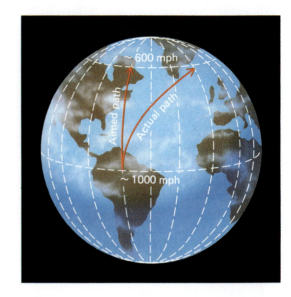

FIGURE 5.2   The coriolis effect on a projectile fired northward from near the equator of the rotating Earth.

the projectile's linear speed to the east must increase if its distance from the axis of rotation decreases. Meanwhile the ground beneath the northbound projectile moves eastward progressively slower, because that ground, closer to the Earth's axis, has less far to move in its daily rotation. We see, then, that the eastward speed of the projectile increases and that of the ground beneath it decreases. Thus, relative to the ground, the missile veers off to the east, that is, to the right for one looking in the direction of its motion.

A similar analysis would show that no matter in what direction a projectile moves, in the Northern Hemisphere it veers off to the right, and in the Southern Hemisphere it veers to the left of its target. This effect must be corrected for in the firing of long-range artillery.

Winds blowing toward a low-pressure area similarly veer off to the right (left in the Southern Hemisphere).

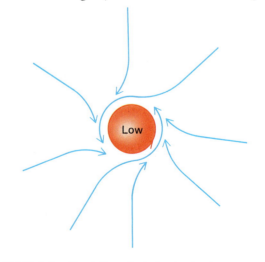

FIGURE 5.3   Circulation of winds about a low-pressure area in the Northern Hemisphere, induced by coriolis forces.

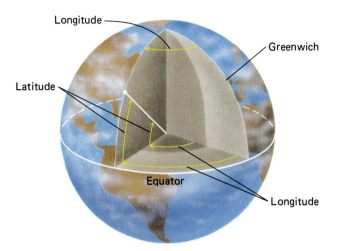

FIGURE 5.4 Latitude and longitude of Washington, D.C.

However, the force continually trying to equalize the pressure of the air accelerates the wind toward the low-pressure area. The wind, rather than "falling" directly into the low center, is caused to circle around the low center by the inertia of the forward-moving air (Figure 5.3). If it were not for the Earth's rotation, winds would blow directly into low-pressure regions, but because the winds veer off and miss the lows, they end up with a circular or **cyclonic** motion. In the Northern Hemisphere, the winds blow around low-pressure centers in a counterclockwise direction, whether they be gentle cyclonic features or violent storms such as hurricanes and tornados. In the Southern Hemisphere the winds are reversed, that is, storms there circulate in a clockwise direction.

(c) Positions on the Earth

To denote positions of places on the Earth, we must set up a system of coordinates on the Earth's surface. The Earth's axis of rotation (that is, the poles and the equator) is the basis for such a system.

A **great circle** is any circle on the surface of a sphere whose center is at the center of the sphere. The Earth's equator is a great circle on the Earth's surface halfway between the North and South Poles. We can also imagine a series of great circles that pass *through* the North and South Poles. These circles are called **meridians;** they intersect the equator at right angles.

Any point on the Earth's surface will have a meridian passing through it (Figure 5.4). This meridian specifies the east-west location of that place—its *longitude*. The zero point of longitudes is arbitrarily defined to be the meridian passing through the old Royal Observatory in Greenwich, England (Figure 5.5). Longitudes are measured either to the east or west of the Greenwich meridian from 0 to 180°.

The other coordinate, the *latitude* of a place, is the angle measured along its meridian between that place and the equator. Latitudes are measured either to the north or south of the equator from 0 to 90°.

Both the latitude and longitude of a location are angles. It is conventional to express these angles in units of degrees (°), minutes ('), and seconds ('') of arc. One minute of arc is defined as 1/60 of a degree, and 1 second of arc is 1/60 of a minute (or 1/3600 of a degree). As it happens, 1 mile on the Earth is about equal to 1 minute of arc. One minute was also about the level of accuracy possible in navigation with the kind of sextant carried on 18th-century sailing ships.

(d) Astronomical Coordinate Systems

In denoting positions of objects in the sky, it is often convenient to make use of the concept of the celestial sphere. We can think of the celestial sphere as being a hollow shell of extremely large radius, centered on the observer (see Chapter 1). The celestial objects appear to be set in the inner surface of this sphere, so we can speak of their *positions on the celestial sphere*. Astronomers have devised coordinate systems, analogous to latitude and longitude, to designate these positions.

First we need to define some terms. In Chapter 1 we discussed the apparent rotation of the celestial sphere about the north and south celestial poles. Halfway between the celestial poles, and thus 90° from each, is the celestial equator, a great circle on the celestial sphere that is in the same plane as the Earth's equator. Great circles passing through the celestial poles and intersecting the celestial equator at right angles (analogous to meridians on the Earth) are called **hour circles.** The great circle passing through the celestial poles and the zenith is called the observer's **celestial meridian.** It coincides with the projection of his terrestrial meridian, as seen from the Earth's center, onto the celestial sphere.

FIGURE 5.5 The Royal Greenwich Observatory, England—internationally agreed upon zero point of longitude on the Earth. (David Morrison)

The celestial meridian intercepts the horizon at the north and south points.

Astronomers specify the apparent position of a star or planet on the celestial sphere by two angles, called **right ascension** and **declination.** Right ascension and declination bear the same relation to the celestial equator and poles that longitude and latitude do to the terrestrial equator and poles. Declination gives the arc distance of a star (or other point on the celestial sphere) along an hour circle north or south of the celestial equator. Right ascension gives the arc distance measured eastward along the celestial equator to the hour circle of the star from a reference point on the celestial equator. That reference point is the **vernal equinox,** one of the two points on the celestial sphere where the celestial equator and the ecliptic intersect.

Several additional celestial coordinate systems are in common use. Each has its advantages for special purposes and is important to astronomers. These systems are defined in Appendix 6.

(e)   The Apparent Motion of the Sky

Imagine an observer at the Earth's North Pole. The celestial north pole is at the zenith, and the celestial equator is at the horizon. As the Earth rotates, the sky turns about a point directly overhead. The stars neither rise nor set; they circle the sky parallel to the horizon. Only that half of the sky that is north of the celestial equator is ever visible to this observer (Figure 5.6). Similarly, an observer at the South Pole would see only the southern half of the sky.

The situation is very different for an observer at the equator (Figure 5.7). There the celestial poles lie at the

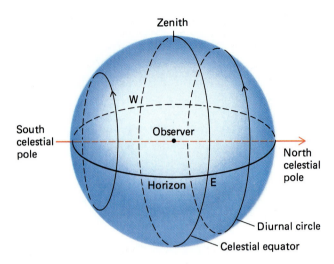

FIGURE 5.7   Sky from the equator; here, stars rise and set perpendicular to the horizon.

north and south points on the horizon. All stars rise and set; they move straight up from the east side of the horizon and set straight down on the west side. During a 24-hr period, all stars are above the horizon exactly half the time (but we can't see them during the day).

For an observer between the equator and North Pole, say at 34° north latitude, the situation is as depicted in Figure 5.8. Here the north celestial pole is 34° above the observer's northern horizon. The south celestial pole is 34° below the southern horizon. As the Earth turns, the whole sky seems to pivot about the north celestial pole, and the stars appear to circle around parallel to the celestial equator. For this observer, stars within 34° of the north celestial pole can never set.

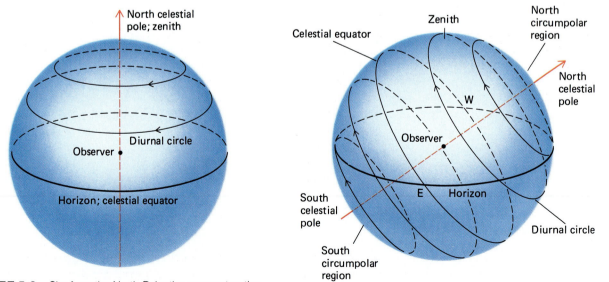

FIGURE 5.6   Sky from the North Pole; the apparent paths of all the stars are parallel to the horizon. The observer is located on a very small Earth at the center; the blue shading is the celestial sphere.

FIGURE 5.8   Sky from latitude 34° N. Stars in the north circumpolar region never set, while those in the south circumpolar region never rise at this latitude.

They are always above the horizon, day and night. This part of the sky is called the *north circumpolar zone* for the latitude 34° N. To observers in the U.S., the Big and Little dippers and Cassiopeia are examples of star groups that are in the north circumpolar zone. On the other hand, stars within 34° of the south celestial pole never rise. That part of the sky is the south circumpolar zone. To most U.S. observers, the Southern Cross is in that zone. At the North or South Pole the entire sky is circumpolar, with half of it above the horizon and half below.

5.2   THE SEASONS

(a)   Proof of the Earth's Revolution

The Earth's revolution about the Sun produces an apparent annual motion of the Sun along the ecliptic, thereby giving rise to our seasons, as we shall see. First, though, let's ask what evidence there is that this motion is real, and that the Earth actually orbits the Sun?

If we adopt Newton's laws of motion and gravitation, it follows simply and directly that the Earth *must* revolve about the Sun and not vice versa. It is obvious that either the Earth goes around the Sun or the Sun goes around the Earth. Thus we have a system of two mutually revolving bodies. The problem is to determine where the common center of revolution is. In Chapter 11 we shall see that the Sun is about 330,000 times as massive as the Earth. Thus the common center of mass (Section 3.3) of the Earth-Sun system must be less than 1/300,000 of the distance from the center of the Sun to the center of the Earth. This puts it well inside the surface of the Sun. Essentially, then, the Earth revolves around the Sun.

There is direct observational evidence that the Earth is not stationary. We have already discussed the retrograde motions of the outer planets (Section 2.1). Other evidence is provided by measurements of parallax and the aberration of starlight. We shall discuss stellar parallax in Chapter 21; aberration is described below.

To understand **aberration,** consider the following analogy. You are walking in the rain, holding a straight drainpipe (Figure 5.9). If the drainpipe is held vertically, and if the raindrops are assumed to fall vertically, they will fall through the length of the pipe only if you are standing still. If you walk forward, you must tilt the pipe slightly forward so that drops entering the top will fall out the bottom without being swept up by the approaching inside wall of the pipe. If the raindrops fall with a speed V, and if you walk with a speed v, the distance by which the top of the pipe precedes the bottom, divided by the vertical distance between the top and bottom of the pipe, must be in the ratio v/V.

Similarly, because of the Earth's orbital motion, if starlight is to pass through the length of a telescope, the

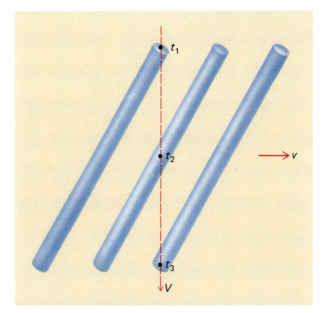

FIGURE 5.9   Raindrops falling through a moving drainpipe—an example of aberration.

telescope must be tilted slightly forward in the direction of the Earth's motion. In other words, the apparent direction of a star is displaced slightly from its geometrical direction, and the displacement is in the direction of the Earth's orbital motion. Analogous to the tilt of the drainpipe, this forward tilt of the telescope is in the ratio of the speed of the Earth to the speed of light. The speed of light is about 10,000 times that of the Earth in its orbit, so the angle through which a telescope must be tilted forward is 1 part in 10,000, or about 20.5 arcsec. The effect is greatest when the Earth is moving at right angles to the direction of the star, and it disappears when the Earth moves directly toward or away from the star. A star that is on the ecliptic appears to shift back and forth in a straight line during the year, since through part of the year the Earth is moving in one direction compared with the star and during the rest of the year the Earth is moving in the opposite direction. A star in a direction perpendicular to the Earth's orbit appears to describe a small circle in the sky, since its apparent direction is constantly displaced in the direction of the Earth's orbital motion from the direction it would have as seen from the Sun. Stars between these extremes appear to shift their apparent directions along tiny elliptical paths of semimajor axis 20.5 arcsec (Figure 5.10).

(b)   The Seasons and Sunshine

The Earth's orbit around the Sun is an ellipse rather than a circle, and our distance varies by about 3 percent from perihelion to aphelion. However, despite what most people think, the changing distance of the Earth from the Sun is not the cause of the seasons. We have seasons because our axis of rotation is tilted with respect

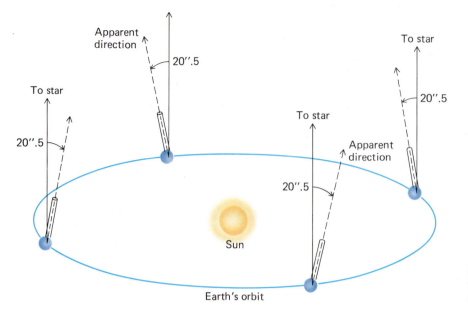

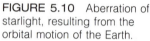

FIGURE 5.10 Aberration of starlight, resulting from the orbital motion of the Earth.

to our orbit. This angle of tilt, which is called the obliquity, is about 23° for the Earth (Section 1.1).

Figure 5.11 shows the Earth's path around the Sun. The line EE' is in the plane of the celestial equator. We see in the figure that on about June 22 (the date of the summer solstice) the Sun shines down most directly upon the Northern Hemisphere of the Earth. It appears 23° north of the equator and thus on that date passes through the zenith of places on the Earth that are at 23° north latitude. The situation is shown in detail in Figure 5.12. To an observer on the equator, the Sun appears 23° north of the zenith at noon. To a person at a latitude 23°

N, the Sun is overhead at noon. This latitude on the Earth, at which the Sun can appear at the zenith at noon on the first day of summer, is called the Tropic of Cancer. We see also in Figure 5.10 that the Sun's rays shine down past the North Pole. In fact, all places within 23° of the pole have sunshine for 24 hr on the first day of summer. The Sun is as far north on this date as it can get; thus, 67° is the southernmost latitude where Sun can ever be seen for a full 24-hr period (the midnight Sun). That circle of latitude is called the Arctic Circle. During this time, the Sun's rays shine very obliquely on the Southern Hemisphere. In fact, all places within 23° of

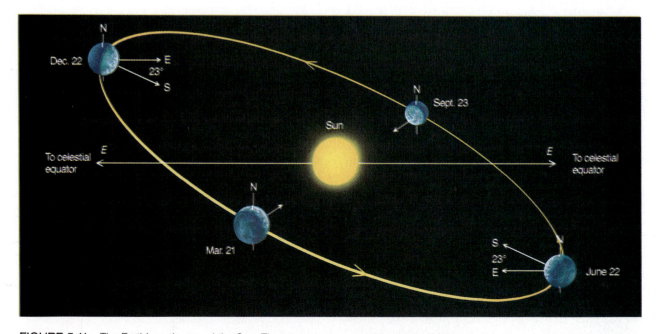

FIGURE 5.11   The Earth's path around the Sun. The seasons are caused by the inclination of the plane of the Earth's orbit to the plane of the equator. (The arrow labeled E marks the Earth's equator; the one labeled S marks the direction toward the Sun.)

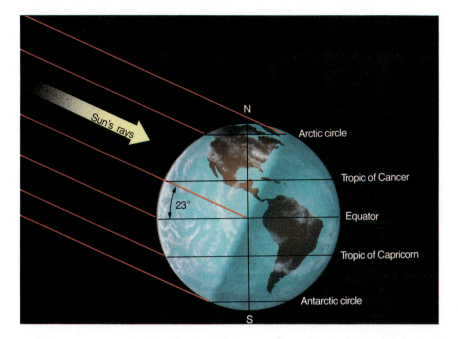

FIGURE 5.12 The Earth on June 22, the summer solstice in the Northern Hemisphere.

the South Pole—that is, south of latitude 67° S (the Antarctic Circle)—do not see the Sun for the entire 24-hr period.

The situation is reversed six months later, about December 22 (the date of the winter solstice), as is shown in Figure 5.13. Now it is the Arctic Circle that has a 24-hr night and the Antarctic Circle that has the midnight Sun. At latitude 23° S, the Tropic of Capricorn, the Sun passes through the zenith at noon. It is winter in the Northern Hemisphere, summer in the Southern.

Finally, we see in Figure 5.9 that on about March 21 and September 23 the Sun appears to be in the direction of the celestial equator. Every place on the Earth then receives 12 hr of sunshine and 12 hr of night. The points where the Sun crosses the celestial equator are the vernal (spring) and autumnal (fall) equinoxes.

Figure 5.14 shows the aspect of the sky at a typical latitude in the U.S. or southern Europe. During the spring and summer, the Sun is north of the equator and is thus up more than half the time. On the first day of summer (about June 22), a typical spot receives about 15 hr of sunshine. Also, notice that the Sun appears high in the sky, and so in these seasons the sunlight is more direct, and thus more effective in heating than in the fall and winter, when the Sun appears at a lower altitude in the sky.

In the fall and winter the Sun is south of the equator, and so it is up less than half the time. On about Decem-

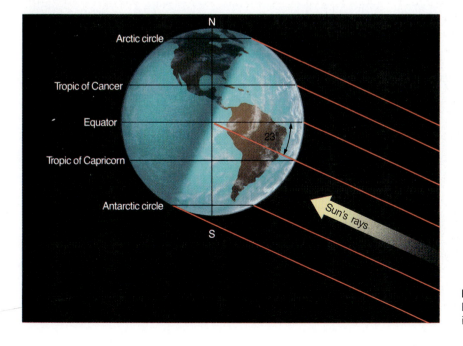

FIGURE 5.13 The Earth on December 22, the winter solstice in the Northern Hemisphere.

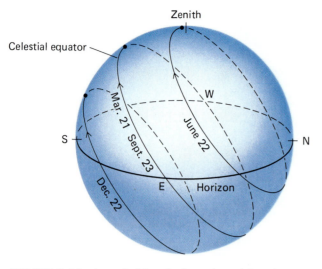

FIGURE 5.14 Aspect of the sky for various dates at a typical place in the temperate Northern Hemisphere.

ber 22, a typical city at, say, 40° north latitude receives only 9 hr of sunshine. At that time the Sun is also low in the sky; a bundle of its rays is spread out over a larger area on the ground (Figure 5.15) than in summer. Because the energy is spread out over a larger area, there is less for each square meter, and so the Sun at low altitudes is less effective in heating the ground.

(c) The Seasons at Different Latitudes

At the equator all seasons are much the same. Every day of the year, the Sun is up half the time, so there are always 12 hr of sunshine at the equator. On about June 22, the Sun crosses the meridian 23° north of the zenith,

while on about December 22, the Sun crosses the meridian 23° south of the zenith.

The seasons become more pronounced as one travels north or south of the equator. At the Tropic of Cancer, on the date of the summer solstice, the Sun is at the zenith at noon. On the date of the winter solstice, the Sun crosses the meridian 47° south of the zenith. At the Arctic Circle, the sun never sets on the first day of summer, but at midnight it can be seen just skimming the north point on the horizon. On about December 22, the Sun does not quite rise at the Arctic Circle, but it just gets up to the south point on the horizon at noon. Between the Tropic of Cancer and the Arctic Circle, the number of hours of sunshine and the noon altitude of the Sun range between these two extremes.

In the Southern Hemisphere, the seasons are reversed from those in the north. While we are having summer in the U.S., it is winter in Australia. Furthermore, in the Southern Hemisphere, the Sun crosses the meridian generally to the north of the zenith. In Buenos Aires, for example, you would want a house with a good northern exposure.

(d) The Effects of the Atmosphere

In the above paragraphs we have been describing the rising and setting of the Sun and stars as they would appear if the Earth had little or no atmosphere. In reality, however, the atmosphere has the curious effect of allowing us to see a little way "over the horizon." This effect is a result of refraction, an optical phenomenon we will discuss in Chapter 8. Because of this atmospheric refraction, the Sun appears to rise earlier, and to

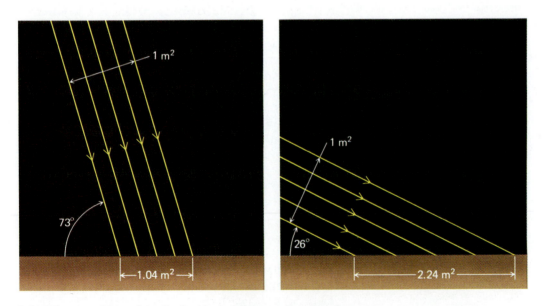

FIGURE 5.15 Effect of the Sun's altitude. When the Sun is low in the sky, its rays are more oblique to the ground and are spread over a larger area than when the Sun is high. This is one of the reasons that winter is colder than summer.

set later, than it would if the atmosphere were not present. In addition, the atmosphere scatters light and provides some twilight illumination even when the Sun is below the horizon. Astronomers define morning twilight as beginning when the Sun is still 18° below the horizon, while evening twilight extends until the Sun sinks more than 18° below the horizon.

These atmospheric effects require the addition of small corrections to many of the statements we have made about the seasons. At the equinoxes, for example, the Sun appears to be above the horizon for a few minutes longer than 12 hr and below the horizon for less than 12 hr. These effects are most dramatic at the Earth's poles, where the Sun actually rises more than a week before it reaches the equator. Also, as a consequence of twilight, the period of real darkness at each pole lasts for only about three months of each year, rather than six.

5.3   TIME

The measurement of time is based on the rotation of the Earth. Over most of human history, time has been reckoned by the positions of the Sun and stars in the sky. Only recently have mechanical and electronic clocks taken over this function in regulating our lives.

(a)   The Length of the Day

The most fundamental astronomical unit of time is the day, measured in terms of the rotation of the Earth. There is, however, more than one way to define the day. Usually, the day is the rotation period of the Earth with respect to the Sun, called the **solar day.** After all, for most people the time of sunrise is more important than the time Arcturus or some other star rises, so we set our clocks to some version of Sun time. However, astronomers also use a **sidereal day,** which is defined in terms of the rotation period of the Earth with respect to the stars.

A solar day is slightly longer than a sidereal day, as a study of Figure 5.16 will show. Suppose we start the day when the Earth's orbital position is at *A*, with the Sun on the meridian of an observer at point *O* on the Earth. When the Earth has completed one rotation with respect to the distant stars (*C*), the Sun will not yet have advanced to the meridian, as a result of the movement of the Earth along its orbit from *A* to *B*. To complete a solar day, the Earth must rotate an additional amount, equal to 1/365 of a full turn. Since there are 360° in a circle, this is very nearly 1°. It takes the Earth about 4 min to rotate through 1°, so the solar day is about 4 min longer than the sidereal day. Each kind of day is subdivided into hours, minutes, and seconds, with each unit of solar time longer than the corresponding unit of sidereal time by about 1 part in 365.

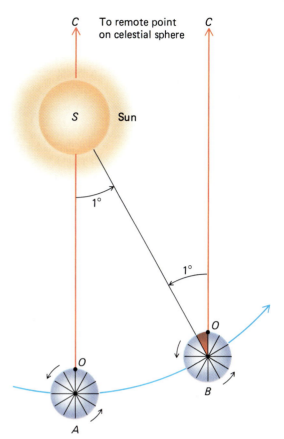

FIGURE 5.16 Sidereal and solar days.

One of the authors of this book remembers a vivid demonstration from his undergraduate days of the slight difference between solar and sidereal time. In the 1950s, before electronic clocks, observatories used pendulum clocks, one keeping standard solar time and the other set to sidereal rate. At his university, these two clocks were in the same room, which was also used by one unfortunate graduate student as his office. Their rates differed by 1 part in 365, so every 365 s they would tick together, then slowly get out of phase until they were again coincident about 6 min later. This constant interplay between the two loudly ticking clocks should have been enough to drive anyone slowly mad, but somehow the students using this peculiar office managed to survive.

The length of the apparent solar day would be constant if the eastward progress of the Sun, in its apparent annual journey around the sky, were precisely constant. However, there are two reasons why the amount by which the Sun shifts to the east is not the same every day of the year.

The first reason is that the Earth's orbital speed varies. In accord with Kepler's second law—the law of areas—the Earth moves fastest when it is nearest the Sun (perihelion) in

2. The length of the apparent solar day is slightly variable because of the Earth's variable orbital speed and the obliquity of the ecliptic.

3. Therefore, the average length of an apparent solar day is defined as the mean solar day.

4. Time based on the mean solar day is mean solar time. It is based on a fictitious "mean Sun" revolving annually on the celestial equator at a perfectly uniform eastward rate.

5. The mean solar time at any one place, called local mean time, varies continuously with longitude, so that two places a few kilometers east and west of each other have slightly different times.

6. Thus we have standardized time, so that each place in a certain zone keeps the same time—the local mean time of the standard meridian in that zone.

7. In many localities, standard time is advanced by 1 hr during part or all of the year to maximize the number of hours of sunshine during the waking hours. This is daylight saving time.

5.4 DATE AND CALENDAR

(a) The Challenge of the Calendar

There are two traditional functions of any calendar. First, it must keep track of the passage of time, allowing people to anticipate the cycle of the seasons and to honor special religious anniversaries. Second, in order to be useful to a large number of people, it must address the problem of the relationships among the basic natural time intervals defined by the motions of the Earth, Moon, and sometimes even the planets. The natural units of the calendar are the *day*, based on the period of rotation of the Earth; the *month*, based on the period of revolution of the Moon about the Earth; and the *year*, based on the period of revolution of the Earth about the Sun. Difficulties have resulted from the fact that these three periods are not commensurable—that is, one does not divide evenly into any of the others.

The rotation period of the Earth is, by definition, 1.0000 day. The period required by the Moon to complete its cycle of phases, called the lunar month, is 29.5306 days. The basic period of revolution of the Earth is 365.2422 days. Clearly, the ratios of these numbers (1.0000/29.5306/365.2422) are not very convenient for calculations. Our natural clocks simply run on different times. This is the historic challenge of the calendar, dealt with in various ways by different human civilizations.

(b) Early Calendars

Even the earliest cultures were concerned with the keeping of time and the calendar. For example, there are Bronze Age monuments in the British Isles, such as Stonehenge. Carbon dating and other studies show that Stonehenge was built in several stages between 2500 and about 1700 B.C. Some of the stones are aligned with the directions of the Sun and Moon during their risings and settings at critical times of the year (such as the beginning of summer and winter), and it is generally believed that at least one function of the monument was connected with the keeping of a calendar.

The Maya in Central America were also concerned with the keeping of time (Figure 5.19). The Mayan calendar was as sophisticated, and perhaps more complex, than contemporary calendars in use in Europe. Apparently, the Maya did not attempt to correlate their calendar accurately with the length of the year or lunar month. Rather, their calendar was a system for keeping track of the passage of days and for counting time far into the past or future. Among other purposes, their calendar was useful for predicting astronomical events—for example, the positions of Venus in the sky.

The ancient Chinese developed an especially complex calendar, largely limited to use by a few privileged court astronomer/astrologers whose positions were hereditary. In addition to the motions of the Earth and Moon, they were able to fit in the approximately 12-yr cycle of Jupiter, which was central to their system of astrology. The Chinese still preserve some aspects of this system in their cycle of 12 "years"—the Year of the Dragon, the Year of the Pig, and so on—that are defined by the position of Jupiter in the zodiac.

FIGURE 5.19 Ruins of the Caracol, a Mayan Observatory at Chichin Itza in the Yucatan, Mexico. (David Morrison)

Our calendar derives from Greek calendars dating from at least the eighth century B.C. They led, eventually, to the Roman calendar introduced by Julius Caesar, which approximated the year by 365.25 days, fairly close to the actual value of 365.2422. The Romans achieved this approximation by declaring years to have 365 days each, with the exception of every fourth year. The **leap year,** which was to have one extra day, bringing its length to 366 days, thus brought the average length of the calendar year to 365.25 days.

The Romans had dropped the almost impossible task of trying to base their calendar on the Moon as well as the Sun, although a vestige of older lunar systems remains, our months having an average length of about 30 days. However, lunar calendars remained in use in other cultures, and Islamic calendars are still primarily lunar rather than solar.

(c) The Gregorian Calendar

Although the Roman (Julian) calendar, which was adopted by the early Christian church, represented a great advance, its average year still differed from the true year by about 11 min, an amount that accumulates over the centuries to an appreciable error. By 1582, that 11 min per year had added up to the point where the first day of spring was occurring on March 11 instead of March 21. If the trend were allowed to continue, eventually Easter would be occurring in early winter. Pope Gregory XIII, a contemporary of Galileo's, felt it necessary to institute a further calendar reform.

The Gregorian calendar reform consisted of two steps. First, ten days had to be dropped out of the calendar to bring the vernal equinox back to March 21; by proclamation, the day following October 4, 1582, became October 15. The second feature of the new Gregorian calendar was that the rule for leap year was changed so that the average length of the year would more closely approximate that of the tropical year. Gregory decreed that three of every four century years, all leap years under the Julian calendar, would be common years henceforth. The rule was that only century years divisible by 400 should be leap years. Thus, 1700, 1800, and 1900—all divisible by four—were not leap years in the Gregorian calendar. On the other hand, the years 1600 and 2000, both divisible by 400, are leap years under both systems. The average length of this Gregorian year, 365.2425 mean solar days, is correct to about 1 day in 3300 years.

The Roman Catholic countries immediately put the Gregorian reform into effect, but countries under control of the Eastern church and most Protestant countries did not adopt it until much later. It was 1752 when England and the American colonies finally made the change. By parliamentary decree, September 2, 1752, was followed by September 14. Although special laws were passed to prevent such breaches of justice as landlords collecting a full month's rent for September, there were still riots and people demanded their 11 days back. Russia did not abandon the Julian calendar until the time of the Bolshevik revolution. The Russians then had to omit 13 days to come into step with the rest of the world.

(d) The Days of the Week

The week is an independent and essentially arbitrary unit of time, although its length may have been based on the interval between the quarter phases of the Moon. In Western cultures the seven days of the week are named for the seven celestial objects that the ancients thought ruled our lives: the Sun, the Moon, and the five planets visible to the unaided eye (Mercury, Venus, Mars, Jupiter, and Saturn.) In English, we can easily recognize Sun-day, Moon-day, and Saturn-day, but the other days come from the Norse equivalents of the Roman gods that gave their names to the planets. In languages more directly related to Latin, the correspondences are clearer. Wednesday, Mercury's day, for example, is Mercoledi in Italian, Mercredi in French, and Miercoles in Spanish. Mars gives its name to Tuesday (Martes in Spanish), Jupiter or Jove to Thursday (Giovedi in Italian), and Venus to Friday (Vendredi in French.) It is interesting to speculate that if we lived in a planetary system where fewer planets were visible, we might well have had a shorter week!

FIGURE 5.20 The Vatican Palace in Rome still appears the way it looked in the time of Pope Gregory XIII. (David Morrison)

In this beautiful image taken with a small amateur's telescope in July 1991, the faint outer atmosphere of the Sun becomes visible during a total eclipse. Note the glowing reddish loops of hot gas, which astronomers call prominences. *The 1991 eclipse was witnessed by millions of people through television coverage from Hawaii and Mexico, two of the sites in the path of the Sun's dark shadow.*
(Courtesy Astronomical Society of the Pacific Archives and Jerry Johnson)

Herodotus, the ancient Greek historian, wrote around 430 B.C. of an earlier period of wars between the Lydians and the Medes, which had gone on without a decisive victory for more than five years. In the sixth year, during one especially fierce battle, a total eclipse of the Sun surprised the combatants, and "the day was on a sudden changed into night." Both armies laid down their weapons, and peace was soon at hand. So awed were earlier civilizations by eclipses of the Sun that they developed a wide range of myths and rituals surrounding them, some of which still survive in folk wisdom. As late as the 19th century, for example, clothes left outside during an eclipse of the Sun in parts of Eastern Europe were burned, lest the poisons associated with eclipses infect the wearer.

The scientific method eventually helped us see the phenomena in the sky not as whims of the gods or the reflections of terrestrial unrest, but merely the results of natural forces that could be understood and foretold through the laws that scientists had discovered. Today, eclipses can be predicted centuries in advance, and the dread that ancient people felt has been replaced by a sense of pride and delight with the power of the human mind. So well do we understand the eclipses we see from Earth, and so uniformly do the principles of celestial motion we have been describing apply elsewhere, that, at the end of this chapter, we are able to discuss eclipses in the system of moons orbiting Jupiter and the mutual eclipses of two stars that are partners in a complex gravitational dance many light years away.

THE MOON AND PLANETS IN THE SKY

Galileo Galilei (1564–1642) advocated that we perform experiments or make observations to ask Nature her ways, rather than deciding how things must be on the basis of preconceived notions. When Galileo turned the telescope to the sky, he found that things were not as philosophers had supposed, discovering sunspots, the mountains of the Moon, the phases of Venus, and the four large satellites of Jupiter— worlds that are still called the Galilean satellites.

Having discussed in Chapter 5 the motions of the Earth and how those motions affect the appearance of the sky, we now turn to the appearance and apparent motions of the Moon and planets. This leads to an examination of eclipses, which take place when one object casts its shadow on another. The physical nature of the Moon is discussed in Chapter 13 and that of the other planets in Chapters 14 through 16.

6.1 OBSERVING THE MOON AND PLANETS

(a) The Planets in the Sky

At almost any time of the night, and at any season, you can spot one or more bright planets visible in the sky. All five of the planets known to the ancients—Mercury, Venus, Mars, Jupiter, and Saturn—are more prominent than any but the brightest of the fixed stars, and they can be seen even from urban locations if you know where and when to look.

Venus, which appears either as an evening "star" in the west after sunset or as a morning "star" in the east before sunrise, is the brightest object in the sky after the Sun and Moon. It far outshines any real star, and under the most favorable circumstances it can even cast a visible shadow. Mars, with its distinctive red color, can be nearly as bright as Venus when it is close to the Earth,

but normally it remains much less conspicuous. Jupiter is most often the second brightest planet, approximately equaling in brilliance the brightest of the stars. Saturn is dimmer, and it varies considerably in brightness, depending on whether its rings are seen nearly edge on (faint) or more widely opened (bright). Finally, Mercury is quite bright, but few people ever notice it because it never moves very far from the Sun and is always seen against bright twilight skies. There is a story (probably apocryphal) that even the great Copernicus never saw the planet Mercury.

True to their name, the planets "wander" against the background of the fixed stars. Although their apparent motions are complex, they reflect an underlying order, which was the basis for the development of the heliocentric model of the solar system.

(b) Moonlight

The amount of light we receive from the Moon varies with its phase. When the Moon is full, its light is nearly bright enough to read by. However, we receive only about 10 percent as much light from the Moon at the first and last quarters.

Despite the brilliance of the full Moon, it shines with less than 1/400,000 the light of the Sun. Even if the entire visible hemisphere of the sky were packed with full Moons, the illumination would be only about one-fifth of that in bright sunlight.

will be bright moonlight in the early evening—a traditional aid to harvesters. The phenomenon of the harvest Moon is most striking in northern latitudes. In the southern hemisphere the same phenomenon occurs in southern autumn (near the time of the vernal equinox).

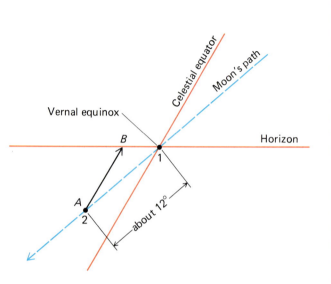

FIGURE 6.2   Explaining the harvest moon.

(d)   The Progression of the Moon's Phases

The phases of the Moon (discussed in Section 1.2b) can be traced using a diagram such as Figure 6.3. We imagine ourselves looking down upon the Earth and the Moon's orbit from the north. The Moon is shown in eight positions in its monthly circuit of the Earth. The Sun is off to the right of the figure at a distance so great that its rays approach the Earth and all parts of the Moon's orbit along essentially parallel paths. The daylight side of the Moon—the side turned toward the Sun—is indicated by shading. For each position of the Moon, its phase, that is, its appearance as viewed from the Earth, is shown just outside its orbit. Several observers are at various places on the Earth, *A*, *B*, *C*, and so on. The time of day, indicated for each observer, depends on the position of the Sun in the sky with respect to the local meridian or, equivalently, on the observer's position on the Earth with respect to the meridian where it is noon.

For person *A*, it is 3:00 P.M. If the Moon is on the meridian, it must be in the waxing crescent phase. (The Moon can be seen easily at noon when in this phase.) If the Moon is in the waning crescent phase, it is setting, for it lies on the western horizon. West is the direction away from which the turning Earth carries the observer, and the horizon lies in a plane tangent to the surface of the Earth at the point where the observer is standing. If

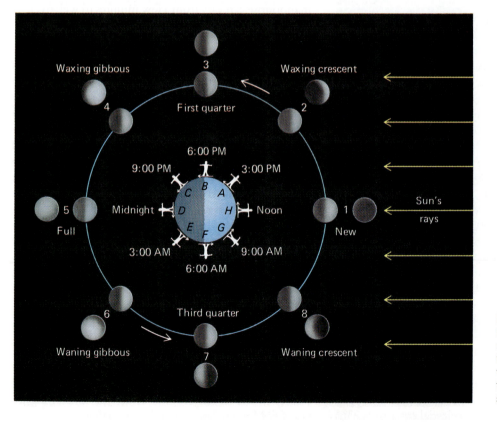

FIGURE 6.3   Phases of the Moon and the time of day. (The outer series of figures shows the Moon at various phases as seen in the sky from the Earth's surface.)

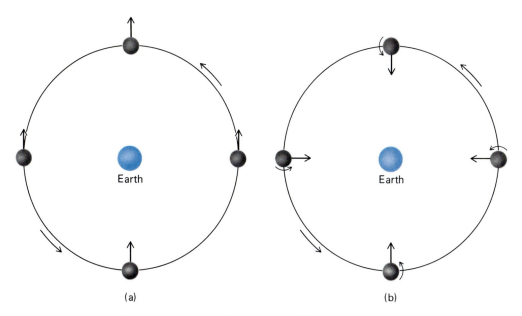

FIGURE 6.4 (a) If the Moon did not rotate, it would turn all its sides to our view. (b) Actually, it does rotate in the same period as it revolves, so we always see the same side. (Thus there is a "back side" or "far side," but certainly not a "dark side.")

the Moon is new, it is in about the same direction as the Sun in the western sky. If it is at first quarter, it is in the eastern sky.

For person B, it is 6:00 P.M. (Person B could be person A 3 hr later. During a period of even a full day, the Moon does not move enough for its phase to change appreciably.) For person B, the Moon is setting if new and rising if full. If it is in the first quarter phase, it appears on the meridian.

For person D, it is midnight. If the Moon is full, it rose at sunset, and it is now on the meridian. The first or third quarter Moon is just setting or rising, respectively.

By studying Figure 6.3, we can tell where the Moon is in the sky at any time of day or night if we know its phase. For example, the full Moon rises at sunset and sets at sunrise. At first quarter the Moon rises at noon and sets at midnight, and so on.

(e)   The Rotation of the Moon

Observations made without optical aid are sufficient to determine that as the Moon goes about the Earth, it keeps the same side toward the Earth. The same facial characteristics of the "man in the Moon" are always turned to our view. If you have never noticed this, start looking carefully at the Moon when it is near full phase and note the unchanging surface pattern, except for the effects of changing phase from night to night.

Although the Moon always presents the same side to us, we must not be misled into thinking that the Moon does not rotate on its axis. In Figure 6.4 the arrow on the Moon represents some lunar feature. If the Moon did not rotate, as in (a), we would see that feature part of the time, and part of the time we would see the other side of

the Moon. Since the Moon rotates on its axis with respect to the stars in the same period as it revolves about the Earth, it always turns the same side toward us (b). The coincidence of the periods of the Moon's rotation and revolution can hardly be accidental. It is believed to have come about as a result of the Earth's tidal forces on the Moon, as discussed in Chapter 4.

We sometimes hear of the back side of the Moon (the side we do not see), called the "dark side." Actually, the back side is dark no more frequently than the front side. Since the Moon rotates, the Sun rises and sets on all sides of the Moon. The back side of the Moon is receiving full daylight at new phase; the dark side is then the one turned toward the Earth.

(f)   The Size and Distance of the Moon

The Moon's distance from the Earth is only about 30 times the diameter of the Earth. Consequently, the direction of the Moon differs slightly as seen from various places on the Earth. The astronomers of ancient Greece and Rome used this fact to determine geometrically the distance to the Moon, and modern astronomers with precision optical observations can do the same thing much better. However, all of these geometrical methods have recently been superseded by new techniques involving radar and laser ranging.

One of the most accurate ways of finding the distance to the Moon is by **radar** (Figure 6.5). The first successful radar contact with the Moon was achieved in 1946. In this technique, radio waves, focused into a beam by a powerful broadcasting antenna, are transmitted to the Moon. Some of this energy is reflected back to the Earth. Since radio waves are a form of electromagnetic

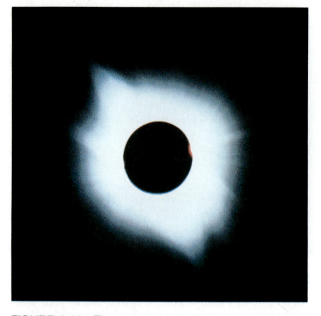

FIGURE 6.15 The corona visible during the July 11, 1991 total solar eclipse photographed from Lomas de Palmira, La Paz, Baja California. Note the two long streamers. (Stephen J. Edberg, ©1991)

uncover the Sun. Gradually the partial phases of the eclipse repeat themselves, in reverse order. At last contact the Moon has completely uncovered the Sun.

In addition to being inspiring to watch, total eclipses of the Sun have considerable astronomical value. Many data are obtained during eclipses that are otherwise not accessible. For example, during an eclipse we can determine the exact relative positions of the Sun and Moon by timing the instants of the four contacts. We can also take direct photographs and make spectrographic observations of the Sun's outer atmosphere.

More than half the time when an eclipse of the Sun takes place, the Moon does not appear large enough in the sky to cover the Sun completely. This means that the umbra of its shadow does not reach all the way to the surface of the Earth. The Moon appears silhouetted against the Sun, with a bright ring of sunlight surrounding it. Such eclipses are called annular. Annular eclipses are of relatively low interest to astronomer and layperson alike, since the Moon does not block enough of the light to reveal the Sun's spectacular corona.

(c) How to Observe a Solar Eclipse

The progress of an eclipse can be observed safely by holding a card with a small (1-mm) hole punched in it several feet above a white surface, such as a concrete sidewalk. The hole in the cardboard produces a pinhole camera image of the Sun (Figure 6.16).

Although there are special filters through which one can safely look at the Sun directly, people are reported to have suffered permanent eye damage by looking at the

Sun through improper filters (or no filter at all!). Common sense (and pain) prevents most of us from looking at the Sun directly on an ordinary day for more than a brief glance. There is nothing about an eclipse that makes the radiation from the Sun more dangerous than it is any other time; on the contrary, we receive less radiation from the Sun when it is partly hidden by the Moon. It is never safe, however, to look steadily at the Sun when it is still in partial eclipse; even the thin crescent of sunlight visible a few minutes before totality has a surface brightness great enough to burn and permanently destroy part of the retina. Unless you have a filter prepared especially for viewing the Sun, it is best to watch the partial phases with a pinhole camera device, as described above.

It is perfectly safe, however, to look at the Sun directly when it is *totally* eclipsed, even through binoculars or telescopes. Unfortunately, unnecessary panic has often been created by well-meaning but uninformed public officials acting with the best intentions. One of us has witnessed two marvelous total eclipses in Australia, during which townspeople held newspapers over their heads for protection and schoolchildren cowered indoors with their heads under their desks. What a cheat to those people to have missed what would have been one of the most memorable experiences of their lifetimes! During totality, by all means look at the Sun.

Nor should you be terrified of accidentally catching a glimpse of the Sun outside totality. How many times have you glanced at the Sun on ordinary days while driving a car or playing ball or tennis? Common sense

FIGURE 6.16 How to watch a solar eclipse safely during its partial phases.

made you look away at once. Do the same if you inadvertently glimpse the Sun directly while it is partially eclipsed.

6.5 PHENOMENA RELATED TO ECLIPSES

Eclipses are not restricted to the Earth-Moon system; they also take place on other planets in the solar system. Also common are situations in which one body passes in front of another, a circumstance that has been used extensively by astronomers to investigate planets, stars, and even distant quasars. These configurations are variously called *eclipses*, *occultations*, and *transits*.

(a) Eclipses in the Planetary System

Eclipses of the Sun are visible from any planet with satellites. In the Jupiter and Saturn systems, with their many satellites, such eclipses are much more common than on Earth. Most of these are annular eclipses, but the larger Jupiter and Saturn satellites produce total solar eclipses. Figure 6.17 shows the umbra of the shadow cast on Jupiter by its innermost large satellite, Io. Within this shadow, a total solar eclipse is in progress. The Viking 1 lander photographed a solar eclipse from the martian surface, produced when the shadow of Phobos fell on the spacecraft.

The equivalents of eclipses of the Moon also take place elsewhere in the solar system, whenever a satellite passes into the shadow cast by its planet. In the Jupiter system, the three inner Galilean satellites (as well as the four small satellites still closer to the planet) undergo an eclipse once in each revolution around the planet. As we will see in Chapter 17, studies of Io carried out during eclipse, when the sunlight is blocked and the surface cools, clearly reveal the presence of hot volcanic sources on Io. Callisto, the outer Galilean satellite, experiences eclipses when its orbit is properly aligned, but much of the time it passes above or below the shadow of Jupiter, just as the Moon usually misses the shadow of the Earth.

A third phenomenon that can occur in systems with multiple satellites is the eclipse of one satellite by another. While very rare, such mutual eclipses do take place in the Jupiter and Saturn systems.

(b) Lunar Occultations

The Moon often passes between the Earth and a star; the phenomenon is called a *lunar occultation*. The stars are so remote that the shadow of the Moon cast in the light of a star is extremely long and sensibly cylindrical. Because a star is virtually a point source, there is no penumbra. During an occultation, a star suddenly disappears as the edge of the Moon crosses the line between the star and observer. Because the Moon moves through an angle about equal to its own diameter every hour, the longest time that a lunar occultation can last is about 1 hr. Geometrically, occultations are equivalent to total solar eclipses, except that they are total eclipses of stars other than the Sun.

The sudden disappearance of a star behind the edge of the Moon during an occultation is evidence that the Moon has no appreciable atmosphere. If there were one, the star would fade gradually, because the starlight would traverse a long path of the lunar atmosphere. Occultations also demonstrate the extremely small angular sizes of the stars. If a star had an appreciable angular size, it would require a perceptible time to disappear behind the Moon, as is true during the partial phases of a total solar eclipse. Actually, the partial phases of occultations have been measured with high-speed detectors and used to measure the angular diameters of stars. But the partial phases are extremely brief, typically a few hundredths of a second or less.

(c) Other Occultations

The term *occultation* is used generally to describe a situation in which one object passes in front of another, while an eclipse takes place when one object moves into the shadow of another. Thus, technically, a solar eclipse is an occultation of the Sun by the Moon. Occultations of stars by solar-system objects have become a powerful tool for investigating planets, satellites, and even small asteroids.

Occultations of stars by planets are much rarer than their occultation by the Moon, because the angular size of planets is so much smaller. Typically, an occultation of a bright star by a planet takes place only once every few decades. The most recent event involving a first-magnitude star was an occultation of Regulus, the brightest star in Leo, by Venus on July 7, 1959. However, with large telescopes, occultations of fainter stars can be used to good purpose,

FIGURE 6.17 The shadow of the satellite Io falling on Jupiter. Within the shadow an observer floating above the clouds of Jupiter would see a total solar eclipse. (NASA/JPL)

Just as water droplets in our atmosphere can bend and disperse sunlight into a rainbow, so instruments called spectroscopes allow astronomers to disperse the light from planets, stars, or galaxies into its component colors, and decipher what conditions are like in those distant objects. This rainbow is seen over one of the rock formations in Monument Valley, Utah.

(Barbara Filet, copyright Tony Stone Worldwide)

In June 1994 a team of astronomers announced the discovery of water molecules in a galaxy 200 million light years away. This is the most distant location where we have ever found water. Its presence was detected not with light waves but from the faint signature that water leaves in *radio* waves. Just as you can tune to your favorite music station amid the static of the radio dial, so astronomers can "tune in" to the waves associated with a variety of chemical substances found among the stars.

Scientists today understand that the light our eyes can detect is a very small sample of the information the universe sends our way. There are many other forms of radiation that are completely invisible to our senses but that nevertheless can help us understand a wide range of phenomena and processes out in space. As the 20th century draws to a close, astronomers have developed instruments to detect all these invisible waves and decode the messages they contain.

In the next few chapters we will examine the nature of light and other radiation, how information is encoded in the waves we receive, and how astronomers have developed instruments to analyze this information. One of the most remarkable facets of the unity of the universe is that the same rules scientists have discovered for the behavior of atoms and radiation on Earth apply on other planets, and even in that distant galaxy where water was found. In this way, the explanation of the rainbow can help us understand the makeup of the stars.

LIGHT: MESSENGER
FROM SPACE

Niels Bohr (1885–1962), a Danish physicist, was awarded the Nobel prize in 1922 for his investigation of the structure of atoms and of the unique patterns of spectral lines that they produce. (AIP Niels Bohr Library, W. F. Meggers Collection)

In astronomy, most of the objects that we study are completely out of reach. If we want to learn about the Sun and stars, we must devise techniques that will allow us to analyze them from a distance. The temperature of the Sun is so high that a spacecraft would be destroyed by its heat long before reaching the solar surface, and the stars are too far away to visit in our lifetimes with the technology now available. Even light, which travels at a speed of 300,000 km/s, takes more than four years to reach us from the *nearest* star.

To study the Sun and stars, we must rely on the messages contained in the light, x rays, radio waves, and other radiation that they emit. In this chapter we shall explore the nature of the radiation that is approaching us with the speed of light from all directions, ready to be sampled by our telescopes. What are the secrets it holds? What are the revelations it will give us about those objects it left years, centuries, even billions of years ago?

7.1 THE NATURE OF LIGHT

With one simple equation, Newton's theory of gravitation accounts for the motions of both the planets and objects on the Earth. Application of this theory to a variety of problems dominated the work of scientists for nearly two centuries. In the 19th century, many physi-

cists turned to the study of electricity and magnetism. The scientist who played a role analogous to that of Newton was physicist James Clerk Maxwell (1831–1879), who was born and educated in Scotland. Maxwell developed a single theory that describes in a small number of elegant equations the way both electricity and magnetism work. It is this theory that allows us to understand the behavior of light.

(a) Maxwell's Theory of Electromagnetism

It was known experimentally in the 19th century that changing magnetic fields could produce electrical currents. Maxwell showed through theoretical calculations that changing electrical currents could also produce magnetic fields, a result that was subsequently confirmed in laboratory experiments. The word **field** is a technical term used in physics to describe the consequences of forces that act on distant objects. For example, the Sun produces a *gravitational field* that controls the Earth's orbit, even though the Sun and the Earth do not come directly into contact. Stationary electric charges produce *electric fields*, and as Maxwell showed, moving electric charges produce *magnetic fields*.

Maxwell found that electric and magnetic fields propagate through space. Maxwell was able to calculate the speed at which an electromagnetic disturbance moves through space and found that it was equal to the speed of light, which had been measured experimentally. On that basis he speculated that light was one form of a broad category called **electromagnetic radiation.** When light enters a human eye, its changing electric and magnetic fields stimulate nerve endings, which then transmit the information contained in these changing fields to the brain.

Since the word *radiation* will be used frequently in this book, it is important to understand what it means. In the modern world, "radiation" is commonly used to describe certain kinds of dangerous subatomic particles released by radioactive materials in our environment. But this is *not* what we mean when we speak of radiation in an astronomy text. Radiation as used in this book is a general term for light, x rays, and other forms of electromagnetic waves. This radiation provides almost our only link with the universe beyond our own solar system.

Energy also has a special meaning in science. In everyday language, we associate "energy" with the fuel we use to run our cars and heat our homes; with the foods we eat, especially just before participating in an athletic event; and with the electricity that lights our streets and houses. Scientists define energy to be the ability to do work, and work in turn is done when force is used to move an object that has mass. Quantitatively, the energy of an object is a measure of its ability to do work.

(b)   The Wave-like Characteristics of Light

Light is produced by regularly repeating changes in electric and magnetic fields. Water waves are another example of a regularly repeating change, and light also has the characteristics of a wave. Some of the features of waves are shown in Figure 7.1. Ocean waves, for example, have alternating crests and troughs. The distance between successive crests is called the **wavelength.** In the Pacific Ocean, waves generated by storms are typically about 200 m long.

The wavelength of visible light lies in the range 0.0000004 to 0.0000007 m. Rather than write all of these zeros, it is customary to express the wavelength of light in nanometers, or nm. One nanometer is one-billionth of a meter (10^{-9} m). Visible light has wavelengths that range from about 400 to 700 nm, with the exact wavelength determining the color of the light. Radiation with a wavelength in the range 400 to 450 nm is perceived by the retina of the eye as the color violet. Radiations of successively longer wavelengths are perceived as the colors blue, green, yellow, orange, and red. (Another unit that is sometimes used to specify wavelengths is the Ångstrom. One nm equals 0.1 Å; wavelengths of 400 nm and 4000 Å are equivalent.)

The **frequency** of a wave is defined to be the number of wave crests (or troughs) that pass a specific point in 1 s. Wind-driven waves break against the coastline at intervals of 5 to 20 s. Their frequency is thus 12 to 3 cycles per minute. The frequency of visible light is much higher—4.3×10^{14} cycles per second for red light to 7.5×10^{14} cycles per second for blue light.

The relationship between the velocity with which a wave propagates, its wavelength, and its frequency is illustrated in Figure 7.1. Imagine a long train of waves moving to the right, past point O, at the speed of light (c). If we measure to the left of O a distance of c km, we arrive at the point P along the wave train that will just reach point O after a period of 1 s. The frequency (f) of the wave train—that is, the number of waves between P and O—times the length of each, λ, is equal to the distance c. Thus we see that for any wave motion, the speed of propagation equals the frequency times the wavelength. That is,

$$c = \lambda f.$$

The Greek letter λ (lambda) is almost always used to denote wavelength.

It is appropriate to compare the propagation of light to the propagation of ocean waves in one more way. While an ocean wave travels forward, the water itself is displaced in a vertical direction. A given water drop does not actually move forward with the wave. The next time you are on a beach, look for a bit of wood or seaweed out beyond where the waves are breaking. You will see that it bobs up and down, staying in about the same place, even though the waves are approaching the shore. Waves that propagate with this kind of motion are called *transverse waves.*

Light also propagates with a transverse wave motion, and it travels with its highest possible speed through a perfect vacuum. In this respect light differs markedly

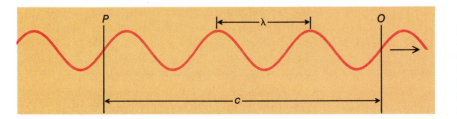

FIGURE 7.1 Electromagnetic radiation has wave-like characteristics. The relationship between the length of the wave (λ), the frequency of the wave, and the speed (c) with which it moves is shown.

from *sound*, which is a physical vibration of matter. Sound does not travel at all through a vacuum. The displacements of the matter that carry a sound impulse are in a *longitudinal* direction, that is, in the direction of the propagation, rather than at right angles to it. Sound is actually a traveling wave of alternate compressions and rarefactions of the matter through which it moves. Sound also travels far more slowly than electromagnetic radiation—only about 1.3 km/s through air at sea level.

(c)   Propagation of Light

As electromagnetic radiation moves away from its source, it spreads out and covers an ever-widening area. The increase in area is proportional to the square of the distance that the radiation has traveled (Figure 7.2). For example, when light from the Sun reaches the Earth, it is spread out over a sphere 1 AU in radius. When it has gone twice as far, to 2 AU from the Sun, that same light is spread over an area four times as great, because the surface area of a sphere is proportional to the square of its radius. When the Sun's radiation reaches Saturn, 10 AU from the Sun, it is spread over an area 100 times that at the Earth's distance. This decrease in energy with increasing distance from the source is called the **inverse-square law,** and it applies to other kinds of energy, including sound. It explains why the inner planets are hot and the outer planets are cold; the farther each planet is from the Sun, the less solar energy it receives on a given area of its surface.

The **apparent brightness** of a light source depends on how much of its energy (that is, light) enters the pupil of our eye or telescope. Since the collecting area of the eye or telescope is constant, the larger the area over which light is spread, the smaller is the fraction observed. We see, then, that the amount of energy passing through a unit area *decreases* with the *square of the distance from the source*. At distances d_1 and d_2 from a light source, the amounts of energy received by a telescope (or other

detecting device), ℓ_1 and ℓ_2 are in the following proportion:

$$\frac{\ell_1}{\ell_2} = \frac{4\pi d_2^2}{4\pi d_1^2} = \left(\frac{d_2}{d_1}\right)^2.$$

The above relationship is known as the inverse-square law of light propagation. In this respect, the propagation of radiation is similar to the effects of gravity, because the force of gravitation between two attracting masses is inversely proportional to the square of their separation.

The inverse-square law for light explains why the stars appear so faint relative to the Sun. A typical star emits about the same total energy as does the Sun, but even the nearest star is about 270,000 times farther away, and so it appears about 73 *billion* times fainter $(73 \times 10^9 = 270,000 \times 270,000)$.

7.2   THE ELECTROMAGNETIC SPECTRUM: MORE THAN MEETS THE EYE

(a)   Types of Electromagnetic Radiation

Light has wavelengths in the range 400 to 700 nm. But what is special about these wavelengths? Is there electromagnetic radiation with wavelengths longer and shorter than these values? It was more than 20 years after Maxwell speculated about this possibility that a new kind of radiation, radio waves, was first generated by the German physicist Heinrich Hertz.

The six broad bands into which we divide electromagnetic radiation are shown in Figure 7.3. Electromagnetic radiation with the shortest wavelengths, not larger than 0.01 nm, is called gamma radiation. **Gamma rays** are often emitted in the course of nuclear reactions and by radioactive elements. They are what is generated in the deep interiors of stars.

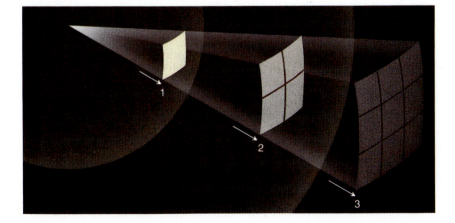

FIGURE 7.2 The inverse-square law for light. As light energy radiates away from its source, it spreads out, so that the energy passing through a unit area decreases as the square of the distance from its source.

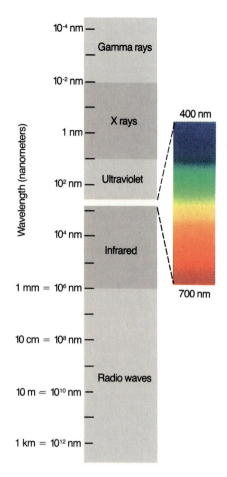

FIGURE 7.3 The electromagnetic spectrum.

Electromagnetic radiation with wavelengths between 0.01 nm and 20 nm is referred to as **x rays,** while radiation intermediate between x rays and visible light is **ultraviolet.** Between visible light and radio waves are the wavelengths of **infrared,** or heat radiation. The **microwaves** used in shortwave communication and in television are radio waves with wavelengths ranging from a few centimeters to a few meters. Other radio radiation can have wavelengths as long as several kilometers.

In 1672, in the first paper that he submitted to the Royal Society, Newton described an experiment in which he permitted sunlight to pass through a small hole and then through a prism. Newton found that sunlight, which gives the impression of being white, is in fact made up of a mixture of all the colors of the rainbow (Figure 7.4). The scientific term for the array of colors produced by visible light is a **spectrum.** The array of radiation of all wavelengths, from gamma rays to radio waves, is called the **electromagnetic spectrum.** Table 7.1 summarizes the types of electromagnetic radiation and indicates the temperatures and types of astronomical objects that emit specific types of electromagnetic radiation.

(b) Radiation Laws

Some astronomical objects emit mostly infrared radiation, others mostly visible light, and still others mostly ultraviolet radiation. What determines the type of electromagnetic radiation emitted by the Sun, stars, and other astronomical objects? The answer is *temperature.*

A *solid* is composed of molecules and atoms that are in continuous vibration. A *gas* consists of molecules that are flying about freely at high speed, continually bumping into one another, and bombarding the surrounding matter. That energy of motion is called heat. The hotter the solid or gas, the more rapid is the motion of the molecules; temperature is a measure of the average energy of those particles.

We now know that electromagnetic radiation is emitted when electric charges *accelerate*: when they change either the speed or the direction of their motion. If an object is hot, then the atoms in that object are moving rapidly, jostling one another, and colliding with electrons, changing their motions at each collision. Each collision therefore results in the emission of electromagnetic radiation—radio, infrared, visible, ultraviolet, and x rays. How much of each type depends on the temperature of the object producing the radiation.

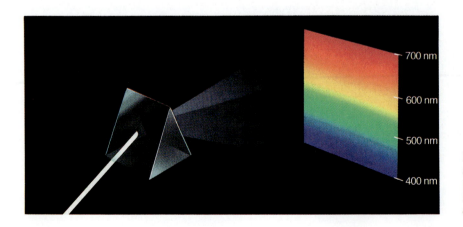

FIGURE 7.4 When Newton passed a beam of white sunlight through a prism, he saw a rainbow-colored band of light that we now call a *continuous* spectrum.

TABLE 7.1 ELECTROMAGNETIC RADIATION			
Type of Radiation	Wavelength Range (nm)	Radiated by Objects at This Temperature	Typical Sources
Gamma rays	Less than 0.01	More than 10^8 K	No astronomical sources this hot; some gamma rays produced in nuclear reactions
X rays	0.01–20	10^6–10^8 K	Gas in clusters of galaxies; supernova* remnants; solar corona
Ultraviolet	20–400	10^5–10^6 K	Supernova remnants; very hot stars
Visible	400–700	10^3–10^5 K	Stars
Infrared	10^3–10^6	10–10^3 K	Cool clouds of dust and gas; planets; satellites
Radio	More than 10^6	Less than 10 K	No astronomical objects this cold; radio emission produced by electrons moving in magnetic fields (synchrotron radiation)

*A *supernova* is the explosion of certain types of stars at the end of their lives.

To understand in more quantitative detail the relationship between temperature and electromagnetic radiation, it is useful to resort to a common tactic used by physicists. Consider an idealized object that absorbs all the electromagnetic energy that impinges on it and does not reflect any of it. Such an object is called a **blackbody.** As it absorbs energy, it heats up until it is emitting energy at the same rate that it is being absorbed.

The radiation from a blackbody has several characteristics that are illustrated in Figure 7.5. First, a blackbody with a temperature higher than absolute zero emits some energy at *all* wavelengths. Second, a blackbody at higher temperature emits *more* energy at all wavelengths

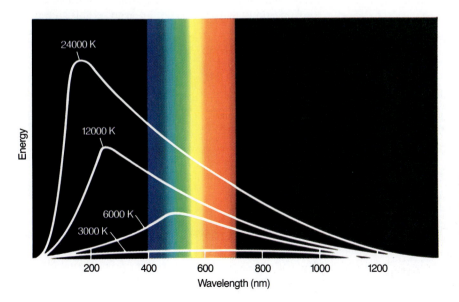

FIGURE 7.5 Energy emitted at different wavelengths for blackbodies at three different temperatures. At hotter temperatures, more energy is emitted at all wavelengths. The peak amount of energy is radiated at shorter wavelengths for higher temperatures (Wien's law).

than does a cooler one. Third, the higher the temperature, the shorter the wavelength at which the maximum energy is emitted.

This third characteristic is one that we have all observed in everyday life. For example, when a burner on an electric stove is turned on low, it emits heat, which is infrared radiation. If the burner is set to a higher temperature, it will glow a dull red. At a still higher setting, it will glow a brighter orange-red. At still higher temperatures, which luckily cannot be reached with ordinary stoves, metal can appear brilliant yellow or even blue-white hot.

The Sun and stars emit energy that approximates that from a blackbody, and so it is possible to estimate temperatures by measuring the energy that they emit as a function of wavelength—that is, by measuring their light at many colors. The temperature at the surface of the Sun, which is where the radiation that we see is emitted, turns out to be 5800 K. (Throughout this text we use the Kelvin or absolute temperature scale. On this scale, water freezes at 273 K and boils at 373 K. All molecular motion ceases at 0 K. The various temperature scales are described in Appendix 5.)

The wavelength at which a blackbody emits its maximum energy can be calculated according to the equation

$$\lambda_{max} = \frac{3,000,000}{T},$$

where the wavelength is in nanometers and the temperature in kelvins. This relationship is called **Wien's law.** For the Sun, the wavelength at which the maximum energy is emitted is 520 nm, which is near the middle of that portion of the electromagnetic spectrum that is called visible light. It is surely no coincidence, but rather a consequence of evolutionary adaptation, that human eyes are most sensitive to electromagnetic radiation at those wavelengths at which the Sun puts out the most energy. Characteristic temperatures of other astronomical objects, and the wavelengths at which they emit most of their energy, are listed in Table 7.1.

If we sum up the contributions from all parts of the electromagnetic spectrum, we obtain the total energy emitted by a blackbody over all wavelengths. That total energy, emitted per second per square meter by a blackbody at a temperature T, is proportional to the

fourth power of its absolute temperature. This relationship is known as the **Stefan–Boltzmann law,** which can be written in the form of an equation as

$$E = \sigma T^4,$$

where E stands for the luminosity and σ is a constant number. If the Sun, for example, were twice as hot, that is, if it had a temperature of 11,600 K, it would radiate 2^4, or 16, times more energy than it does now.

7.3   THE STRUCTURE OF THE ATOM

(a)   Spectral Lines

The visible spectrum of a blackbody is a *continuous* rainbow of colors (Figure 7.6). If the Sun behaves like a blackbody, then, when sunlight passes through a prism, we would expect to see radiation of all the possible colors. In fact, this is what Newton, who used very simple equipment, did see. In 1802, the English chemist William Wollaston (1766–1828) repeated Newton's experiment, but he used a slit instead of a round hole. After the sunlight passed through the prism, Wollaston used a lens to form an image of the slit at every wavelength—and some of the colors were missing! We now know that there are thousands of places (wavelengths) where light is missing. Sunlight has thousands of dark **spectral lines** (Figure 7.7).

During the 19th century, scientists learned that hot gases on Earth produce bright spectral lines. Each gas—sodium, sulfur, carbon, and so on—produces its own unique pattern of spectral lines (Figure 7.8). Just as we can identify an individual person from fingerprints, so it is possible to identify a gas purely by examination of its spectrum. In 1860, Gustav Kirchhoff concluded from his analysis of the solar spectrum that sodium is present in the Sun's atmosphere. (Later in this section there is an explanation of why spectral lines are sometimes bright and sometimes dark.)

Analysis of spectral lines is the key to modern astrophysics. In 1835, the French philosopher Auguste Comte concluded that it would be possible to measure the sizes, distances, and motions of stars, but that it would never be possible by any means to learn their chemical composition. Only 25 years later, Kirchhoff

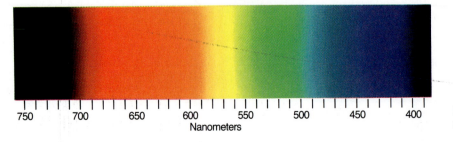

FIGURE 7.6 A continuous spectrum of visible light.

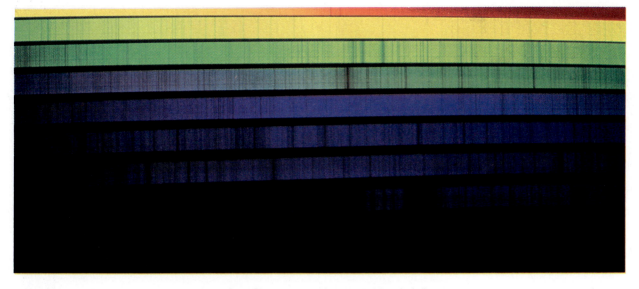

FIGURE 7.7 The visible spectrum of the Sun. The spectrum is crossed by dark lines produced by atoms in the solar atmosphere that absorb light at certain wavelengths. (National Solar Observatory/National Optical Astronomy Observatories)

proved him resoundingly wrong. In the years following Kirchhoff's discovery of sodium in the Sun, astronomers identified many other chemical elements in the Sun and stars, but it was only in the 20th century, with the development of a model for the atom, that scientists learned how spectral lines are formed.

(b) Structure of the Atom

The idea that matter is composed of tiny particles—atoms—massed together is at least 25 centuries old. It was not until the 20th century, however, that scientists invented instruments that permitted them to probe inside an atom and discover that it was not, as Newton thought, hard and indivisible. Instead, it is a complex structure composed of still smaller particles.

The first of these smaller particles was discovered by J.J. Thomson in 1897. Named the **electron,** this particle is negatively charged. Since an atom in its normal state is electrically neutral, each electron in an atom must be balanced by the same amount of positive charge. The obvious next problem was to determine where in the atom the positive and negative charges are located. In 1911 British physicist Ernest Rutherford devised an experiment that provided part of the answer to this question. What he did was to bombard a piece of gold foil, which was about 400 atoms thick, with a beam of *alpha particles* emitted from a radioactive material (Figure 7.9). We now know that alpha particles are helium atoms that have lost all of their electrons. Most of the alpha particles passed through the gold foil just as if it and the atoms composing it were nearly empty

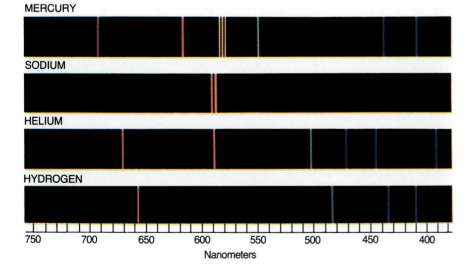

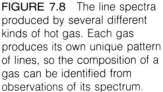

FIGURE 7.8 The line spectra produced by several different kinds of hot gas. Each gas produces its own unique pattern of lines, so the composition of a gas can be identified from observations of its spectrum.

FIGURE 7.9 (a) When Rutherford allowed alpha particles from a radioactive source to strike a target of gold foil, he found that some of the alpha particles rebounded back in the direction from which they came. (b) From this experiment, he concluded that the atom must be constructed like a miniature solar system, with the positive charge concentrated in the nucleus. The negative charge was assumed to orbit in the large volume around the nucleus.

space. About 1 in 8,000 of the alpha particles, however, completely reversed direction and bounced backward from the foil. Rutherford wrote, "It was quite the most incredible event that has ever happened to me in my life. It was almost as incredible as if you fired a 15-inch shell at a piece of tissue paper and it came back and hit you."

The only way to account for the alpha particles that reversed direction when they hit the gold foil is to assume that nearly all of the mass and all of one type of charge, either positive or negative, in each individual gold (and helium) atom is concentrated in a tiny **nucleus.** We now know that it is the positive charge that is located in the nucleus. When an alpha particle

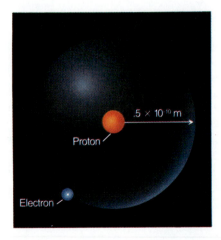

FIGURE 7.10 Schematic diagram of a hydrogen atom in its lowest energy state, which is also called the ground state. Although the orbit of the electron is drawn as if it were well defined, in fact the orbit shown is only the most probable one. The electron may move in a variety of orbits all relatively close to the one shown. The proton and electron have equal but opposite charges, which exert an electromagnetic force that binds the hydrogen atom together.

strikes a nucleus, the two positive charges repel and the particle reverses direction. Rutherford's model placed the other type of charge—the electrons as we now know—in orbit around this nucleus.

Rutherford's model requires that the electrons be in motion. Since positive and negative charges attract each other, stationary electrons would simply fall into the nucleus. Because most of the atom is empty, nearly all of Rutherford's alpha particles were able to pass right through the gold foil without colliding with anything.

(c)   The Atomic Nucleus

The simplest atom is hydrogen, and the nucleus of ordinary hydrogen contains a single positively charged particle called a **proton.** Moving around this proton is a single electron. The mass of an electron is nearly 2000 times smaller than the mass of a proton, but the electron carries a charge that is exactly equal in magnitude but opposite in sign to that of the proton. The electron's charge is negative instead of positive (Figure 7.10). Opposite charges attract one another, and so it is the **electromagnetic force** that binds the proton and electron together, just as gravity is the force that keeps the planets in orbit around the Sun.

There are many other types of atoms. Helium, for example, is the second most abundant element in the Sun. Helium has two protons in its nucleus, instead of the single proton that characterizes hydrogen. In addition, the helium nucleus contains two **neutrons,** particles with a mass slightly larger than that of the proton but with no electric charge. Moving around this nucleus are two electrons, so that the total net charge of the helium atom is also zero (see Figure 7.11).

From this description of hydrogen and helium, perhaps you have guessed the pattern for building up all of

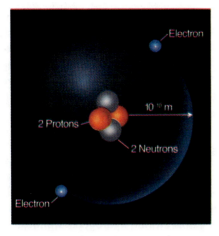

FIGURE 7.12 *Schematic diagram of the nuclei of the isotopes of hydrogen. A single proton in the nucleus defines the atom to be hydrogen, but there may be zero, one, or two neutrons. By far the most common isotope is the one with only a single proton. A hydrogen nucleus with one neutron is called deuterium; one with two neutrons is called tritium.*

FIGURE 7.11 Schematic diagram of a helium atom in its lowest energy state. Two protons occur in the nucleus of all helium atoms. In the most common variety of helium, the nucleus also contains two neutrons, which have nearly the same mass as the proton but carry no charge.

the elements that we find in the universe. The specific element is determined by the number of protons in the nucleus. Carbon has 6 protons, oxygen has 8, iron has 26, and uranium has 92. In its normal state, each atom has the same number of electrons as protons, and these electrons follow complex orbital patterns around the nucleus.

Although the number of neutrons in the nucleus is usually approximately equal to the number of protons, the number of neutrons is not necessarily the same for all atoms of a given element. For example, most atoms of hydrogen contain no neutron at all. There are, however, hydrogen atoms that contain one proton and one neutron, and others that contain one proton and two neutrons. The various types of nuclei of hydrogen are called **isotopes** of hydrogen (Figure 7.12), and other elements have isotopes as well.

7.4 SPECTROSCOPY

(a) The Bohr Atom

There is one serious problem with Rutherford's model for atoms. As we have already seen, Maxwell's theory says that when electrons change either their speed or their direction of motion, they emit energy in the form of electromagnetic radiation. Since orbiting electrons constantly change their direction of motion, they should emit a constant stream of energy. Earth-orbiting satellites spiral back toward Earth as they lose energy through friction with the Earth's atmosphere. So, too, should electrons spiral into the nucleus of the atom as they radiate electromagnetic energy.

It was the Danish physicist Niels Bohr (1885–1962) who solved the mystery of the electrons. He suggested

that the spectrum of hydrogen can be understood if it is assumed that *only orbits of certain sizes are possible* for the electron. Bohr further postulated that so long as the electron moves only in one of these allowed orbits, it radiates no energy. If the electron moves from one orbit to another closer to the atomic nucleus, then it must give up some energy in the form of electromagnetic radiation. Conversely, energy is required to boost the electron from a smaller orbit to one farther from the nucleus, and one way to obtain the necessary energy is to absorb electromagnetic radiation if some is streaming past the atom from an outside source.

Note that Bohr's bold postulate is not in accord with everyday experience. A satellite in orbit around the Earth can take any orbit we choose to give it. This is not so in the atom, where nature permits only certain orbits and forbids all others. Strange as it seems, the predictions of this postulate have been confirmed again and again by all our experiments.

When an electron moves from a larger to a smaller orbit, it emits a discrete packet of electromagnetic energy, which is called a **photon.** The energy of this photon is directly proportional to the frequency of the electromagnetic radiation. Higher frequency corresponds to higher energy. Since higher frequency also corresponds to shorter wavelength, photons of violet and blue light have higher energy than those of red light. The highest energy photons are gamma rays; those of lowest energy are radio waves. Photons have no mass.

(b) Particles and Waves

In this description of how the hydrogen atom absorbs and emits energy, we have talked about light as though it were made up of little energetic particles—photons. But we have already seen that electromagnetic energy also propagates as waves. So what is light really—is it made up of particles or waves? The answer is neither and both. In some situations it is easier to describe the way light interacts with the material world by treating it as if

to the average bowler. But an electron has a tiny mass, and if its velocity is uncertain by the same amount as our bowling ball, its position is uncertain by more than 1 m!

It may seem that we are just talking about our inability to make a precise measurement, and that the real momentum and real position exist with perfect precision. But how can we define something like perfect precision if there is no possible experiment, even theoretically, by which we can determine it? Since the limitation is theoretical, we must regard the uncertainty principle as inherent to nature.

Differences in energy of the electron orbits in atoms are usually expressed in *electron volts*. An electron volt (abbreviated eV) is the small amount of energy acquired by an electron after being accelerated through a potential difference of 1 volt. The energy needed to raise the hydrogen atom from its lowest energy level to the next lowest one is about 10.2 eV. One eV = 1.602×10^{-19} joules. One joule of energy expended in 1 s is a *watt* of power. Typical electric light bulbs use about 100 watts.

Ordinarily, an atom is in the state of lowest possible energy, its **ground state.** In the Bohr model, the ground state corresponds to the electron being in the innermost orbit. However, an atom can absorb energy, which raises it to a higher energy level (corresponding, in the Bohr picture, to the movement of an electron to a larger orbit). The atom is then said to be in an **excited state.** Generally, an atom remains excited for only a very brief time. After a short interval, typically a hundred-millionth of a second or so, it drops back down to its ground state, with the simultaneous emission of light. The atom may return to its lowest state in one jump, or it may make the transition in steps of two or more jumps, stopping at intermediate levels on the way down. With each jump, it emits a photon of the wavelength that corresponds to the energy difference between the levels at the beginning and end of that jump.

An energy-level diagram for a hydrogen atom and several possible atomic transitions are shown in Figure 7.15; compare this figure with the Bohr model, shown in Figure 7.13.

Because atoms that have absorbed light and have become excited generally de-excite themselves and emit that light again, you might wonder why *dark* spectral lines are ever produced. In other words, why doesn't this re-emitted light "fill in" the absorption lines? Some of the re-emitted light actually is received by us, but this light fills in the absorption lines only to a slight extent. The reason is that the atoms re-emit light in mostly random directions, and only a small fraction of the re-emitted light is directed toward the observer. On the other hand, we can observe the re-emitted light as emission lines if we can view the absorbing atoms from a direction from which little or no background light is coming—as we do, for example, when we look at clouds of hot gas located in the space between the stars. Figure 7.16 illustrates the situation.

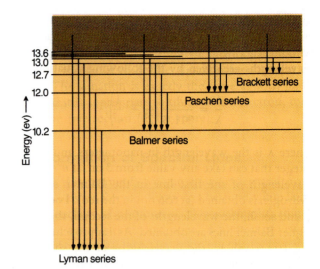

FIGURE 7.15 Energy-level diagram for hydrogen. The shaded region represents energies at which the atom is ionized.

Atoms in a gas are moving at high speeds and continually collide with one another and with electrons. They can be excited and de-excited by these collisions as well as by absorbing and emitting light. The velocity of atoms in a gas depends on its temperature, and so the higher the temperature, the higher is the velocity and the higher the energy of the collisions. The hotter the gas, therefore, the more likely it is that electrons will occupy the outermost orbits, which correspond to the highest energy levels.

(e) Ionization

We have described how certain discrete amounts of energy can be absorbed by an atom, raising the atom to an excited state and moving one of its electrons farther from its nucleus. If enough energy is absorbed, the electron can be removed completely from the atom. The atom is then said to be **ionized.** The minimum amount of energy required to ionize an atom from its ground state is called its *ionization energy* or *ionization potential*. The ionization potential for hydrogen is about 13.6 eV (Figure 7.15).

Still greater amounts of energy must be absorbed by the ionized atom (called an **ion**) to remove an additional electron. Successively greater energies are needed to remove the third, fourth, and fifth electrons from the atom, and so on. If enough energy is available (in the form of photons with very short wavelengths or in the form of a collision with a very fast-moving electron or another atom), an atom can become completely ionized, losing all of its electrons. A hydrogen atom, having only one electron to lose, can be ionized only once; a helium atom can be ionized twice, and an oxygen atom, eight times.

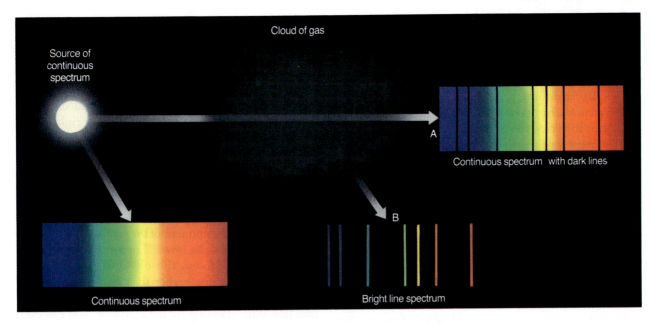

FIGURE 7.16 Production of bright and dark spectral lines. The atoms in the gas cloud produce absorption lines in the continuous spectrum of the white light source when viewed from direction *A*, but they produce emission lines (of the light they re-emit) when viewed from direction *B*.

An atom that has become ionized has lost a negative charge—that carried away by the electron—and thus is left with a net positive charge. It has, therefore, a strong affinity for a free electron. Eventually, an electron will be captured, and the atom will become neutral (or ionized to one less degree) again. During the capture process, the atom emits one or more photons, depending on whether the electron is captured at once to the state corresponding to the lowest energy level of the atom or whether it stops at one or more intermediate levels on its way to the ground state.

Just as the excitation of an atom can result from a collision with another atom, ion, or electron (collisions with electrons are usually most important), so also can ionization. The rate at which such collisional ionizations occur depends on the atomic velocities and hence on the temperature of the gas. The rate of recombination of ions and electrons also depends on their relative velocities, that is, on the temperature. In addition, it depends on the density of the gas; the higher the density, the greater the chance for recapture, because the different kinds of particles are crowded closer together. From a knowledge of the temperature and density of a gas, it is possible to calculate the fraction of atoms that have been ionized once, ionized twice, and so on. In the surface layer of the Sun, for example, we find that most of the hydrogen and helium atoms are neutral, whereas most of the atoms of calcium, as well as many other metals, are once ionized.

The energy levels of an ionized atom are entirely different from those of the same atom when it is neutral. In each degree of ionization, the energy levels of the ion, and thus the wavelengths of the spectral lines it can produce, have their own characteristic values. In the Sun, therefore, we find lines of neutral hydrogen and helium, but of ionized calcium. (Ionized hydrogen, having no electron, can produce no absorption lines.) Examining such absorption lines is the main way astronomers have of determining what the stars are made of.

(f) Spectra of Molecules

Molecules, combinations of two or more atoms, can also absorb or emit light. Molecular lines are seen in the spectra of cool stars. In addition to undergoing electronic transitions, molecules can also rotate and vibrate, all of which involve energy. Only certain energy levels of vibration and rotation can occur, just as only certain energy levels are allowed for atoms. The vibrational and rotational energies are generally low, but they add to or subtract from the energies corresponding to transitions of electrons from one energy level to another. Consequently, in place of each single energy level found in atoms there is a series of closely spaced levels, each one corresponding to a different mode of vibration or rotation of the molecule. Many more different transitions between energy levels are possible, therefore, differing from one another only slightly in energy or wavelength. Molecules, in other words, produce series of closely spaced lines known as molecular bands. In spectra of those stars in which molecules exist, these many molecular lines within a band are often not resolved as separate, and only a single broad absorption feature is observed.

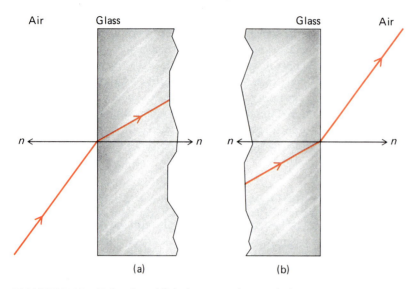

FIGURE 7.19   Refraction of light between glass and air.
(a) When light passes from air to glass. (b) When light
passes from glass to air.

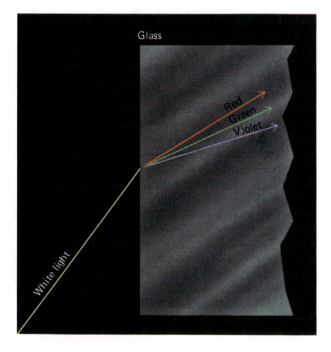

FIGURE 7.20   Dispersion at the interface of two
transparent substances.

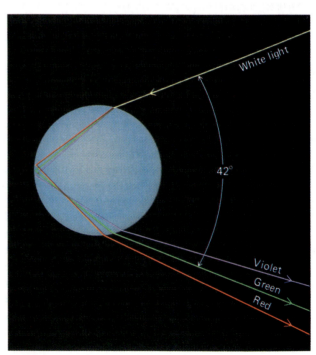

FIGURE 7.21   Dispersion in a raindrop.

is bent different amounts in passing from air into glass. Figure 7.4 shows how light can be separated into its colors with a prism, a triangular piece of glass. Upon entering one face of the prism, light is refracted once, the violet light more than the red, and upon leaving the opposite face, the light is bent again, and so is further dispersed. Greater dispersion can be obtained by passing the light through a series of prisms. We will return to this topic in Chapter 9.

(d)   Weather Optics

Nature provides an excellent example of the dispersion of light in the production of a rainbow. Raindrops, tiny spherical droplets of water in the air, act like prisms. Light from the Sun entering a raindrop is bent, the blue and violet light being bent the most. This bent light strikes the inside rear surface of the drop, and some of it is reflected back toward the front surface. This reflected light leaves the raindrop by passing through the same side that it enters. But when it leaves the drop, it is again refracted, and again dispersed, just as when light leaves a glass prism. Thus sunlight is spread into the rainbow of colors—the rainbow is nothing more than the spectrum of sunlight. Figure 7.21 shows how a raindrop produces a spectrum.

The light that emerges from a raindrop is most intense at an angle of about 42° from the direction at which it enters. Thus to see a rainbow, the observer must have the Sun behind him, at an altitude of less than 42° above the horizon (Figure 7.22). The rainbow appears as an arc, with an angular radius of 42°, centered about a point exactly opposite the Sun. Since the Sun must be above the horizon to illuminate the raindrops, the center of the rainbow must be below the horizon. An observer on the ground can never see a rainbow as more than half a complete circle, although observers in airplanes or on mountains may occasionally see more.

Although the arc of a rainbow has a radius of 42°, the different colors of sunlight are refracted and reflected by the droplets back to the observer in slightly different directions, so the band of color has a finite width. It may be seen in Figure 7.22 that from the upper drops it is the red light, bent the least, that enters the observer's eye, and from the lower drops it is the violet light, bent the most, that the observer sees. Thus, the top (or outside) of the arc of the rainbow appears red, and the bottom (or inside) of the arc appears violet. The other colors come from the drops in between.

A little of the light is reflected a second time before leaving the drops. This light emerges at an angle of about 51° from that of the incoming light and produces a fainter secondary rainbow in an arc of 51° radius, thus lying outside the primary rainbow. In the secondary rainbow, the red is on the inside of the arc and the violet on the outside.

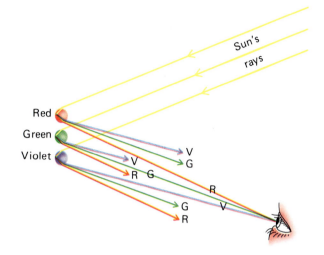

FIGURE 7.22 Formation of a rainbow.

Another natural phenomenon that involves the refraction and dispersion of light is a halo about the Sun or Moon. A halo is a faint ring of light, of angular radius 22°, caused by the bending of light as it passes through the tiny ice crystals that form cirrus clouds. As in the rainbow, the violet light is refracted most in passing through the crystals. Consequently, the outer edge of a solar or lunar halo appears violet, the inner edge red.

S U M M A R Y

7.1 Light is one form of **electromagnetic radiation.** The **wavelength** of light determines the color of visible radiation. Wavelength (λ) is related to **frequency** (f) and the speed of light (c) by the equation $c = \lambda f$. The **apparent brightness** of a source of electromagnetic energy decreases with increasing distance from that source in proportion to the square of the distance. The mathematical equation describing this relationship is known as the **inverse-square law.**

7.2 **Gamma rays, x rays,** and **ultraviolet radiation** are forms of electromagnetic radiation with wavelengths shorter than that of visible light. **Infrared, microwave,** and longer wave radio radiation have wavelengths longer than that of light. The higher the temperature of a **blackbody,** the shorter the wavelength at which the maximum amount of electromagnetic radiation is emitted. The mathematical equation describing this relationship ($\lambda_{max} = 3,000,000/T$) is known as **Wien's law.** The total energy emitted per square meter

increases with increasing temperature. The relationship between emitted energy and temperature ($E = \sigma T^4$) is known as the **Stefan-Boltzmann** equation.

7.3 Atoms consist of a **nucleus** containing one or more positively charged particles known as **protons.** All atoms except hydrogen also contain one or more **neutrons** in the nucleus. Neutrons have no charge but have about the same mass as the proton. Negatively charged **electrons** orbit the nucleus, and the number of electrons normally equals the number of protons. The number of protons defines the element (hydrogen, helium, and so forth) of the atom. Nuclei with the same number of protons but different numbers of neutrons are different **isotopes** of the same element.

7.4 When an electron moves from one orbit to another closer to the atomic nucleus, a **photon** is emitted, and a **spectral emission line** is formed. **Absorption lines** are formed when an electron moves to an orbit farther from the nucleus. Since

thought to influence terrestrial events, knowledge of astronomy—or rather astrology—in the wrong hands might carry with it the power to overthrow the ruling dynasty.

Telescopes were a surprisingly late addition to the complement of instruments housed within observatories—surprising because the technology for making glass is an ancient one. The Mesopotamians were the first to fuse sand and ash to form glass, and by 1200 B.C. the Babylonians knew how to make objects of blown glass. Chinese mirrors, including curved mirrors and burning mirrors, were manufactured long before the birth of Christ. The Chinese did not, however, make either a telescope or a microscope until after the information on how to do so reached them from Europe (Figure 9.2).

(a)   The First Telescopes

The telescope was almost surely invented by a manufacturer of eyeglasses, who by chance aligned and looked through two lenses, one concave and the other convex. This was an accident that took nearly 300 years to happen, since spectacles were used in Europe by the year 1300. Once the first telescope was made, word traveled quickly throughout Europe, so quickly that no one was able to establish unambiguous claim to having been the inventor.

Galileo himself claimed after the fact that he had invented the telescope on the basis of profound considerations of both perspective and the doctrine of refraction. In fact, Galileo had done no research in optics, and his telescope was the product of trial and error combined with knowledge that such a device had already been constructed by others. Galileo was, of course, not the first—and surely not the last—scientist to claim that results achieved by accident or luck were, in fact, the product of a deep understanding of the implications of a new theory.

On August 25, 1609, Galileo demonstrated one of his first telescopes, which had a magnification of 9×, to officials of the Venetian government. By a magnification of 9×, we mean that the linear dimensions of the object being viewed appeared 9 times larger or, equivalently, that the objects appeared 9 times closer than they really were. There were obvious military advantages associated with a device for seeing distant objects. For his invention Galileo's salary was nearly doubled, and he was granted lifetime tenure as a professor. His colleagues were outraged, particularly since the invention was not even an original one.

In yet another example of how it often takes a very long time to do what in hindsight seems so obvious, Galileo did not use his telescope for astronomy for several months. First, he had to devise a stable mount, and he also improved the optics to provide a magnification of 30×. Galileo also had to acquire confidence in the telescope. At that time, the eyes were believed to be

(a)

(b)

FIGURE 9.2   Two surviving pretelescopic observatories. (a) The Jantar Mantar, built in 1724 by Maharaja Jai Singh in Delhi, India. (b) Seventeenth-century bronze instruments from the old Chinese imperial observatory, Beijing. (David Morrison)

the final arbiter of the truth about sizes, shapes, and colors. Lenses, mirrors, and prisms were known to distort distant images by enlarging them, reducing them, or even inverting them. Galileo undertook repeated experiments to convince himself that what he saw through the telescope was identical to what he saw up close. Only then could he begin to believe that the miraculous phenomena that were revealed in the heavens were real. While Galileo was convinced of the validity of what he saw, others were not. One unbelieving colleague said that he ". . . tested this instrument of Galileo's in a thousand ways, both on things here below and on those above. Below, it works wonderfully; in the sky it deceives one, as some fixed stars are seen double." Another scholar refused even to look through the telescope because doing so gave him a headache.

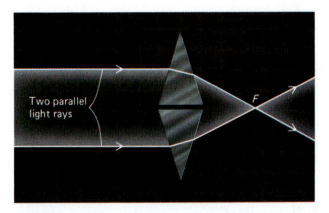

FIGURE 9.3 Principle of image formation.

(b) Formation of an Image by a Lens or a Mirror

Now let's see how a telescope works. The first task is to form an image of a source of radiation. The image can then be detected, measured, reproduced, and analyzed in a host of ways.

One important point to bear in mind as you read this section is that the planets and stars are so far away that by the time their light reaches us, the rays are essentially parallel to each other. (This is not true for nearby objects on Earth.) We will use the laws of reflection and refraction described in Section 7.2 to understand the formation of images by optical systems. Images were first produced by simple convex lenses. To illustrate the principle, let us imagine two triangular prisms, base to base, as in Figure 9.3. Suppose we select two of the parallel rays of light from a distant object and allow one ray to enter each prism. The light rays are refracted by the prisms and meet at a point F.

This is the principle underlying the formation of an image by a lens. In Figure 9.4 a simple convex lens is shown. Parallel light from a distant star or other light source is incident on the lens from the left. A convex lens is thicker in the middle than at the edges. In cross section it is no more than a series of segments of prisms piled one on another, with slightly different slopes to their sides. If the curvatures of the surfaces of the lens are

correct, light passing through the lens will be refracted in such a way that parallel rays converge toward a point. Convex lenses whose surfaces are portions of spherical surfaces are easiest to manufacture. Such lenses will refract a parallel beam of light to a point as shown in Figure 9.4.

The point where light rays come together is called the **focus** of the lens. At the focus, an image of the light source appears. The distance of the focus, or image, behind the lens is called the **focal length** of the lens. A lens or other device that forms an image is called an **objective.** A telescope objective collects incoming radiation from a distant object and brings all of this radiation to a focus. Think of telescopes as light buckets that collect all the light that falls on them. The wider the bucket, the more light is collected and the fainter is the object that can be seen. The amount of light collected is proportional to the area of the objective, and therefore to the square of its diameter or **aperture.** Astronomical telescopes are generally built with as large an aperture as possible, in order to collect the maximum amount of light.

In Figure 9.5 we see how an image is formed of an extended source like the Moon. From each point on the Moon, light rays approach the lens along parallel lines. However, from different parts of the Moon, the parallel rays of light approach the lens from different directions. The rays of light from each point on the Moon are focused at a point. The distance between the lens and the image is the focal length of the lens.★ If a screen, such as a white card, is placed at this distance behind the lens, an inverted (upside down) image of the Moon appears on it.

Note that if part of the lens is covered up, or if the middle is cut out, an entire image will still be formed. *All* parts of the lens contribute to *each* part of the image. Covering up part of the lens cuts down the total amount of light that can strike each portion of the image and thus makes the image fainter, but nevertheless the whole image is formed. An ordinary camera lens produces an image at the focal plane (where the film is placed), just as is shown in Figure 9.5. Every photographer knows that it is possible to cover up part of a camera lens by "stopping it down" with an iris diaphragm. Reducing the aperture in this way will cut down the brightness of the image (and hence the effective exposure on the film), but the outer parts of the image will not be removed. The part of the lens that remains uncovered still produces the entire image.

Rays of light can also be focused to form an image with a concave mirror—one hollowed out in the middle. Parallel rays of light, as from a star, fall upon the

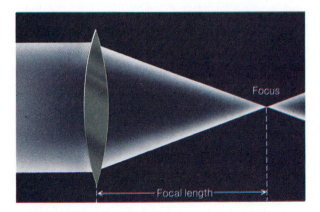

FIGURE 9.4 Formation of an image by a simple convex lens.

★ We discuss here only the case in which the object whose image is formed is so distant that light from any point on it can be regarded as approaching the lens along parallel rays. This is always true when any astronomical body is observed. Nearby terrestrial objects may be so close that the assumption is not valid. Then the image is formed at a point farther from the lens than the focal length.

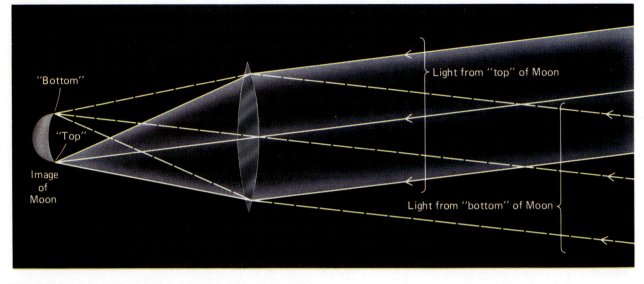

FIGURE 9.5   Formation of the image of an extended object.

curved surface of the mirror (Figure 9.6), which is coated with silver or aluminum to make it highly reflecting. Each ray of light is reflected according to the law of reflection. If the mirror has the correct concave shape, all the rays are reflected back through the same point, the focus of the mirror. The image of the star thus appears at the focus. As in a lens, the distance from the mirror to the focus is called its focal length.

Rays of light from an extended object are focused by a mirror, exactly as they are by a lens, into an inverted image of the object. The principal difference between image formation by a lens and that by a mirror is that the mirror reflects the light back into the general direction from which it came, so that the image forms in front of the mirror. One way to inspect the image is to allow light to illuminate a white card at the focus of the mirror. The card, of course, will block off part of the incoming light. However, the presence of the card will not produce a "hole" in the image but will merely reduce its overall brightness.

The best image is formed by a concave mirror whose cross-section is a parabola rather than a circle. However, even a spherical surface produces an image of fairly good quality.

The most important properties of an image are its scale or size, its brightness, and its resolution. We shall consider these in turn. In this section, we ignore the limitations imposed by the Earth's atmosphere (Section 9.4a).

(c)   Scale of an Image

The *scale* of an image is a measure of its size. In all astronomical applications, we are dealing with objects whose sizes can be expressed in angular units, and it is generally convenient to express the scale of an image as the linear distance in the image that corresponds to a certain angular distance in the sky. For example, suppose that an image of the Moon were produced that is exactly 1 cm across. The Moon has an apparent or angular size of $\frac{1}{2}°$ in the sky. The scale of the image is thus $\frac{1}{2}°$ per centimeter, or 2 cm per degree.

The scale of an image depends on only the focal length of the lens or mirror that produces it. Numerically, the distance s in an image, corresponding to 1° in the sky, is given by the equation

$$s = 0.01745f,$$

where f is the focal length of the lens or mirror. For example, the 5-m (200-in.) mirror of the Hale telescope on Palomar Mountain has a focal length of 16.8 m. It produces images with a scale of 0.29 m per degree, which corresponds to 12 arcsec/mm. Thus the image of the Moon is about 15 cm in diameter—the size of a saucer. Most astronomical telescopes have image scales ranging from 10 to 200 arcsec/mm.

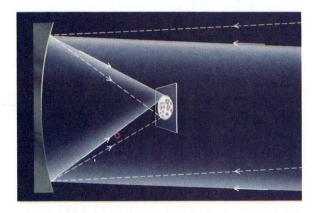

FIGURE 9.6   Formation of an image by a concave mirror.

(d) Brightness of an Image

The primary purpose of a telescope is to collect light and thus make faint sources appear bright. The *brightness* of an image is a measure of the amount of light energy that is concentrated into a unit area of the image, such as a square millimeter. The brightness of an image determines whether it is above the threshold of visibility or, alternatively, how long a time would be required to record the image photographically.

The brightness of an image of an *extended* object (such as the Moon, a nebula, or even the faint illumination of the night sky) is greater the greater the amount of light flux that passes through the objective (lens or mirror) to form the image. The brightness is less the larger the image area over which that amount of flux must be spread. The amount of flux reaching the image is proportional to the area of the objective and hence to the square of its diameter, or aperture. The area the flux is spread over is proportional to the square of the focal length of the objective, for as we have seen, the image diameter varies directly with the focal length. Hence the brightness B of an extended image is given by

$$B = \text{constant} \times \left(\frac{a}{f}\right)^2,$$

where a is the aperture and f is the focal length. The constant of proportionality is a number whose value depends on the units chosen to measure the various quantities and also on the amount of light actually leaving each unit area of the object.

(e) Resolution

The angular size of the image of a point source depends on the wavelength of light used and the diameter of the lens or mirror. If the wavelength and aperture are measured in the same units (for example, both in meters), the smallest angle (alpha, α) in arcseconds that can be resolved by a lens or mirror of aperture d is given approximately by the equation

$$\alpha = 2.1 \times 10^5 \times \frac{\lambda}{d},$$

where λ is the wavelength. If λ is chosen as 550 nm—near the middle of the visible spectrum—and if d is measured in meters, the formula becomes

$$\alpha = 0.116/d.$$

The above formula, called Dawes' criterion, represents an average of empirical studies of the ability of telescopes of small aperture to resolve double stars. As we shall see, atmospheric turbulence further degrades the actual resolving power of large telescopes, so that the resolution of a 4-m telescope may be, in practice, no greater than that of a 1-m telescope.

(f) Magnification

An eyepiece is a small magnifying lens that is used to view the image formed by a large objective. When an extended celestial object is viewed through a telescope equipped with an eyepiece, it appears enlarged, that is, closer than when viewed with the unaided eye. The factor by which an object appears larger (or nearer) is called the **magnifying power** of the telescope. For example, the Moon has an angular diameter of $\frac{1}{2}°$ when viewed with the naked eye. If, when viewed through a particular telescope, the Moon appears to have a diameter of 10°, the magnifying power of the telescope is 20.

To read fine print one must use a magnifying glass of higher power than when reading the print in a newspaper. A higher power magnifying glass has a shorter focal length. Similarly, eyepieces of different focal lengths, used in conjunction with the same telescope objective, produce different image magnifications. It is the purpose of the objective of a telescope to produce an image; it is the purpose of the eyepiece to magnify the image to the point where details in it can be viewed. In principle, any desired magnification can be obtained if an eyepiece of sufficiently short focal length is used. Therefore, it does not make sense to ask an astronomer what the "power" of a telescope is. The power can be changed at will by using different eyepieces. The term magnifying power loses its significance completely when, as in most modern telescopes, the image is not viewed through an eyepiece at all, but is displayed on a television screen.

(g) The Complete Telescope

Telescopes have two primary purposes: to collect and concentrate light and to form an image that can be studied in a variety of ways. Now that we have seen how an image is produced, we are able to understand the operation of a telescope. There are two general kinds of optical astronomical telescopes in use: (1) refracting telescopes, which utilize lenses to produce images, and (2) reflecting telescopes, which utilize mirrors to produce images.

The **refracting telescope** is probably the most familiar, although it is rarely used in modern research observatories. Galileo's telescopes were refractors, and ordinary binoculars are two refracting telescopes mounted side by side (Figure 9.7). A lens is usually mounted at the front end of an enclosed tube. The tube is not really essential—its purpose is merely to block out scattered light; an open framework would suffice. In a refracting telescope, the objective is the lens at the front of the tube. The image is formed at the rear of the tube.

All of the large research telescopes in the world are **reflecting telescopes,** and this design is also more popular among amateur astronomers. The reflecting telescope was first conceived of by James Gregory in 1663, and the first successful model was built by New-

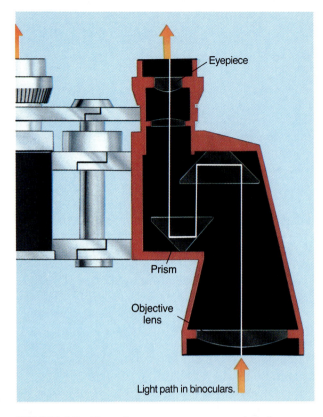

FIGURE 9.7 Binoculars are common examples of refracting telescopes.

FIGURE 9.8 The observer's cage at the prime focus of the Hale 5-m telescope. (Caltech)

ton in 1668. A concave mirror (usually parabolic in shape) is used as an objective. The mirror is placed at the bottom of a tube or open framework. The mirror reflects the light back up the tube to form an image near the front end.

With a reflecting telescope, the problem of image accessibility arises because a concave mirror produces the image in front of the mirror, in the path of the incoming light. There are various arrangements for getting at the focus. Which one is adopted depends on the type of telescope and on the purpose for which the image is to be used.

The place where the image is formed by the mirror is the *prime focus* (Figure 9.8). If the image is to be recorded, a photographic film holder or electronic detector can be mounted at the prime focus in the middle of the mouth of the telescope tube. Since the detector system blocks a small fraction of the incoming light, the brightness of the image is slightly dimmed.

Figure 9.9 illustrates a variety of other focus arrangements for reflecting telescopes. In the Newtonian design, which is commonly used on smaller amateur telescopes, a flat mirror is mounted diagonally in the middle of the tube so that it intercepts the light just before it reaches the focus, diverting it to an eyepiece outside the tube (Figure 9.9b). The most popular design for modern research telescopes is the Cassegrain system, in which a small convex mirror (rather than a flat mirror) is suspended in the telescope tube. The convex mirror inter-

cepts the light before it reaches the prime focus and reflects it back down the tube of the telescope. Usually a hole is provided in the center of the objective mirror so that the light reflected from the convex mirror can form an image behind the objective (Figure 9.9c). In the coudé arrangement, Figure 9.9e, a convex mirror intercepts the light just before the prime focus is reached. The light is reflected back down the tube until it reaches one of the pivot points about which the telescope tube can be rotated to point to various parts of the sky. There it is intercepted by a flat mirror that reflects the light outside the tube to a fixed observing station. Because the station is not attached to the moving part of the telescope, heavy equipment can be used there.

9.2 OPTICAL DETECTORS

The popular view of the astronomer is a person in a cold observatory peering through a telescope all night. Most astronomers do not live at observatories but near the universities or laboratories where they work. Even when they are at an observatory, astronomers seldom inspect telescopic images visually, except to center the telescope on a desired region of the sky or to make adjustments. On the contrary, photographic plates or electronic detectors are used to record the image permanently for detailed analysis after the observations are completed. Typically, one successful night at the telescope yields enough data to keep an astronomer busy for weeks of analysis and interpretation (Figure 9.10).

(a) Telescopic Photography

Through most of the 20th century, photographic plates and films have served as the prime astronomical detectors, whether for direct imaging or for photographing spectra. To photograph an astronomical object, the image of that object is allowed to fall on a light-sensitive coating that, when developed, provides a permanent

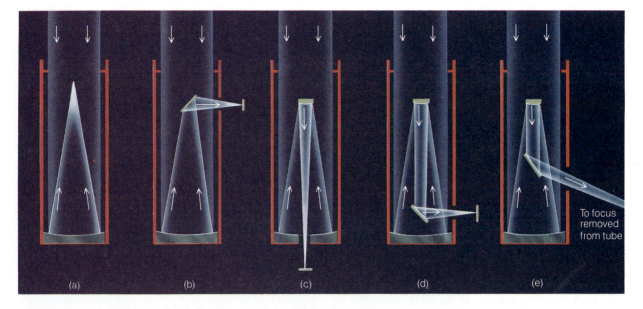

To focus
removed
from tube

(a)   (b)   (c)   (d)   (e)

FIGURE 9.9 Various focus arrangements for reflecting telescopes: (a) prime focus, (b) Newtonian focus, (c,d) two types of Cassegrain focus, and (e) coudé focus.

record of the image—one that can be measured, studied, enlarged, published, and inspected by many individuals. When used for photography, a telescope becomes nothing more than a large camera; the lens or mirror of the telescope serves as the camera lens.

One important advantage of photography is that a photographic plate can build up an image during a long exposure. Astronomical exposures often run hours in length. The longer the exposure, the more faint light gradually accumulates to help build up the photographic image. Objects can be detected that are more than 100 times too faint to see by just looking through a telescope. The layperson is often disappointed by a first

look through an astronomical telescope, for the sight is nothing compared with the spectacular photographs (such as those reproduced in this book) that are the result of long time exposures.

The photographic plate is an excellent device for collecting and storing a large amount of information, and plates have been used by astronomers for more than a century. Photographic plates do, however, have limitations. Photography, for example, is *nonlinear*—that is, equal differences in exposure do not produce equal differences in the blackening of the plate. Moreover, the photographic plate *saturates* (turns black) with long enough exposure, and no additional data can be recorded. Consequently, it is far from ideal for measuring the brightness of astronomical objects. In addition, photographic plates are much less sensitive than electronic detectors. At best the **quantum efficiency** of a photographic plate is only about 1 percent. This means that it requires, typically, a hundred photons falling on a small portion of the plate to produce a measurable blackening when the plate is developed. It has long been desired, therefore, to find a more ideal detector—one that is linear, does not saturate, and has high quantum efficiency.

FIGURE 9.10 Astronomers check the instrumentation mounted at the Cassegrain focus of the 2.1-m telescope on Kitt Peak before beginning their observations. (National Optical Astronomy Observatories)

(b)   Electronic Detectors

There are two basic types of electronic detectors that are widely used by optical astronomers. The first type is the so-called *photoemissive detector*, in which a photon strikes a light-sensitive surface with enough energy to free an electron completely from it. One example of such a device is a photomultiplier tube (Figure 9.11). The primary application of photomultipliers in astronomy is

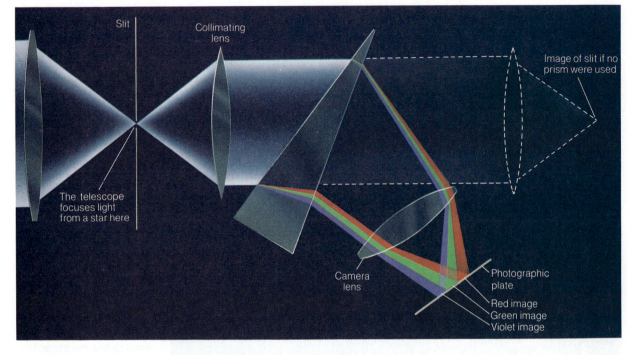

FIGURE 9.14   Design of a simple prism spectrometer for astronomy.

often measured and recorded electronically rather than photographically, but the optical principle is the same.

An image of the spectrum shows the star's light spread into a streak, with short wavelengths (violet) at one end and long wavelengths (red) at the other. Absorption lines in the spectrum manifest themselves as narrow ranges of wavelength where the light intensity is low. What we need to know is exactly where in terms of wavelength the dark lines and other spectral features occur. Thus it is also necessary to record some wavelength reference. Astronomers usually do this several times while recording the star's spectrum. They temporarily interrupt the exposure on the star and allow the light from a laboratory source (whose spectrum consists only of bright emission lines) to pass through the slit. The spectrum of the laboratory reference source thus appears as a series of bright lines. The wavelengths of these lines are precisely known, and the wavelengths of stellar spectral lines can be determined precisely, relative to the lines of the laboratory source.

9.3   ASTRONOMICAL OBSERVATORIES

We have discussed the basic principles that explain how astronomical telescopes and detectors work. However, to be useful, all of these different technologies have to be brought together in an *astronomical observatory*—a place where large telescopes and their auxiliary instrumentation are able to operate every clear night to collect

the data required by astronomers to help them understand the universe.

(a)   Telescope Mounts

Binoculars or a small spotting telescope can be held in your hands. Larger telescopes must be steadied by a tripod or some similar sturdy mounting. At an astronomical observatory, where the individual telescopes typically weigh 100 tons or more, proper mounting of the optical instrumentation becomes a matter of major concern.

The telescope must be mounted so that it can be pointed toward any direction in the sky and can move smoothly to follow the apparent motion of the source under study. Until very recently, almost all astronomical optical telescopes had **equatorial mounts.** An equatorial mount (Figure 9.15) allows the telescope to turn to the north and south about one axis and to the east and west about another. The axis for the east-west motion of the telescope is parallel to the axis of the Earth's rotation, and the other axis, about which the telescope can rotate to north or south, is perpendicular to this axis.

The two axes of motion of a telescope with an equatorial mount allow the telescope to be turned directly in right ascension and declination, the celestial coordinates in which astronomical positions are generally tabulated. An important advantage of the equatorial mount is that a simple slow motion of the telescope about its axis parallel to the Earth's axis—the polar axis of the tele-

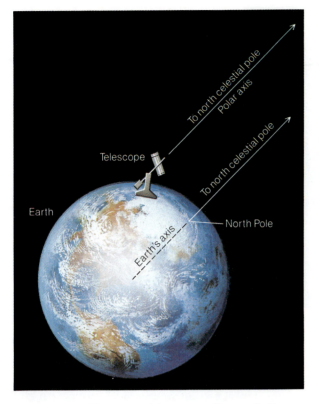

FIGURE 9.15 The equatorial mount, until recently the standard design for astronomical telescopes.

FIGURE 9.16 The 4-m telescope of Cerro Tololo Inter-American Observatory in Chile (operated by the U.S. National Optical Astronomy Observatories) uses the standard type of "horseshoe" equatorial mounting first pioneered by the Hale 5-m telescope at Palomar Observatory. (National Optical Astronomy Observatories)

scope—is sufficient to compensate for the apparent motions of the stars across the sky that result from the Earth's rotation. Even when a moving source such as a planet or asteroid is being tracked, only small adjustments with respect to the basic motion around the polar axis are needed (Figure 9.16).

An equatorial mount, however, is a disadvantage for a very large telescope because it is difficult to adjust for the gravitational stress on its heavy mirror. For a given telescope aperture, an **altitude–azimuth mount,** in which the telescope rotates simply about one vertical and one horizontal axis, is smaller, easier to construct, and much cheaper than an equatorial mount. All of the new large optical telescopes, such as the 10-m Keck telescope on Mauna Kea, have altitude–azimuth mounts, as do most large radio telescopes (Chapter 10). It is more complicated for them to compensate for the Earth's rotation and track the stars, but today that problem can be handled with computer-controlled driving mechanisms, while the advantages of reduced size and mechanical simplicity are considerable.

To point and track with the high accuracy required by modern telescopes, the motion of the telescope is controlled by a computer, which senses the position of the telescope and provides commands to electric motors connected to both axes. Many telescopes today can be pointed, on command, to within an accuracy of a few arcseconds, thus greatly speeding the location of faint sources. Once the telescope is in position, the computer-controlled drives will maintain pointing to within a fraction of an arcsec.

The telescope computer stores the coordinates of thousands of stars in its memory, and it can calculate at any moment the apparent positions of solar-system objects. In practice (and if the system is working properly), the astronomer needs only to enter the name of the object to be studied, and a minute or so later it will appear in the center of the field of view, ready for measurements to begin.

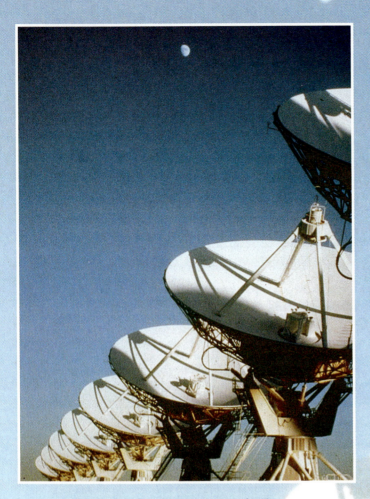

Part of the Very Large Array of radio antennae near Socorro, New Mexico. Each of the 27 dish-shaped antennae is 25 meters in diameter.
(National Radio Astronomy Observatory)

One of the most dramatic scenes in the science fiction film *2010* takes place not in space but on Earth, when the American protagonist meets his Russian counterpart in a surrealistic landscape of giant white machines. This remarkable vision is not a movie set or computer animation, but rather the Very Large Array radio telescope in New Mexico. Even while the movie scenes were being filmed, the telescope was carrying out its primary task of receiving the faint whispers of radio radiation from the universe and synthesizing "radio pictures" of distant objects.

A dazzling and otherworldly scene more familiar to Americans in the mid-1990s was the repair of the Hubble Space Telescope. For a week in December 1993, millions of people watched in fascination as video images were broadcast to Earth of astronauts repairing this giant scientific satellite and replacing some of its instruments. Like the Very Large Array, the Hubble Space Telescope is a tool of modern astronomy, designed to study the invisible radiation from the universe.

As astronomers have developed techniques for detecting the long-wavelength and short-wavelength radiation that our eyes cannot make out, their perspective on the universe has expanded dramatically. Before we saw "through a glass, darkly"; today we can begin to appreciate the glory of the universe in all its spectral bands, from high-energy gamma rays to long-wavelength radiowaves.

Grote Reber (b. 1911), an amateur astronomer and electronics expert, built in his own backyard the first radio telescope specifically designed to observe radio waves from space. From 1937 until after World War II, he was the world's only active radio astronomer. (Ohio State University)

NEW WINDOWS ON THE UNIVERSE

Near the middle of the 20th century, astronomy moved squarely into the forefront of science. The dramatic advances in astronomy were spurred in large part by the opening of new windows in the electromagnetic spectrum. Before 1931, we recorded information from space only in optical wavelengths. Radio astronomy had its humble beginning in that year, but it did not really develop until after World War II. Then, beginning in the 1960s, astronomers began to use the new tools of space flight to reach ultraviolet, x-ray, and other wavelengths blocked by the Earth's atmosphere.

In less than three decades our vision, formerly sensitive only to visible light, broadened to encompass the entire electromagnetic spectrum. It was predictable that there would be new discoveries, but no one could have foreseen just how far-reaching those discoveries would be. The new astronomy stimulated by access to the full electromagnetic spectrum—from gamma rays to low-frequency radio waves—has been one of the most productive areas of modern science. Here we summarize briefly how we detect this hitherto invisible radiation.

10.1 RADIO ASTRONOMY

(a) Early History

In 1931 Karl G. Jansky (1905–1950), an American radio engineer at the Bell Telephone Laboratories, built a rotating radio antenna array designed to operate at a wavelength of 14.6 m (Figure 10.1). With this array he attempted to investigate the sources of shortwave interference. In addition to temporary intermittent interference due to such phenomena as thunderstorms, he encountered a steady hiss-like static coming from an unknown source. He discovered that this radiation came in most strongly about 4 min earlier on each successive day and correctly concluded that since the Earth's sidereal rotation period is 4 min shorter than a solar day, the radiation must be originating from some region of the celestial sphere. Investigation showed that the source of the radiation was the Milky Way.

It was more than a decade before the astronomical community paid serious attention to Jansky's important discovery. Jansky has, however, received belated honors; today the standard unit of radio flux received from space is the Jansky (Jy). (One Jansky is 10^{-26} watt/m^2 striking the Earth's surface in one unit of frequency.)

Actually, the new radio astronomy did not go totally unnoticed, thanks to the American amateur astronomer, electronics engineer, and radio ham, Grote Reber (b. 1911). In 1936 Reber built the first radio telescope—an antenna specifically designed to receive cosmic radio waves (Figure 10.2). He constructed it of wooden two-by-fours and galvanized iron, in his backyard in Wheaton, Illinois. Subsequently Reber built other improved antennas and remained active in the field for more than 30 years.

Figure 10.1   The rotating radio antenna used by Jansky in his serendipitous discovery of radio radiation from the Milky Way. (Bell Laboratories)

In his first decade of radio observations, Reber worked practically alone. By 1940 he confirmed Jansky's conclusion that the Milky Way is a source of radio radiation, and in 1944 he published in *The Astrophysical Journal* the first contour maps of the radio brightness of the Milky Way as it appears at a wavelength of 1.87 m. He also discovered discrete sources of radio emission in the galactic center, Cygnus, and Cassiopeia, as well as radio waves from the Sun. From 1937 until after World War II, Reber was the world's only active radio astronomer.

Meanwhile, in 1942 radio radiation from the Sun was picked up by radar operators in England. After the war, the technique of making astronomical observations at radio wavelengths developed rapidly, especially in Australia, the Netherlands, England, and later in the U.S. Radio waves have now been received from many astronomical objects—the Sun, Moon, planets, gas clouds in our Galaxy, other galaxies, and many other objects. The technique of radio astronomy has become an essential tool of modern observational astronomy.

(b)   Detection of Radio Energy from Space

It is important to understand that radio waves are not "heard"; they have nothing whatever to do with sound. In commercial radio broadcasting, radio waves are modulated or coded to carry sound information; the sound itself is not transmitted. The radio waves merely carry the information that a radio receiver must decode and convert into sound by means of a loudspeaker or earphones. Sound is a physical vibration of matter; radio waves, like light, are a form of electromagnetic radiation. We can also code visible light to carry sound information, as is done, for example, by the sound track on a movie film or by telephone systems that use fiberoptic cables.

Many astronomical objects emit all forms of electromagnetic radiation—radio waves as well as light, infrared and ultraviolet radiation, and so on. The radio waves we receive from space are those that can penetrate the ionized layers of the Earth's atmosphere—those with wavelengths in the range from a few millimeters to about 20 m. The human eye and photographic plates are not sensitive to radio waves. We detect this form of

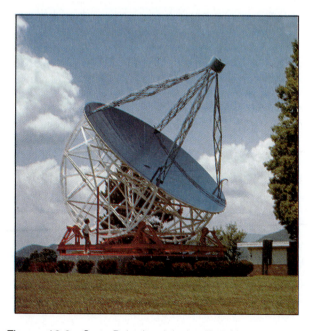

Figure 10.2   Grote Reber's original radio telescope, now reconstructed at the National Radio Astronomy Observatory in Greenbank, West Virginia. (National Radio Astronomy Observatory)

radiation by making use of the fact that radio waves induce a current of electricity in conductors. An *antenna* is such a device; it intercepts radio waves, which induce a feeble current in it. The current is then amplified in a radio receiver until it is strong enough to measure or record.

An antenna by itself can't tell us the source of the radiation we receive. We can understand why through an analogy. If we lay a photographic plate out on the ground in daylight, it will be exposed by sunlight, indicating the presence of a light source in the sky. We can place various color filters in front of such plates to detect the presence of various colors in the light that exposes them. Such an experiment, however, does not indicate the direction in the sky of the light source. Similarly, a radio antenna can be strung up outside, and currents induced in it indicate the presence of a source of radio radiation. Electronic filters in the radio receiver can be tuned to amplify only one frequency at a time and thus determine what frequencies or wavelengths are present in the radio radiation (just like optical filters). As in the case of a photographic plate laid out in sunlight, however, a single antenna does not indicate the direction of the source. What is needed is something that can focus the waves—a radio telescope.

(c)   Radio Reflecting Telescopes

Radio waves are reflected by conducting surfaces just as light is reflected from an optically shiny surface. Furthermore, the same law of reflection applies to both optical and radio waves. A radio reflecting telescope consists of a parabolic reflector, analogous to a telescope mirror. The reflecting surface can be solid metal or a fine wire mesh. The reflecting paraboloid is called a "dish" (Figure 10.3). Radio dishes are usually mounted so that they can be steered to point to any direction in the sky and gather up radio waves, just as an optical reflecting telescope can be pointed in any direction to gather up light. The radio waves collected by the dish are reflected to the focus of the paraboloid, where they form a radio image.

At the telescope's focus, the radio energy induces a weak current, which is amplified by a receiver, not unlike ordinary home receivers in principle. Receivers can be tuned to select a single frequency, but today it is more common to use sophisticated data processing techniques to allow thousands of separate frequency bands to be detected. Thus the modern radio receiver operates much like a spectrometer on an optical telescope. The signals, after computer processing, are recorded on magnetic disk or tape.

One advantage of astronomical observations at radio wavelengths is that some of the important atmospheric effects discussed in Section 9.4 are not bothersome. In particular, radio observations are not affected by atmospheric seeing. They are less affected by weather and sky brightness, and at some wavelengths observations can even be made throughout the entire 24-hr day. Radio interference from transmitters in Earth satellites, from radar, and even from such ubiquitous devices as automobile ignition systems, however, is a serious problem.

Figure 10.3   The 100-m radio telescope near Bonn, West Germany. (Max-Planck Institut für Radioastronomie)

The ability of a radio telescope to gather radiation depends on its aperture. The radio energy received from most astronomical bodies is very small compared with the energy in the optical part of the electromagnetic spectrum. Hence radio dishes are usually built in large sizes; few are under 6 m across. At first thought, the problem of constructing a large parabolic reflecting surface to sufficient accuracy might seem prohibitive. The 5-m mirror of the Hale telescope has a surface accurate to about five-millionths of a centimeter, about one-eighth of the wavelength of visible light. However, a radio dish designed to receive radio waves of a length of 25 cm, for example, need be accurate to only about 3 cm, to achieve the same relative level of precision.

(d)   Resolving Power of a Radio Telescope

The resolution of a radio telescope depends on its diameter relative to the wavelength of observation. Remember that, in the absence of atmospheric turbulence, the resolution of an optical telescope is given by

$$\alpha = 2.1 \times 10^5 \frac{\lambda}{d}$$

where d is the aperture of the telescope (Section 9.1c). Since atmospheric turbulence is not a problem at radio wavelengths, this formula works well for radio telescopes. Thus we can easily compute the resolving power of a radio telescope.

The wavelength of radio radiation, λ in the above formula, is far greater than that of visible light, so the resolving power for a telescope of a given size is correspondingly less. Radio waves of 20-cm wavelength, for example, are some 400,000 times longer than waves of visible light. To resolve the same angle, a radio telescope would have to be 400,000 times larger than an optical telescope. To resolve 1 arcsec at 20-cm wavelength requires a radio telescope nearly 40 km across. The largest steerable radio telescopes in use today are only about 100 m in diameter. At a wavelength of 20 cm, their resolution is only about 400 arcsec. The largest radio telescopes have far poorer resolving power than even the human eye.

Angular resolution at radio wavelengths can be greatly enhanced with the technique of **interferometry,** which uses two or more separate radio telescopes. By electronically combining the signals from different dishes, it is possible to achieve a resolution that depends on the *separation* of the antennas rather than their individual sizes. If the antennas are 40 km apart, for

Figure 10.4   A radio interferometer. Millimeter-wave telescopes of the Owens Valley Radio Observatory, operated by the California Institute of Technology. (Caltech)

example, their resolution at 20-cm wavelength is 1 arcsec (Figure 10.4).

(e) Radio Interferometry

Let us see how a radio interferometer works. Suppose, for example, two radio dishes are placed some distance apart. Unless the source of radio radiation happens to lie along a perpendicular bisector of the line between the antennas, the radio waves will strike one antenna a brief instant before the other, so that the two antennas will receive the same waves at slightly different times and thus become "out of phase" with each other—that is, each antenna receives a different part of a given wave (Figure 10.5). The difference in phase between the waves detected at the two antennas can be measured electronically. Because this phase difference depends on the angle that the direction to the source makes with the line between the antennas, that angle can be determined. If the two antennas are due east and west of each other, an observation of this sort gives a much more accurate measure of the east-west position of the source in the sky than could be obtained with a single radio telescope. If the antennas are placed due north and south of each other, the other coordinate of the source can be found.

The farther apart·the components of a two-element interferometer are placed (the longer the *baseline*), the higher the resolution is and the more accurately we can pinpoint the direction of the source. Most major radio astronomy observatories, therefore, have several telescopes so that pairs of them can be operated together as interferometers. A multi-element interferometer is also called an *array*. The overall sensitivity to radio radiation is then the sum of the collecting areas of the telescopes so combined, and the resolution (or *beam width*) in a direction along the line between the telescopes is that of a single telescope of total aperture equal to that separation.

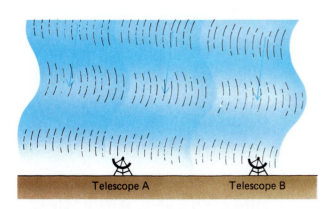

Figure 10.5 The principle of the radio interferometer. Two radio telescopes simultaneously observe the same source, which is not in a direction perpendicular to the line connecting the two telescopes. Thus the waves reach telescope A slightly out of phase with (behind) those reaching telescope B. The phase lag depends on the direction of the source and hence determines it.

(f) Interferometer Arrays

The most powerful radio telescopes are made up of several separate receivers. In these **arrays** the radiation detected at each receiver can be compared, by computer, with that from each other one in the array. In this way an image of the source can be reconstructed as it would appear with the resolution of a single telescope with aperture equal to the maximum separation in the array. The technique is called *aperture synthesis* and was largely pioneered by astronomer Martin Ryle at Cambridge, England, working in the 1960s.

The world's most impressive radio telescope array is the Very Large Array (VLA) near Socorro, New Mexico (Figure 10.6). The VLA, operated by the National Radio Astronomy Observatory (financed by the National Science Foundation), was dedicated in 1980. It has 27 telescopes, each of aperture 25 m, that can all be moved along rails laid out in a large "**Y**" configuration with a total span of about 36 km (Figure 10.7). The VLA normally operates at four wavelengths: 3 cm, 6 cm, 12 cm, and 21 cm. By electronically combining the signals from all of its individual telescopes, this array permits the radio astronomer to make "pictures" of the sky at radio wavelengths that are comparable to those obtained with an optical telescope. The resolution of the VLA was intentionally matched to that of ground-based optical telescopes to facilitate comparison of images of the same object at optical and radio wavelengths.

By tracking a particular area of the sky for several hours, the VLA can produce a radio map or image of 1-arcsec resolution over a field of view of several arcminutes. Alternatively, and again like an optical telescope, the VLA can be used for shorter exposures of less sensitivity. Much of our modern understanding of astrophysical phenomena, particularly high-energy processes such as those associated with active galaxies, is now being derived in part from observations made with this facility.

But even more can be done to increase the resolution of radio telescopes. Now that time standards can be coordinated to high precision, we can extend the interferometer principle to *very long baseline interferometry* (VLBI). Two different radio telescopes, thousands of kilometers apart, can simultaneously observe the radio waves from the same source and record them on tape, along with the time from a very accurate standard clock. Later, these two tapes can be analyzed by computer to find the phase difference between the radio radiation at the two stations, and hence the direction of the source. Baselines as long as from California to Parkes (in Australia) and from Greenbank, West Virginia, to the Crimea (in Ukraine) have been used. The resulting angular resolution of the sources observed is as great as a few ten-thousandths of an arcsec—far surpassing the angular resolution of optical telescopes.

Naturally, astronomers want to be able to utilize an

Figure 10.6 Part of the Y-shaped Very Large Array (VLA) near Socorro, New Mexico. The individual antennas are 25 m in aperture. (National Radio Astronomy Observatory)

array of very long baseline interferometers, just as they have constructed arrays of ordinary radio telescopes like the VLA. To achieve this goal, the U.S. National Radio Astronomy Observatory built the Very Long Baseline Array (VLBA), consisting of ten linked telescopes each of 25 m aperture distributed over 9000 km of U.S. territory, from the Virgin Islands to Hawaii (Figure 10.8). Dedicated in 1993, the VLBA is the largest single integrated astronomical instrument in the world. Its resolution is nearly 1000 times better than that of the VLA, or about 0.001 arcsec—greater than that of any other astronomical instrument on Earth or in space. At this resolution, features as small as 10 astronomical units (AU) could be distinguished at the center of our Galaxy.

The eventual addition of one or more telescopes in space has been suggested as a means to increase further

Figure 10.7 Aerial view of the VLA. This telescope consists of 27 movable antennas spread over a total span of 36 km. (National Radio Astronomy Observatory)

Figure 10.8 The distribution of antennas that constitute the U.S. VLBA (Very Long Baseline Array). (National Radio Astronomy Observatory)

the resolution of the ground-based VLBA, beyond the limits presently set by the dimensions of the Earth. Russia, in collaboration with the U.S. and several European nations, is developing an orbiting receiver called Radioastron that could be launched in the late 1990s to link up with the VLBA.

(g) Radar Astronomy

Radio astronomy, like optical astronomy, is based on the measurement of the electromagnetic radiation naturally emitted by objects in the universe. In contrast, **radar** is the technique of *transmitting* radio waves to an object and then detecting the radiation that the object *reflects* back to the transmitter. The time required for the radio waves to make the round trip can be measured electronically, and because radio waves travel with the known speed of light, the distance of the object is

determined. Astronomers have also learned to use radar techniques to determine properties of targets, such as size, shape, and composition.

In recent decades the radar technique has been applied to the investigation of the solar system. Radar observations of the Moon and planets have yielded our best knowledge of the distances of these worlds and have played an important role in navigating spacecraft throughout the solar system. In addition, as will be discussed in later chapters, radar observations have determined the rotation periods of Venus and Mercury, probed the tiny Earth-approaching asteroids and the nuclei of comets, analyzed the rings of Saturn, and investigated the surfaces of Mercury, Venus, Mars, and the large satellites of Jupiter and Saturn. Radar is particularly critical for the study of Venus, since it can penetrate the otherwise opaque atmosphere and clouds (Chapter 15).

One of the special advantages of radar as an astronomical tool is that, in contrast to other techniques, the scientist can control the properties of the transmitted beam. In most of astronomy, our role is entirely passive; all we can do is to detect and try to understand the radiation that nature sends to us. The transmitted radar signal, however, is entirely within our control. We can vary its strength, frequency, polarization, and other properties. In effect, radar astronomy is an *experimental* rather than an *observational* science. The radar beam serves as a tool with which to probe remote members of the planetary system.

The largest radar telescope (in fact, the largest reflecting telescope of any kind in the world) is the 1000-ft (305-m) bowl near Arecibo, Puerto Rico, a facility of the National Astronomy and Ionosphere Center, funded by the National Science Foundation and operated by Cornell University (Figure 10.9). The telescope was completed in 1963, and in 1974 its reflecting surface was rebuilt with such high precision that it can

Figure 10.9 The 305-m (1000-ft) dish at the National Astronomy and Ionosphere Center, Arecibo, Puerto Rico, operated by Cornell University and sponsored by the National Science Foundation. (Cornell University)

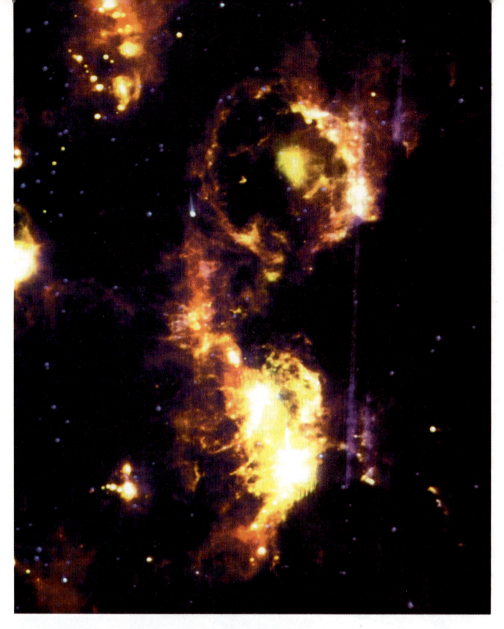

Figure 10.14 A mosaic map of part of star-forming clouds in the constellation of Orion, made by IRAS. The rotation of the satellite continuously swept its telescope around the sky. Each line across the picture shows the change in signal strength recorded in a scan with one of IRAS's detectors. In this picture, data with a wavelength of 12 μm are color-coded blue, at 60 μm are coded green, and at 100 μm are coded red. (NASA/JPL)

two other wavelength bands, at about 300 and 700 μm (0.3 and 0.7 mm). Spectroscopy has been particularly productive in the submillimeter, since a number of compounds (such as CO and other carbon molecules) in the interstellar gas emit primarily in this part of the spectrum or at adjacent millimeter wavelengths.

Two telescopes have been constructed at Mauna Kea that are optimized for submillimeter observations. One, with a 15-m aperture, is a joint project on the Netherlands, Britain, and Canada. The other, 10 m in aperture, is operated by Caltech (Figure 10.15). For wavelengths between 1 mm and 3 mm, the most important instruments are the 30-m IRAM telescope in Europe and the 25-m Nobeyama telescope in Japan (Table 10.1).

10.3 HIGH-ENERGY ASTRONOMY

(a) Early History

The atmosphere is completely opaque to electromagnetic radiation of wavelength less than about 300 nm. Consequently, ultraviolet, x-ray, and gamma-ray observations must be made from space. Such observations first became possible in 1946, when the U.S. acquired several V2 rockets captured from the Germans. Herbert Friedman (b. 1916) and his colleagues from the U.S. Naval Research Laboratory instrumented these rockets for a series of pioneering flights, used initially to detect far-ultraviolet radiation from the Sun. Since then, many

Figure 10.15 The California Institute of Technology 10-m submillimeter telescope on Mauna Kea, Hawaii, shown at its dedication in 1986. Very dry sites such as Mauna Kea are required for ground-based observations in this part of the electromagnetic spectrum. (Janet Morrison)

rockets have been launched to make x-ray and ultraviolet observations of the Sun and other celestial objects.

Beginning in the 1960s, a number of Earth satellites have been launched to carry out astronomical observations. Scientist astronauts and cosmonauts have also made astronomical observations from Skylab, from the Spacelab facility of the U.S. Shuttle, and from the Russian Mir space station.

(b)   Orbiting Ultraviolet Observatories

Ultraviolet telescopes are similar to optical telescopes, except that their optical surfaces need special coatings with high ultraviolet reflectivity. The first successful OAO (Orbiting Astronomical Observatory), OAO 2, was launched by the U.S. in 1968 and carried instruments developed by the University of Wisconsin and by the Smithsonian Astrophysical Observatory at Harvard for obtaining ultraviolet spectra of astronomical objects in the spectral range 120 to 400 nm. The next important ultraviolet observatory was OAO Copernicus, launched in 1972, which carried a 0.8-m ultraviolet telescope and three small x-ray telescopes. Copernicus is best remembered for its many discoveries about the composition, temperature, and structure of the interstellar gas.

The most productive orbiting ultraviolet observatory to date (still in operation in 1994) is the International Ultraviolet Explorer (IUE) satellite, launched in January 1978 and carrying instruments developed by NASA and by the United Kingdom's Science Research Council. The IUE can obtain high-quality spectra in the range 115 to 320 nm with exposure times of up to 15 hours. The IUE is in a synchronous orbit (that is, its period of revolution about the Earth is equal to the period of Earth's rotation), so that it is always in view of its

control headquarters at NASA's Goddard Space Flight Center in Greenbelt, Maryland, and the European control center in Madrid. Astronomers from all over the world go to Greenbelt or Madrid as guest investigators. Also still operating in 1994 is the Extreme Ultraviolet Explorer of the University of California at Berkeley, which yielded the first all-sky survey at shorter ultraviolet wavelengths (less than 120 nm).

(c)   X-Ray Observatories

X rays are electromagnetic radiation of wavelength less than about 10 nm. The shorter the wavelength, the higher the energy of the photons (Section 7.1), and x-ray and gamma-ray astronomers often speak of the *energies* of the photons they observe rather than of their wavelengths. A photon of wavelength 1.24 nm has an energy of 1000 eV, abbreviated 1 keV. Those of lower energy are called soft x rays, and those of higher energy (shorter wavelength) are called hard x rays.

We saw earlier that radio waves are characteristic of the energy emitted from cold bodies, while visible light and the ultraviolet are characteristic of the energy from hot bodies of temperature 10^3 to 10^5 K (like the surfaces of stars). X rays are generally emitted from gas at very high temperatures—10^6 to 10^8 K. Thus in different spectral regions we preferentially observe parts of the universe that are at different temperatures.

X rays from space were first observed with instruments flown on balloons and rockets. By 1967, about 30 discrete sources of x rays had been discovered. X-ray astronomy made a sudden advance in December 1970, when NASA launched the first orbiting x-ray observatory. Since the launch was carried out from Kenya, that observatory was named Uhuru, the Swahili word for "freedom." Uhuru systematically scanned the sky for x-ray sources and charted more than 200 of them during its lifetime (to 1973). These sources are named numerically according to the constellation in which they are found; for example, the famous source Cygnus X-1 is the first x-ray source to be found in Cygnus.

Beginning in 1977 the United States launched a series of High-Energy Astronomy Observatories. X-ray astronomy received a truly spectacular boost with the launching of HEAO 2, the Einstein Observatory, in November 1978 (Figure 10.16). Although x rays are easily absorbed in ordinary optical systems, they can be reflected from polished surfaces that they strike at a grazing angle—like stones skipping across water. The Einstein satellite had an x-ray telescope consisting of a complex set of concentric parabolic and hyperbolic cylindrical surfaces that used grazing reflection to focus x rays into an actual image that could be detected electronically and transmitted to Earth. The size of the telescope aperture was only 0.6 m, but because of the grazing angles of the reflecting surfaces to the incoming

Figure 10.16 Artist's impression of the Einstein X-Ray Observatory. (NASA)

x-ray photons, the actual mirror surface was equivalent to that of the 2.5-m (100-in.) telescope.

The Einstein telescope was designed to record x rays of wavelengths from 0.3 to 5 nm and had a field of view of about 1°. In the first few months of its operation, it was an unqualified and spectacular success, with a sensitivity for detecting weak sources 1000 times as great as anything that preceded it. This is equivalent to changing from a small amateur telescope to the 200-in. Hale reflector on Palomar Mountain. In later chapters we will have more to say about the thousands of x-ray sources discovered with the Einstein telescope and what they mean.

(d) Gamma-Ray Astronomy

Gamma rays were first discovered among the radiation emitted during the decay of radioactive elements. We know today that atomic nuclei have excited energy states, analogous to those of an atom when it changes its configuration of electrons. The differences in energy between the nuclear states, however, are very much higher, and the photons that can be absorbed or emitted by nuclei changing states are gamma rays. The boundary between x rays and gamma rays is usually taken as about 0.01 nm (about 100 keV).

Gamma rays of cosmic origin were first discovered by the Vela satellites, which were launched by the U.S. Department of Defense to carry out worldwide surveil-lance for possible explosions of nuclear bombs, which would emit gamma rays. In 1967 the Vela detected bursts of gamma radiation that investigation showed could not originate from within the solar system. Such bursts have also been detected by Uhuru and other orbiting observatories equipped with detectors of high-energy radiation, including some of the OSOs. Each lasts anywhere from about a tenth of a second to several seconds.

The bursts detected by Vela were of relatively low energy, as gamma rays go. Sources of far higher energy gamma rays—greater than 10 million eV (10 MeV)—were found subsequently. The first certain detection of high-energy gamma rays of cosmic origin was with an experiment on OSO 3, in which the plane of our Galaxy was detected in the light of gamma rays with energies greater than 50 MeV. Other satellites, especially the second Small Astronomy Satellite, SAS 2, extended this picture of the gamma-ray sky. In high-energy gamma radiation, the Milky Way appears somewhat as it does to the unaided eye, only narrower—a belt around the sky about 2° wide. There are also discrete gamma-ray sources, mostly along the Milky Way; some have been identified with old supernovae.

(e) Solar Observatories

Some of the first rocket observations were of the Sun, which is the brightest source in the sky. Later, many satellites were launched by both NASA and the U.S. Navy (as well as the U.S.S.R.) to monitor the Sun and help predict the effects of solar activity on the Earth's atmosphere, including possible disruption of military communications. Solar observations were also carried out from the shortlived NASA Skylab space station in the 1970s.

Eight NASA solar satellites were launched during the years 1962 through 1975. These orbiting solar observatories (OSOs) obtained thousands of ultraviolet spectra of the Sun. The last and most successful observatory, OSO 8 carried two instruments (one American and one French) for obtaining ultraviolet spectra of light from tiny regions of the Sun, and six were x-ray and gammay-ray detectors for exploring sources of this high-energy radiation from other directions in space. The OSO satellites were especially productive in studies of the high-temperature outer atmosphere of the Sun (called the solar corona), particularly in discovering the structure and variability of the x-ray emission from the corona.

The two most important solar space observatories since the OSO series were the U.S. Solar Maximum satellite, launched in 1980, and the Japanese Yohkoh (Sunbeam) x-ray satellite, launched in 1991. The Solar Maximum (SMM) failed a few months after launch, but in 1984, Space Shuttle astronauts docked the errant satellite, brought it on board, and made the necessary electronic repairs before releasing it back into an independent orbit (Figure 10.17). Solar Max finally ended its mission in 1989, when it re-entered the Earth's atmosphere. Yohkoh is primarily an x-ray satellite, and it has produced thousands of beautiful, high-resolution x-ray images of the Sun that are permitting detailed study of solar flares and other energetic phenomena in the corona.

10.4 NEW OBSERVATORIES IN SPACE

(a) Changing Strategies

Most of the orbiting observatories we have described were launched in the 1970s. During the 1980s there was a substantial gap in this sequence of space observations, with only the reliable IUE continuing its operations. A similar gap existed in planetary exploration, with not one U.S. planetary launch between 1978 and 1989, compared with an average of almost one per year going to the Moon and planets during the previous two decades. This interruption can be attributed in large part to the development of the Space Shuttle, which absorbed a disproportionate share of the NASA funding during the 1970s. The problem became worse after the explosion of the Shuttle Challenger in 1986, just when the Shuttle was becoming operational and scientific flights were about to resume. It is also true, however, that as the cost of individual science missions goes up, their frequency of flight must decline unless there are major infusions of new funding into the NASA science program.

In the 1990s only a few astronomical satellites are being built, and because of their high cost most of them represent international collaborations. This new generation of space observatories, however, has capabilities far greater than those of the satellites of the 1970s, more than compensating for their smaller numbers. Unfortunately, these may be the last of the large space observatories for some time. The collapse of the Soviet Union and cutbacks in space science funding in the U.S. and Europe will limit us in the future to smaller spacecraft with more modest capabilities.

(b) Hubble Space Telescope

The first of the new generation of large space observatories was the Hubble Space Telescope (HST) (Table 10.2; Figure 10.18). This optical-ultraviolet telescope, with its aperture of 2.4 m, is named for Edwin Hubble, the California astronomer who discovered the expan-

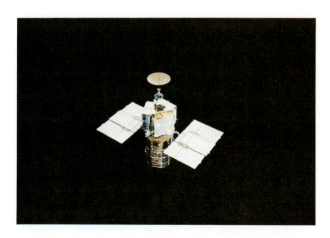

Figure 10.17 The Solar Max Satellite in 1984 at the time of its on-orbit repair by astronauts on the U.S. Space Shuttle. (NASA)

TABLE 10.2 HUBBLE SPACE TELESCOPE	
Telescope	
Aperture	2.4 m
Focal ratio	f/24
Resolution	0.05 arcsec
Pointing precision	0.01 arcsec
Tracking precision	0.007 arcsec
Electrical power	4 kW
Spacecraft length	13.1 m
Spacecraft mass	11.6 tons

Figure 10.18   The Hubble Space Telescope (HST) at the time of its launch in 1990. (NASA)

sion of the universe in the 1920s. HST was launched by the Space Shuttle in April 1990 and is operated from Goddard Space Flight Center, working with the NASA-funded Space Telescope Science Institute in Baltimore. It is instrumented for direct imaging and spectroscopy in the visible and ultraviolet, but in the late 1990s one or more of these instruments will be replaced, and it is expected that the HST will then also work in the near infrared part of the spectrum. Because it can be serviced by Shuttle astronauts, HST can be

reconfigured and improved over time, unlike any of its smaller predecessors.

The primary objective of the HST is to take advantage of the absence of atmospheric seeing to obtain high-resolution images and thereby to probe deeper into space than is possible with any ground-based telescope. The HST also can operate in the ultraviolet, at wavelengths too short to penetrate to the surface of the Earth. The instruments on the telescope are summarized in Table 10.3. Unfortunately, scientists discovered

TABLE 10.3 HST INSTRUMENTS	
Wide Field and Planetary Camera	8 CCD cameras, each 800 × 800 pixels. Wavelengths 115 to 1100 nm. (JPL/Caltech)
Faint-Object Camera	TV with image intensifier, 512 × 1024 pixels. Wavelengths 115 to 650 nm. (ESA/Dornier/Matra)
Faint-Object Spectrograph	Wavelength range 115 to 850 nm. Spectral resolving power 250 to 1300. (Martin Marietta)
High-Resolution Spectrograph	Wavelength range 105 to 320 nm. Spectral resolving power 2000 to 100,000. (NASA/Ball Aerospace)

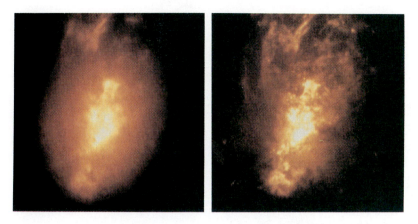

Figure 10.19 Improvement in the clarity of images taken with the Hubble Space Telescope. Both images show the central region of the active galaxy NGC 1068, about 60 million light years away, the left image before the repair, the right one after. (NASA/STScI)

after launch that a major manufacturing error had been made in the shape of the primary mirror, which deviates significantly from a parabola. Consequently, the images were blurred and the effective resolution of the telescope was about five times worse than the design goal of 0.1 arcsec. This error could not be corrected until the Shuttle revisited HST in December 1993 to install a new camera containing compensating optics. Figure 10.19 illustrates the improvement achieved. At the same time the astronauts also added an external optical correction device that permitted the other instruments to function at nearly their original capability (Figure 10.20). Images and other information discovered by HST appear throughout this book.

Following its refurbishment, HST is the premier astronomical instrument of the world, used by hundreds of astronomers from the U.S. and Europe. It is also the most expensive telescope in the world, with an annual operating cost of about a quarter of a billion dollars—ten times as much as any ground-based observatory. In spite of its great success, it seems unlikely that other space observatories of this scale will be orbited in the foreseeable future, and no successor to HST is currently planned.

(c) Rosat X-Ray Observatory

The primary x-ray observatory of the 1990s was a German satellite (with British and U.S. collaboration) called the Roentgensatellit, named for the German scientist who discovered x rays, Wilhelm Roentgen. It is more commonly called simply Rosat. Rosat, launched into orbit in June 1990 (2 months after HST), had a mass of 2.4 tons. It was still operating in 1994, producing both an x-ray survey of the entire sky and a series of more detailed images of individual sources. In its survey mode, Rosat was able to catalogue more than 100,000 x-ray sources. Its x-ray telescope had both higher sensitivity and greater resolving power than the Einstein satellite that had pioneered x-ray imaging in the 1970s. Some of the discoveries from Rosat will be presented in chapters on stellar evolution and external galaxies.

Toward the end of the decade, NASA plans to launch

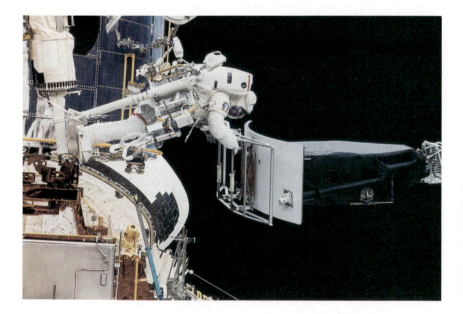

Figure 10.20 Astronaut/astronomer Jeffrey Hoffman is holding the old Wide Field and Planetary Camera during the December 1993 repair of the Hubble Space Telescope as he rides the remote manipulator arm above the payload bay of the Space Shuttle. The telescope extends above his body at the left. (NASA)

a larger x-ray satellite called the Advanced X-Ray Astrophysics Facility (AXAF). AXAF has a mass of more than 5 tons and is designed for imaging with still greater sensitivity and resolution.

(d) Compton Gamma-Ray Observatory

Another of the large NASA space observatories was the Compton Gamma Ray Observatory, launched in April 1991. Its objectives are to conduct an all-sky survey for gamma-ray sources as well as to study individual objects in detail at a variety of energies. The spacecraft is named for American physicist Arthur Holly Compton, an early 20th-century pioneer in the laboratory study of gamma rays. With a mass of more than 16 tons, the Compton observatory is one of the largest scientific payloads ever launched into space (Figure 10.21).

The Compton observatory carries four instruments designed to detect gamma rays over a broad range of energies from 20 thousand to 30 billion eV. The Burst and Transient Source Experiment (BATSE) studies the enigmatic gamma-ray bursts first discovered by the Vela satellites. Two telescopes (Comptel and EGRET) produce images of gamma ray sources with resolutions of about 0.1°, and the Oriented Scintillation Spectrometer (OSSE) measures the energy (spectrum) of incoming gamma rays.

By the end of 1994 the BATSE instrument had pinpointed more than 1000 gamma-ray bursts and demonstrated that the sources are distributed uniformly across the sky. Astronomers still do not know what causes these events, but the evidence suggests that the sources are at great distances and may be near the limits of the observable universe. Continuous sources of gamma rays have been identified at the center of our Galaxy and from many active galaxies. We will discuss some of these results near the end of this book.

(e) Infrared Space Observatories

With the launch of HST, Rosat, and Compton and the construction of AXAF, the most obvious part of the electromagnetic spectrum not covered by current space observatories is the infrared. In spite of the success of IRAS and COBE in the 1980s, as well as great strides in development of infrared detectors made subsequently, no successor has been launched. This situation will be corrected in 1995 or 1996 when the European Infrared Observatory in Space (ISO) is launched. With a larger

Figure 10.21 The deployment of the Compton Gamma-Ray Observatory from the Shuttle on April 1991. In this view from the Shuttle payload bay, the remote manipulator arm is used to move the satellite away from the Shuttle Atlantis. (NASA)

aperture and more sensitive detectors than IRAS, ISO should deliver a major improvement in our ability to detect faint sources at longer infrared wavelengths.

Beyond ISO, two proposals have been made for the next generation of infrared space telescopes, but neither has yet received funding. The U.S. project is called SIRTF, or Space Infrared Telescope Facility. In Europe, a telescope called Edison is being studied with a larger aperture than SIRTF but operating a higher temperature. Both SIRTF and Edison will, if funded, take advantage of a new generation of more capable infrared detectors.

S U M M A R Y

10.1 In the 1930s, radio astronomy was pioneered by Jansky and Reber. A radio telescope is basically a radio antenna (often a large parabolic dish) connected to a receiver and (possibly) a

multichannel analyzer to derive spectral information. The resolution of a single dish depends on its diameter relative to the wavelength, but much enhanced resolution is obtained by

interferometry, including the development of interferometer **arrays** like the 27-element VLA. Expanding to very long baseline arrays (with antennas separated by thousands of km, as in the VLBA), radio astronomers can achieve resolutions as good as 0.001 arcsec. **Radar** involves transmitting as well as receiving. The largest radar telescope is the 305-m dish at Arecibo.

10.2 Infrared includes wavelengths from about 1 μm to 1 mm, where the wavelengths beyond 100 μm are also called the submillimeter. Short of 2.5 μm, infrared techniques are similar to those of optical astronomy. Longward of 2.5 μm, the atmosphere and telescope are strong thermal emitters and special techniques must be used to cool and shield infrared detectors. In addition to ground-based facilities, infrared observations are made from the Kuiper Airborne Observatory and from space. IRAS and COBE provided the first infrared all-sky surveys from space.

10.3 High energy astronomy refers to observations made at ultraviolet, x-ray, and gamma-ray frequencies. All these observations must be made from above the atmosphere. Many orbiting ultraviolet observatories have been flown, notably a series of solar observatories and the International Ultraviolet Explorer (IUE). X rays (wavelength shorter than 10 nm) originate in hot, energetic sources. Uhuru carried out an all-sky survey, and much progress was made with the Einstein x-ray observatory and the more recent Yokhoh solar x-ray satellite. Gamma-ray bursts were discovered by the Vela military satellites.

10.4 Few space observatories were launched during the 1980s, but several major facilities are becoming available in the 1990s. These include the Hubble Space Telescope (HST) for visible and ultraviolet work, the Infrared Space Observatory (ISO), the Rosat and AXAF x-ray telescopes, and the Compton Gamma Ray Observatory.

E X E R C I S E S

THOUGHT QUESTIONS

1. Suppose you are looking for a site for an optical observatory, an infrared observatory, a submillimeter observatory, and a radio observatory. What are the main criteria of excellence for each of these cases? What sites on Earth are actually thought to be the best today?

2. Radio and radar observations are often made with the same antenna, but otherwise they are very different techniques. Compare and contrast radio and radar astronomy, in terms of the equipment needed, the methods used, and the kind of results that are obtained.

3. Suppose there were a stationary array of radio telescopes lined up in an east-west direction and located at the equator of the Earth. Would the system determine more accurately the right ascension or the declination of a cosmic radio source? Why?

4. Compare infrared telescopes with those built primarily for observations in the visible part of the spectrum. What are the main differences? Should it be possible to construct a telescope so that it can operate in an optimum fashion in both wavelength regions?

5. Infrared detectors are usually operated at very low temperatures, typically 4 K. Explain the advantages of using cold detectors. In space, the entire telescope can also be cooled. What additional advantages does this bring? Why can't we also cool the telescope optics on a ground-based telescope?

6. What are the advantages of putting a telescope in an airplane, such as the NASA Kuiper Airborne Observatory? How does the 1-m KAO telescope compare in capability with larger ground-based telescopes, for both infrared and visible observations?

7. What are the advantages and disadvantages of an observatory on the Moon, relative to (a) a ground-based observatory, or (b) an orbiting observatory?

PROBLEMS

8. The VLA has a maximum baseline of 36 km. What is the angular resolution possible with the system operating at a wavelength of 6 cm?

9. If, with a baseline of 6000 km, a VLB interferometer just resolved an angle of 0.0042 arcsec, at what wavelength would the observations have been made?
Answer: 12 cm

10. If a radio telescope array has four separate antennas, how many different pairs can be selected among them for interferometry? Actually, the number of different pairs among n objects is $n(n - 1)/2$. How many pairs exist among the 27 antennas in the VLA?
Answer: 351

11. What is the wavelength of 50 MeV gamma-ray photons?
Answer: About 2.5×10^{-5} nm

★**12.** Suppose at some northern latitude a straight-line array of radio telescopes is laid out in an arbitrary direction (not necessarily north-south or east-west, but not excluding these possibilities). Explain how the use of the array as an interferometer at different times during the day (or night) can yield two-dimensional information about the structure of the source and its direction in the sky.

The Saturn system, showing the planet, its rings, and several satellites, as photographed by the Voyager spacecraft. This figure is a collage, constructed of individual close-up images of each world, with the sizes adjusted for perspective. Note the variety in the appearances of Saturn's satellites.

(NASA/JPL)

To the unaided eye, the planets look just like stars. Only their slow movement against the stellar background identifies them as members of our solar system. Even seen through a telescope, only four planets—Venus, Mars, Jupiter, and Saturn—appear as any more than tiny featureless disks. The great majority of solar system objects, including more than 60 planetary moons and thousands of asteroids or minor planets, are no more than faint, star-like points of light when observed through even the largest telescopes.

How have we learned so much about these distant worlds, given the limitations of our telescopes? In part, our understanding has emerged from the patient collection and analysis of their radiation by astronomers observing with optical and infrared telescopes. Spectroscopy, for example, has revealed the presence of atmospheres on many objects and yielded the composition and temperatures of the gases in these atmospheres. Even more important, however, have been the robot space probes that have explored the planets directly.

The spacecraft exploration of the solar system, beginning in 1959 with the first successful lunar probes, is one of the great historic themes of 20th-century science. Within less than one human generation, our "robot representatives" have visited eight planets, dozens of moons, four ring systems, comets, and asteroids—the most ambitious and productive period of exploration since the time of Columbus. One by one, these faint points of light have been transformed into real worlds. The next dozen chapters of this book deal with the planets as places, each with its own unique characteristics, sights, and history.

THE PLANETARY SYSTEM

Gerald P. Kuiper (1905-1973), Dutch-born American astronomer, made many contributions to the theory of the origin of the solar system, discovered the atmosphere of Titan, carried out pioneering work in infrared astronomy, founded several observatories, initiated the NASA airborne astronomy program, and was an influential architect of the early NASA program of lunar and planetary exploration. (University of Arizona)

The ancient observer, who considered the Earth to be central and dominant in the universe, regarded the Sun, Moon, and planets as luminous orbs that moved about on the celestial sphere through the zodiac. While our solar system is indeed dominated by one body, it is the Sun, not the Earth. Relative to the Sun, the planets are tiny and inconspicuous objects. But the planets are important to us, since we live on one of them.

We turn now to those worlds of the planetary system. They will be considered individually in detail in later chapters, and their properties are summarized in Appendices 9, 10, and 11. We begin, however, by looking at some of the general characteristics of the whole system and noting a few of the properties that its constituent worlds have in common.

11.1 INVENTORY OF THE SOLAR SYSTEM

(a) Overview

The Sun is the dominant member of the solar system, as shown in Table 11.1. The Sun is a star, generating 4×10^{26} watts of power and holding the planetary system in its gravitational field. Its radiation provides most of the energy to heat the planets. In addition to its heat and light, the Sun is the origin of most of the thin gas that fills the solar system. A **solar wind** of ionized atoms and electrons constantly streams away from the hot upper atmosphere of the Sun at speeds of several hun-

TABLE 11.1	MASS OF MEMBERS OF THE SOLAR SYSTEM
Object	Percentage of Mass
Sun	99.80
Jupiter	0.10
Comets	0.05
All other planets	0.04
Satellites and rings	0.00005
Asteroids	0.000002
Dust and debris	0.0000001

dred kilometers per second. The solar wind strikes the planets, interacting with their atmospheres and magnetic fields.

We shall describe the Sun, a typical star, in Chapters 27 and 28. In this and the following chapters, our attention is focused instead on the other members of the planetary system, which include, in addition to the planets themselves, satellites, rings, comets, asteroids, and dust orbiting the Sun.

(b) The Planets

Most of the material in the planetary system is concentrated in the largest planet, Jupiter. The nine planets are the Earth, five other planets known to the ancients (Mercury, Venus, Mars, Jupiter, and Saturn), and three discovered since the invention of the telescope (Uranus, Neptune, and Pluto).

The planets all *revolve* about the Sun in approximately the same plane, like marbles rolling on a table. Each of the planets also *rotates* about an axis running through it, and in most cases the direction of rotation is the same as that of its revolution about the Sun. The similarity in motions of the planets is one of the basic facts of the solar system, presumably related to the way the planets formed.

The four planets closest to the Sun (Mercury through Mars) are called the inner or **terrestrial planets;** often the Moon is also discussed as a part of this group, bringing the total of terrestrial bodies to five. The terrestrial planets are relatively small worlds, composed primarily of rock and metal. The remaining five planets (Jupiter through Pluto) are called the outer planets. Jupiter, Saturn, Uranus, and Neptune are also referred to as giant or **jovian planets.** The jovian planets are large—at least ten times more massive than the Earth—and they are composed primarily of lighter ices, liquids, and gases. Little Pluto is neither a terrestrial nor a jovian planet; it is similar to the satellites of the outer planets. Table 11.2 summarizes some of the main facts about these nine planets.

(c) Other Members of the Planetary System

Most of the planets are accompanied by one or more satellites; only Mercury and Venus are alone. There are 62 known satellites (see Appendix 11), and undoubtedly many other very small satellites remain undiscovered. Saturn has 19, Jupiter 16, Uranus 15, Neptune 8, Mars 2, and Earth and Pluto 1 each. Six of these satellites are comparable in size to our Moon or to the planet Pluto.

Each of the jovian planets has a system of rings, discussed in Chapter 17. The best known are the rings of Saturn, which were first seen by Galileo. The rings of Uranus were discovered in 1977, those of Jupiter in 1979, and those of Neptune not until 1985.

The asteroids (Chapter 18) and comets (Chapter 19) are small objects orbiting the Sun. Comets differ from asteroids in both composition and orbits. **Comets** are chunks of frozen gases, ice, and dust that revolve about the Sun in highly elongated orbits. Although they are only a few kilometers in diameter, they become visible to us when they are heated by the Sun to produce sometimes spectacular atmospheres and tails millions of kilometers long (Figure 11.1). In contrast, **asteroids** are rocky objects located primarily between the orbits of Mars and Jupiter. They are never spectacular, being visible only through a telescope.

The number of still smaller objects revolving about the Sun—objects too small to observe with telescopes—is great indeed. These particles range from boulders to dust grains. When one of them encounters the Earth, it is heated by friction with the atmosphere and at least partially vaporized. Looking up at the night sky, we see a brief flash of light and perhaps a trail of luminous vapor that persists for a few seconds before fading back into blackness. This "falling star" or "shooting star" is properly called a **meteor.** If a part of the particle should survive its fiery plunge through the atmosphere to reach the surface, we call it a **meteorite.** Meteorites (discussed in Chapter 20) are valuable to scientists because they are samples of cosmic material that can be analyzed in detail in the laboratory.

(d) Origin of the Solar System

One of the objectives of studying the solar system is to understand how it formed, a topic we discuss in some detail in Chapter 20. One of the important things we have learned in our studies of the planets is that all members of the solar system formed together about 4.5 billion years ago. The Sun, the planets, and the smaller bodies of the system are the same age and share many basic properties. Nor is there anything special about the Earth; our planet shared the same birth as the other members of the system.

From the study of our own solar system, and by analogy with stars that we observe to be forming in other parts of our Galaxy, we conclude that the solar system formed from a rotating, hot cloud of gas and dust called the **solar nebula.** This solar nebula was composed primarily of hydrogen and helium, the two most abundant elements in the universe (and today the main constituents of our Sun). The central parts of this solar nebula condensed to form the Sun, while the planets and smaller bodies formed in a large rotating disk that surrounded the Sun. In the following chapters, we will refer repeatedly to the existence of this solar nebula, and in Chapter 20 we will discuss how the observed properties of the planets reveal something of the nature of the solar nebula itself.

11.2 DISCOVERIES OF THE PLANETS

Most ancient civilizations knew the five planets that are easily seen with the unaided eye—Mercury, Venus, Mars, Jupiter, and Saturn. The initial recognition of these objects is lost in the haze of prehistory. However, the three outer planets—Uranus, Neptune, and Pluto—are all *discovered* objects. In their times, each discovery of a new planet created widespread public interest and brought fame to the discoverer. Even today, we frequently read newspaper articles speculating on the possible existence of a tenth planet, or "Planet X."

(a) Discovery of Uranus

Uranus was discovered on March 13, 1781, by the German-English musician and amateur astronomer William Herschel, who was making a systematic telescopic survey of the sky in the constellation of Gemini. Herschel noted that through his telescope one object did not appear as a stellar point but seemed to be a small

TABLE 11.2 THE PLANETS

Name	Distance from Sun (AU)	Revolution Period (yr)	Diameter (km)	Mass (10^{23} kg)	Density (g/cm³)	Reflectivity
Mercury	0.39	0.24	4,878	3.3	5.4	0.1
Venus	0.72	0.62	12,102	48.7	5.3	0.7
Earth	1.00	1.00	12,756	59.8	5.5	0.5
Mars	1.52	1.88	6,787	6.4	3.9	0.2
Jupiter	5.20	11.86	142,800	18,990	1.3	0.5
Saturn	9.54	29.46	120,540	5,686	0.7	0.5
Uranus	19.18	84.07	51,200	869	1.2	0.5
Neptune	30.06	164.82	49,500	1,024	1.6	0.5
Pluto	39.44	248.6	2,200	0.01	2.1	0.4

disk. He believed it to be a comet and followed its motion for some weeks. Later, its orbit was computed and found to be nearly circular, lying beyond that of Saturn. Thus Herschel's "comet" was unquestionably a new planet.

Uranus can be seen by the unaided eye on a clear, dark night, but it is near enough to the limit of visibility that it is indistinguishable from a very faint star. It is so inconspicuous that its motion escaped notice until after its telescopic discovery. However, it turned out that Uranus had been plotted as a star on charts of the sky on at least 20 previous occasions between 1690 and 1781. These earlier observations were useful later in determining how perturbations were altering the planet's orbit.

Herschel proposed naming the newly discovered planet Georgium Sidus, in honor of George III, En-

FIGURE 11.1 Comet Halley, photographed from Mauna Kea Observatory in 1986. (University of Hawaii, courtesy Dale Cruikshank)

gland's reigning king. Others suggested the name Herschel, after the discoverer. The name finally adopted, in keeping with the tradition of naming planets for gods of Greek and Roman mythology, was Uranus, father of the Titans and grandfather of Jupiter.

The discovery of Uranus brought Herschel great fame and a full-time career in astronomy. As we shall see later in this book, Herschel became one of the most successful and influential astronomers in history, and he is usually credited with being the "father" of stellar astronomy. Often it is said that Herschel's discovery of Uranus was an accident. However, he disputed this accusation. At that time he was using the finest telescopes in the world, and he was undertaking a systematic survey of the sky. If a planet were there, he would have found it.

The discovery of Uranus also brought satisfaction to the German astronomer Johann Bode (1749–1826), because it fit beautifully into a sequence of numbers announced in 1766 by Daniel Titius that describe the approximate distances of the planets from the Sun. Bode had been so impressed with Titius' progression that he published it in his own introductory astronomy text. This sequence is known as the Titius–Bode law. It is obtained by writing down the numbers: 0,3,6, 12, . . . , each succeeding number in the sequence (after the first) being double the preceding one. If 4 is added to each number and the sum divided by 10, the resulting numbers are the approximate radii of the orbits of the planets in astronomical units, as can be seen in Table 11.3. The rule breaks down completely for Neptune and Pluto, but these planets were unknown in Bode's time. Now we recognize this "law" as just a coincidence, without physical basis. But it played an important role in the history of astronomy.

(b) Discovery of Neptune

Whereas the discovery of Uranus was unexpected, Neptune was found as the result of mathematical prediction. By the first decade of the 19th century, it was apparent that all was not well with Uranus. Even after allowing for the perturbing effects of the known

scope that could record a 12° by 14° area of the sky on a single photograph. The new camera went into operation in 1929, and the search was continued for the ninth planet.

Unfortunately, Gemini lies near the Milky Way, and some 300,000 star images were recorded on each exposure. It was an immense task to compare all the star images on each of two or more photographs of the same field in the hope of finding one image that changed position with respect to the rest, revealing itself as the new planet. The job was facilitated by the invention of the blink microscope, a device for comparing two different photographs of the same region of the sky. The operator's vision is automatically shifted back and forth between corresponding parts of the two photographs. If the star patterns are the same on the two plates, the observer sees a constant, although flickering, picture. However, if one object has moved slightly in the interval between the times the two plates were taken, the image of that object appears to jump back and forth as the view alternates between the two plates. In this way, moving objects can quickly be picked out from among the many thousands of star images.

In February 1930, Clyde Tombaugh (b. 1906) (Figure 11.4), comparing photographs made on January 23 and 29 of that year, found an object whose motion appeared to be about right for a planet far beyond the orbit of Neptune (Figure 11.5). It was within 6° of the position Lowell predicted for the unknown planet. Announcement of the discovery was made on Lowell's birthday, March 13, 1930. The new planet was named for Pluto, the god of the underworld. (Appropriately, the first two letters of Pluto are the initials of Percival Lowell; this is

about as close as one can come to naming a planet for a person.)

Although in 1930 the discovery of Pluto appeared to be a vindication of gravitational theory similar to the 19th-century triumph of Adams and Leverrier, we now know that Lowell's calculations were wrong. When the mass of Pluto was finally measured, it was found to be much less than that of the Moon. It could not possibly have exerted any measurable pull on either Uranus or Neptune. Recently, the Pioneer and Voyager spacecraft have penetrated beyond the orbit of Pluto, and they show no drift that might be attributed to an undiscovered mass. Further, a survey of the entire sky in the infrared region, carried out in 1983 by the Infrared Astronomical Satellite (IRAS), revealed no hidden "Planet X." Today it is generally accepted that the supposed perturbations of Uranus and Neptune are not, and never were, real.

11.3 SOME BASIC IDEAS IN PLANETARY SCIENCE

The study of the planetary system has gone through three distinct phases. From the most ancient times through the 18th century, the mystery of the motions of the planets dominated astronomy. As we have seen, Tycho's observations of planetary positions led to Kepler's determination of the laws of planetary motion and thus to Newton's brilliant synthesis and the foundation of modern science. This emphasis on celestial mechanics persisted until the evolution of the telescope led to the second, or astronomical, phase. For about the last century, planetary astronomers have concentrated their efforts on the study of the physical and chemical nature of the planets as determined by telescopic observation. The third phase is that of direct exploration of the planetary system by spacecraft, in which the traditional interests are supplemented by new fields of planetary geology, planetary meteorology, and space physics.

In the following chapters we shall consider the results of both telescopic and space-probe exploration of the individual planets and their satellite and ring systems. But first we introduce, in the remainder of this chapter, some basic ideas that are needed to understand the processes at work in the planetary system.

(a) Fundamental Characteristics of Planets

The most fundamental characteristics of a planet are its mass, its chemical composition, and its distance from the Sun. From these data alone, we can predict to some extent many of its other features, such as its surface temperature or whether or not it is likely to have an atmosphere. We have already described how to measure the mass and location of a planet, and Section 11.4 is devoted to an overview of their chemical compositions.

FIGURE 11.4 Clyde Tombaugh, at the time of his discovery of the planet Pluto in 1930. (Lowell Observatory, courtesy Robert Millis)

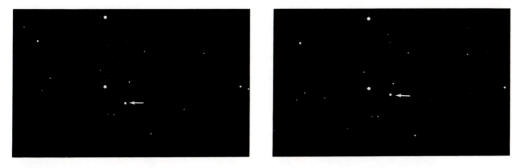

FIGURE 11.5 Two photographs of Pluto showing its motion among the stars in a 24-hr period, photographed with the 200-in. Hale telescope. (Caltech/Palomar Observatory)

(b) Rotation

All the planets are observed to rotate. Astronomers use several different techniques to determine their rotation rates.

1. The most direct method is to watch permanent surface features move across the disk. Even naked-eye observations are sufficient to show that the Moon always keeps the same face toward the Earth, that is, that the rotation period of the Moon is equal to its period of revolution about the Earth. Mars' rotation period of a little over 24 hr was observed telescopically more than a century ago, and today most of the known rotation rates for planets and satellites have been derived by this technique, using either telescopic or spacecraft data. If the object is too small to show individual surface features, one can look for a periodic variation in brightness as it rotates. This approach has given us our value for the rotation period of Pluto, and it yielded rotation periods for many of the satellites of Jupiter and Saturn long before the arrival of the first spacecraft.

2. Doppler radar was first used to derive the rotation periods of Mercury and Venus, although the results were later verified using method 1. Radar waves at a single frequency are beamed to a planet, and the reflected signal is measured. These reflected waves are Doppler-shifted according to the line-of-sight velocity of each part of the target with respect to the transmitter on Earth. Waves reflected from the approaching edge are thus shifted to shorter wavelengths than those that are reflected from the center of the disk, and the signal from the receding part of the planet's surface is shifted to longer wavelengths. Measurement of the range of wavelengths in the reflected signal (its bandwidth) thus gives a measure of the rotation rate of the target (Figure 11.6).

3. The gaseous giant planets do not reflect radar waves, and observations of the motions of markings yield only the wind velocities in the atmosphere. The true rotation rate of the underlying planet can be found, however, if there is a magnetic field generated in the core, since this field rotates with the core. The rotation rates of Jupiter, Saturn, Uranus, and Neptune are all defined by the measured rotation rates of their magnetic fields.

(c) Surface Temperature

Since the planets obtain most of their energy from sunlight, their temperatures depend primarily on their distance from the Sun. The closer to the source of energy, the warmer the planet should be. It is relatively straightforward to calculate a characteristic or **effective temperature** for any particular distance from the Sun, and these values provide at least a first approximation to the expected temperatures on individual planets or satellites.

The effective temperature corresponds to a balance between sunlight striking a planet and emission of infrared radiation back into space. At the Earth (a distance of 1 AU from the Sun), the power received from the Sun is 1,370 watts for each square meter of surface. The corresponding temperature for a dark, perfectly absorbing material (a blackbody) to radiate an exactly equal quantity of energy is just under 400 K. However, the Earth is not a blackbody but a real rotating planet with oceans and atmosphere. It reflects part of the sunlight back into space, and it distributes heat from the sunlit regions to the cooler poles and to the night side. The average surface temperature on the Earth is actually about 280 K, a little above the freezing point of water (273 K). Without the atmosphere, the temperature would be nearer to 260 K and the oceans would freeze.

The incident sunlight varies with distance according to the inverse-square law. As we saw in Section 7.2b, the emission of energy from a hot surface is proportional to the temperature to the fourth power (the Stefan-Boltzmann law). A little algebra should convince you, therefore, that the effective temperature varies as the inverse square root of the distance from the Sun:

those encountered in the Sun and stars. Matter at high temperatures is relatively easy to understand, since it consists of individual atoms or fragments of atoms in the gaseous state. But under planetary conditions, atoms interact with one another to produce molecules and minerals, and we must deal with the complexities of *chemistry*.

(a)   Five Kinds of Matter

In a quick overview of planetary compositions, we can simplify the situation by considering just five characteristic kinds of matter:

1. *Fluid.* The gaseous form of matter is well known to us. It forms the atmospheres of planets, and we often speak of the giant planets as composed primarily of the gases hydrogen and helium. At the pressures found within the giant planets, however, these gases undergo a transformation to the liquid state. Most of the matter in the planetary system is fluid: either gas or liquid.

2. *Plasma.* As we saw in Chapter 7, a plasma is a dilute hot gas composed of ionized atoms: basically, positively charged ions and negatively charged electrons. Unlike the motion of electrically neutral gas, a plasma's motion is responsive to magnetic and electric fields. The solar wind that streams through interplanetary space is a plasma, as are the charged atomic particles trapped in the magnetic fields of planets.

3. *Ice.* The cosmically abundant elements hydrogen, oxygen, carbon, and nitrogen all form simple compounds that freeze into solids at the temperatures of the outer solar system. The most important of these is water ice (H_2O), but significant quantities are also expected of ices of NH_3, CH_4, CO_2, and CO. Ices form the main building blocks of comets, and they may also contribute most of the mass to the cores of the giant planets.

4. *Rock.* The next most abundant materials are the rocks, which consist of more complex compounds of silicon, oxygen, magnesium, calcium, sulfur, carbon, iron, and other elements. Rock constitutes the main building blocks of the inner planets and the asteroids.

5. *Metal.* Most metallic elements readily form compounds with oxygen and thus contribute to the rocky material. However, there are places in the planetary system, particularly in the cores of planets, where metal is separated from rock and exists in the pure state. The two most abundant metals in the core of the Earth are iron and nickel.

As we look at individual planetary bodies we shall see that they are composed of various proportions of metal, rock, ice, and fluid, all immersed in a sea of interplanetary plasma.

(b)   Oxidized and Reduced Environments

Because oxygen and hydrogen are abundant and chemically reactive elements, they tend to dominate the chemistry of the solar system. Much of the chemical evolution of a planet therefore depends on the relative proportions of these two elements.

On the Earth, hydrogen is relatively rare, since this light gas escapes easily from the upper atmosphere (Section 11.5). Oxygen therefore dominates, and we live in an **oxidized** environment. Most of the rocks that make up the crust of the Earth are composed of various compounds of oxygen, and there is oxygen to spare in our atmosphere. If hydrogen or hydrogen-rich compounds are introduced on Earth, they are quickly broken down by chemical interactions with oxygen. Pure metal is also unstable and oxidizes, as we see by the rusting of iron or the tarnishing of silver.

All of the terrestrial planets are chemically oxidized to various degrees, but only the Earth has free oxygen in its atmosphere. As we will see in Chapter 12, the presence of oxygen is the direct result of photosynthetic life on our planet.

When there are two or more hydrogen atoms present for each atom of oxygen, the hydrogen dominates and the chemical environment is said to be **reduced.** Any available oxygen combines with hydrogen to produce water (H_2O), and the leftover hydrogen combines with other elements to produce an entirely different set of compounds. Among these are ammonia (NH_3) and the hydrocarbons (compounds of hydrogen and carbon).

The giant planets all have chemically reduced atmospheres with plentiful free hydrogen. Their visible clouds are composed of ammonia (NH_3) or methane (CH_4) crystals, and their spectra show the presence of many hydrocarbons. Thus most of the material in the planetary system is characterized by a reducing chemistry.

(c)   Rocks and Minerals

Rocks are mixtures of compounds composed in part of the elements silicon and oxygen. They are made up of assemblages called *minerals.* Unlike a mineral, which consists of a single compound, a rock is typically much more heterogeneous. Examples of minerals include silicates such as quartz (SiO_2), metallic oxides such as hematite (Fe_2O_3), sulfides such as iron pyrite (FeS_2), and carbonates such as calcite ($CaCO_3$).

Rocks can be classified by their minerals, but a much simpler system is to categorize them according to their history. On Earth, three such classes are commonly used:

1. *Igneous rock*, which formed by the cooling of molten material (Figure 11.8). The two most abundant kinds of igneous rock on the surface of the Earth are **basalt,** the lava that makes up the ocean floors, and **granite,** the most common continental rock.

2. *Sedimentary rock*, which formed by the deposition of fragments of igneous rock or of living organisms. On Earth, these include the common sandstones, shales, and limestones.

3. *Metamorphic rock*, produced by the chemical and physical alteration of igneous or sedimentary rock at high temperature and pressure. Metamorphic rock is produced on Earth because geological activity carries surface rock to considerable depths and then brings it back up to the surface. On less active planets, metamorphic rock should be rare.

There is a fourth very important category of rock not represented on the Earth or Moon that can tell us much about the early history of the planetary system— **primitive** rock:

4. *Primitive rock*, which has largely escaped modification by heating. Primitive rock represents the original material out of which the planetary system was made. There is no primitive material left on the terrestrial planets because these planets were heated above the melting point of rock early in their history. To find primitive rock, we must look to smaller objects, such as comets and asteroids. Fragments of primitive material also reach the Earth in the form of some meteorites.

A piece of marble on Earth is composed of materials that have gone through all four of these stages. Beginning as primitive material, it was heated in the early Earth to form igneous rock, subsequently eroded and redeposited (perhaps many times) to form sedimentary rock, and finally transformed several kilometers below the Earth's surface into the hard, white metamorphic stone we see today.

(d) Planetary Interiors

The study of the interiors of planets is a difficult subject. Our knowledge of the interior of the Sun is more advanced than our knowledge of the interior of our own planet. It is easier to explore the surfaces of the other planets than to penetrate even a few kilometers toward the interior of the Earth.

There are indirect ways, however, to probe the interiors of our own and other planets. One of the simplest clues to the composition of the interior, for example, is given by a calculation of the *density* of the planet (mass/volume). Consider the five kinds of matter discussed in Section 11.4a. Ices have densities typically near 1.0 g/cm^3. Rocks are much more dense, usually in the range 2.5 to 3.5 g/cm^3. And metals are denser yet, often more than 8 g/cm^3. A knowledge of the density of a planet may be enough to estimate the relative proportions of these three components.

As an example, we can look at the densities of the Earth, the Moon, and the planet Pluto. The Moon's density is 3.3 g/cm^3, within the range for rock. Since we know the Moon is too warm to contain ice, we conclude that it is predominantly composed of rock, with at most a small metal core. The Earth's higher density

FIGURE 11.8 A flow of lava, in the process of "freezing" to create new basaltic rock. (David Morrison)

TABLE 12.1	BASIC PROPERTIES OF THE EARTH
Semimajor axis	1.00 AU
Revolution period	1.00 yr
Diameter	12,756 km
Mass	5.98×10^{24} kg
Density	5.5 g/cm³
Uncompressed density	4.5 g/cm³
Surface gravity	9.8 m/s²
Escape velocity	11 km/s
Rotation period	$23^h\, 56^m$
Surface area	5.1×10^8 km²
Atmospheric pressure	1.00 bar
Atmospheric composition	N_2(78%), O_2(21%)

experience is with the outermost skin of the Earth's crust, a layer no more than a few kilometers in depth. All of the information we have about the bulk properties of the Earth has been deduced indirectly. It is important to remember that we know less about our own planet a few kilometers beneath our feet than we do about the surfaces of Venus or Mars.

The surface rocks of the Earth have densities mostly in the range of 2.5 to 3.0 g/cm³. However the density of the planet as a whole is 5.5 g/cm³. Because of this high overall density, we conclude that the interior of the Earth is very dense indeed, and therefore that its composition is probably quite different from the familiar rocks of the crust.

The above argument is not sufficient, however, to demonstrate beyond a doubt that the composition of the interior of the Earth is different from that of the observable crust. The weight of the various layers of the Earth causes the pressure to increase inward, compressing the materials in the interior and increasing their density. It is necessary, therefore, to use experimental measurements of the physical properties of rocks to see if they can be compressed sufficiently to explain the apparent increase in density of the Earth with depth.

Such studies show that rock is not sufficiently compressible, and they lead us to an interesting property called the **uncompressed density** of the Earth. The uncompressed density, which is equal to 4.5 g/cm³, is the density that an average piece of our planet would have if it were not under high pressure. It is the uncompressed density that should be compared with the densities of various materials to estimate composition. To achieve the average uncompressed density of 4.5 g/cm³, the interior of the Earth must include high-density material as well as rock. Since the cosmically most abundant such material is metallic iron, we conclude that the interior of the Earth is enriched in iron and perhaps other metals relative to crustal rocks.

(b) Structure

The structure of the planet can be studied using the transmission of seismic waves—large-scale vibrations—through its interior. As we shall see in Chapter 27, similar methods are being used to investigate the interior structure of the Sun. On Earth, seismic waves are produced by earthquakes, which generate vibrations from the sudden slippage of parts of the crust. Some of these vibrations travel along the surface; others pass directly through the interior. The seismic vibrations are recorded by delicate instruments and analyzed to determine the paths and velocities of the waves through the Earth's interior.

Seismic studies have shown that most of the interior of the Earth is solid, and that it consists of several distinct layers of different composition (Figure 12.2).

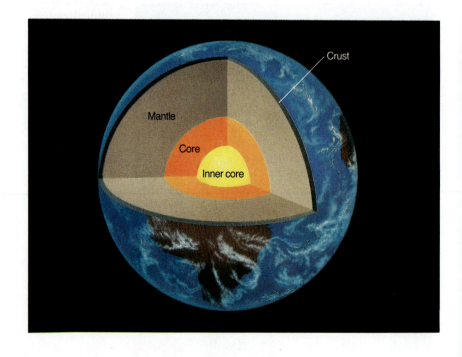

FIGURE 12.2 Interior structure of the Earth.

The uppermost layer is the **crust.** The crust under the ocean, which covers 55 percent of the surface, is typically about 8 km thick and is composed of *basalts.* The continental crust, which covers 45 percent of the surface, is from 20 to 70 km thick and is predominantly made of *granite.* On both ocean floors and continents, these igneous rocks are often buried by sedimentary and metamorphic rocks produced by weathering and erosion of the surface material. Although the crust is the part of the Earth we know best, it is important to remember that it makes up only about 0.3 percent of the mass of the Earth.

The major part of the Earth is the **mantle,** which stretches from the base of the crust down to a depth of 2900 km. The density in the mantle increases downward from about 3.5 g/cm^3 to more than 5 g/cm^3; however, its uncompressed density is everywhere about 3.5 g/cm^3. Its composition is believed to be igneous silicate rocks. Samples of material from the upper mantle are occasionally ejected from volcanoes, permitting a detailed analysis of its chemistry.

Within the mantle is the **core** of the Earth, a high-density region with a diameter of 7000 km, substantially larger than the planet Mercury. We surmise that the outer part of the core acts like a liquid, for it does not transmit certain kinds of seismic waves. The innermost part of the core (about 2400 km in diameter) is extremely dense and probably solid. The primary constituent of both parts of the core is believed to be iron, probably also containing substantial quantities of nickel, sulfur, and other cosmically abundant elements.

At the base of the mantle, the temperature is 4500 K and the pressure is 1.3 million bars, where 1 **bar** is defined as 10^5 pascals (the metric unit of pressure), or approximately the atmospheric pressure at the surface of the Earth. At the center of the planet, the pressure is nearly 4 million bars (4×10^9 pascals), but the temperature is about the same, approximately 5000 K. The interior is heated by the decay of radioactive elements in the mantle and crust. In the mantle, this heat is carried upward by **convection,** the slow rising of currents of hotter material. In the crust, heat escapes upward to the surface by **conduction** through the solid rock or by the release of molten lava in volcanic eruptions. We are all familiar with conduction as the process that transmits heat from the bottom of a metal pan on the stove to its handle. We will examine some consequences of this release of energy from the interior in Section 12.2.

Table 12.2 summarizes the interior structure of the Earth.

(c)   The Earth's Magnetism

Additional clues concerning the interior are provided by the Earth's magnetic field. This field is similar to that produced by a bar magnet. Many students are familiar with the way iron filings align themselves along the lines of force that extend between the north and south poles of a bar magnet. Between the magnetic poles of the Earth stretch similar lines of force along which compass needles align. The magnetic poles, however, are not coincident with the rotational poles, but are tilted by a few degrees. The overall strength of the Earth's magnetic field is fairly weak, averaging about one-half gauss at the surface (where the gauss is the metric unit of magnetic field strength).

The Earth's magnetism results from electric currents moving in the core of the planet. Being composed of metal, the core is electrically conducting. The rotation of the Earth generates slow motions in the metallic core that act like a giant dynamo, generating the observed field. Because these motions are turbulent, the strength and alignment of the field vary, and the magnetic poles wander about with respect to geographical position.

In addition, the polarity of the field reverses itself completely from time to time. We detect these changes in polarity from the direction of magnetism preserved in igneous rocks. When the rock is molten, its iron-bearing compounds are weakly magnetized by the Earth's field; when it solidifies, this alignment is "frozen in." We measure the strength and polarity of magnetism in rocks of different ages to trace the magnetic history of the Earth and find that its field has reversed polarity about 100 times in the past 50 million years. The cause of these field reversals is not well understood.

The Earth's magnetic field extends into surrounding space. Above the atmosphere, this field is able to trap small quantities of plasma (mostly from the solar wind), creating the Earth's magnetosphere (Figure 12.3). The **magnetosphere** is defined as that region surrounding the planet within which our magnetic field dominates over the weak interplanetary field that originates in the Sun. Since the size of the magnetosphere depends on the strength of the Earth's field relative to the strength of the interplanetary field, it can expand or contract, depending on the value for the Earth's field strength and the level of solar activity. Typically, the Earth's magnetosphere extends about 60,000 km, or ten Earth radii, in the direction of the Sun. It is shaped like a wind sock pointing away from the solar wind. In the downstream direction, the magnetosphere can reach as far as the orbit of the Moon.

TABLE 12.2	STRUCTURE OF THE EARTH		
Region	Radius (km)	Composition	Mass (%)
Core	3500	Iron, nickel	33
Mantle	6370	Silicates	67
Crust	6378	Granite, basalt	0.3

The existence of the magnetosphere was discovered in 1958 by instruments on the first U.S. Earth satellite, Explorer 1, which recorded the plasma trapped in its inner part. The regions of high-energy trapped plasma are often called the Van Allen belts in recognition of the University of Iowa professor, James Van Allen, who built the scientific instrumentation for Explorer 1 and correctly interpreted the satellite measurements. Van Allen's name was thus given to the largest feature of our planet. Audacious students have been known to ask him for his belt so that they can own a "Van Allen belt."

Since 1958, hundreds of satellites have explored various regions of the magnetosphere, and a scientific discipline called space plasma physics, which is devoted to understanding magnetospheric phenomena, has developed. We will discuss planetary magnetospheres in more detail in Chapter 16, when we look at the much larger magnetospheres of the giant planets.

12.2   CRUST OF THE EARTH

(a)   Geological Processes

The study of the Earth's crust and the processes that modify it is called *geology*. Most geological structures can trace their origin to the effects of heat escaping from the interior of the planet. Geologists call these **endogenic** processes, meaning that their source is from the inside. One familiar example of an endogenic process is volcanism. Internal activity can also result in the compression or expansion of the crust. The endogenic activity associated with such crustal forces is called **tectonic.** Compression or expansion can cause fractur-

ing of the surface or can pile up material to build mountains.

The dominance of endogenic geological processes is a special property that sets the Earth apart from most other planetary bodies, which are much more influenced by **exogenic,** or external, processes. The most important exogenic features on other planets are impact craters, produced by collisions with comets, asteroids, or other space debris.

Impact craters are also produced on Earth. If the Earth's level of internal geological activity were as low as that of the Moon, for instance, it would have as many impact scars as does our cratered satellite. There are few identifiable impact craters on Earth, primarily because of its high level of endogenic geological activity with associated water and ice erosion. This activity has destroyed all evidence of craters produced early in the history of the planet. Even relatively young craters are erased within a few million years of their formation by a combination of erosion and weathering.

The most prominent impact crater on the Earth is Meteor Crater (Figure 12.4), a mile-wide scar in the arid plains of northern Arizona. Meteor Crater was produced 50,000 years ago by the impact of an iron meteoroid with a mass of about a million tons, creating an explosion equivalent to a 20-megaton nuclear bomb. Larger craters are older and more eroded. Altogether, several hundred craters have been found on Earth, but most are so badly eroded that they can be identified only by an expert geologist. We will defer the detailed discussion of impact cratering to the next chapter, when we investigate the Moon. Additional information on the cratering of the Earth, past and present, is in Chapter 18. In this chapter, we concentrate on the internally driven geology that is so characteristic of our planet.

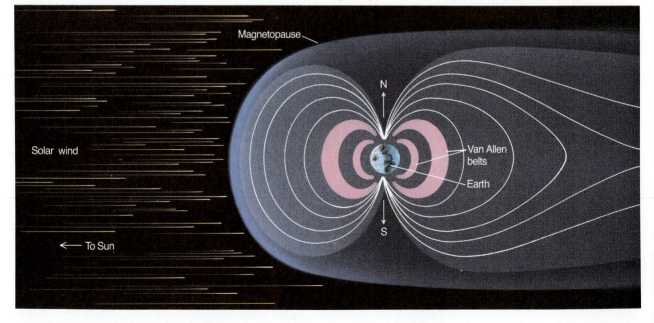

FIGURE 12.3 Cross-section of the Earth's magnetosphere and the Van Allen belts, as revealed by numerous spacecraft.

FIGURE 12.4 Meteor Crater in Arizona. This relatively fresh terrestrial impact crater about a mile in diameter was formed 50,000 years ago. Older craters, especially in wetter climates, are quickly eroded and usually become unrecognizable after a few million years. (Meteor Crater, Northern Arizona)

(b) Plate Tectonics: A Geological Revolution

Many schoolchildren, in studying maps or globes of the Earth, notice that North and South America, with a little juggling, look as if they could almost be nestled up against Europe and Africa. It seems as if these great land masses were once together and somehow tore apart. The same idea also occurred to the German meteorologist Alfred L. Wegener (1880–1930) early in the 20th century. He looked at the matter in considerable detail, making a good empirical case that the continents had drifted apart. Wegener based his arguments on detailed geological similarities between the east and west shores of the Atlantic Ocean. At that time, however, the mantle and crust of the Earth were known to be solid, and Wegener could propose no plausible mechanism by which the continents could move. His ideas of *continental drift* were therefore dismissed and even ridiculed by nearly all of the scientific community, even though additional evidence accumulated over the following decades that supported past connections between land masses that are now widely separated.

Wegener had proposed that the continents somehow moved through the fixed, underlying basaltic crust. This idea doesn't work. Continental drift could be accepted only when a new theory of the crust of the Earth was developed, called **plate tectonics.** In this theory, we recognize that the oceanic crust also moves, driven by the slow convection currents that transport heat within the mantle. The crust and upper mantle (to a depth of about 60 km) are divided into more than a dozen major plates that fit together like the pieces of a jigsaw puzzle, but these plates are also capable of moving slowly with respect to one another. This mobile part of the Earth is called the **lithosphere** (Figure 12.5.)

The direct evidence for moving lithospheric plates was first supplied in the Atlantic Ocean, where geologists in the 1960s demonstrated that fresh lava was being injected along the Mid-Atlantic Ridge, a line of volcanic mountains running approximately north and south along the center of the ocean basin. At the same time, geologists found that on either side of the ridge the ocean floor was gradually separating at a speed of a few meters per century. At this rate, the entire Atlantic

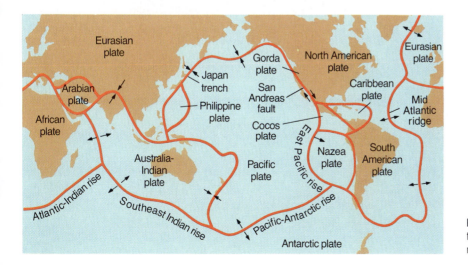

FIGURE 12.5 Tectonic plates on the Earth. Arrows indicate the motion of the major plates.

Ocean, which is about 4000 km across, could have formed within 100 million years, strongly in agreement with the evidence for geological continuity across the Atlantic accumulated by Wegener and others. Very quickly, other plate boundaries were located where similar sea-floor spreading was taking place. A great many previously unconnected geological phenomena were seen to make better sense in the context of the new plate tectonics. Within a few years, a fundamental change in our geological perspective took place—a true "scientific revolution."

But how can a system of interlocking plates covering the entire surface of the Earth move about? If plates in the Atlantic and elsewhere are spreading apart, they must be jamming together somewhere else. The analysis of the interactions of moving plates provides the basis for understanding most large-scale geological activity on Earth.

Four basic kinds of interactions between plates are possible: (1) they can pull apart, (2) one plate can burrow under another, (3) they can slide alongside each other, or (4) they can jam together. Each of these activities is important in determining the geology of the Earth.

(c)   Rift and Subduction Zones

Plates pull apart from each other along **rift zones,** such as the Mid–Atlantic Ridge, driven by upwelling convection currents in the mantle (Figure 12.6a). A few rift

zones are found on land, the best known being the central African rift, an area in which the African continent is slowly breaking apart. Most rift zones, however, are in the oceans. The new material that rises through the crust to fill the space between the receding plates is basaltic lava, the kind of igneous rock that forms most of the ocean basins.

Considerable heat is released at the midocean rifts, not all of it in the form of volcanic eruptions. One of the most dramatic discoveries about our planet in recent years has been the existence of colonies of remarkable lifeforms that cluster around the hot, mineral-laden springs that result when seawater circulates through the superheated rift rocks. These hydrothermal vents in the dark ocean depths represent one of the few places on our planet where life has learned to obtain its nutrients and energy directly from geological activity, without dependence on sunlight or the organic products of other living creatures.

From a knowledge of sea-floor spreading, we can calculate the average age of the oceanic crust. About 60,000 km (6×10^7 m) of active rifts have been identified, with average separation rates of about 4 m per century (4×10^{-2} m per year). The new area added to the Earth each year is therefore 6×10^7 m $\times 4 \times 10^{-2}$ m per year, or just over 2 km^2. The total area of ocean crust is 260 million km^2. Dividing the total area by the new area added each year, we obtain about 100 million years as the average age of the oceanic crust. This is a very

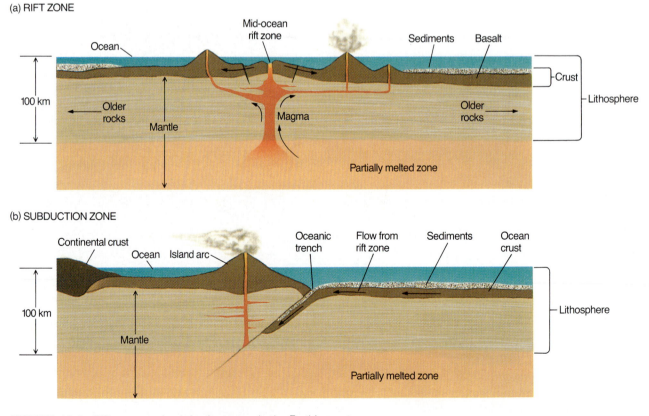

(a) RIFT ZONE

(b) SUBDUCTION ZONE

FIGURE 12.6 Rift zones and subduction zones in the Earth's crust.

short interval in geological time, less than 3 percent of the age of the Earth. The present ocean floors are among the youngest features on the planet!

If plates are spreading at rift zones, they must also be colliding at other locations. When one plate at such a boundary slides under another, the region is called a **subduction zone** (Figure 12.6b). Generally, continental masses cannot be subducted, but the thin oceanic plates can be rather readily forced down into the upper mantle. Often a subduction zone is marked by an oceanic trench, a fine example being the deep Japan Trench along the coast of Japan. Approximately the same total area—2 km²—is subducted each year as is created by sea-floor spreading.

The subducted plate is forced down into regions of high pressure and temperature, eventually melting several hundred kilometers below the surface. Its material is recycled into a downward-flowing convection current, ultimately balancing the material that rises along rift zones.

A part of the subducted material reaches the surface more directly through volcanic eruptions. All along the subduction zone, earthquakes and volcanoes mark the death throes of the plate. Some of the most destructive earthquakes in history have taken place along subduction zones. These include the 1923 Yokohama earthquake and fire, which killed 100,000 Japanese, and the 1966 quake that killed half a million Chinese in the city of Tangshan.

When you think about it, the concept of plate tectonics is quite remarkable. The maps of the Earth's continents that you learned in school are not permanent. The ground beneath your feet moves. On geological time scales, the planet is in a constant state of change, responding to awesome interior forces that dwarf anything generated by human engineering.

(d) Fault Zones

Along much of their lengths, the crustal plates slide along parallel to each other. These plate boundaries are marked by cracks or **faults.** Along active fault zones, the motion of one plate with respect to the other is several meters per century, about the same as the spreading rates along rifts.

One of the most famous faults is the San Andreas Fault, lying on the boundary between the Pacific Plate and the North American Plate. This fault runs from the Gulf of California in the south to the Pacific Ocean just west of San Francisco in the north (Figure 12.7). The Pacific Plate, to the west, is moving northward, carrying Los Angeles, San Diego, and parts of the southern California coast with it. In several million years, Los Angeles will be an island off the coast of San Francisco.

Unfortunately for us, the motion along most fault zones does not take place smoothly. The creeping

motion of the plates against each other builds up tectonic stresses in the crust that are released in sudden, violent slippages, generating earthquakes. Since the average motion of the plates is constant, the longer the interval between earthquakes, the greater the stress and the larger the energy released when the surface finally moves. For example, the part of the San Andreas Fault near the central California town of Parkfield has been slipping about every 22 years, moving an average of about 1 m (5 cm/yr × 22 years) (Figure 12.8).

Potentially the worst situation occurs in southern California. The average interval between major earthquakes in the Los Angeles region is about 140 years, and the average motion is about 7 m. The last time the San

FIGURE 12.7 The San Andreas Fault, a very active region where one crustal plate is sliding sideways with respect to the other. The fault is marked by the valley running up the center of this photo; the dark line to the left is tumbleweeds piled along a fence line. (USGS)

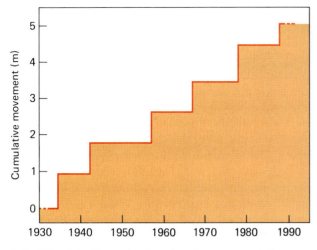

FIGURE 12.8 Example of earthquake movement along the San Andreas Fault. The measured displacement of the Pacific Plate relative to the North American Plate is shown in the Bear Valley area south-east of Monterey, California. The typical interval between earthquakes is about a decade on this section of the fault. (Data courtesy of William Ellsworth, USGS)

FIGURE 12.9 The Himalayan Mountains and the Tibetan Plateau are the highest region on the Earth, produced by tectonic forces resulting from the collision of India with Asia. (David Morrison)

Andreas Fault slipped in this area was in 1857; tension has been building ever since, and sometime soon it is bound to be released. For all of the damage it caused, the 1994 Los Angeles quake did little to relieve this accumulating stress.

(e)   Mountain Building

Subduction can take place only at boundaries involving thin oceanic plates. When two continental masses are brought together by the motion of the crustal plates, neither can slip under the other. The Earth buckles and folds, forcing some rock deep below the surface and raising other folds to heights of many kilometers. This is the way most of the mountain ranges on Earth were formed. As we will see, however, quite different processes produced the mountains on other planets.

The highest mountain range on Earth, the Himalayas, is still being formed as the Indian subcontinent is forced against the Asian mainland (Figure 12.9). India was formerly a large island near Australia in the Indian Ocean, but during the past few million years it has been forced up against the Eurasian Plate. The Alps are the result of the African Plate's bumping into Europe. In North America, the Rocky Mountains probably also are the product of pressure between two plates, but this boundary is not very active at present, and there is some debate among geologists concerning exactly what is happening here.

At the same time a mountain range is being formed by upthrusting of the crust, its rocks are subject to the erosional force of water and ice (Figure 12.10). The

sharp peaks and serrated edges characteristic of our most beautiful mountains have little to do with the forces that make them, but are instead the result of the processes that tear them down. Ice is an especially effective sculptor of mountains. In a planet without moving ice or running water, mountains will tend to remain smooth and dull. This is exactly the case on our Moon, as we shall see in the next chapter.

(f)   Volcanoes

Terrestrial volcanoes mark the locations where subsurface molten rock, called **magma,** rises to the surface. One example is the midocean ridges, which are volcanic features formed by mantle convection currents at plate boundaries. A second, major kind of volcanic activity is associated with subduction zones, and volcanoes sometimes also appear in regions where continental plates are colliding.

Another location for volcanic activity on our planet is found above so-called mantle hot spots, areas far from plate boundaries where heat is nevertheless rising from the interior of the Earth. Perhaps the best-known such hot spot is under the island of Hawaii, where it supplies the energy to maintain three currently active volcanoes, two on land and one under the ocean. The Hawaii hot spot has been active for at least a hundred million years, and it has generated a 3500-km-long chain of volcanic islands (Figure 12.11). The tallest Hawaiian volcanoes are among the largest single volcanic features on Earth, up to 100 km in diameter and rising 9 km above the ocean floor. However, as we shall see, other planets have volcanoes that considerably surpass these size records.

The fact that the Hawaiian volcanoes form a long chain of islands rather than one gigantic volcano is the

FIGURE 12.10 The Alps, a young region of the Earth's crust where sharp mountain peaks are being sculpted by glaciers. (David Morrison)

direct result of the motion of the Pacific Plate, which slides over the hot spot at a rate of several centimeters per year. In 1 or 2 million years, the plate moves a distance equal to the average separation of the Hawaiian Islands. Thus, the island chain represents a time sequence, with each major island a few million years older than the next. A couple of million years from now, all the currently active volcanoes will have fallen silent, and a new volcanic island will be forming to the southeast of the present island of Hawaii.

Volcanic eruptions on Earth can take several forms. The Hawaiian volcanoes are called **shield volcanoes** because of their shape, characterized by long, gradual slopes built up by successive flows of fluid lavas (Figure

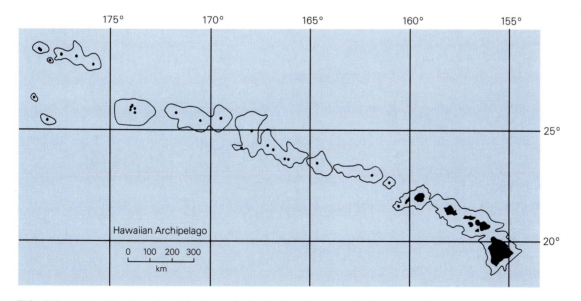

Hawaiian Archipelago

0 100 200 300
km

FIGURE 12.11 The Hawaiian Islands, a chain of volcanic mountains formed during the past 100 million years by movement of the Pacific Plate over a mantle hot spot. The figure shows islands and the contour at a 10,000-ft depth. Three volcanoes are currently active at the southeast end of the chain.

FIGURE 12.12   Eruption of Mauna Loa, a shield volcano in Hawaii. Highly fluid basaltic lava in thin, overlapping flows produces the gentle slopes and smooth profiles of shield volcanoes. The constellation of the Southern Cross is visible to the left of the volcanic plume. (Dale P. Cruikshank)

FIGURE 12.13   A "fire-fountain" of lava, building up a steep-sided cinder cone. Even more explosive eruptions are responsible for forming such cone-shaped volcanoes as Fuji in Japan or Etna in Italy. (USGS photo by J.D. Griggs)

12.12). In contrast, there are other more explosive volcanoes (Fuji in Japan, Vesuvius in Italy, and Mt. St. Helens in Washington State are well-known examples) that are characterized by steep-sided cones created by the fall-back from violent fire-fountains of molten lava (Figure 12.13). Various types of conical volcanoes are called cinder cones, stratovolcanoes, or composite volcanoes. Both shield volcanoes and conical volcanoes have been identified on other planets as well.

If the lava is very fluid and is erupted rapidly, no mountain is formed, but instead vast plains of basalt called **flood basalts.** These flood basalt eruptions are the largest in terms of volume of lava produced, and they also have their counterparts on other planets, particularly in the formation of the lunar lava plains. A terrestrial example of such an eruption is the 60 million-year-old Columbia River flood basalts that cover eastern Washington State and have a total volume estimated at 400,000 km^3.

(g)   Weathering, Erosion, and Sedimentation

While most terrestrial landforms can ultimately trace their origin to the deep internal processes discussed above, their details are often the result of local weathering and erosion. Water, wind, and ice all mod-ify the surface of our planet, as does the chemical interaction with the atmosphere and oceans. In particular, almost all sharply sculpted features, from mountain peaks to deep canyons, are the product of water and ice erosion.

FIGURE 12.14   Sand dunes in Death Valley National Monument. (David Morrison)

Water and ice do their work in arid climates as well as wet ones. Indeed, the lack of vegetation in arid lands often makes the ground more susceptible to erosion when occasional big rainstorms or floods occur. Do not be surprised, therefore, that the spectacular rock formations of the American Southwest or the Arabian Desert are primarily the result of water erosion. Even in such desert climates, wind rarely has a major effect on rock, compared with water. However, wind is effective in transporting dust and sand, which are produced by weathering of rock. The most common landforms sculpted by wind are sand dunes in the great sand seas of the world (Figure 12.14).

The fine material eroded away from the mountains of the Earth or produced by chemical weathering of the crust must go somewhere. A part ends up in the desert sand seas, but most is carried by water into low-lying areas or into the sea itself. These sediments cover much of the igneous rock on both the continents and the oceans, and eventually they fuse into new sedimentary rock. Typically, the sedimentary deposits on the ocean floors are more than a kilometer thick. Of course, this sediment is recycled along with the oceanic basalt in the subduction process, some to replenish the mantle and some to reappear on the surface by volcanic eruption.

12.3   THE EARTH'S ATMOSPHERE AND OCEANS

(a)   Structure of the Atmosphere

We live at the bottom of the ocean of air that envelops our planet. The atmosphere, weighing down upon the surface of the Earth under the force of gravitation, exerts a pressure at sea level of 1 bar, which equals the weight of 1.03 kg over each square centimeter, or 10.3 tons/m^2. If the mass of the air over 1 m^2 is 10.3 tons, the total mass of the atmosphere may be found by multiplying this figure by the surface area of the Earth in square meters. We find, thus, that the total mass of the atmosphere is about 5×10^{15} tons, or about one-millionth the total mass of the Earth.

At higher and higher altitudes, the air thins out until it disappears into the extremely sparse gases of the magnetosphere, at an altitude of several hundred kilometers (Figure 12.15). In the thin upper reaches of the atmosphere, we can observe the **aurora,** or polar lights, which are produced when ions and electrons from the magnetosphere bombard the upper air. Most auroras are found to occur at heights of 80 to 160 km, but a few are as high as 1000 km.

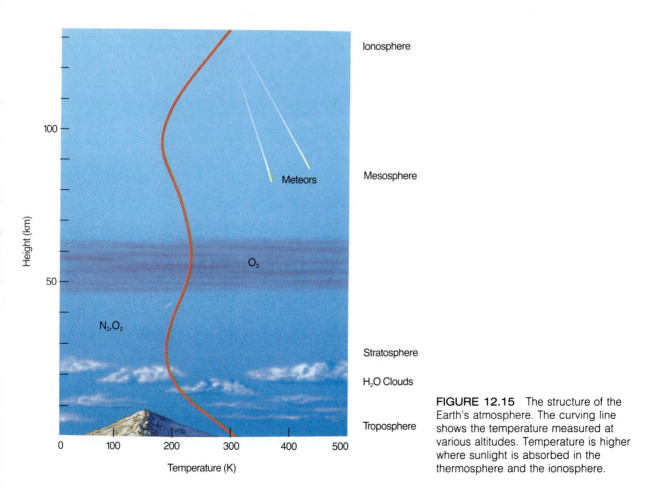

FIGURE 12.15   The structure of the Earth's atmosphere. The curving line shows the temperature measured at various altitudes. Temperature is higher where sunlight is absorbed in the thermosphere and the ionosphere.

Most of the atmosphere is concentrated in the **troposphere.** That is where clouds form and airplanes fly. The troposphere is characterized by convection currents produced as warm air, heated by the surface, rises and is replaced by descending currents of cooler air. Within the troposphere, temperature drops rapidly to values near $-50°C$ (which is also, coincidentally, about $-50°F$) at its upper boundary, called the tropopause. The altitude of the tropopause is about 10 km.

Above the troposphere, and extending to a height of about 80 km, is the **stratosphere.** Most of the stratosphere is cold and free of clouds, but in its upper part the temperature rises again. The hot layer is due to the absorption of solar energy by ozone. **Ozone** (abbreviated O_3) is a heavy form of oxygen, having three atoms per molecule instead of the usual two. It absorbs solar ultraviolet light, protecting the lower atmosphere and surface; this absorbed energy heats the ozone and warms the parts of the atmosphere where it is present. Even a modest reduction in the ozone is sufficient to increase human skin cancer rates and perturb the ecological balance on our planet. Beginning in the early 1980s, significant ozone loss was recorded, first over the Antarctic and later spreading to temperate latitudes in both hemispheres. Most of this loss is due to release into the atmosphere of the manufactured chemicals called chlorofluorocarbons (CFCs). Although production of CFCs is being banned or severely restricted, the ozone loss is expected to persist for at least another century.

From 65 to 80 km, the temperature drops to below $-50°C$ again (see Figure 12.15). Above 80 km the temperature rises rapidly through a region called the **thermosphere,** and at 400 to 500 km it reaches values above 1000°C. This is the region where many satellites, including the Space Shuttle, orbit the Earth. In spite of the high temperature of the individual atoms, the astronauts are in no danger of burning, because there are too few atoms in the atmosphere to cause significant heating. The atmosphere above about 400 km is called the **exosphere.** In these upper layers of the atmosphere (the thermosphere and exosphere), molecules of oxygen and nitrogen break up into individual atoms of those elements. Ultraviolet radiation from the Sun ionizes many of these atoms, giving rise to the Earth's **ionosphere.**

Earth, by allowing animals to oxidize their food to produce energy. Oxygen is also required for all forms of combustion (rapid oxidation); thus it is necessary for most of our heat and power production. In the process of photosynthesis, green plants absorb carbon dioxide and release oxygen, which helps to replenish the oxygen consumed by humans and other animals.

A complete census of the Earth's atmosphere, however, should consider more than the gas now present. Suppose, for example, that our planet were heated to above the boiling point of water (100°C, or 373 K). The oceans would boil, and their water vapor would become a part of the atmosphere. To estimate how much water vapor would be released, we note that there is enough water to cover the entire Earth to a depth of about 3000 m. Since the pressure exerted by 10 m of water is about equal to 1 bar, the average pressure at the ocean floor is about 300 bars (3000 m divided by 10 m/bar). Water (or any material) weighs the same whether it is in liquid or vapor form. Therefore, the atmospheric pressure of water if the oceans boiled away would also be 300 bars. Water would dominate the Earth's atmosphere, with nitrogen and oxygen reduced to the status of trace constituents.

On a warmer Earth, another source of additional atmosphere would be found in the sedimentary carbonate rocks of the crust. These minerals contain abundant carbon dioxide, which, if released by heating, would generate about 70 bars of CO_2, far more than the current CO_2 pressure of only 0.0005 bar. Thus the atmosphere of a warm Earth would be dominated by water vapor and carbon dioxide, with a surface pressure close to 400 bars.

We do not know how the Earth first acquired an atmosphere. Today we see that CO_2, H_2O, and other gases are released from volcanoes. Much of this apparently new gas, however, is probably recycled material that has been subducted through plate tectonics. The atmosphere today is controlled in part by geological and biological activity. As to the original source of the atmosphere and oceans, there are three possibilities: (1) the atmosphere could have been formed with the rest of the Earth, as it accumulated from the gas and dust in the solar nebula; (2) it could have been released from the interior through volcanic activity, subsequent to the

(b) Composition

At the Earth's surface, the atmosphere consists of about 78 percent nitrogen (N_2) and 21 percent oxygen (O_2), with traces of argon (Ar), water vapor (H_2O), carbon dioxide (CO_2), and other gases. Variable amounts of dust particles and water droplets are also found suspended in the air. The composition of the dry atmosphere is summarized in Table 12.3.

The gases nitrogen and argon are relatively inert chemically. It is oxygen that sustains animal life on

TABLE 12.3	COMPOSITION OF THE DRY ATMOSPHERE	
Element or Compound		**Amount (%)**
Nitrogen (N_2)		78.1
Oxygen (O_2)		21.0
Argon (Ar)		0.93
Carbon dioxide (CO_2)		0.03
Neon (Ne)		0.002

FIGURE 12.16 A large tropical storm, marked by clouds swirling in a cyclonic direction around a low-pressure region. (NASA)

formation of the Earth; or (3) the atmosphere may be derived from impacts by comets or other icy materials from the outer parts of the solar system. Current scientific opinion favors the third possibility, the cometary hypothesis, but all three mechanisms may have contributed.

(c)   Weather and Climate

The convection of the troposphere, particularly the evaporation and condensation of water, gives rise to the phenomena we call *weather*. The energy that powers the weather is derived primarily from the sunlight that heats the surface. As the planet rotates, and as slower seasonal changes take place in the deposition of sunlight, the atmosphere and oceans try to redistribute the heat from warmer to cooler areas. Weather on any planet represents the response of its atmosphere to changing inputs of energy from the Sun.

The weather on Earth is closely tied to the presence of the oceans and of water vapor in the atmosphere. As water evaporates, it stores large quantities of energy, which can be released later through condensation. The violence of a summer thunderstorm or the awesome power of a hurricane is the result of energy released by the condensation of atmospheric water. On a planet without abundant water, the weather should be calmer and more predictable.

The rotation of the Earth also has a major influence on the circulation of the atmosphere. On a slowly rotating planet, the simplest circulation pattern would involve warm air rising near the equator and cool air descending near the poles. This is approximately the kind of cir-

culation observed on Venus. The Earth's rotation, however, breaks up this simple flow pattern to create the large cyclonic weather systems that dominate the temperate regions of the planet (Figure 12.16). On an even more rapidly spinning planet, such as Jupiter, we will see that north-south motion of the atmosphere is almost impossible, and the circulation is dominated by extremely strong eastward-blowing winds.

Climate is a term used to refer to the effects of the atmosphere on a longer time scale, measured in decades or centuries. Changes in climate are difficult to detect, but as they accumulate, their effect can be devastating. Modern farming practice, in particular, is highly sensitive to the temperature and rainfall. Calculations show, for instance, that a drop of only 2°C throughout the growing season would cut the wheat production of Canada and the U.S. in half. In Section 12.4 we will return to the problem of climate change, and in Chapter 27 we will discuss evidence for the association of climate changes with the activity level of the Sun.

(d)   The Ice Ages

The best-documented changes in the climate of the Earth are the great ice ages, which have periodically lowered the temperature of the Northern Hemisphere of the Earth over the past million years or so. During an ice age, snow accumulates over the continental masses of North America, Europe, and Asia, eventually building up great ice deposits or glaciers. At the same time the sea level drops as more and more water builds up on land. Such conditions can persist for tens of thousands of years. Today we are in a relatively warm period, interpreted by many scientists as a fairly short-lived interglacial interval between major ice ages.

It is generally believed today that the ice ages are the result of changes in the tilt of the Earth's rotation axis (its obliquity) produced by the gravitational influences of other planets, as originally calculated in 1920 by the Serbian scientist Milutin Milankovitch (1879–1958). With modern computing techniques, these changes can be predicted for the Earth and for other planets. As we will see in Chapter 14, one of the exciting discoveries about Mars is that it also seems to experience periodic ice ages. Soon we may have data on both the Earth and Mars to use in testing and refining our understanding of the astronomical causes of ice ages.

12.4   LIFE AND THE CHEMICAL EVOLUTION OF THE EARTH

(a)   The Uniqueness of Life

Earth is the only inhabited planet in the solar system. We can now make this dramatic assertion with some confidence. About a century ago, it was widely believed that Mars harbored not only life, but also intelligent creatures with an advanced civilization. As recently as the 1960s, many astronomers still thought that some form of life probably existed on Mars. But by now the data from numerous space probes have almost completely ruled out the presence of life on other planets in our solar system. However, there are likely to be many planets elsewhere in the Galaxy, including perhaps some with intelligent inhabitants. But that is another story, one that we save for Chapter 28.

Terrestrial life has played an important role in the story of our planet. Life arose early in Earth's history, and living organisms have been interacting with their environment for billions of years. We all recognize that lifeforms have evolved to adapt themselves to the environment on Earth, and now we are beginning to realize that the Earth itself has been changed in important ways by the presence of living matter.

At the basic molecular level, all life on Earth is similar. The molecules of deoxyribonucleic acid (DNA) and ribonucleic acid (RNA) that are fundamental to life are helical—corkscrew shaped—and all have the right-hand thread; that is, in all living organisms, the helical molecules twist the same way. It's easy enough to understand why this should be so. In the early development of life, some molecule with the ability to reproduce itself got formed. It had either the right-hand or left-hand thread—it couldn't be both at once. These self-replicating molecules developed cells and organisms around them. They fed and reproduced, underwent mutations, and were modified by natural selection. Eventually we ended up with something holding an astronomy book in its hand. If life had sprung up independently all over the Earth, we might expect by chance to find some forms with right-hand helices and others with left-hand helices in their chromosomes. But all of us, from the lowliest bacterium to the biggest elephant, are made of organic molecules twisting the same way, and all are based on the same genetic code. We apparently have a single common ancestor; if other, independent lifeforms originated during the long history of the Earth, their descendants do not survive today.

(b)   Origin of Life

The record of the birth of life on Earth has been lost in the restless motions of the crust. During the first few hundred million years of its existence, the Earth was battered by impacts from space, some of them large enough to destroy any life that might have formed. Yet by 3.6 billion years ago, less than a billion years after the formation of the planet, life was abundant in the oceans and had evolved to the stage of large colonies of micro-

(a)

(b)

FIGURE 12.17   Cross-section of stromatolites, both (a) fossil and (b) contemporary. These colonies of microorganisms date back more than 3 billion years. (NASA/ARC, courtesy David DesMarais)

organisms, called stromatolites, whose fossils can still be seen in ancient rocks (Figure 12.17).

What was it like on our planet about 3.9 billion years ago, when our distant ancestors apparently made their first appearance? We cannot observe those conditions directly, but we can look for clues from laboratory experiments. Experiments can tell us, for example, what kinds of conditions were necessary for the formation of the organic molecules that are the chemical building blocks that make up all living things. Such studies indicate clearly that an oxygen-rich atmosphere like that of the Earth today could not have led to the proper chemical reactions. However, if the experiments start with a less oxidizing atmosphere (no free oxygen) and abundant water, it turns out to be easy to produce a wide variety of organic molecules in the laboratory. These include such complex compounds as the amino acids and simple proteins. Thus the presence of organic molecules provides evidence that our atmosphere has evolved over time.

Scientists are not surprised to find that the Earth must have had a less oxidizing atmosphere at the time life was forming. We know that we have oxygen in our atmosphere today only because it is produced by green plants through the process of photosynthesis, followed by burial of the organic material so that it cannot recombine with the atmospheric oxygen. In the absence of life, any free oxygen quickly combines with surface rocks to produce oxides. Before life began, our atmosphere presumably was dominated by carbon dioxide and perhaps carbon monoxide, like the atmospheres of Venus and Mars today.

To form abundant organic compounds in the laboratory, however, requires more than the absence of oxygen. Also needed are such hydrogen-rich compounds as methane (CH_4) and ammonia (NH_3). Either the Earth must once have had an atmosphere containing methane and ammonia, or else the necessary organic materials came from elsewhere. Both reduced atmospheric compounds like methane and complex organic compounds like alcohol and hydrogen cyanide are abundant in comets, so they may have been carried to the Earth as part of the early cometary bombardment.

(c) Evolution of the Atmosphere

If reduced compounds were abundant in the early atmosphere, they probably did not last long. Before the atmosphere contained ozone, ultraviolet light could penetrate to the surface and break such compounds apart. The hydrogen released from the destruction of methane and ammonia would have escaped, so that such compounds could not be put back together again. Thus there was an inevitable transition toward an oxidizing atmosphere: from NH_3 to N_2, and from CH_4 to CO.

Although photosynthetic plants existed in the seas more than 3 billion years ago, studies of the chemistry of ancient rocks show that the atmosphere of the Earth lacked substantial oxygen until at least 2 billion years ago. Apparently the oxygen gas was removed by chemical reactions with the crust as quickly as it formed. Slowly, however, the increasing evolutionary sophistication of life led to a growth in plant population, finally reaching the point where oxygen was produced faster than it could be removed, and the atmosphere became more and more oxidizing.

The appearance of free oxygen eventually led to the formation of the Earth's ozone layer. Before that, it was unthinkable for life to venture outside the protective oceans, and the land masses of Earth were barren. The presence of oxygen allowed the colonization of the land. Oxygen dissolved in the oceans also made possible a tremendous proliferation of animals, creatures who lived off of the organic materials produced by plants and the oxygen in the seas. Eventually the animals too moved to the land, as evolution adapted them to breathing oxygen directly from the atmosphere.

(d) Carbon Dioxide and Global Warming

One of the most important consequences of life has been a reduction in atmospheric CO_2. In the absence of life we would have a much more massive atmosphere dominated by CO_2, but life has effectively taken most of this gas out of the atmosphere.

Most of the carbon is currently found in sediments composed of carbonate minerals such as limestone. Carbonates are an essential material for the shells of marine creatures, which have evolved techniques for extracting CO_2 from the water and manufacturing intricate carbonate structures for their protection. When they die, their shells sink to the ocean floor to build thick layers of limestone. The carbon remains trapped on the ocean floor until it is subducted, when much of it returns to the atmosphere in the form of CO_2 released by volcanic eruptions. But evidently the marine organisms quickly cycle it back into sediments, so that at any time only a minute fraction of the Earth's carbon is present in the atmosphere. Another way that life removes CO_2 is by producing deposits of fossil fuels, predominantly coal and oil. These substances are primarily carbon, for the most part extracted from atmospheric CO_2 hundreds of millions of years ago.

We have a special interest in the CO_2 content of the atmosphere because of the role this gas plays in the atmospheric greenhouse effect (Section 11.4). CO_2 is the most important greenhouse gas in the Earth's atmosphere, although water vapor and the CFCs also contribute. Even the relatively small amount of CO_2 in our atmosphere today is sufficient to raise the average surface temperature more than 20°C higher than it would be if none of this gas were present. Note that contrary to many press reports stimulated by public

Apollo 11 astronaut Buzz Aldrin on the lunar surface. Because there is no atmosphere, ocean, or geological activity on the Moon today, the footprints you see in this image will likely be preserved in the lunar soil for millions of years.

(NASA)

The Moon is the nearest planetary body. It is the only other world that we can recognize as a globe with our unaided eyes, and the only one whose geological history we can decipher from telescopic views alone. We might expect that scientists would have understood the main outlines of lunar geology and evolution for centuries, but this is not the case. The Moon is too different, too alien a place, for easy comparisons with the Earth. Not until the second half of the 20th century did we begin to find realistic answers to the questions of its origin and the forces that have made the Moon we see today.

The Moon also provides an accessible record of the past. Because it is geologically inactive, its surface preserves evidence of events that took place billions of years ago. This ability to probe our past is one of the most important goals of modern astronomy, whether we are trying to establish the history of the solar system or of the entire universe. Further, the Moon demonstrates the role played by violent impacts in the history of the solar system. Nearly every feature we see on the lunar surface was produced by an impact, and the Moon itself was born in an interplanetary collision of extraordinary violence. Similar catastrophic events have influenced the development of other planets, as we will see in later chapters.

Mercury, the second smallest of the planets, is very similar to the Moon. It too experienced catastrophic impacts in its past, and its arid surface is covered with impact craters just like those on the Moon. By discussing both Mercury and the Moon in the same chapter, we emphasize comparative planetology—the study of solar system processes through comparing their actions in different locales.

Grove K. Gilbert (1843–1918), a founding member and Chief Geologist of the U.S. Geological Survey, was among the first scientists to recognize the importance of impact cratering on the Earth and Moon. In the 1890s his careful arguments for the impact origin of the lunar craters, although decades ahead of their time, laid the foundations for the modern science of lunar geology and the development of a chronological sequence for the evolution of the terrestrial planets. (USGS)

CHAPTER

13

CRATERED WORLDS: MOON AND MERCURY

The Moon, the only celestial object to have been visited by humans, is the best-known member of the solar system beyond the Earth. Unlike our planet, the Moon is geologically dead, a small world that has long since exhausted its internal energy sources. Mercury, the smallest of the terrestrial planets, is in many ways similar to the Moon, which is why both are discussed in this chapter. These two objects are both lacking in atmospheres, deficient in geological activity, and dominated by the effects of impact cratering (Figure 13.1). Neither would be a very pleasant place for humans to live, but they are fascinating worlds for the scientist to study.

13.1 GENERAL PROPERTIES OF THE MOON

(a) Some Basic Facts

The Moon has only $1/80$ of the mass of the Earth and a surface gravity too low to retain an atmosphere. If, early in its history, the Moon outgassed an atmosphere from its interior or collected a temporary envelope of gases from impacting comets, such an atmosphere was lost before it could leave any recognizable evidence of its short existence. All signs of water are similarly absent. Indeed, the Moon is dramatically deficient in a wide range of **volatiles,** those elements and compounds that evaporate at moderate temperatures. Something in its

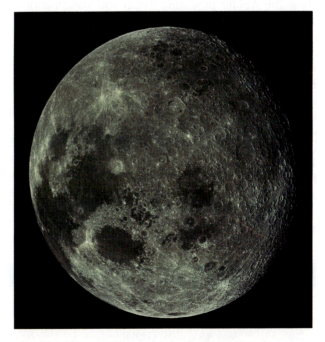

FIGURE 13.1   The Moon as photographed from a spacecraft near the Earth. The hemisphere shown consists primarily of heavily cratered highlands. The resolution of such an image is several kilometers, similar to that of high-powered binoculars or a small telescope. (NASA)

past removed whatever volatiles might have been accumulated initially from the solar nebula or from cometary impacts.

An impact explosion of the sort described above leads to a characteristic kind of crater. We will describe here a crater 50 km or so in diameter, the sort that would be produced by a small asteroid or comet impact, but most of the same features are also present for both larger and smaller craters.

The central cavity is initially bowl-shaped ("crater" comes from the Greek word for "cup"), but the gravitational rebound of the crust partially fills it in, producing a flat floor and sometimes creating a central peak. Around the rim, landslides create a series of terraces. The floor is generally below the level of the surrounding terrain. Terraces, flat depressed floors, and central peaks (Figure 13.18) are all characteristics of impact craters.

The rim of the crater is turned up by the force of the explosion and bent back like the pages of a book. Surrounding the rim is a blanket of ejecta, consisting of material thrown out by the explosion that falls back to create a rough, hilly apron, typically about the same width as the diameter of the crater. Additional, higher speed ejecta fall at greater distances from the crater, often digging small craters (called secondary craters) where they strike the surface. Some of these streams of ejecta can extend for hundreds or even thousands of kilometers from the crater, creating bright crater rays that are prominent in photographs taken near full phase (Figure 13.19).

(c)   Using Crater Counts to Date Planetary Surfaces

If a planet has little erosion or internal activity, like the Moon during the past 3 billion years, it is possible to use the numbers of impact craters counted on its surface to estimate the age of that surface. By "age" we mean the

FIGURE 13.18 King Crater on the far side of the Moon, a relatively fresh lunar crater 75 km in diameter, clearly showing most of the features associated with large lunar impact craters. (NASA)

FIGURE 13.19 This view of Mare Imbrium includes numerous secondary craters and other ejecta from the large crater Copernicus, on the upper horizon. Copernicus is almost 100 km in diameter. (NASA)

time since there was a major disturbance such as the volcanic eruptions that produced the lunar maria.

This technique works because the rate at which impacts have occurred has been roughly constant subsequent to the end of the heavy bombardment about 3.8 billion years ago. Therefore, in the absence of forces to eliminate craters, the number of craters is just proportional to the length of time the surface has been exposed.

Estimating ages from crater counts is a little like the experience you might have walking along the sidewalk in a snowstorm, when the snow has been falling steadily for a day or more. You may notice that in front of some houses the snow is deep, while next door the sidewalk may be almost clear. Do you conclude that less snow has fallen in front of Mr. Jones' house than Ms. Smith's? No, of course not. Instead, you conclude that Jones has recently swept the walk clean while Smith has not. Similarly, the numbers of craters indicate how long since a planetary surface was last "swept clean."

Except for the Moon and Earth, we have no planetary samples to permit exact ages to be calculated from radioactive decay (Section 11.2d). Without samples, the planetary geologist cannot tell if a surface is a million years old or a billion. But the number of craters provides an important clue. On a given planet, the more heavily cratered terrain will always be the older (as we have defined age above). And if we can calibrate the relation-

ship between crater numbers and ages using what we know about the Moon, we may be able to estimate the ages of surfaces on other cratered planets, such as Mars or Mercury.

(d) Cratering Rates

The rate at which craters are being formed in the vicinity of the Earth and Moon cannot be measured directly, since the interval between large crater-forming impacts is longer than the span of human history. Remember that Meteor Crater is 50,000 years old. However, the cratering rate can be estimated from the number of craters on the lunar maria, or it can be calculated from the numbers of potential projectiles (asteroids and comets) present in the solar system today (Figure 13.20). Fortunately, both lines of reasoning lead to about the same answers.

For the entire surface area of the Moon, these calculations indicate that a crater of 1 km diameter should be produced about every million years, a 10-km crater every 10 million years, and several 100-km craters every billion years. Comets and asteroids appear to contribute about equally to these statistics. For the Earth, the average frequencies of impact are about 20 times greater, primarily because Earth is a larger target.

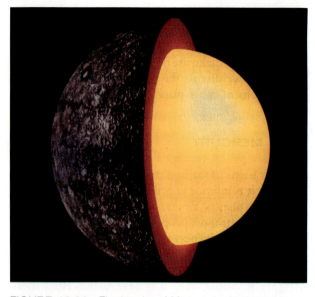

FIGURE 13.22 The interior of Mercury is dominated by a metallic core about the same size as our Moon. (University of Arizona, courtesy Robert Strom)

This high uncompressed density, larger than that of any other planet or satellite, tells us that the composition of Mercury is different from that of either the Earth or the Moon.

The high uncompressed density of Mercury implies a large proportion of metals. The most likely models for the interior suggest an iron-nickel core with a mass amounting to 60 percent of the total, with the rest of the planet made up primarily of silicates. The core has a diameter of 3500 km and extends to within 700 km of the surface. We could think of Mercury as a metal ball the size of the Moon surrounded by a rocky crust 700 km thick (Figure 13.22).

The escape velocity is too low and the surface temperature too high for Mercury to retain any substantial atmosphere. In 1985, however, an extensive but extremely thin atmosphere of sodium was detected spectroscopically. Apparently, atoms of this metal are ejected from the surface through bombardment by the solar wind.

One of the surprises of the Mariner 10 mission was the discovery of a magnetic field on Mercury. Although less than 1 percent as strong as the Earth's field, Mercury's magnetic field is likewise intrinsic to the planet. Such planetary magnetism was not expected because the interior of Mercury was thought to be solid, so that a field would not be generated. It is difficult to see how such a small planet could retain enough interior heat to have a liquid metal core. The most likely suggestion is that there is considerable sulfur mixed with the iron in the core, sufficient to lower its freezing temperature and maintain at least part of the iron-sulfur core in the liquid state.

(c) Rotation

Visual studies of Mercury's indistinct surface markings were once thought to indicate that the planet kept one face to the Sun, and for many years it was widely believed that Mercury's rotation period equaled its period of revolution about the Sun, which is 88 days. Doppler radar observations of Mercury in the mid-1960s, however, showed conclusively that Mercury does rotate with respect to the Sun. Its sidereal period of rotation (that is, with respect to the distant stars) is about 59 days. The Italian physicist Giuseppe Colombo (1920–1984) first pointed out that this is very nearly two-thirds of the planet's period of revolution, and subsequently it was found that there are theoretical reasons for expecting that Mercury can rotate stably with a period of exactly two-thirds that of its revolution—58.65 days.

If Mercury were not perfectly spherical, but were deformed so that one dimension through the equator were longer than the others, the Sun's force on that bulge should force the long axis of the planet to point to the Sun when it is at perihelion, where the Sun's differential force on Mercury is strongest. This condition would be met if the planet rotated with its revolution period, but also with certain other rotation periods. The most likely alternative rotation period is two-thirds of the revolution period, so that at successive perihelions alternate ends of the long axis of the planet are pointed toward the Sun. This is the observed situation.

(d) The Surface of Mercury

The first closeup data on Mercury were obtained in March 1974, when the Mariner 10 spacecraft passed within 9500 km and photographed the surface with resolution down to 150 m. Mariner 10, in a planned orbit about the Sun with exactly twice the orbital period of Mercury itself, passed the planet again in September 1974 and March 1975 for additional photography and other measurements.

Although the temperature of Mercury had been measured from the Earth, better data came from the Mariner 10 flybys. The daylight temperature on the surface ranges up to about 700 K at noontime, nearly hot enough to melt lead. Just after sunset, however, the temperature drops quickly (as it does on the Moon) to about 150 K, and it then slowly cools to about 100 K just before dawn. The range in temperature on Mercury is thus 600 K (more than 1000°F), greater than on any other planet.

Mercury strongly resembles the Moon in appearance (Figure 13.23). It is covered with thousands of craters and larger basins up to 1300 km in diameter. There are also scarps (cliffs) over a kilometer high and hundreds of kilometers long, as well as ridges and plains. Some of the brighter craters are rayed, like Tycho and Copernicus on the Moon, and many have central peaks.

Most of the mercurian features have been named in commemoration of artists, writers, composers, and other contributors to the arts and humanities, in contrast to the scientists commemorated on the Moon. One of us (Morrison) is the head of the International Astronomical Union commission that assigns names to features on Mercury. Among the most prominent craters are Bach, Shakespeare, Tolstoy, Mozart, and Goethe. A large basin has been named Caloris (Latin for "heat"), since the Sun shines directly down on the basin when Mercury is at perihelion, and Caloris is therefore probably the place on the planet with the highest noon temperature.

The Caloris basin (Figure 13.24) resembles the Orientale lunar basin in both size and appearance. Evi-

dently it began with a large impact. This basin shows evidence of much flooding from lava flows; however, these flows do not have the distinctive dark color that characterizes the lunar maria. Thus there is no equivalent on Mercury of the "Man in the Moon." Since Mercury is so difficult to study telescopically and has been visited by only one spacecraft, we actually know very little about the chemistry of its surface or the probable source of these lavas.

Although the Caloris basin is extremely hot, Mercury has cold surface areas as well. Because its axis of rotation is almost exactly perpendicular to the orbit plane, there are no seasons on Mercury, and the north and south poles remain in eternal shadow. In 1992 astronomers found that there were circular regions of high radar reflectivity near both poles, similar in their radar reflection properties to the polar caps of Mars and the Earth. These highly reflective regions appear to consist of ice in the shadowed interiors of polar craters. While it is no

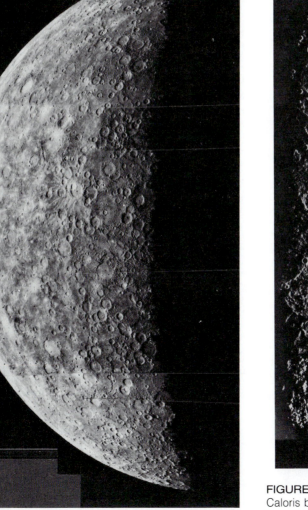

FIGURE 13.23 General view of the planet Mercury as photographed by Mariner 10 in 1974. (NASA/JPL)

FIGURE 13.24 In the left half of this photomosaic is the Caloris basin, 1300 km in diameter. This partially flooded impact basin is the largest structural feature on Mercury seen by Mariner 10. Compare this photo with that of the Orientale basin on the Moon (Figure 13.10). (NASA/JPL)

FIGURE 13.25 Discovery scarp on Mercury. This tectonic feature, nearly 1 km high and more than 100 km long, cuts across several craters, proving that the compression took place after the craters were formed. The width of the frame is about 100 km. (NASA/JPL)

surprise that temperatures are currently low enough in these dark craters to maintain water ice, astronomers do not understand the origin of this ice on a planet that is otherwise exceedingly hot and dry.

Mercury also differs from the Moon in having tectonic features, apparently due to compression of the crust. Mercury's distinctive long scarps, which sometimes cut across craters, are compressional tectonic features (Figure 13.25). Apparently the planet shrank, wrinkling the crust, and it did so after most of the craters on its surface were formed. If the standard cratering chronology applies to Mercury, this shrinkage must have taken place sometime during the first 500 million years of Mercury's history.

If we more fully understood the interior structure and composition of Mercury, we could probably calculate how internal changes in temperature might have led to this global compression, which has no counterpart elsewhere in the solar system. In general, it appears that a very early episode of expansion may have led to cracking of the surface and the release of deep magmas to produce the volcanic plains of the planet. Later, the internal expansion was replaced by shrinking of the core. The surface cracks closed, volcanism stopped, and the compressional scarps were formed. But the details of this sequence remain quite speculative. Many planetary scientists feel that a new space mission to Mercury would be justified to explore further these and other mysteries of this unusual planet.

S U M M A R Y

13.1 The Moon has $\frac{1}{80}$ the mass of the Earth and a density of only 3.3 g/cm³. It is composed mostly of rock, with little (if any) metal core. Relative to the Earth, it is depleted of both metal and **volatiles.** Most of what we know of the Moon is derived from nine Apollo flights (1968–1973), six of which landed. About 400 kg of samples were returned, and ALSEP stations monitored the lunar surface until 1978.

13.2 The lunar surface is dominated by exogenic impact craters, which are densest on the old lunar **highlands.** The highlands, with their anorthositic composition, are more than 4 billion years old and represent the original crust of the Moon. About 17 percent of the surface is covered with younger basaltic plains (the **maria**), erupted between 3.8 and 3.3 billion years ago from deep magma sources. This endogenic activity then ceased. Most maria occupy ancient impact basins on the side of the Moon facing the Earth. The lunar mountains (most ringing mare basins) are of impact origin. The Moon is covered with a deep soil (the **regolith**) produced by constant bombardment of cosmic debris, much in the form of **micrometeorites.** This bombardment also creates the **breccias** that are among the most common types of lunar rock.

13.3 Gilbert defended the impact origin of the large lunar craters in the 1890s, but many scientists still supported the volcanic alternative into the 1950s. Impact craters have characteristic forms that are different from those of volcanic craters. The lunar surface, especially in the maria, constitutes an excellent scorecard for recording cosmic impacts. Impact rates have been roughly constant for more than 3 billion years, and the lunar data permit us to determine the frequency of impacts of various sizes. Prior to 3.8 billion years ago, however, there was a period of heavy bombardment (thousands of times current rate). There is evidence that similar conditions obtained throughout the inner solar system, permitting us to use the lunar rates to determine ages of other planetary surfaces from their crater densities.

13.4 Mercury is the closest planet to the Sun; its orbital period is 88 days, and its rotation period (first determined by radar) is 57 days. Mercury is similar to the Moon in being a small, airless body with an old surface dominated by impacts. Its interior is very different, however. The uncompressed density of Mercury is 5.3 g/cm³, indicating a very large core of iron. Mercury also has a weak magnetic field. As revealed by Mariner 10, the surface displays many craters, including the large Caloris basin, with evidence also of ancient flood basalts. Unlike the Moon, it has tectonic scarps, indicating some interior shrinkage of the planet in the past.

E X E R C I S E S

THOUGHT QUESTIONS

1. One of the primary scientific objectives of the Apollo program was the return to Earth of samples of lunar material. Why was this so important? What can be learned from samples, and are they still of value now?

2. Apollo astronaut David Scott dropped a hammer and a feather together on the Moon, and both reached the ground at the same time. There are two reasons why this experiment on the Moon had distinct advantages over the same experiment as performed by Galileo on the Earth. What are these advantages?

3. Galileo thought the lunar maria were seas of water. If you had no better telescope than Galileo's, could you prove that the maria are not composed of water?

4. Why did it take so long for geologists to recognize that the lunar craters had an impact origin rather than being volcanic?

5. Explain the evidence for a period of heavy bombardment on the Moon about 4 billion years ago. What might have been the source of this high flux of impacting debris?

6. How would a crater made by the impact of a comet on the Moon differ from a crater made by the impact of an asteroid?

7. Explain why the presence of flooded craters (the ghost craters) in the lunar maria is evidence for a long time interval between the formation of the mare basins and the filling of those basins by lunar magma.

8. Why are the lunar mountains smoothly rounded rather than having sharp, pointed peaks (the way they had almost always been depicted before the first lunar landings)?

9. The lunar highlands have about ten times more craters on a given area than do the maria. Does this mean that the highlands are ten times older? Explain your reasoning.

10. Give several reasons why Mercury would be a particularly unpleasant place to live.

11. Only one spacecraft, Mariner 10, has been sent to Mercury. Can you think of any reasons why there have not been additional missions to follow up on the discoveries of Mariner 10?

PROBLEMS

12. The Moon was once closer to the Earth than it is now. When it was at half its present distance, how long was its period of revolution?

13. In any one mare there are a variety of rock ages, typically spanning about a hundred million years. The individual lava flows as seen in Hadley Rill by the Apollo 15 astronauts were about 4 m thick. Estimate the average interval between lava flows if the total depth of the lava in the mare is 2 km.

14. If the typical depth of the regolith on the lunar maria is about 10 m, how much new regolith accumulates, on the average, each century?

15. The Moon requires about one month (0.08 year) to orbit the Earth. Its distance is about 400,000 km (0.0027 AU). Use Kepler's third law, as modified by Newton, to calculate the mass of the Earth relative to the Sun.

★16. Suppose an isolated mountain is observed on the Moon 100 km from the terminator. Its shadow is 40 km long. How high is the mountain?

★17. Suppose a mountain peak on the night side of the Moon rises just high enough to catch some of the rays of the rising Sun and shine like a bright spot of light. If the mountain is just 100 km from the terminator, what is its height?

These heavily eroded canyonlands on Mars, tantalizingly similar to those on our own planet, testify to a long history of tectonics and water erosion on Mars. This Viking photograph shows an area about 60 km across.
(NASA/U.S. Geological Survey, courtesy of Alfred McEwen)

Mars is the only other planet in the solar system where humans might someday feel at home. Although it has no oxygen in its atmosphere and the temperatures are almost always below the freezing point of water, Mars resembles our planet in other ways; its 25-hour day and its progression of seasons are similar, the minerals on its surface are like those on Earth, and its soaring volcanic mountains and deeply eroded river beds look familiar to terrestrial geologists. It has an atmosphere, clouds, weather, and seasons, and temperatures on a summer day can climb above freezing. Many scientists would love to travel there in person to explore its wonders, and perhaps some of us alive today will actually have that opportunity.

Although Mars is not yet within the reach of human explorers, it has been visited by many robotic spacecraft. In this chapter we will see how a succession of such missions have flown past, orbited, and ultimately landed on Mars, returning a wealth of data on this sister world.

We will compare Mars with the Earth and try to understand the processes that shape its geology and control its environment. It is not enough to describe what Mars is like; we must look for the reasons it differs from the Earth. We also have the opportunity to look backward in time to a dimly understood era when Mars was even more Earth-like: a time when the skies were blue, rain fell, and life may even have flourished. This is the aspect of Mars that most fascinates us: its "lost youth" and the possible evidence from those early days that Mars once supported life, perhaps life very different from the biology of our own planet.

MARS: THE PLANET MOST LIKE EARTH

Thomas A. Mutch (1931–1980) of Brown University was one of the first geologists who sought to carry out comparative studies of the planets. In 1976, as the Team Leader for the Viking lander imaging experiment, he served as an articulate and enthusiastic advocate for the exploration of Mars. Following the completion of the Viking program, Mutch became the NASA Associate Administrator for Space Science, but his leadership of the U.S. space science program ended little more than a year later when he was killed in a mountain-climbing accident in the Himalayas. (Brown University)

Mars is the planet most like the Earth in its surface conditions. In spite of its smaller size and greater distance from the Sun, Mars has experienced a geological history that is similar in many ways to that of our own planet, in contrast to the static cratered surface of the Moon. Mars also differs from the Moon in having abundant supplies of H_2O, an essential ingredient for life. Unfortunately, almost all of the H_2O is frozen beneath the surface or in polar caps. The red planet now seems frozen in a terminal ice age. Although life may once have flourished there, deterioration in the martian environment probably led to its extinction billions of years ago.

Whether or not life once existed on Mars, we are confident that humans with their advanced technology could establish a foothold there. The subsurface ice could yield both water and oxygen, the soil could support crops under enclosed greenhouses, and rocket fuels could be synthesized from the atmosphere. Indeed, Mars is the only planet where we think a self-sufficient human colony could eventually be established.

14.1 GENERAL PROPERTIES OF MARS

(a) Basic Facts

The median distance of Mars from the Sun is 227 million km, but its orbit is somewhat eccentric (0.093), and its heliocentric distance varies by 42 million km. The sidereal period of revolution is 687 days, and the synodic period is 780 days. Thus approximately once every 26 months, Mars is at opposition, when it is above the horizon all night and is most favorably placed for observation. The same 26-month interval separates the best launch opportunities for sending spacecraft to Mars.

Seen through a telescope, Mars is both tantalizing and disappointing. The planet is distinctly red, owing to the presence of iron oxides in its soil (Figure 14.1). Its reflectivity (about 30 percent) is more than twice that of the Moon. At its nearest, Mars has an apparent diameter of about 25 arcsec, and the best resolution obtainable is about 100 km, or about the same as the Moon seen with the unaided eye. At the resolution of 100 km, however, no hint of topographical structure can be detected: no mountains, no valleys, not even impact craters (Figure 14.2). On the other hand, the bright polar caps of the red planet can easily be seen, together with dusky surface markings that gradually change in outline and intensity from season to season. Mars alone of all the planets has a surface that can be seen clearly from Earth, and this surface exhibits changes that bespeak a dynamic atmosphere and were once thought to be hints of the presence of life.

The existence of permanent surface markings enables us to determine the rotation period of Mars with great accuracy; its sidereal day is $24^h 37^m 23^s$, just a little greater than the rotation period of the Earth. This high preci-

FIGURE 14.1 Mars as photographed from the Viking spacecraft in 1976. The red color is due to a pervasive dust of clay soil containing iron oxides. Darker areas have similar composition but different texture.(NASA/USGS)

sion is not obtained by watching Mars for a single rotation, but by noting how many turns it makes in a long period of time. Good observations of Mars date back more than 200 years, a period during which tens of thousands of martian days have passed. The rotation period is known now to within a few hundredths of a second.

The rotational pole of Mars has a tilt, or obliquity, of 25°, similar to that of the Earth's pole. Thus Mars experiences seasons very much like those on Earth. Because of the longer martian year, however, seasons there each last about six of our months. Bright polar caps of water or carbon dioxide ice form at each pole during the winter and largely evaporate during spring and summer.

As a planet, Mars is rather small, with only 11 percent the mass of the Earth. It is larger than either the Moon or Mercury, however, and unlike them it retains a thin atmosphere. This atmosphere is composed almost entirely of CO_2, but its pressure is less than 1 percent that of Earth. Mars is also large enough to have supported considerable endogenic geological activity, some apparently persisting to the present day. On the whole, it is a much more interesting and attractive place than either of the objects studied in the last chapter.

Table 14.1 summarizes some of the basic data for Mars.

(b) The Canal Controversy

Approximately 100 years ago Mars became an object of great public interest and the subject of a controversy that persisted for decades. At the favorable opposition of

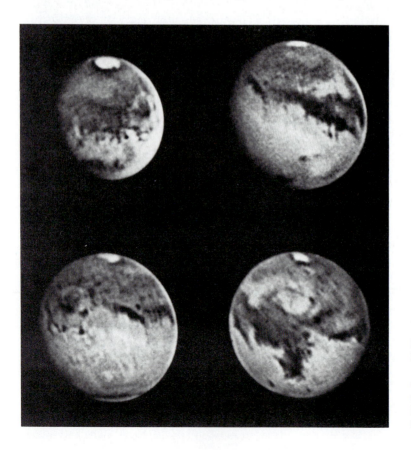

FIGURE 14.2 These are among the best Earth-based photographs of Mars, taken in 1988 when the planet was exceptionally close to the Earth. The polar caps and detailed dark surface markings are evident, but no topographical features are seen. (Steve Larson, University of Arizona)

TABLE 14.1	PROPERTIES OF MARS AND EARTH	
	Mars	**Earth**
Semimajor axis (AU)	1.52	1.00
Revolution period (yr)	1.88	1.00
Diameter (km)	6794	12756
Mass (Earth = 1)	0.11	1.00
Density (g/cm³)	3.9	5.5
Uncompressed density (g/cm³)	3.8	4.5
Surface gravity (Earth = 1)	0.38	1.00
Escape velocity (km/s)	5	11
Rotation period (hr)	24.6	23.9
Surface area (Earth = 1)	0.28	1.00
Surface reflectivity	0.2	0.5
Atmospheric pressure (bar)	0.006	1.00
Atmospheric composition	CO_2(95%)	N_2(78%)

1877, the Italian astronomer Giovanni Schiaparelli (1835–1910) announced the discovery of long, faint, straight lines on Mars that he called *canale*, or channels. In English-speaking countries, the term *canale* was translated as canals, a word that implies an artificial origin. After the existence of these markings was confirmed by other observers at the opposition of 1879, there was widespread speculation in both scientific and public circles that artificial waterways existed on Mars, presumably the work of an advanced race of Martians. The public interest in (and concern over) Martians was considerably enhanced in 1898 with the publication of H. G. Wells' popular novel *War of the Worlds*, which depicts an invasion of Earth by aliens from Mars.

The most effective proponent of intelligent life on Mars was Percival Lowell (1855–1916), self-made American astronomer and member of the great Lowell family of Boston (Figure 14.3). Lowell was fascinated by the discovery of the canals, and he decided to devote his talents and fortune to pursuing studies of the red planet. After an extensive search for the best available site, in 1894 he constructed the Lowell Observatory at Flagstaff in the Territory of Arizona. An ardent observer, Lowell saw hundreds of martian canals through his telescopes. He even spotted similar linear markings on Venus and the large satellites of Jupiter. In addition, however, Lowell was an effective author and public speaker. During the first decade of the 20th century, he made a convincing case for intelligent Martians, who, he believed, had constructed huge canals to carry water from the polar caps in an effort to preserve their existence in the face of a deteriorating climate. (Later, as we saw in Section 11.2, Lowell was instrumental in early searches for the ninth planet.)

The martian canals were always difficult to study, glimpsed only occasionally as "seeing" caused the tiny image of Mars to shimmer in the telescope. Lowell saw them everywhere, but many other observers were skeptical. When the new, larger telescopes constructed at Mt. Wilson failed to confirm the existence of canals, the skeptics seemed to be vindicated. Astronomers had given up the idea by the 1930s, although it persisted in the public consciousness until the first spacecraft photographs clearly showed that there were no martian canals. Now it is generally accepted that the canals were an optical illusion, the result of the human mind's tendency

FIGURE 14.3 Percival Lowell in about 1910, observing with his 24-in. telescope at Flagstaff. (Lowell Observatory)

elevation several kilometers lower than the older southern uplands. In many places the boundary between the two terrains is relatively abrupt, with the surface sloping down 4 or 5 km in a span of a few hundred kilometers. There are no mountain rings to mark the borders of the lowland plains, however, and they did not originate, like the lunar mare basins, in catastrophic impacts. Some other process, still unknown, destroyed the ancient crust over half the planet and lowered the surface levels by about 5 km on the average.

Mars does have impact basins, but they lie in the old southern uplands. The largest basin, called Hellas, is about 1800 km in diameter and 6 km deep, larger than the Imbrium basin on the Moon. Smaller but better preserved is the Argyre basin (Figure 14.7), with a diameter of 700 km. Both basins are surrounded by mountains formed from the impact explosion.

In addition to the north-south division of the planet into old uplands and younger volcanic plains, Mars displays an impressive east-west asymmetry. On one side of the planet, straddling the boundary between uplands and plains, is an immense bulge the size of North America that rises nearly 10 km above its surroundings. This is the Tharsis bulge, a volcanically active region crowned by four great volcanoes that rise another 15 km into the pink martian sky. We will discuss Tharsis and its volcanoes in some detail in Section 14.2.

The total range in elevations on Mars is very large, amounting to 30 km between the tops of the highest volcanoes and the bottom of the Hellas basin. In comparison, the maximum elevation range on Earth (from Mt. Everest to the deep ocean trenches) is only about 17 km. As we will see in Chapter 15, Venus has about the same elevation range as the Earth. Why are the mountains of Mars so much larger? We will answer this question in Section 15.4.

(e)   The Moons of Mars

In the early 1600s, Kepler, hearing of Galileo's discovery of four satellites of Jupiter, speculated that Mars should have two satellites, based on numerological considerations: If Earth had one moon and Jupiter had four, then Mars (being between them) should have two moons. English author Jonathan Swift picked up this idea in his satire *Gulliver's Travels,* published in 1726. He described Gulliver's fictional visit to the land of Laputa, where he found that Laputian astronomers had discovered

> . . . [two] satellites, which revolve about Mars, whereof the innermost is distant from the centre of the primary planet exactly three of the diameters, and the outermost five; the former revolves in the space of ten hours, and the latter in twenty one and a half; so that the squares of their periodical times are very near in the same proportion with the cubes of their distance from the centre of Mars, which evidently shows them to be governed by the same law of gravitation that influences the other heavenly bodies.

It is an interesting coincidence that in 1877, 150 years later, Asaph Hall (1829–1907) of the U.S. Naval Observatory actually discovered two small satellites of Mars that closely resemble those described by Swift. (More significant than Swift's "prediction" of the martian moons, perhaps, is that he was aware of and understood Kepler's third law. Swift was not a scientist. How many intellectuals of today are this knowledgeable about modern science?)

The satellites are named Phobos and Deimos, meaning "fear" and "panic"—appropriate companions of the god of war, Mars. Phobos is 9380 km from the center of Mars and revolves about it in $7^h 39^m$; Deimos has a distance of 23,500 km and a period of $30^h 18^m$. The "month" of Phobos is less than the rotation period of Mars; consequently, Phobos would appear to an observer on Mars to rise in the west.

Phobos and Deimos were studied at close range by the Viking orbiters in 1976, while Phobos was the target of a Soviet mission in 1989. They are both rather irregular, somewhat elongated, and heavily cratered. Since these two small satellites are thought to be captured asteroids, we will discuss them in more detail in Chapter 18, which is devoted to the asteroids.

FIGURE 14.7 The Argyre impact basin, about 700 km in diameter, photographed by the Viking orbiter. Note also the haze layers in the planet's atmosphere. (NASA/JPL)

(a)

FIGURE 14.8 Olympus Mons, the largest volcano on Mars, and probably the largest in the solar system. This shield volcano may still be intermittently active. (a) General view; also note the surrounding clouds, which are controlled in part by air rising over the 25-km-high mountain. (b) The summit caldera, about 80 km in diameter. It consists of several interconnected, flat-floored collapse craters, similar to calderas of terrestrial volcanoes. (NASA/JPL)

(b)

14.2 VOLCANOES AND OTHER GEOLOGICAL FEATURES

(a) The Martian Shield Volcanoes

About a dozen very large volcanoes have been found on Mars, most of them associated with the Tharsis bulge and its immediate surroundings. Many more small volcanoes also dot the surface, primarily in the younger northern half of the planet. Three of the most dramatic of these volcanoes lie along the crest of the Tharsis bulge. Each is about 400 km in diameter, and all rise to the same height. The fourth and largest volcano, Olympus Mons (Mt. Olympus), is on the northwest slope of Tharsis (Figure 14.8a). It is more than 500 km in diameter, with a summit 25 km above the surrounding plains. The volume of this immense volcano is nearly 100 times greater than the largest terrestrial volcano, Mauna Loa in Hawaii.

These great martian mountains are shield volcanoes, similar in shape to their terrestrial counterparts in such locations as Hawaii or the Galapagos Islands. Presum-

ably they consist of many overlapping flows of fluid basaltic lavas. Indeed, the detailed patterns of these flows can be traced in high-resolution Viking photographs of their slopes. The summits of shield volcanoes are marked by **calderas,** which are broad, flat-floored collapse craters (Figure 14.8b). The caldera on Olympus Mons is 80 km across, while that of Arsia Mons, another Tharsis volcano, is 120 km in diameter.

The Viking imagery permits a detailed examination of the calderas and slopes of these volcanoes to search for impact craters. As described in Section 13.3c, the number of such impact craters is a direct measure of the age of a surface. Many of the volcanoes show fair numbers of such craters, suggesting that they ceased activity a billion years or more ago. However, Olympus Mons has very, very few impact craters. Its present surface cannot be more than about a hundred million years old, and it could be much younger yet. Some of the fresh-looking lava flows we see might have been formed a hundred years ago, or a thousand, or a million, but geologically speaking they are young. It is quite probable that these great volcanoes remain intermittently active today.

(b) Canyons and Tectonic Features

The Tharsis bulge consists of more than a collection of huge volcanoes. In this part of the planet, the surface itself has bulged upward, forced by great pressures from below. Thus the Tharsis bulge (as opposed to the volcanoes that lie on it) is primarily tectonic in origin. Other tectonic features on Mars include extensive cracks in the crust that criss-cross and surround the Tharsis area, produced by the same forces that raised the bulge (Figure 14.9).

On the Earth we see abundant evidence of plate tectonics, in which the crust responds to internal forces by dividing into plates that move with respect to each other. There is no evidence of plate tectonics on Mars, however. For example, there is no martian equivalent of a San Andreas Fault, where one piece of the crust slides alongside another. Instead, there is the single feature of Tharsis, forced upward but not induced to shift sideways. It is as if crustal forces began to act but then subsided before full-scale plate tectonics could begin.

The most spectacular tectonic features on Mars make up a great canyon system called the Valles Marineris, which extends for about 5000 km (nearly a quarter of the way around Mars) along the slopes of the Tharsis bulge (Figure 14.10). The main canyon is about 7 km deep and up to 100 km wide. It is so large that the Grand Canyon of the Colorado River would fit comfortably into one of its side canyons.

The term canyon is somewhat misleading, because the Valles Marineris canyons were not cut by running water. They have no outlets. They are basically cracks, produced by the same crustal tensions that caused the Tharsis uplift. However, water is believed to have played a later role in shaping the canyons, primarily through the undercutting of the cliffs by seepage from deep springs. This undercutting led to landslides, gradually widening the original cracks into the great valleys we see today. The material from recent landslides is clearly visible on the valley floors, while earlier landslides have probably been eroded away by windstorms sweeping down the canyons (Figure 14.11).

The Valles Marineris has been compared with the African Rift Valley on Earth, which is the result of mantle convection that has pulled apart the crust. It may be that similar forces were at work on Mars 2 to 3 billion years ago, which is when we estimate both Tharsis and the canyons were formed. However, we must conclude that full-scale plate tectonics never developed on Mars, even though the planet may have been headed in that direction early in its geological history.

(c) The Cratered Uplands

Most of the martian southern hemisphere consists of cratered upland plains. The martian craters are clearly of

FIGURE 14.9 Tectonic features in the Tharsis region of Mars, produced by tension in the crust, and two large shield volcanoes. The width of this frame is about 400 km. (NASA/JPL)

FIGURE 14.10 The multiple canyons that comprise the Valles Marineris. These valleys are primarily tectonic in origin; that is, they represent splitting of the crust rather than erosion. Three large volcanoes on the Tharsis bulge appear on the left. The width of this frame is about 5000 km, equal to the width of North America. (NASA/JPL)

impact origin, although they have been modified more than lunar craters by subsequent erosion. There are no rays, and most craters have also lost their ejecta blankets. Some of the oldest appear to have been partially filled in, perhaps by windblown dust.

Among the most curious features of the equatorial plains are multitudes of small, sinuous (twisting) channels that appear to have been cut by running water. Typically, these channels are simple valleys, a few meters deep, some tens of meters wide, and perhaps 10 or 20 km long. They are called *runoff channels*, because they appear to have carried the surface runoff of ancient rainstorms (Figure 14.12). As we will discuss in the next

section, there are no rainstorms today on Mars. These runoff channels seem to be telling us that the planet had a very different climate long ago. How can we tell when rain might have last fallen on Mars?

The best that we can do is to count impact craters and use the theory described in Section 13.3. Such crater counts show that the uplands of Mars are more cratered than the lunar maria but less so than the lunar highlands. Thus they are older than the maria, presumably at least 3.9 billion years. Most scientists would guess that they were formed during the final heavy bombardment about 4 billion years ago. Interestingly, there are no runoff channels in the younger terrain to the north.

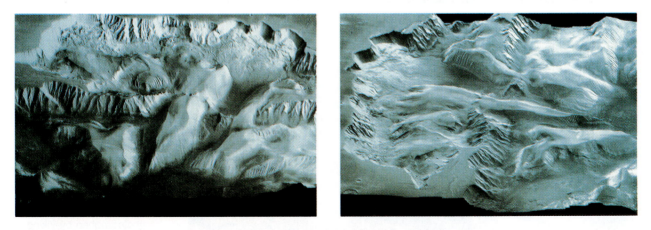

FIGURE 14.11 Oblique view of cliffs and landslides in the Valles Marineris, shown from two different perspectives. These computer-generated scenes of Ophir and Candor Chasmas are exaggerated in vertical relief by a factor of two. Individual landslides are as much as 50 km in width. (Image processing by L.M. Bertolini and A.S. McEwen, USGS)

FIGURE 14.15 Surface frost photographed at the Viking 2 landing site during late winter. (NASA/JPL)

new fragments found in Antarctica in an excellent state of preservation (Figure 14.16).

The most obvious special characteristic of the SNC meteorites is that they are basalts, and they are relatively young, with ages of 1.3 billion years (since the lava solidified). We know they are not from the Moon, and in any case there was no lunar volcanic activity as recently as 1.3 billion years ago. Their composition eliminates the Earth as a possible source. By process of elimination, the only reasonable origin seems to be Mars, where the Tharsis volcanoes were certainly active at that time. This guess has been confirmed by analysis of tiny bubbles of gas trapped in one of these meteorites, the composition of which match almost perfectly the atmospheric properties of Mars as measured directly by Viking.

These pieces of Mars were presumably ejected in some large impact, probably on one of the Tharsis volcanoes. The scar from that impact must be visible, but of course we do not know which crater it was. In regard to helping us understand Mars, the work on these meteorites has just begun, and we do not know how much they will tell us. Already, analyses of the SNC meteorites, together with other information on the density and interior structure of the planet, have yielded our best estimates of the bulk composition of Mars. These values are summarized in Table 14.2.

14.3   POLAR CAPS, ATMOSPHERE, AND CLIMATE

(a)   The Martian Atmosphere

The atmosphere of Mars is composed primarily of carbon dioxide (average value 95 percent), with about 3 percent nitrogen and 2 percent argon. The predominance of CO_2 over N_2 is not too surprising when you remember (Section 12.3b) that the Earth's atmosphere

therefore, that scientists have recently concluded there are samples of martian material already available for study on Earth.

These recently identified martian rocks are a rare class of meteorites called **SNC meteorites.** Four SNC meteorites have been in our collections for a number of years, and they have recently been supplemented by several

FIGURE 14.16 One of the SNC meteorites, believed to be fragments of basalt ejected from Mars. (NASA)

TABLE 14.2	AVERAGE SURFACE COMPOSITION OF MARS*	
Mineral		**Percent**
SiO_2		46
Fe_2O_3		19
Al_2O_3		8
SO_3		7
MgO		6
CaO		6
Na_2O		1
H_2O		1

* Adapted from a summary by Heinrich Wanke, Max Planck Institut für Chemie, based on data from the Viking landers, the Phobos orbiter, and laboratory analyses of the SNC meteorites.

TABLE 14.3	COMPOSITION OF THE MARTIAN ATMOSPHERE	
Element or Compound		**Percent**
Carbon dioxide (CO_2)		95.3
Nitrogen (N_2)		2.7
Argon (Ar)		1.6
Oxygen (O_2)		0.15
Water (H_2O)		0.03
Neon (Ne)		0.0003

would also be mostly CO_2 if the CO_2 were not locked up in marine sediments. Water vapor is not expected to be a major component of the martian atmosphere, since temperatures are almost always well below the freezing point. Table 14.3 lists the exact abundances as measured by the Viking landers. Note, however, that these are average values; the actual abundance of CO_2 varies with the season.

Like the Earth, Mars has a troposphere, in which convection takes place, and above it a more stable cold stratosphere. The martian troposphere typically has a height of about 10 km during the daytime, when it is heated from below by sunlight absorbed at the surface. At night or over the polar regions, however, the surface temperature is much lower, and the troposphere disappears.

Several types of clouds can form in the atmosphere. First there are dust clouds, raised by winds, which can sometimes grow to cover a large fraction of the surface. Second are water ice clouds similar to those on Earth. These often form around mountains, just as happens on our planet (see Figure 14.8); low-lying fog also is possible near dawn, when the surface temperature drops to −100°C (Figure 14.17). Finally, the CO_2 of the atmosphere can itself condense at high altitudes to form hazes of dry ice crystals. The CO_2 clouds have no counterpart on Earth, since temperatures never drop low enough here for this gas to condense.

Most of the time, the atmosphere of Mars is much clearer than that of the Earth, with relatively few clouds. The exceptions to this rule are associated with the major dust storms that occur during southern hemisphere summer. Once begun, these storms seem to feed upon themselves, spreading outward across the surface and carrying more and more dust high into the atmosphere. On occasion, such storms can envelop the entire planet. When this happens, sunlight is absorbed primarily in the dust clouds rather than at the surface, and the atmo-

sphere is heated from above. The troposphere disappears, temperatures all over the planet tend to become uniform, and the circulation of the atmosphere changes. Eventually, after a month or two, the dust settles out and the atmosphere returns to its normal state.

With the exception of the great dust storms, martian weather is largely predictable. The thinness of the atmosphere and the absence of water lead to a situation in which the atmosphere responds very quickly to changing surface temperature, and the wind patterns tend to repeat almost exactly each day. The situation can vary substantially from one surface location to another, however, influenced by local topography. Martian conditions provide an interesting and useful test of models developed to predict weather on the Earth, since the martian case is simpler. If a meteorological computer program fails for Mars, it is unlikely to work properly for Earth.

Although there is water vapor in the atmosphere and occasional clouds of water ice can form, liquid water is

FIGURE 14.17 Morning fog in the martian canyonlands. The width of this frame is about 1000 km. (NASA/JPL)

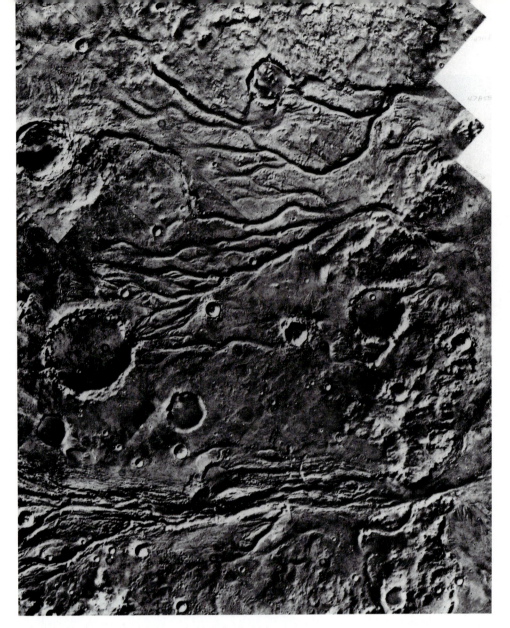

FIGURE 14.21 Large outflow channels photographed by Viking. These features appear to have been formed in the distant past from massive floods of water. The width of this frame is about 400 km. (NASA/JPL)

ington State called the Channeled Scablands. These were carved 18,000 years ago when a large glacial lake in Montana burst its natural dam and emptied within a matter of hours, generating flows 120 m deep and cutting new channels up to 10 km wide through the bedrock.

Presumably, the martian outflow channels were also formed by sudden floods. But where did the water come from? As far as we can tell, the source regions contained abundant water frozen in the soil as **permafrost.** Most of the subsurface of Mars remains below freezing temperature at all times, so that permafrost is stable indefinitely. In the past, however, some local source of heating must have released this water, leading to catastrophic flooding. Perhaps this heating was associated with the formation of the volcanic plains, which apparently happened at about the same time the channels formed.

The two kinds of martian channels thus provide evidence of two periods in the past when water was present in liquid form. The first, about 4 billion years ago, was a time when rain fell and the atmosphere was probably much larger and warmer than it is at present. The second, perhaps a billion years later, represented the release of frozen groundwater by volcanic heating. Since then, the planet has probably been as cold and dry as we see it today, although occasional shorter periods of warmer and wetter climate are not impossible in more recent times.

(d) Climate Change

The evidence cited above indicates that the climate of Mars has varied on at least two different time scales. The layering in polar regions suggests climatic changes similar to our ice ages, with time scales of tens of thousands of years. In addition, there are longer term trends. Billions of years ago temperatures were warmer, rain fell, and the atmosphere must have been much more substantial than it is today.

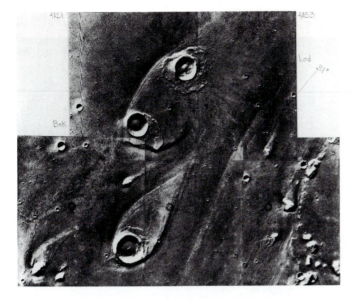

FIGURE 14.22 Flow features in Chryse, near the place where several major outflow channels once emptied into the basin. The sculpted islands are up to 100 km in length. (NASA/JPL)

The long-term cooling of Mars and loss of its atmosphere are a result of both its small size (and low escape velocity) relative to the Earth and its greater distance from the Sun. Presumably, Mars formed with a much thicker atmosphere, and the atmosphere maintained a higher surface temperature because of the greenhouse effect. Escape of the atmosphere to space, however, gradually lowered the temperature. Eventually it became so cold that the water froze out of the atmosphere, further reducing its ability to retain heat. The result is the cold, dry planet we see today. Probably this loss of atmosphere took place within a few hundred million years; from the absence of runoff channels in the northern plains, it seems that rain has not fallen for at least 3 billion years on Mars.

14.4 THE SEARCH FOR LIFE ON MARS

(a) Early Ideas

Even after scientists early in this century rejected the martian canals as evidence of intelligent life, there was still a widespread feeling in the astronomical community that some simpler kind of life was possible, or even probable, on Mars. Of all the planets, Mars had conditions most similar to those on Earth, with surface temperatures not much lower than those in the polar regions of our planet. Before the 1960s, its atmospheric pressure was thought by scientists to be about $1/10$ of the terrestrial value, rather than $1/100$ as we now know. In addition, there was direct evidence that argued for plant life on that planet. Observed changes in the dusky markings on Mars seemed to follow a seasonal pattern. In many cases these markings darkened with the coming of spring, as if in response to higher temperatures and the availability of water vapor from the shrinking polar cap. Some observers argued that the markings were green, suggesting the presence of photosynthetic plant life. There were even hints of spectral features that matched those of sunlight reflected from green leaves on Earth.

With the advantages of hindsight, we now recognize that much of this attitude represented wishful thinking. The spectral features turned out to have been an observational error, and later measurements showed that the dark markings are orange, not green. The apparent seasonal changes, while real, are the result of the deposition and stripping away of light-colored dust from darker terrain, not the growth of plant life. Further, the Mariner 4 spacecraft revealed in 1965 a desolate cratered surface and showed that the atmosphere was too inconsequential to permit liquid water to exist on the martian surface.

Although the new data from the first spacecraft and from more sophisticated telescopic studies were not encouraging for the idea of widespread plant life on Mars, they by no means excluded the possibility of life of *some* kind. Conditions on Mars were still the most favorable to be found beyond the Earth. Most scientists felt that the only way to settle the issue was to go to Mars and look.

(b) The Viking Life-Detection Experiments

The primary objectives of the Viking mission were to land on the martian surface and search for life, using an automated, miniaturized biological laboratory. The entire spacecraft was carefully sterilized to ensure that no terrestrial microorganisms were accidentally transported to Mars, thus protecting the integrity of the martian environment as well as guarding against a false-positive signal from the spacecraft experiments.

The Viking strategy was to search for microorganisms, on the grounds that these were likely to be the most abundant lifeforms as well as the easiest to detect.

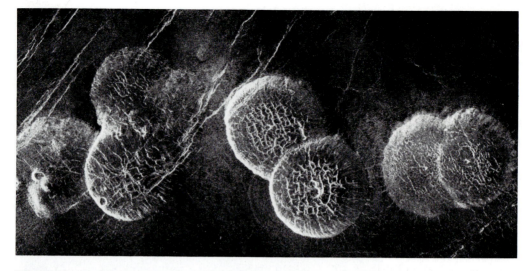

FIGURE 15.6 Pancake-shaped volcanoes on Venus. These remarkable circular domes, each about 25 km across, are the result of eruptions of highly viscous lava. (NASA/JPL)

of extremely long lava channels or rivers. These lava rivers sometimes reach lengths of several thousand kilometers, comparable to the longest water courses on Earth, such as the Mississippi or Nile Rivers. Some of the lavas erupted on Venus must be extremely fluid and fast flowing to be able to form such long channels before they solidify into rock. It would be fascinating to have information on the chemical composition of the lavas on Venus in order to understand the differences between the viscous flows of the pancake domes and the fluid flows of the long channels.

All of the volcanism described above is the result of eruption of lava onto the surface of the planet. But the hot magma rising from the interior of a planet does not always make it to the surface. On both the Earth and Venus, this magma can collect to produce bulges in the crust. Many of the granite mountain ranges on Earth, such as the Sierra Nevada in California, involve subsurface collections of magma. Such features are common on Venus, and in the absence of plate tectonics or surface erosion, they are much more visible than on the Earth. Their characteristic visible expression is a large circular or oval feature called a *corona* (Figure 15.7). Coronae are typically several hundred kilometers across, with slightly raised interiors surrounded by a depressed ring or moat. They are a unique feature of venerian geology, never seen on any other planet or satellite in the solar system.

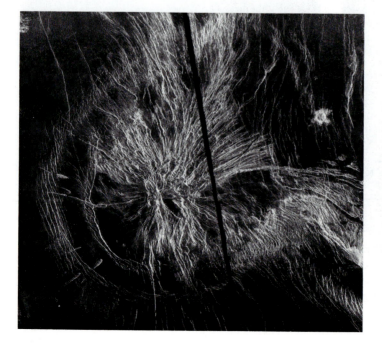

FIGURE 15.7 Pandora Corona, a circular feature 350 km in diameter produced when magma pressure built up below the surface of Venus. This Magellan radar image illustrates the extensive radial and concentric cracks that resulted from this subsurface pressure. The black streak is due to missing radar data. (NASA/JPL)

(d) Tectonic Activity

The mantle convection currents on Venus do more than bring magma to the surface. As on the Earth, these convection currents exert tectonic forces on the crust. Although full-scale plate motion has not occurred on Venus, its crust is constantly subjected to pushing and stretching, creating a variety of tectonic features. The geology of Venus is dominated by the forces associated with mantle convection, and this unique geology is sometimes called "blob tectonics" to distinguish it from terrestrial plate tectonics.

On the lowland plains, tectonic forces have broken the lava surface to create remarkable patterns of ridges and cracks (Figure 15.8). In a few places the crust has been torn apart to generate great rift valleys. The circular features associated with coronae are tectonic ridges and cracks, and most of the mountains of Venus also owe their existence to tectonic forces.

The Ishtar continent, which has the highest elevations on Venus, is the product of tectonic forces. In many ways Ishtar and its high Maxwell Mountains resemble the Tibetan Plateau and Himalayan Mountains on the Earth. Both are the product of compression of the crust, and both are maintained by the continuing forces of mantle convection. Figure 15.9 illustrates the steep flanks of the Ishtar plateau, where folded mountains rise from the lowland plains to the south. These features look much more like the mountains of the Earth than anything seen on the Moon, Mars, or other planets—except, of course, for the lack of erosion and the absence of overlying soil or vegetation.

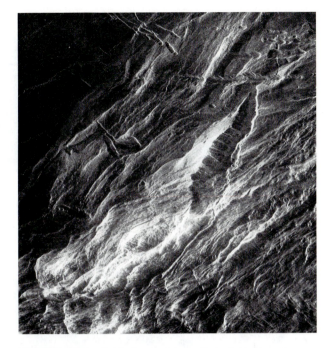

FIGURE 15.9 Folded mountains with individual peaks rising about 2 km high. This region of the Danu Mountains in the Ishtar Continent may be similar in origin to the Himalayan Mountains on the Earth. (NASA/JPL)

(e) On the Surface

The successful Venera landers of the 1970s found themselves on an extraordinarily inhospitable planet, with a surface pressure of 90 bars and a temperature of 730 K. The surface of Venus is hot enough to melt lead and zinc. Despite these unpleasant conditions, the Russian spacecraft were able to photograph their surroundings and perform chemical analyses of surface rocks. They found that the crust in the landing areas was igneous, primarily basalts. Examples of the Venera photographs are shown in Figure 15.10. Each wide-angle panorama shows a desolate, flat landscape with a variety of rocks, some of which may be ejecta from impacts. Other areas show flat, layered lava flows, much as we might expect on such a volcanically active planet as Venus.

The Sun cannot shine directly through the heavy, opaque clouds, but the surface is fairly well lit by diffused sunlight. The illumination is about the same as that on Earth under a very heavy overcast, but with a strong red tint, since the massive atmosphere blocks blue light. (Some of the NASA radar images have been tinted orange to match this color, although radar images have no intrinsic color.) The weather is unchanging—hot and dry with calm winds. Because of the heavy blanket of clouds and atmosphere, one spot on the surface of Venus is like any other as far as weather is concerned. The explanation for the uniform conditions and blistering temperatures is to be found in the planet's atmosphere.

FIGURE 15.8 Tectonic features on Venus. This region of the Lakshmi Plains has been fractured to produce a grid of cracks and ridges. This width of the radar image is 40 km. (NASA/JPL)

15.3   The massive atmosphere is 96 percent CO_2. Thick clouds at an altitude between 30 and 60 km are composed of H_2SO_4 and perhaps elemental sulfur. Stratospheric winds blow at 100 m/s. The high surface temperature is maintained by the greenhouse effect. Venus may have evolved to its present state from more Earth-like conditions by the **runaway greenhouse effect.** There is evidence of former water, now escaped.

15.4   The chapter concludes with a comparison of the terrestrial planets (Earth, Venus, Mars, Mercury, Moon). All are rocky, differentiated objects. Earth, Venus, and Mars have similar composition and internal structure, but the Moon is depleted in metals and Mercury is depleted in silicates. The level of endogenic activity is (as expected) proportional to mass: greatest for Earth and Venus, less for Mars, and absent for Moon and Mercury. Mountains can be the result of impacts, volcanism, or tectonics. Whatever their origin, higher mountains can be supported on smaller planets, where the surface gravity is less. All the terrestrial planets may have acquired their atmospheric volatiles from comet impacts. Moon and Mercury lost their atmospheres; most volatiles on Mars are frozen owing to its greater distance from the Sun; Venus retained CO_2 but lost H_2O when it developed a massive greenhouse effect. Only Earth still has liquid water and hence can support life.

E X E R C I S E S

THOUGHT QUESTIONS

1. Would astronomers be likely to learn more about the Earth from observatories on Venus or Mars? Explain.

2. Explain some of the problems that would be encountered in trying to build a spacecraft that could operate on the surface of Venus for a full Venus year.

3. What are the advantages of using radar imaging rather than ordinary cameras to study the topography of Venus? What are the relative advantages of these two approaches to mapping the Earth or Mars?

4. Discuss the main ways in which the geology of Venus is similar to that of the Earth, and the main differences between the two planets.

5. Why is there so much more carbon dioxide in the atmosphere of Venus than in that of the Earth? Why so much more than in the martian atmosphere?

6. How could Venus have avoided a runaway greenhouse effect? (In formulating your answer, consider the effects of changing its distance from the Sun.)

7. Venus is now a very dry planet. Why is this troubling? Discuss why it seems likely that Venus should once have had more water, and what evidence there is from Pioneer Venus for the existence of more water in the past.

8. Compare the basic data on orbits, mass, size, density, and rotation for Venus, Earth, and Mars.

9. Summarize the main reasons for differences among the terrestrial planets in their: (1) geological activity; (2) mountain heights; (3) atmospheric pressure and composition; and (4) amount and state of water.

10. Estimate the maximum height of the mountains on a hypothetical planet similar to the Earth but with twice the surface gravity of our planet.

11. We have seen how Mars can support greater elevation differences than Earth or Venus. According to the same arguments, the Moon should have higher mountains than any of the other terrestrial planets, yet we know it does not. What is wrong with this line of reasoning?

PROBLEMS

12. Venus requires 440 days to move from greatest western elongation to greatest eastern elongation, but only 144 to move from greatest eastern elongation to greatest western elongation. Explain why. (A diagram will help.)

13. At its nearest, Venus comes within about 40 million km of the Earth. How distant is it at its farthest?

14. Venus has no satellite. But suppose that it had a satellite just like our Moon, orbiting at the same distance as the Moon. What would be the orbital period of this satellite?

15. Magellan scientists estimate that the volcanic eruptions on Venus generate about 2 cubic km of new lava per year. At this rate of eruption, estimate how long it would require to cover the entire surface of the planet uniformly to a depth of 1 km? Generally, if we wish to obliterate an impact crater, we need to bury it to a depth of at least

2 km. How long would it take to "resurface" all of Venus to a depth of 2 km?

16. The surface of Venus is very nearly twice as hot as the surface of the Earth. At what wavelength is the maximum of its infrared radiation (*Hint*: Use the Wien Law)? How hot would the surface have to be to shift its thermal radiation down to wavelengths less than 1000 nm, where the CO_2 atmosphere is more transparent?

17. Calculate the relative *land* areas of the Earth, Moon, Venus, and Mars. (Note: 70 percent of the Earth is covered with water).

18. Mariner 2 to Venus was the first successful interplanetary spacecraft. It traveled to Venus from the Earth on an elliptical orbit in which the Earth was at aphelion and Venus at perihelion. Calculate the total period of revolution about the Sun for a spacecraft on such an orbit. Then calculate the one-way travel time from Earth to Venus on this same trajectory.

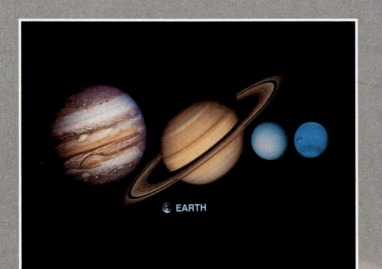

Earth with the four giant planets—Jupiter with its colorful cloud bands, Saturn with its magnificent rings, featureless Uranus, and blue Neptune—shown to the same scale. These images were all obtained by the two Voyager spacecraft as part of their grand tour of the outer solar system.
(NASA/JPL)

The history of astronomy is a story of expanding horizons. In the 1970s, planetary exploration experienced a dramatic growth, as we learned to build rockets and spacecraft that allowed us to probe beyond the inner planets. As a practical matter, the most important advances that brought the outer solar system within our reach were in electronics and computers. The same technologies that created digital watches, microwave ovens, and VCRs also enabled us to construct such compact and multifaceted spacecraft as the Pioneers and Voyagers to probe the solar system beyond Mars.

In 1974 Pioneer 10 was the first spacecraft to reach Jupiter. Fifteen years later Voyager 2 flew past Neptune in the finale of its grand tour of the giant planets. Within the short span of 15 years, we became acquainted, in a way never before possible, with the four giant planets, their more than 50 moons (24 of them discovered by Voyager), and four systems of rings.

When we look at these giant worlds in detail, we find that they are very different from the smaller inner planets. Composed primarily of hydrogen and helium, they are mostly fluid, with no solid surfaces. Spacecraft could not land *on* them but only *in* them (as will the Galileo Jupiter Probe in 1995). There are no craters here, no volcanoes, no geological records of solar system history. In fact, the giant planets have changed very little over the past 4.5 billion years. We see them as they were when they formed, in the dawn of planetary evolution, and they have much to tell us about these ancient times.

EARTH

THE OUTER PLANETS

James Van Allen (b. 1914), American physicist from the University of Iowa, was one of the originators of the discipline of space plasma physics. Van Allen extended his studies from the Earth to the larger and more complex magnetospheres of the outer planets as an experimenter on the Pioneer spacecraft, and his strong advocacy was critical in approval of the Galileo mission to Jupiter. He is also a frequent critic of NASA's policy of emphasis on manned space flight at the expense of robotic missions.

The members of the outer solar system are very different from the inner planets. The four giant planets are much larger, distances between them are vastly increased, and they are accompanied by extensive systems of satellites and rings. From the perspective of an objective observer, the outer solar system is where the action is, and the giant planets are the important members of the Sun's family. In contrast, the little cinders of rock and metal that orbit closer to the Sun seem like an insignificant afterthought. In this chapter we describe the outer planets, and in Chapter 17 we shall investigate their satellite and ring systems.

16.1 FOUR GIANTS AND PLUTO: AN OVERVIEW

(a) Census of Objects

There are five planets in the outer solar system: Jupiter, Saturn, Uranus, Neptune, and Pluto. The first four of these are often called the giant or jovian planets while Pluto is by far the smallest planet—more similar to the satellites of the jovian planets.

Jupiter and Saturn are so large that in spite of their great distances they appear as bright as the brightest stars. They were both well known to ancient peoples, who revered them as important gods. In contrast, both Uranus and Neptune were discovered after the invention of the telescope, one as the result of detailed theoretical calculations (Section 11.2). Pluto is also a discovered planet, the product of a careful, systematic telescopic search.

Each of the giant planets has a system of orbiting satellites and rings. Many of these have been discovered recently as a result of the first spacecraft studies of these systems, and undoubtedly our present census is incomplete. Saturn appears to have the most extensive system, consisting of its beautiful and complex rings plus 19 known satellites. Jupiter is a close second, with 16 satellites and a very faint ring. Uranus has an intricate system of narrow, dark rings and 15 satellites, while Neptune has faint rings and 8 satellites.

(b) Chemistry in the Outer Solar System

In moving outward beyond Mars and the asteroid belt, we enter a region of different planetary composition. Beyond about 4 AU from the Sun, water ice was able to condense and thus to become available as a raw material, in addition to the silicates and metals present in the inner solar system. Since the atoms that constitute water—hydrogen and oxygen—are among the most abundant in the universe, a great deal of water ice formed. Beyond

TABLE 16.1	ABUNDANCES IN THE OUTER SOLAR SYSTEM	
Material	**Percent (by mass)**	
Hydrogen (H_2)	77	
Helium (He)	22	
Water (H_2O)	0.6	
Methane (CH_4)	0.4	
Ammonia (NH_3)	0.1	
Rock (including metal)	0.3	

10 AU, additional ices could also condense, but none of these is nearly as plentiful as water ice. Table 16.1 lists the main materials available to form a planet, based on the abundances we expect to have characterized the solar nebula.

The second major chemical distinction in the outer solar system is a result of the larger spacing between planets and the accumulation of more massive cores of rock and ice. The developing cores of Jupiter and Saturn grew large enough before the dissipation of the gaseous solar nebula to attract and hold the hydrogen and helium from large volumes of space. Uranus and Neptune captured much less hydrogen and helium; this is why these two planets are both smaller than, and different in composition from, Jupiter and Saturn.

With so much hydrogen available, the chemistry of the outer solar system is *reducing*. Most of the oxygen present is chemically combined with hydrogen to make H_2O, and it is therefore unavailable to form many oxidized compounds with other elements. The compounds detected in the atmospheres of the giant planets are thus hydrogen-based gases, such as methane (CH_4) and ammonia (NH_3), or more complex hydrocarbons, such as ethane (C_2H_6) and acetylene (C_2H_2).

(c)   Exploration of the Outer Solar System

Four spacecraft, all from the U.S., have penetrated beyond the asteroid belt to initiate the exploration of the outer solar system. The challenges of probing so far from Earth are considerable. Flight times to the outer planets are measured in years to decades, rather than the few months required to reach Venus or Mars. Spacecraft must be highly reliable, and they must also be capable of a fair degree of independence and autonomy, since the light-travel time between Earth and the spacecraft is several hours. If a problem develops near Saturn, for example, the spacecraft computer must deal with it directly. To wait hours for the alarm to reach Earth and instructions to be routed back to the spacecraft could spell disaster. These spacecraft also must carry their own electrical energy sources, since sunlight is too weak to supply energy through solar

cells. Heaters are required to keep instruments at proper operating temperatures, and spacecraft must have more powerful radio transmitters and large antennas if their precious data are to be transmitted to receivers on Earth a billion kilometers or more distant.

The first spacecraft to the outer solar system were Pioneers 10 and 11, launched in 1972 and 1973 as pathfinders to Jupiter. Their main objectives were to determine whether a spacecraft could navigate through the asteroid belt without collision with small particles and to measure the radiation hazards in the magnetosphere of Jupiter. Both spacecraft passed through the asteroid belt without incident, but the energetic plasma associated with Jupiter nearly wiped out their electronics, providing information necessary for the design of subsequent missions. Pioneer 10 flew past Jupiter in 1974, after which it sped outward toward the limits of the solar system. Pioneer 11 undertook a more ambitious program, using the gravity of Jupiter during its 1975 encounter to divert it toward Saturn, which it reached in 1979.

FIGURE 16.1   Voyager scientists working during one of the six major planetary encounters that took place between 1979 and 1989. (NASA/JPL)

The primary scientific missions to the outer solar system were Voyagers 1 and 2, launched in 1977 (Figure 16.1). The Voyagers carried 11 scientific instruments, including cameras and spectrometers as well as devices to measure the magnetic field and plasma characteristics of planetary magnetospheres. Voyager 1 reached Jupiter in 1979 and used a gravity assist from that planet to take it on to Saturn in 1980. The second Voyager, arriving at Jupiter four months later, followed a different path to accomplish a full grand tour of the outer planets: Saturn in 1981, Uranus in 1986, and Neptune in 1989. Most of the information in this chapter and in Chapter 17 is derived from the Voyager missions.

Voyager followed a trajectory made possible by the alignment of the four giant planets on the same side of the Sun. About once every 175 years, these planets are in a position where a single spacecraft can visit them all, using gravity-assisted flybys to adjust its course for the next encounter. We are fortunate that this opportunity was seized. Because of this alignment, every planet in the outer solar system except Pluto has been visited by spacecraft; otherwise, it would probably have been well into the next century before this basic reconnaissance of the planetary system was accomplished.

The next steps in the exploration of the outer solar system involve extended study of Jupiter, Saturn, and their satellites. The Galileo mission to Jupiter was launched in 1989 and will reach the planet in 1995. It will deploy an entry probe into the planet for direct studies of the atmosphere, before beginning a three-year orbital tour, during which there will be repeated close flybys of the four large Galilean satellites. The similar Cassini mission to Saturn is under development as a cooperative venture between NASA and the European Space Agency. Cassini is planned for launch in 1996 and arrival at Saturn in 2002.

Table 16.2 summarizes the encounter dates for the Pioneer and Voyager missions to the outer solar system.

16.2 THE JOVIAN PLANETS

(a) Basic Properties

The median distance of Jupiter from the Sun is 778 million km, 5.2 times that of the Earth (5.2 AU); its period of revolution is just under 12 years. Saturn is

TABLE 16.2	MISSIONS TO THE JOVIAN PLANETS	
Planet	Spacecraft	Encounter Date
Jupiter	Pioneer 10	Dec 73
	Pioneer 11	Dec 74
	Voyager 1	Mar 79
	Voyager 2	Jul 79
	Galileo	Dec 95
Saturn	Pioneer 11	Sep 79
	Voyager 1	Nov 80
	Voyager 2	Aug 81
Uranus	Voyager 2	Jan 86
Neptune	Voyager 2	Aug 89

about twice as far away as Jupiter, at an average distance from the Sun of 1427 million km (9.6 AU). Saturn completes one revolution in very nearly the standard human generation of 30 years, providing the longest natural time interval available to ancient peoples. Uranus and Neptune are more distant yet. Uranus orbits at 19 AU with a period of 84 years, and Neptune at 30 AU requires 165 years for each circuit of the Sun. Not until 2010 will a full Neptune "year" have passed since its discovery in 1845. Some of the main properties of these four planets are summarized in Table 16.3.

Jupiter truly deserves the nickname giant planet. It is 318 times as massive as the Earth, a value that is very close to $\frac{1}{1000}$ the mass of the Sun. Its diameter is about 11 times the Earth's diameter and about $\frac{1}{10}$ the diameter of the Sun. Jupiter's density is 1.3 g/cm^3, much lower than that of any of the terrestrial planets. The mass of Saturn is 95 times that of the Earth, and its density is only 0.7 g/cm^3—the lowest of any planet and less than that of water. In fact, Saturn would be light enough to float, if an ocean large enough to contain it existed.

Uranus and Neptune are substantially smaller, with masses only about 15 times that of the Earth and hence only 5 percent as great as Jupiter. They are intermediate in size between Jupiter and Saturn and the terrestrial planets. Their densities of 1.2 g/cm^3 and 1.6 g/cm^3, respectively, are higher than that of Saturn, in spite of their smaller mass and weaker internal compression due to gravity. This must be because their composition is fundamentally different, consisting, for the most part, of

TABLE 16.3	BASIC PROPERTIES OF THE JOVIAN PLANETS					
Planet	Distance (AU)	Period (yr)	Diameter (km)	Mass (Earth = 1)	Density (g/cm³)	Rotation (hr)
Jupiter	5.2	11.9	142,800	318	1.3	9.9
Saturn	9.5	29.5	120,540	95	0.7	10.7
Uranus	19.2	84.1	51,200	14	1.2	17.2
Neptune	30.1	164.8	49,500	17	1.6	16.1

heavier materials than the hydrogen and helium that are the primary constituents of Jupiter and Saturn. We will discuss the details of their composition below.

(b)   Appearance and Rotation

When we look at each of the giant planets, we see only its atmosphere, composed primarily of hydrogen and helium gas (Figure 16.2). If any solid surface existed, it would be invisible to us, hidden by opaque clouds. The uppermost cloud deck of Jupiter and Saturn, and therefore the part of the planet we see when looking down from above, is composed of ammonia crystals. On Neptune, the upper cloud deck is methane. On Uranus, we detect no obvious cloud deck at all, but only a deep and featureless haze.

Seen through a telescope, Jupiter is a colorful and dynamic planet. Distinct details in its cloud patterns allow us to determine the rotation rate of the atmosphere at the cloud level, although such an apparent rotation of the atmosphere may have little to do with the spin of the underlying planet. Much more fundamental is the rotation of the mantle and core, as indicated by periodic variations in the magnetic field. This period of $9^h 56^m$ gives Jupiter the shortest "day" of any planet.

The rotation period of Saturn is $10^h 40^m$, as derived from variations in its radiation at radio wavelengths, which are in turn linked to its magnetic field. Uranus and Neptune have slightly longer rotation periods ($17^h 14^m$ and $16^h 6^m$, respectively), also determined from the rotation of their magnetic fields.

The axis of rotation of Jupiter is tilted by only 3°, so there are no seasons to speak of. However, Saturn does have seasons, since its axis of rotation is inclined at 27° to the perpendicular to its orbit. Neptune has about the same tilt as Saturn (29°); therefore, it experiences similar seasons. However, Uranus has an axis of rotation that is tilted by 98° with respect to the north direction. This unusual tilt creates very strange seasons, with each pole alternately tipped toward the Sun for about 40 years at a time.

(c)   Composition and Structure

Astronomers are confident that the interiors of Jupiter and Saturn are composed primarily of hydrogen and helium. Of course, these gases have been measured only in their atmospheres, but calculations first carried out nearly 50 years ago by Yale University astronomer Rupert Wildt (1905–1976) have shown that these two gases are the only possible materials out of which a planet with the observed masses and densities of Jupiter and Saturn could be constructed. There remain some uncertainties in calculating models, however, primarily because of our incomplete knowledge of the compressibility of liquid hydrogen and helium at the temperatures

FIGURE 16.2 Jupiter, as photographed by Voyager. The banded structure on the clouds represents strong east-west winds in the atmosphere. (NASA/JPL)

and pressures that exist inside Jupiter and Saturn. The best models for Jupiter predict a central pressure of over 100 million bars and a central density of about 31 g/cm³.

The internal structures of the giant planets are very different from those of the rocky inner planets, and the materials of which they are composed can take on strange forms. At depths of only a few thousand kilometers below the visible clouds of these planets, pressures become so high that hydrogen changes from a gaseous to a liquid state. Still deeper, this liquid hydrogen can act like a metal, if the pressure is great enough. The greater part of the interior of Jupiter is liquid metallic hydrogen. Because Saturn is less massive, it has only a small volume of metallic hydrogen, but most of its interior is liquid. Uranus and Neptune are probably too small to reach internal pressures sufficient to liquefy hydrogen.

Each of these planets has a core composed of heavier materials, as demonstrated by detailed analyses of their gravitational fields. Presumably these cores are the original rock-and-ice bodies that formed before the capture of gas from the surrounding nebula. The cores exist at pressures of tens of millions of bars, compared with a pressure of 4 million bars at the center of the Earth. While scientists speak of the giant planet cores being composed of rock and ice, we can be sure that neither rock nor ice assumes any familiar forms at such pressures and the accompanying high temperature. What is really meant by "rock" is any materials made up primarily of iron, silicon, and oxygen. By "ice" is meant

materials composed primarily of the elements carbon, nitrogen, and oxygen in combination with hydrogen.

Figure 16.3 illustrates the interior structures of the four jovian planets. It appears that all four have similar cores of "rock" and "ice." On Jupiter and Saturn, the cores constitute only a few percent of the total mass, consistent with the initial composition of raw materials shown in Table 16.1. However, most of the mass of Uranus and Neptune resides in these cores, demonstrating that these two planets were unable to attract massive quantities of hydrogen and helium.

It is interesting that Jupiter has very nearly the maximum possible size for a body of "cold" hydrogen, that is, one that is not generating energy as does a star. Less massive bodies than Jupiter would occupy a smaller volume (like Saturn). More massive bodies, by virtue of their greater gravitation, would also be compressed to a smaller volume than Jupiter's. Such an object, with a mass larger than Jupiter's but not large enough to maintain nuclear reactions, is called a brown dwarf or infrared dwarf (Chapter 28).

(d) Internal Heat Sources

Because of their large sizes, each of the giant planets was strongly heated during its formation by the collapse of surrounding nebular gas onto its core. Jupiter, being the largest, was by far the hottest. In addition, it is possible for giant, largely gaseous planets to generate heat after

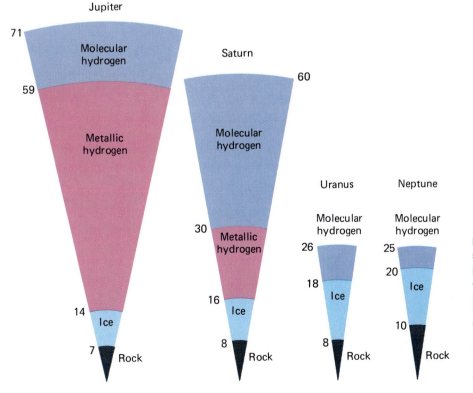

FIGURE 16.3 Internal structures of the four jovian planets, drawn to scale. Jupiter and Saturn are composed primarily of hydrogen and helium, but the mass of Uranus and Neptune consists, in large part, of compounds of carbon, nitrogen, and oxygen. Radii of regions of differing composition are labeled in units of 1000 km.

FIGURE 16.4   Jupiter's complex and colorful clouds. Also shown in this Voyager photograph are two of Jupiter's satellites: Io (*left*) and Europa (*right*). (NASA/JPL)

formation by slow contraction. A shrinkage of as little as 1 mm per year can liberate substantial gravitational energy. Nevertheless, the discovery in 1969 by Frank J. Low of the University of Arizona of an internal source of energy for Jupiter came as a considerable surprise to the astronomical community. Low made his observations with a small (12-in.) telescope carried above atmospheric water vapor in a NASA Lear Jet aircraft. Only later, after a similar source had been found for Saturn, did theorists carry out the calculations required to show that such energy sources were to be expected for the giant planets.

Jupiter has the largest internal energy source, amounting to 4×10^{17} watts. It is glowing with the equivalent of 4 million billion hundred-watt light bulbs. This level of energy is about the same as the total solar energy absorbed by Jupiter. The atmosphere of Jupiter is therefore somewhat of a cross between a normal planetary atmosphere, which obtains most of its energy from the Sun, and the atmosphere of a star, which is entirely heated from below. Most of the internal energy of Jupiter is primordial heat, left over from the formation of the planet 4.5 billion years ago.

Saturn has an internal energy source about half as large as that of Jupiter, which means (since its mass is only about one-quarter as great) that it is producing twice as much energy per kilogram of material as is Jupiter. Since Saturn is expected to have much less primordial heat, there must be another source at work generating most of this 2×10^{17} watts of power. This source is believed to be the separation of helium from hydrogen in the interior. In the liquid hydrogen mantle, the heavier helium forms drops that sink toward the core, releasing gravitational energy. In effect, Saturn is still differentiating. This precipitation of helium is possible in Saturn because it is cooler than Jupiter; at the temperatures in Jupiter's interior, hydrogen and helium remain well mixed.

Uranus and Neptune are different. Neptune has a small internal energy source, while Uranus does not emit enough internal heat for it to have been measured. As a result, these two planets have almost the same effective temperatures, in spite of the greater distance of Neptune from the Sun. No one knows why these two planets differ in their internal heat.

16.3 ATMOSPHERES OF THE JOVIAN PLANETS

(a) Atmospheric Composition

Spectroscopic observations of the jovian planets began in the 19th century, but for a long time the observed spectra could not be interpreted. For a time these planets were suspected of being self-luminous, and their spectra were thought to consist of emission bands rather than absorption features. As late as the 1930s, the most prominent absorption bands photographed in these spectra remained unidentified. Then methane (CH_4) was identified in the atmospheres of Jupiter and Saturn, followed by ammonia (NH_3).

At first it was thought that CH_4 and NH_3 might be the primary constituents of these atmospheres, but now we know that hydrogen and helium are actually the dominant gases. But neither hydrogen nor helium has easily detected spectral features, and it was not until the Voyager spacecraft measured the far-infrared spectra of Jupiter and Saturn that a reliable abundance of the elusive helium could be found on either planet. Table 16.4 summarizes the compositions of these two atmospheres.

The compositions of the two atmospheres are generally similar, except that on Saturn there is only about half as much helium—the result of the precipitation of helium that contributes to Saturn's internal energy source (Section 16.2d). The measurement by Voyager of this depletion of atmospheric helium represents an impressive confirmation of the theory that helium can precipitate in Saturn but not in Jupiter. The atmospheres of Uranus and Neptune have about the same abundance of helium relative to hydrogen as is found on Jupiter.

TABLE 16.4	ATMOSPHERIC COMPOSITIONS OF JUPITER AND SATURN (NUMBER OF MOLECULES RELATIVE TO HYDROGEN)	
Gas	Jupiter	Saturn
H_2	1	1
He	0.12	0.06
CH_4	2×10^{-3}	2×10^{-3}
NH_3	2×10^{-4}	2×10^{-5}
C_2H_2	8×10^{-7}	1×10^{-7}
C_2H_6	4×10^{-5}	8×10^{-6}
PH_3	4×10^{-7}	3×10^{-6}

(b) Clouds and Atmospheric Structure

The clouds of Jupiter are among the most spectacular sights in the solar system, much beloved by makers of science fiction films. They range in color from white to orange to red to brown, swirling and twisting in a constantly changing kaleidoscope of patterns (Figure 16.4). Saturn shows similar but very much subdued cloud activity; instead of vivid colors, its clouds are a nearly uniform butterscotch hue (Figure 16.5).

At the temperatures and pressures of the upper atmospheres of Jupiter and Saturn, methane remains a gas, but ammonia can condense, just as water vapor condenses in the Earth's atmosphere, to produce clouds. The primary clouds that we see when we look at these planets, whether from a spacecraft or through a telescope, are composed of crystals of frozen ammonia. The ammonia cloud deck marks the upper edge of the convective troposphere; above it is the cold stratosphere.

On both planets the temperature near the cloud tops is about 140 K (only a little cooler than the polar caps of Mars). On Jupiter this cloud level is at a pressure of about 0.1 bar, but on Saturn it occurs at about 1 bar of pressure. Because the ammonia clouds lie so much deeper on Saturn, they are more difficult to see, and the overall appearance of the planet is much more bland than that of Jupiter.

Within the tropospheres of these planets, the temperature and pressure both increase with depth. Through breaks in the ammonia clouds, we can see other layers of cloud that exist in these deeper regions of the atmosphere—regions that will be sampled directly by the Galileo probe. In 1995 this probe (Figure 16.6) will enter the atmosphere of Jupiter and descend to a pressure level between 10 and 20 bars before its battery power is exhausted. Below the thin ammonia clouds it should pass through a clear region, but at a pressure of about 3 bars we expect the probe to enter another thick deck of condensation clouds, composed of ammonium hydrosulfide (NH_4SH). The ammonium hydrosulfide clouds probably also contain some sulfur particles, which color them a darker yellow or brown.

As it descends to a pressure of 10 bars and ever-higher temperatures, the Galileo probe should pass next into a region of frozen water clouds, and below that into clouds of liquid water droplets perhaps similar to the common clouds of the terrestrial troposphere. This region corresponds almost to a "shirt sleeve" environment, in which astronauts could exist quite comfortably if they carried scuba gear for breathing. But with no solid surface to stop it, the probe will continue to descend, penetrating to dark regions of higher and higher pressure and temperature. No matter how strongly it was built, eventually the probe will be crushed and swallowed in the black depths, where the

FIGURE 16.5 Saturn and its rings, photographed by Voyager. The clouds of Saturn are less colorful than those of Jupiter, but the structure and dynamics of the atmosphere are similar. (NASA/JPL)

great pressures finally transform the atmospheric hydrogen into a hot, dense liquid.

Above the visible ammonia clouds the atmosphere of Jupiter is clear and cold, reaching a minimum temperature of near 120 K. At still higher altitudes temperatures rise again, just as they do in the upper atmosphere of the Earth, owing to the absorption of solar ultraviolet light. In this region photochemical reactions create a variety of fairly complex hydrocarbon compounds by the action of solar radiation on the gases in the atmosphere. A thin layer of organic aerosols—a *photochemical smog*—lies far above the visible clouds.

There is one other mystery of the jovian clouds that should be mentioned, and that concerns their colors. The ammonia condensation clouds identified on the planet should be white, like water clouds on Earth, yet we see beautiful and complex patterns of red, orange, and brown. Some additional chemical or chemicals must be present to lend the clouds such colors, but we do not know what they are. Various photochemically produced organic compounds have been suggested, as well as sulfur and red phosphorus. But there are no definite identifications, nor any immediate prospects of solving this mystery.

The atmospheric structure of Saturn is similar to that of Jupiter. Temperatures are somewhat colder, and the atmosphere is more extended as the result of Saturn's lower surface gravity, but qualitatively the same atmospheric regions, and the same condensation clouds and photochemical reactions, should be present. Figure 16.7 compares the atmospheric structures of the four jovian planets.

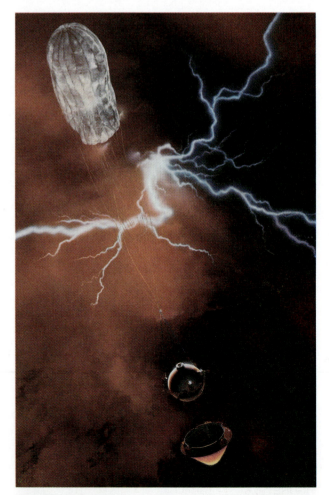

FIGURE 16.6 Artist's impression of the Galileo Probe descending through the jovian clouds in 1995. (NASA/ARC)

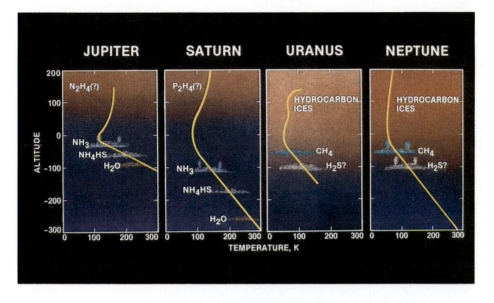

FIGURE 16.7 The structure of the atmosphere and clouds of the jovian planets. (NASA/JPL)

Unlike Jupiter and Saturn, Uranus is almost entirely featureless as seen at wavelengths that range from the ultraviolet to the infrared (Figure 16.8). Calculations indicate that the basic atmospheric structure of this planet should resemble that of Jupiter and Saturn, although the upper condensation clouds (at the 1-bar pressure level) are composed of methane (CH_4), rather than ammonia. However, the absence of an internal heat source suppresses convection and leads to a stable atmosphere with little visible structure. Also, the troposphere is hidden from view by a deep, cold, hazy stratosphere.

Neptune differs dramatically from Uranus in its appearance (Figure 16.9), although the basic atmospheric structures (and effective temperatures) are almost iden-

tical. The upper clouds are composed of methane, which forms a thin cloud near the top of the troposphere at a temperature of 70 K and a pressure of 1.5 bars. Most of the atmosphere above this level is clear and transparent, with less haze than on Uranus. Scattering of sunlight lends Neptune a deep blue color similar to that of the Earth's atmosphere. Another cloud layer exists at a pressure of 3 bars and is perhaps composed of hydrogen sulfide ice particles.

The primary difference between Uranus and Neptune is the presence on Neptune of convection currents from the interior, powered by the planet's internal heat source. These currents carry warm gas above the 1.5-bar cloud level, forming additional clouds at elevations

FIGURE 16.8 The planet Uranus as photographed by Voyager in 1986, when its rotation pole was tipped toward the Sun. (NASA/JPL)

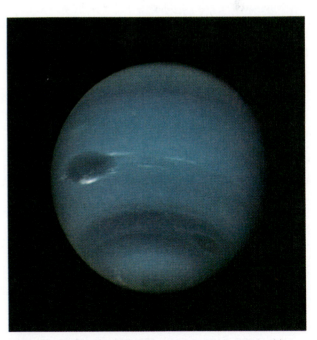

FIGURE 16.9 The planet Neptune as photographed in 1989 by Voyager. (NASA/JPL)

FIGURE 16.10 The atmosphere of Neptune. The long bright cirrus clouds are made of crystals of methane ice injected into the lower stratosphere. These clouds cast their shadows on the solid cloud layer about 75 km beneath. (NASA/JPL)

about 75 km higher. The high-altitude clouds form bright white patterns against the blue planet beneath. They can even cast distinct shadows on the methane cloud tops, permitting their altitudes to be calculated (Figure 16.10).

(c)   Winds and Weather

Observations of the changing cloud patterns in the atmospheres of the jovian planets permit us to measure wind speeds and track the circulation of the atmosphere. The atmospheric dynamics observed on these planets differ fundamentally from those of the terrestrial planets. There are three primary reasons for these differences: (1) These planets have much deeper atmospheres, with no solid lower boundary; (2) they spin faster than the terrestrial planets, suppressing north-south circulation patterns and accentuating east-west airflow; and (3) on all except Uranus, internal heat sources (Section 16.2d) contribute about as much energy as sunlight, forcing the atmospheres into deep convection to carry the internal heat outward.

The main features of the visible clouds of Jupiter are alternating dark and light bands that stretch around the planet parallel to the equator. These bands are semipermanent features, although they shift in intensity and position from year to year. Consistent with the small obliquity of Jupiter, there are no detectable seasonal effects.

More fundamental than these bands are the underlying east-west wind patterns in the atmosphere, which do not appear to change at all, even over many decades (Figure 16.11). The main such feature on Jupiter is an eastward-flowing equatorial jet stream with a speed of 300 km/hr, similar to the speed of jet streams in the Earth's upper atmosphere. At higher latitudes there are alternating east- and west-moving streams, with each hemisphere an almost perfect mirror image of the other. Saturn shows a similar pattern, but with a much stronger equatorial flow at a speed of 1300 km/hr (almost 400 m/s).

Generally, the light zones on both Jupiter and Saturn are regions of upwelling air, capped by white ammonia cirrus clouds. They apparently represent the tops of upward-moving convection currents. The darker belts are regions where the cooler atmosphere moves downward, completing the convection cycle; they are darker because there are fewer ammonia clouds and it is possible to see deeper in the atmosphere, perhaps down to the ammonium hydrosulfide clouds.

In spite of the strange seasons induced by the 98° tilt of its axis, Uranus' basic circulation is east to west, just as it is on Jupiter and Saturn. The mass of the atmosphere and its capacity to store heat are so great that the alternating 40-year periods of sunlight and darkness have little effect; in fact, Voyager measurements show that the atmospheric temperatures are a few degrees higher on the dark, winter side than on the hemisphere facing

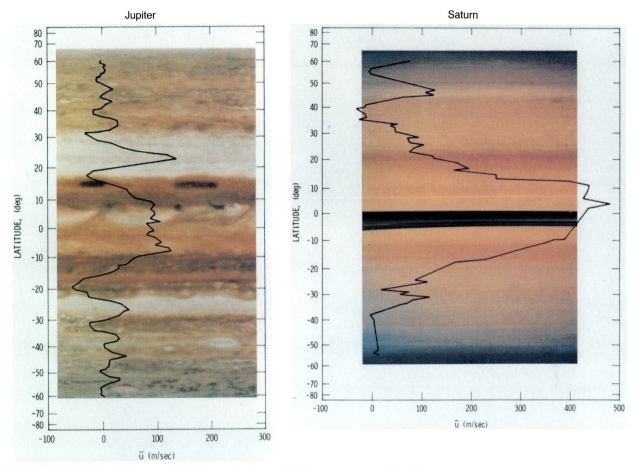

FIGURE 16.11 Zonal (east-west) winds on Jupiter and Saturn as measured by Voyager. The wind speeds are referred to the rotation of the core as determined from magnetic field and radio measurements. (NASA/JPL)

the Sun. The dynamics of these atmospheres are complex, and we do not understand the seasonal effects in detail.

Neptune's weather is characterized by strong east-west winds generally similar to those observed on Jupiter and Saturn. The highest wind speeds near the equator reach 2100 km/hr (600 m/s), nearly twice as fast as the peak winds on Saturn. The Neptune equatorial jet stream actually approaches supersonic speeds. Deep convection carries warmer gas up from the interior, also contributing to an appearance almost as striking as that of Jupiter.

(d) Storms

Superimposed on the regular atmospheric circulation patterns described above are many local disturbances—weather systems or storms, to borrow terrestrial terminology. The most prominent of these are large oval high-pressure regions on both Jupiter and Neptune.

The largest and most famous "storm" on Jupiter is the Great Red Spot, or GRS, a reddish oval in the southern hemisphere that is almost 30,000 km long—big enough to hold two Earths side by side (Figure

16.12). First seen 300 years ago, the GRS is clearly much longer-lived than storms in our own atmosphere. The GRS also differs from terrestrial storms in being a high-pressure region characterized by anticyclonic motion. The counterclockwise rotation has a period of six days. Three similar but smaller disturbances on Jupiter, called the "white ovals," formed about 1940; these are only about 10,000 km across.

We don't know what causes the GRS or the white ovals, but it is possible to understand how they can last so long once they do form. On Earth, a large oceanic hurricane or typhoon typically has a lifetime of a few weeks, or even less when it moves over the continents and encounters friction with the land. On Jupiter, there is no solid surface to slow down an atmospheric disturbance, and furthermore the sheer size of these features lends them stability. It is possible to calculate that on a planet with no solid surface the lifetime of anything as large as the GRS should be measured in centuries, while lifetimes for the white ovals should be measured in decades. These time scales are consistent with the observed lifetimes of jovian storms.

In spite of its smaller size and different cloud composition, Neptune has an atmospheric feature surprisingly

FIGURE 16.12   The Great Red Spot of Jupiter. Below it and to the right is one of the "white spots," which are similar smaller anticyclonic features. (NASA/JPL)

similar to the jovian GRS. Neptune's Great Dark Spot (Figure 16.13) is nearly 10,000 km long. Like Jupiter's GRS, it is found at latitude 20° S, and its size and shape are similar relative to the size of the planet. This Great Dark Spot rotates in an anticyclonic direction with a period of 17 days. Finally, just as with the GRS, we do not yet understand the origin of the Great Dark Spot.

16.4   MAGNETOSPHERES

Among the most dramatic features of the giant planets are their magnetospheres. Like the magnetosphere of the Earth, these regions are defined as the large cavities within which the planet's magnetic field dominates over the interplanetary magnetic field. Inside the magnetosphere, higher density plasma can be contained, and ions and electrons can be accelerated to high energies. These physical processes are similar to those dealt with by astrophysicists in many distant objects, from pulsars to quasars. The magnetospheres of the giant planets and the Earth provide the only nearby analogs of these cosmic processes that can be studied directly.

(a)   Planetary Magnetic Fields

In the late 1950s, radio energy was observed from Jupiter that is more intense at longer than at shorter wavelengths—just the reverse of what is expected from thermal radiation. It is typical, however, of the radiation emitted by electrons accelerated by a magnetic field, called **synchrotron radiation** (Chapter 30). Later observations showed that the radio energy originated from a region surrounding the planet whose diameter is several times that of Jupiter itself. The evidence sug-

FIGURE 16.13   The Great Dark Spot of Neptune. The spot is accompanied by streamers of bright methane cirrus clouds that form around it at a higher elevation. (NASA/JPL)

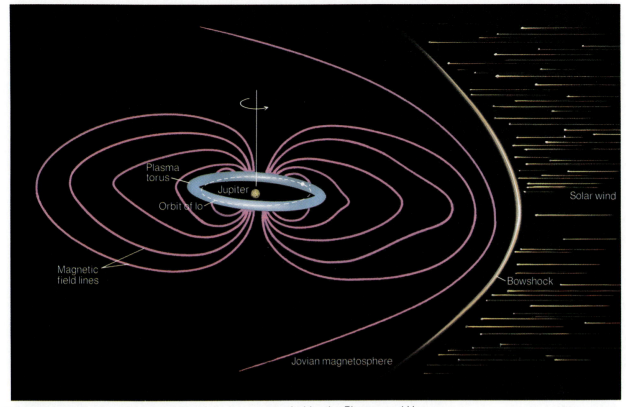

FIGURE 16.14 The magnetosphere of Jupiter as revealed by the Pioneer and Voyager missions.

gested, therefore, that there are a vast number of charged atomic particles circulating around Jupiter, spiraling through the lines of force of a magnetic field associated with the planet. This phenomenon is like the Van Allen belt around the Earth (Section 12.1).

The Pioneer and Voyager spacecraft supplemented these indirect measurements with direct studies of the magnetic field and magnetosphere of Jupiter (Figure 16.14). They found Jupiter's surface magnetic field to be 20 to 30 times as strong as the Earth's field. Because of Jupiter's great size, moreover, its total magnetic energy is enormous compared with the Earth's.

The jovian magnetic axis, like that of the Earth, is not aligned exactly with the axis of rotation of the planet, but is tipped at some 15°. The magnetic axis also does not pass exactly through the planet's center but is offset by about 18,000 km. In addition, the jovian field has the opposite polarity of the Earth's current value. However, the Earth's field is known to reverse polarity from time to time, and the same may be true of Jupiter's field.

Saturn does not emit strong synchrotron radiation, because its magnetosphere is depleted in electrons by collisions between electrons and its rings and inner satellites. It does have a substantial magnetic field, however, as discovered by Pioneer 11 and the Voyagers. Unlike the fields of the Earth and Jupiter, Saturn's field is almost perfectly aligned with its rotation axis.

The magnetic field of Uranus was not discovered until the Voyager flyby in 1986. The strength of the field is comparable to that of Saturn, about what would be expected from the size of the planet. However, the orientation of the magnetic field of Uranus is very different (Figure 16.15). Like Jupiter's field, it is offset from the center of the planet, but to a greater degree (by about one-third of the planet's radius). In addition, the magnetic field of Uranus is tilted by 60° with respect to the axis of rotation—the extreme opposite of Saturn's field.

Neptune's magnetic field was not discovered until 1989, when the Voyager 2 spacecraft reached this distant world. Its configuration is similar to that of Uranus, with a magnetic axis tilted by 55° from the rotational axis. The offset of the neptunian field is the greatest of any planet, amounting to nearly half the planet's radius. The magnetic fields of the four giant planets are compared with that of the Earth in Table 16.5.

Presumably the magnetic fields of the outer planets are generated in much the same way as the field of the Earth. All of these planets spin rapidly, so there is a ready source of energy to power their internal magnetic generators. Jupiter and Saturn have large interior regions of metallic liquid hydrogen that act like the liquid iron core of the Earth. In the case of Uranus and Neptune, however, the metallic region may be in the hydrogen-water mantle, possibly accounting for the large offset of the field from the center of the planet.

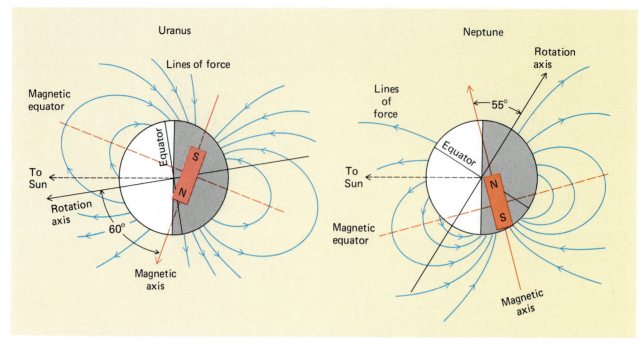

FIGURE 16.15 The magnetic fields of Uranus and Neptune as revealed by Voyager; note the large offset from the center of the planet and the tilt with respect to the planet's axis of rotation.

Although the detailed mechanisms may not be well understood, these planets seem to meet the conditions required for the generation of a planetary magnetic field in a spinning metallic core.

(b) Magnetosphere of Jupiter

The jovian magnetosphere is one of the largest features in the solar system. It is actually much larger than the Sun and completely envelops the innermost satellites of Jupiter. If we could see the magnetosphere, it would appear the size of our Moon. The total mass of the ions and electrons in the magnetosphere, however, is less than the mass of the Great Pyramid of Giza in Egypt.

On its upstream side (facing toward the solar wind), each magnetosphere is bounded by a pressure balance between the plasma inside and the solar wind streaming toward it at about 400 km/s. At the outer planets, the magnetic fields are stronger than Earth's, and the solar wind is weaker, contrib-

uting to the large size of their magnetospheres. The actual borders vary, however, with the changing pressure of the solar wind.

The magnetosphere is characterized as much by the plasma trapped within it as by the planetary magnetic field. The two primary sources of plasma in the Earth's magnetosphere are the solar wind (mostly protons and electrons) and atmospheric atoms (mostly nitrogen) that escape upward from the planet. On Jupiter both of these sources also apply, but they are supplemented by a much stronger source on the large Galilean satellites. Unlike our Moon, these satellites are enveloped by the magnetosphere. Some atoms are ejected or *sputtered* from their surfaces by the impact of energetic magnetospheric ions. An even larger source of ions is provided by oxygen and sulfur from the active volcanoes of Io. In the inner magnetosphere of Jupiter, the dominant ions are sulfur and oxygen.

The ultimate fate of ions in the jovian magnetosphere is similar to that of the ions of the terrestrial magnetosphere. For both planets, some magnetospheric particles escape (mostly down the magnetotail), and some are lost by colli-

TABLE 16.5	PLANETARY MAGNETIC FIELDS			
Planet	Average Surface Field (gauss)	Dipole Moment (Weber-m)	Tilt	Offset (planet radii)
Jupiter	4	1×10^{19}	10°	0.1
Saturn	0.2	3×10^{17}	1°	0.0
Uranus	0.3	3×10^{16}	60°	0.3
Neptune	0.2	2×10^{16}	55°	0.5
Earth	0.3	6×10^{14}	11°	0.0

sion with the planet's atmosphere (where they generate auroral discharges). In addition, Jupiter loses some by collision with its satellites, but in general more new ions are released by sputtering than old ones destroyed. Since the Earth's Moon is far outside our magnetosphere, it does not contribute as either a source or a sink of magnetospheric ions.

The ions and electrons within Jupiter's magnetosphere are accelerated by the spinning magnetic field of the planet, eventually reaching extremely high energies. It is these energetic particles that generate synchrotron radiation at radio wavelengths by processes that are similar, whether they take place at Jupiter or in a distant galaxy. One of the main challenges of space plasma physics is to understand the processes that produce such high energies. Experience with the data on the magnetospheres of the planets acquired during recent decades has demonstrated the extreme complexity of these processes, and it is clear that much additional theoretical work will be required before we can interpret observations of astrophysical sources with great confidence.

(c) The Io Plasma Torus

One of the major features of the magnetosphere of Jupiter is associated directly with its innermost Galilean satellite, Io. Io is volcanically active, erupting large quantities of sulfur and sulfur dioxide into the space surrounding it. While most of this material falls back to the surface, it is estimated that about 10 tons per second is lost to the magnetosphere. These ions of oxygen and sulfur form a donut-shaped plasma torus surrounding Jupiter approximately at Io's orbit, at a distance of five Jupiter radii from the planet. In addition to being investigated directly by spacecraft, the Io plasma torus has been imaged telescopically from Earth in the glow emitted by oxygen and sulfur atoms as they recapture electrons from the plasma.

The energetic ions of the plasma torus and its surroundings would be very dangerous to both spacecraft and humans, if any should ever venture close to Jupiter. When these ions strike a solid surface, they damage it directly and also generate lethal x rays. On or near Io, a human could survive for only a few minutes. Special shielding would be required for spacecraft electronics, and no spacecraft has been built that could last more than a few hours in this environment. Thus an Io lander or orbiter is far beyond our present capabilities, and it is probably safe to predict that human exploration of the inner jovian system will never be possible.

As Io orbits Jupiter in this sea of energetic plasma, it generates an electric current that flows along magnetic field lines between the satellite and the planet. This magnetic flux tube acts like a wire carrying a current estimated at 5 million amperes. Where it reaches the atmosphere of Jupiter, this current stimulates strong radio noise emissions. These emissions were picked up by radio astronomers many years before spacecraft reached Jupiter, and their association with the orbital position of Io was established.

Ions from the Io plasma torus also strike the upper atmosphere of Jupiter at high latitudes, producing auroral glows that were measured by the Voyager spacecraft (Figure

FIGURE 16.16   Jupiter at night. This long-exposure Voyager picture shows auroral glows in the upper atmosphere due to the magnetosphere as well as lightning flashes illuminating the clouds on the planet below. (NASA/JPL)

16.16) and should be observable by the Hubble Space Telescope.

(d) Magnetospheres of Saturn, Uranus, and Neptune

The magnetospheres of the other three jovian planets are generally similar to that of Jupiter. The physical dimensions of Saturn's magnetosphere are about one-third as great, and those of Uranus and Neptune are still smaller, approximately in proportion to the sizes of the planets themselves.

The density of the solar wind declines with distance from the Sun, and therefore the pressure exerted on each magnetosphere by the solar wind also decreases with distance. The upstream boundary of the jovian magnetosphere is usually between 50 and 100 times the radius of Jupiter, or roughly 5 million km. That of Saturn is at 20 to 40 Saturn radii, or about 2 million km. The boundary for Uranus is near 20 Uranus radii, or about 0.5 million km, and that of Neptune is slightly smaller yet. On the downstream side all of the magnetospheres extend much farther in a magnetotail, and the boundary is even more variable.

All four outer-planet magnetospheres differ with respect to the plasma composition, and each has different sources and sinks of atomic particles. All of the magnetospheres obtain a part of their ions from the solar wind (mostly protons and helium nuclei, since these are the primary constituents of their solar wind). Some of the ions also originate from the atmosphere of the planet (also mostly protons, since hydrogen is the dominant gas in all four giant planets). Only Jupiter has a large ion source from the surfaces of its satellites (the sulfur and oxygen ejected by the volcanoes of Io). Saturn, however, has a major source of ions in the atmosphere of its large satellite Titan, which is constantly losing nitrogen to the saturnian magnetosphere. Both Uranus and Neptune derive some oxygen ions from sputtering of ice on the surfaces of their icy satellites, but their magnetospheres consist primarily of protons and electrons, derived from the planet's atmosphere and the solar wind.

The ultimate fates of ions in the magnetospheres of these four planets are more nearly the same. All magnetospheres lose some ions by escape and some by collision with the planet's atmosphere. All also lose ions by collision with satellites and rings. These effects are most important for Saturn. The large surface areas of its inner satellites and especially its rings absorb almost all of the plasma in the inner magnetosphere. Therefore, the magnetosphere of Saturn is almost empty in comparison with that of Jupiter. The situation is more complicated for Uranus and Neptune, because of the large tilt of their magnetic fields, which amplifies the loss rate from collisions with rings and inner satellites.

Table 16.6 summarizes the main properties of the magnetospheres of the giant planets.

16.5   PLUTO AND ITS MOON

(a)   Orbit and General Characteristics of Pluto

Pluto is neither a jovian planet nor a terrestrial planet. It is a small object (diameter about 2300 km), perhaps one of many such bodies that once orbited the Sun beyond the giant planets. Several other objects have recently been found in similar orbits (see Chapter 18), and many others probably still remain undiscovered on the outer edges of the solar system. Because it is so much smaller than the other planets, some have even suggested that Pluto itself should be classed as a minor planet or asteroid.

Pluto's orbit has the highest inclination to the ecliptic (17°) of any planet and also the largest eccentricity (0.248). Its median distance from the Sun is 40 AU, or 5.9 billion km, but its perihelion distance is under 4.5 billion km, within the orbit of Neptune. Pluto passed its perihelion in 1989, and it remains inside the orbit of Neptune until 1999. Even though the orbits of these two planets cross, there is no danger of collision; because of its high inclination, Pluto's orbit clears Neptune's by 385 million km. Pluto completes its orbital revolution in a period of 248.6 years. Since its discovery in 1930, the planet has traversed less than one-quarter of its long path around the Sun.

(b)   Discovery of Charon

Pluto has a satellite, discovered in 1978 by James W. Christy of the U.S. Naval Observatory. He noticed a peculiarity on the images of Pluto obtained with the observatory's 1.5-m telescope at Flagstaff. The images of Pluto were slightly elongated, while those of stars on the same photograph were not (Figure 16.17). A check of the observatory records revealed that some of the other images taken under excellent seeing conditions also showed this image distortion, although most did not. Such an effect could take place if there were a second, unresolved source in a periodic orbit around Pluto. A better image of Pluto and Charon, obtained by the Hubble Space Telescope, is shown in Figure 16.18.

Christy followed up on this hunch and found that all of the observations could be matched if the satellite had a period of 6.387 days and a maximum separation from Pluto of just under 1 arcsec. This is the same as Pluto's rotation period, showing that the planet keeps the same side always turned toward its satellite. Since the satellite also almost surely has the same rotational period, we have here the only planet-satellite system in which both members are tidally locked together.

The exact nature of the orbit of Pluto's satellite was not confirmed until 1985, when the system had turned to the point at which the satellite and the planet began to **occult** each other—that is, to pass alternately in front of one another—on each satellite orbit. These occultation observations established the satellite orbit and indicated that it had a diameter of about 1200 km, more than half the size of Pluto itself. In 1985 the

TABLE 16.6	MAGNETOSPHERES OF THE GIANT PLANETS		
Planet	Radius (R_{planet})	Ion Source	Ion Composition
Jupiter	50–100	Solar wind	H^+
		Atmosphere	H^+
		Sputtering	H^+, O^+
		Io volcanoes	O^+, S^+
Saturn	15–25	Solar wind	H^+
		Atmosphere	H^+
		Titan	N^+
		Rings and satellites	H^+, O^+
Uranus	15–20	Solar wind	H^+
		Atmosphere	H^+
Neptune	15–20	Solar wind	H^+
		Atmosphere	H^+

FIGURE 16.17 Highly enlarged negative image of Pluto on a photograph made at the U.S. Naval Observatory at Flagstaff, Arizona. The "bump" on the upper right is Pluto's satellite, Charon. (U.S. Naval Observatory)

International Astronomical Union named it Charon for the boatman of Greek mythology who transported souls to Hades, the realm of Pluto.

Confirmation of the orbit of Charon also clarified what had been suspected for some time—that Pluto's rotation is retrograde. The obliquity of Pluto's rotational axis is 112°, similar to that of Uranus. At present, the equator of Pluto faces approximately toward the Sun, but in about 60 years the pole will nearly point to the Sun, the way Uranus' pole does now.

(c) Surfaces of Pluto and Charon

Pluto has not been visited by spacecraft, and it is so faint that studies require the use of the largest telescopes in the world. However, thanks to modern astronomical detectors and the mutual occultations of Pluto and Charon, astronomers have acquired a surprising amount of information about these objects.

As described in Chapter 11, the search for a ninth planet was motivated by the desire of Percival Lowell and others to explain observed discrepancies in the orbit of Uranus. When Pluto was discovered, astronomers assumed that it had a mass several times larger than the mass of the Earth, as indicated by the apparent perturbations of Uranus. However, this large mass was not consistent with the apparent small size of Pluto, unless Pluto had an unreasonably large density (greater than 25 g/cm^3). Was the mass of Pluto (and hence the basis for its prediction) wrong? The discovery of Charon provided a means to measure the mass of Pluto accurately using Kepler's laws. Its mass is $1/400$ the mass of the Earth, insufficient to have caused observable perturbations in Uranus or Neptune. A recent reanalysis of the observations shows that the discrepancies were not real in any case, and both Uranus and Neptune are following the orbits they should in the absence of any unknown planet.

The diameter of Pluto is only 60 percent as large as the Moon. From the diameter and mass, we find a

density of 2.1 g/cm^3, suggesting that Pluto is composed in part of water ice. We will see in the next chapter that this density is higher than that of most of the satellites of Jupiter and Saturn, but similar to that of Neptune's satellite Triton. Detailed calculations indicate that the composition of Pluto is about 75 weight-percent rock and metal with 25 weight-percent H_2O ice.

Pluto's surface is highly relective, suggestive of ice or frost. Spectra of reflected sunlight obtained by NASA astronomer Dale P. Cruikshank and his colleagues show that the primary surface frost is frozen nitrogen, with small (about 1 percent each) quantities of frozen carbon monoxide (CO) and methane (CH_4). The temperature of most of the surface is between 40 and 45 K; if it were higher, nitrogen would not be stable as ice but would rapidly evaporate. Pluto is now in its warmest period, near perihelion, and its surface temperature is expected to drop by about 10 K over the next few decades as it moves farther from the Sun.

Charon has about half the diameter of Pluto (1200 km), making it the largest satellite in the solar system relative to its primary planet. Unlike Pluto, the surface

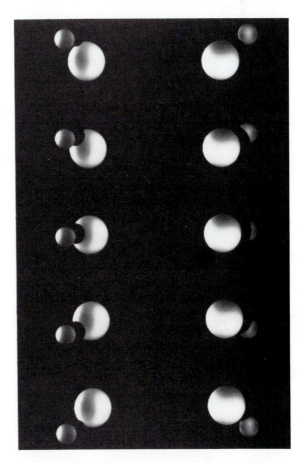

FIGURE 16.18 The appearance of Pluto and Charon as modeled in their computer by Marc Buie (Lowell Observatory) and David Tholen (University of Hawaii), based on observations of mutual occultations. On the left, Charon moves in front of Pluto; on the right, it moves behind the planet. (Space Telescope Science Institute, courtesy of Marc Buie)

Ground based   Hubble Space Telescope

FIGURE 16.19   The Hubble Space Telescope's image showing Pluto and Charon as distinct objects, compared with a ground-based image of Pluto and Charon taken with the Canada-France-Hawaii Telescope on Mauna Kea.

TABLE 16.7 BASIC PROPERTIES OF PLUTO	
Distance (average)	39.5 AU
Period of revolution	248.6 yr
Diameter	2310 km
Mass (Earth = 1)	0.0025
Density	2.1 g/cm^3
Period of rotation	6.39 days
Surface composition	Nitrogen ice
Atmospheric composition	N$_2$
Atmospheric surface pressure	6 × 10^{-5} bar

of Charon shows the spectral signature of H$_2$O ice. Astronomers do not know whether it contains frozen nitrogen as well, nor is it clear why there is no spectral evidence of CH$_4$. Perhaps Charon once contained CH$_4$ but was not able to retain this gas, because of its lower surface gravity or higher surface temperature.

The probable appearance of the Pluto-Charon system is illustrated in Figure 16.19. These images, based on a detailed model derived from the mutual occultation observations of 1985–89 by Mark Buie of Lowell Observatory and David Tholen of the University of Hawaii, illustrate a specific pair of occultation events.

(d)   Pluto's Atmosphere

The presence of frozen nitrogen on the surface of Pluto suggests that an atmosphere might be present, since nitrogen evaporates at low temperatures. This atmosphere was first measured in 1988 during an occultation of a star by Pluto, when the light rays from the star were bent slightly by the presence of a tenuous envelope of gas. The atmosphere consists mostly of nitrogen with trace quantities of CO and CH$_4$. Its surface pressure is only 60 millionths of a bar. Calculations suggest that the atmosphere is continuously escaping from the planet and being lost to space. Thus Pluto constantly loses mass, but at a very slow rate.

Since Pluto is near perihelion, its surface temperature is relatively high and the atmosphere has its maximum extent and rate of loss. As it moves farther from the Sun, the atmosphere will shrink and the loss of nitrogen will decrease. Calculations indicate that it will shrink by about a factor of two every decade from now until the middle of the 21st century.

The basic facts about Pluto are summarized in Table 16.7. It is unlike any other planet, but as we will see in the next chapter, Pluto resembles several of the icy satellites of the outer planets, particularly Neptune's satellite Triton. Triton seems to have the same size and composition as Pluto, and currently they are nearly at the same distance from the Sun. Both also have thin, continually escaping atmospheres made primarily of nitrogen. When we look at the Voyager discoveries about Triton in the next chapter, we should consider that in this satellite we may be seeing a twin of Pluto.

S U M M A R Y

16.1   The outer solar system contains the four jovian planets and Pluto. The chemistry is generally reducing, and Jupiter and Saturn have an overall composition similar to that of the Sun. Exploration has been carried out by Pioneers 10 and 11 and by the two Voyager spacecraft. Voyager 2, perhaps the most successful of all space science missions, successfully explored Jupiter (1969), Saturn (1981), Uranus (1986), and Neptune (1989)—a grand tour of the jovian planets.

16.2   Jupiter is 318 times more massive than the Earth. Saturn is about 25 percent, and Uranus and Neptune are only 5 percent, as massive as Jupiter. All four have deep atmospheres and opaque clouds, and all rotate quickly (periods from 10 to 17 hr). Jupiter and Saturn have extensive mantles of liquid hydrogen. Uranus and Neptune are depleted in hydrogen and helium relative to Jupiter and Saturn (and the Sun). Each jovian planet has a core of "ice" and "rock" of about ten Earth masses. Jupiter, Saturn, and Neptune have major internal heat sources, obtaining as much (or more) energy by convection from their interiors as by radiation from the Sun. Uranus has no measurable internal heat.

16.3   The four jovian planets have generally similar atmospheres, composed mostly of hydrogen and helium. There are small quantities of methane (CH_4) and ammonia (NH_3) gas, both of which also condense to form clouds. Deeper (invisible) cloud layers consist of H_2O and possibly NH_4SH (Jupiter and Saturn) and H_2S (Neptune). We do not know what colors the clouds of Jupiter are. Atmospheric dynamics are dominated by east-west circulation. Jupiter displays the most active cloud patterns, with Neptune second. Saturn is generally bland, and Uranus is featureless (perhaps owing to its lack of an internal heat source). The two major storms (the Great Red Spot on Jupiter and the Great Dark Spot on Neptune) are similar anticyclonic, oval-shaped, high-pressure systems.

16.4   The jovian planets have substantial magnetic fields, approximately in proportion to their sizes. Within their large magnetospheres, trapped ions and electrons are accelerated to high energies and emit **synchrotron radiation.** Jupiter has the most active magnetosphere, partly because of the ions provided by the volcanoes of Io, which also produce the Io plasma torus.

16.5   Pluto was discovered in 1930, and its satellite Charon was discovered in 1978. Pluto is the smallest planet, and its orbit has the largest inclination and eccentricity. Its mass is only ¼₀₀ that of the Earth. The Pluto-Charon pair is remarkable, however, in having two objects of so nearly comparable size orbiting each other and tidally locked together. The mutual **occultations** of 1985 to 1990 have yielded much information, including the sizes of Pluto and Charon and their surface compositions. Pluto also has an atmosphere, which is near its maximum extent in the late 20th century, since the planet is near perihelion and hence relatively warm (45 K).

E X E R C I S E S

THOUGHT QUESTIONS

1. We often speak of the giant planets having approximately solar or cosmic composition. Is this strictly true? To what degree does each of these planets depart from cosmic composition?

2. What would you expect a planet with five times the mass of Jupiter to be like?

3. Jupiter is denser than water, yet it is composed, for the most part, of two light gases, hydrogen and helium. How can it be so dense?

4. Would you expect to find oxygen gas in the atmospheres of the giant planets? Why or why not?

5. The water clouds believed to be present on Jupiter and Saturn exist at temperatures and pressures similar to those at locations in the clouds in the terrestrial atmosphere. What would it be like to visit such a location on Jupiter or Saturn? In what ways would the environment differ from that in the clouds of Earth?

6. Describe the different processes that lead to substantial internal heat sources for Jupiter and Saturn. Since these two objects generate much of their energy internally, should they be called stars instead of planets? Justify your answer.

7. Describe the seasons of Uranus, and compare them with the seasons of the Earth.

8. It has been suggested that Pluto should be called an asteroid (minor planet) rather than a "real" planet. Why (or why not) is this a good idea?

PROBLEMS

9. Calculate the flight times to Saturn, Uranus, and Neptune for simple spacecraft orbits in which the Earth is at perihelion and the target planet is at aphelion. Compare these with the flight times for the Voyager 2 mission. Why are they so different?

10. Jupiter's Great Red Spot rotates in 6 days and has a circumference equivalent to a circle with radius of 10,000 km. Neptune's Great Dark Spot is one-quarter as large and rotates in 17 days. For each, calculate the wind speeds at the outer edges of the spots. How do these compare with the winds in terrestrial hurricanes?

11. As the Voyager spacecraft penetrated the outer solar system, the illumination from the Sun declined. Relative to the situation on Earth, how bright is the sunlight at each of the jovian planets?

★12. Estimate how frequently all four giant planets are approximately in alignment (for example, with celestial longitudes differing by no more than 60°), permitting a "grand tour" trajectory like that of Voyager 2.

Among the wonders of the outer solar system is the eery, multihued surface of Jupiter's moon Io, shown here in a Voyager photograph. The dark features and craters are volcanic, while the origin of mountains (such as the large mountain in the lower right) remains mysterious even today.
(NASA/JPL)

Unlike some of the inner planets, the giant outer planets do not orbit the Sun alone. They are accompanied by busy systems of satellites and rings, whose investigation by Voyager is thought by many people to represent the high point of the space program. In just one decade, from 1979 to 1989, Voyager dazzled us with discoveries undreamed of by scientists, such as the huge volcanic plumes on Jupiter's moon Io, braided and kinky rings around Saturn, and nitrogen geysers on Neptune's moon Triton. As a result we recognize that the solar system is teeming with fascinating worlds, presenting a much richer variety of places and processes than we find if we limit ourselves to the nine planets.

In contrast with the liquid giants, the satellites of the outer solar system have solid surfaces that have recorded their evolutionary history, providing insights into the formation and evolution of the solar system. Because they were formed so far from the Sun, they are composed in large part of water ice. We all know that an ice cube is different from a pebble, and we would not expect these icy satellites to have the same properties as rocky worlds. But if it is sufficiently cold, an ice cube can be as hard as a pebble. Perhaps it is not too surprising, then, to see the similarities between the icy satellites and inner planets: both have experienced their own versions of tectonics and volcanism. Much of the focus of this chapter is upon comparing the geology and chemistry of the outer satellites with those of more familiar rocky planets, including our own Earth.

SATELLITES AND RINGS

Edward C. Stone (b. 1936) of Caltech is a space plasma physicist who has played a central role in the study of the outer planets and their satellites and rings. As Project Scientist for Voyager, Stone led NASA's most successful mission of exploration since Apollo, revealing for the first time the ring and satellite systems of the outer planets. Stone, who is now Director of the Jet Propulsion Laboratory, was also a central player in the development of the 10-m Keck telescope at Mauna Kea, Hawaii.

The giant planets are accompanied by satellites, which orbit them like planets in a miniature solar system. Although small in comparison to the planets they accompany, several of these satellites are larger than the planet Mercury, and many of them show evidence of a surprising degree of geological and atmospheric evolution. Equally interesting in the outer solar system are the ring systems, which provide fascinating examples of physical processes with applications ranging from the formation of planetary systems to the spiral structure of galaxies.

17.1 RING AND SATELLITE SYSTEMS

(a) General Properties

The rings and satellites of the outer solar system are chemically distinct from objects in the inner solar system, as is to be expected from the fact that they formed in regions of lower temperature. The primary difference, as we have frequently noted, was the availability of large quantities of water ice as building material for bodies beyond the asteroid belt. Notable, in addition, is the presence of dark, organic compounds formed in the solar nebula. Mixed with the ice that is present in these objects, this dark, primitive material often results in low reflectivities. Paradoxically, therefore, the ring and satellite systems contain many objects that are both icy and black.

Most of the satellites in the outer solar system are in direct or regular orbits; that is, they revolve about their parent planet in an east-to-west direction and very nearly in the plane of the planet's equator. Ring systems are also in direct revolution in the planet's equatorial plane. Such objects probably formed at about the same time as the planet by processes similar to those that formed the planets in orbit around the Sun.

In addition to the regular satellites, there are irregular satellites that orbit in a retrograde (west-to-east) direction or have orbits of high eccentricity or inclination. These are usually smaller satellites, located relatively far from their planet, and they were probably formed between the planets and subsequently captured into orbit.

We have said that ice is a major component of most of these objects, but how do we know this to be the case? There are two ways, and they complement each other. First, we can analyze the spectrum of reflected sunlight to determine if any of the prominent infrared ice absorption bands are present. Using this approach, investigators have found that many satellites and the rings of Saturn have icy surfaces. However, such spectra refer only to the surface, and they may not be representative of the bulk composition of the satellites.

Estimating the composition of planetary interiors is never easy, since they cannot be sampled directly. But as we saw for the inner planets, a useful indication of bulk composition is provided by a measurement of density.

Most of the measured densities for outer-planet satellites are less than 2.0 g/cm^3, in contrast to the larger densities of terrestrial bodies. Since water ice is the main low-density material expected from condensation in the solar nebula, it is natural to try to reproduce these densities with combinations of rock and ice. Most of these measured densities are matched by compositions that include 30 to 50 percent ice (by mass).

Planet-sized objects composed in part of ice are unlikely to experience the same sort of geological evolution as the rocky worlds studied in Chapters 12 to 15. One difference results from the low melting temperature of ice. Only a little internal heating is required to melt these objects, resulting in rapid differentiation. Ice also expands and contracts differently from rock as its temperature changes, producing different tectonic stresses in the planetary crust. Even before the Voyager flights, there was reason to expect some geological surprises in the outer solar system.

(b)   The Jupiter System

Jupiter has 16 satellites and a faint ring. The 16 satellites include the 4 large Galilean satellites (Figure 17.1) dis-

FIGURE 17.1 The four large Galilean satellites of Jupiter shown with the planet Jupiter in a combination of Voyager images. (NASA/JPL)

covered in 1610 by Galileo: Callisto, Ganymede, Europa, and Io. The smallest of these, Europa and Io, are about the size of the Moon. The largest, Ganymede and Callisto, are larger than Mercury.

The other 12 jovian satellites are much smaller. They divide themselves conveniently into three groups of four each. The inner four all circle the planet inside the orbit of Io; one of these, Amalthea, has been known for about a century, but the other three were discovered by Voyager. The outer satellites consist of four in direct but highly inclined orbits and four farther out in retrograde orbits. These eight are believed to be captured objects. The two groupings may indicate two parent bodies that were broken up in a collision early in the history of the jovian system. These eight outer satellites are dark, apparently primitive objects.

(c)   The Saturn System

Saturn has 19 known satellites in addition to its magnificent rings. The largest of the satellites, Titan, is almost as big as Ganymede in the jovian system, and it is the only satellite with a substantial atmosphere. The composition of six other regular satellites, with diameters between 400 and 1600 km, is about half water ice. Saturn also has two distant irregular satellites, one of which (Phoebe) is in a retrograde orbit.

The rings of Saturn are broad and flat, with only a few gaps. Individual ring particles are composed of H_2O ice and are typically the size of tennis balls. Gravitational interactions between various small inner satellites and the rings are responsible for much of the detailed ring structure observed by Voyager.

(d)   The Uranus System

The regular ring and satellite system of Uranus shares the 98° tilt of the planet. It consists of 11 rings and 15 regular satellites. The five largest satellites are similar in size to the regular satellites of Saturn, with diameters of 500 to 1600 km, while the ten smaller satellites and the ring particles are very dark, reflecting only a few percent of the sunlight that strikes them.

The rings of Uranus, discovered in 1977, are narrow ribbons of material with broad gaps between—fundamentally different from the broad rings of Saturn. Presumably the ring particles are confined to these narrow paths by the gravitational effects of small satellites.

(e)   The Neptune System

Neptune has eight satellites: six regular satellites close to the planet and two irregular satellites. The most interesting of these is Triton, a relatively large satellite in a retrograde orbit. Triton has an atmosphere, and active volcanic eruptions were discovered there by Voyager in its 1989 flyby.

TABLE 17.1	THE LARGEST SATELLITES			
Name	Diameter (km)	Mass (Moon = 1)	Density (g/cm³)	Reflectivity (percent)
Ganymede	5280	2.0	1.9	40
Titan	5150	1.9	1.9	20
Callisto	4820	1.5	1.8	20
Io	3640	1.2	3.5	60
Moon	3476	1.0	3.3	12
Europa	3130	0.7	3.0	70
Triton	2710	0.3	2.1	80

The rings of Neptune are narrow and faint. Like those of Uranus, they are composed of dark materials. One ring is distinguished by the presence of three bright regions that represent unexplained concentrations of ring material.

17.2   LARGE SATELLITES

In this section we discuss the six large satellites of the outer solar system. Table 17.1 summarizes their properties, with the Moon listed for comparative purposes.

(a)   The Three Largest Satellites

The three largest satellites are Ganymede and Callisto in the jovian system and Titan in the saturnian system. All three of these have about the same diameter (from 5280 km for Ganymede down to 4820 km for Callisto) and nearly identical density (1.9 g/cm³). Each therefore appears to have the same composition, and we would expect them to have experienced parallel, and perhaps nearly identical, evolution. It is thus with considerable interest that we note that the three are different in several fundamental ways.

Since they have the same size and composition, these three largest satellites probably have the same general interior structure. M.I.T. chemist John Lewis first argued in about 1970 that such objects are almost surely differentiated into a central, Moon-sized core of rock and mud, surrounded by a thick mantle of ice or, possibly, liquid water. The crust would be hard, brittle ice at the temperature prevalent on these satellites (Figure 17.2).

Both Callisto and Ganymede apparently conform to this expectation. Titan may be geologically similar, but unfortunately we have limited knowledge of its surface, since Titan, like Venus, is hidden under opaque clouds. One of the most obvious questions to ask when comparing these three objects is why Titan developed an extensive atmosphere, while Callisto and Ganymede have none. First, however, let us look at the geological

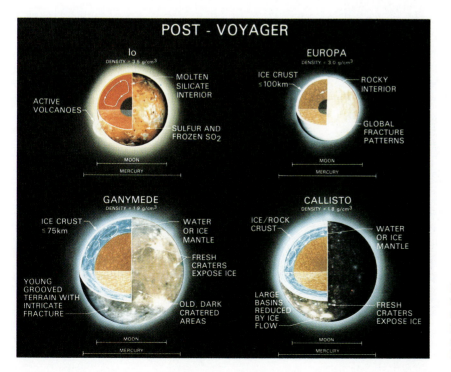

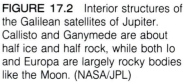

FIGURE 17.2   Interior structures of the Galilean satellites of Jupiter. Callisto and Ganymede are about half ice and half rock, while both Io and Europa are largely rocky bodies like the Moon. (NASA/JPL)

FIGURE 17.3 The heavily cratered surface of Jupiter's outermost Galilean satellite, Callisto. (NASA/JPL)

record preserved on the surfaces of Ganymede and Callisto.

(b)   Geology of Ganymede and Callisto

Callisto and Ganymede provide an excellent introduction to the geology of icy worlds. We begin with Callisto, the simpler of the two. The entire surface of Callisto is covered with impact craters, like the lunar highlands (Figure 17.3). The existence of this heavily cratered surface tells us three important things not known before Voyager: (1) An icy planet retains impact craters in its surface if its temperature is low enough; (2) there was a heavy bombardment by debris in the outer solar system as well as nearer the Sun; and (3) Callisto has experienced little, if any, geological activity other than impacts for a long time—probably billions of years.

The craters of Callisto do not look exactly like their counterparts in the inner solar system. They tend to be much flatter, as if the surface did not have the strength to support much vertical relief. Such subdued topography is to be expected for an ice crust at the temperatures of 130 to 140 K measured near local noon on Callisto, since ice loses some of its strength as it is warmed. Farther from the Sun, in the Saturn system, temperatures are so low that ice is as strong as rock.

Ganymede, the largest satellite in the solar system, is also cratered, but less so than Callisto. About one-third

of its surface seems to be contemporary with Callisto; the rest formed later, after the end of the heavy bombardment period. This younger terrain on Ganymede is probably about as old as the lunar maria or the martian volcanic plains, judging from crater counts.

The younger terrain on Ganymede (Figure 17.4) was produced when tectonic forces (probably due to expansion of ice in the mantle) cracked the crust, flooding many of the craters with water from the interior and forming extensive parallel mountain ridges. This mountainous terrain, with its ridges evenly spaced about 15 km apart, covers more than one-quarter of the surface. There is even evidence that blocks of the older, heavily cratered terrain may have rotated or slipped at the time the younger crust was forming, providing a surprising analog of Earth's plate tectonics on this icy satellite.

Ganymede experienced expansion and consequent resurfacing during the first billion years after its formation, while Callisto did not. Apparently the small difference in size between the two led to this difference in their evolution.

(c)   The Atmosphere of Titan

Titan's atmosphere was discovered in 1944 by Gerard P. Kuiper (1905–1973), one of the few professional astronomers who worked in the area of planetary studies during the middle part of this century. His spectra, obtained at the McDonald Observatory in Texas, showed absorptions due to methane gas. Subsequent observations established the presence of dense clouds, obscuring the surface from our view. By the time of the Voyager flyby in 1980, many scientists suspected that the atmosphere of Titan might be as substantial as that of the Earth.

The Voyager 1 flyby of Saturn was designed to yield as much information as possible about Titan. Voyager passed within 4000 km, and it also flew behind the satellite as seen from the Earth, producing an occultation. In this way, its radio signal traversed successive paths through Titan's atmosphere, generating data from which scientists could reconstruct the atmospheric profile all the way down to the invisible surface. The measured surface pressure was 1.5 bars, higher than that on any of the terrestrial planets except Venus.

The composition of Titan's atmosphere is primarily nitrogen, another respect in which Titan resembles the Earth. Methane and argon amount at most to a few percent each. Additional compounds detected spectroscopically in Titan's upper atmosphere include carbon monoxide (CO), various hydrocarbons, and nitrogen compounds such as hydrogen cyanide (HCN) (Table 17.2). The discovery of HCN was particularly interesting, since this molecule is the starting point for formation of some of the components of deoxyribonu-

FIGURE 17.4 Jupiter's largest satellite, Ganymede. (a) Global view. (b) Detail of a complex region that includes both old and modified terrain. The width of the frame (b) is 1000 km. (NASA/JPL)

cleic acid (DNA), the fundamental genetic molecule essential to life on Earth.

There are multiple cloud layers on Titan (Figure 17.5). The lowest clouds are in the troposphere, within the bottom 10 km of the atmosphere; these are condensation clouds composed of methane. Methane plays the same role in Titan's atmosphere as water does on Earth;

the gas is only a minor constituent of the atmosphere, but it condenses to form the major clouds in the troposphere. Much higher, photochemical reactions have produced a dark reddish haze or smog consisting of complex organic chemicals. Formed at an altitude of several hundred kilometers, this aerosol slowly settles downward, where it presumably has built up a deep

TABLE 17.2	COMPOSITION OF TITAN'S ATMOSPHERE
Gas	Abundance
Major Components (bulk atmosphere)	
Nitrogen (N_2)	90–98%
Argon (Ar)	0–10%
Methane (CH_4)	1–5%
Hydrogen (H_2)	0.2–0.6%
Hydrocarbons (stratospheric)	
Ethane (C_2H_6)	13 ppm
Acetylene (C_2H_2)	2 ppm
Ethylene (C_2H_4)	0.1 ppm
Propane (C_3H_8)	0.7 ppm
Diacetylene (C_4H_2)	0.002 ppm
Methylacetylene (CH_3C_2H)	0.004 ppm
Nitriles (stratospheric)	
Hydrogen cyanide (HCN)	0.2 ppm
Cyanoacetylene (HC_2CN)	0.002 ppm
Cyanogen (C_2N_2)	0.002 ppm
Oxygen compounds (stratospheric)	
Carbon monoxide (CO)	60 ppm
Carbon dioxide (CO_2)	0.01 ppm

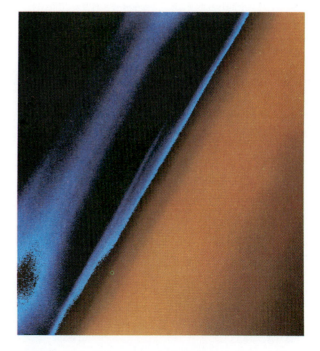

FIGURE 17.5 Enhanced color photograph of the upper atmosphere of Titan, showing multiple haze layers. (NASA/JPL)

FIGURE 17.6 Europa has a surface of water ice, crossed by complex cracks and low ridges. The width of the frame is 1000 km, the same as Figure 17.4b. (NASA/USGS)

layer of tar-like organic chemicals on the surface of Titan.

Titan's surface temperature is about 90 K, held uniform by the blanketing atmosphere. At such a low temperature, there may be seas of liquid methane and ethane. Organic compounds are chemically stable at Titan's temperatures, unlike the situation on the warmer, oxidizing Earth. Therefore Titan's surface probably records a chemical history that goes back billions of years. Many people believe that this satellite

will provide more insights into the early history of Earth's atmosphere, and even into the origin of life, than any other object in the solar system.

Why does Titan have an atmosphere while Ganymede and Callisto do not? Partly it is because Titan is farther from the Sun and therefore colder. At low temperatures the molecules in the atmosphere move more slowly and are less likely to escape. But the primary reason must be that Titan outgassed from its interior more gas than was ever present on the two jovian satellites. At the distance from the Sun where Titan formed, small but significant amounts of methane and ammonia were present, mixed with water ice, while Ganymede and Callisto apparently had none. Subsequently, photochemical reactions dissociated most of the ammonia (NH_3) in Titan's atmosphere into hydrogen and nitrogen. The light hydrogen molecules escaped into space, leaving behind the heavier molecules of nitrogen.

(d)   Europa and Io

Europa and Io, the inner two Galilean satellites, are not icy worlds like most of the satellites of the outer planets. Similar to our Moon in density and size, they appear to be predominantly rocky objects. How did they fail to acquire the ice that must have been plentiful at the time of their formation? The most probable cause is Jupiter itself, which became hot and radiated a great deal of infrared energy during the first few million years after its formation. Temperatures therefore rose in the disk of material near the planet, and the ice evaporated, leaving Europa and Io with compositions more appropriate to bodies in the inner solar system.

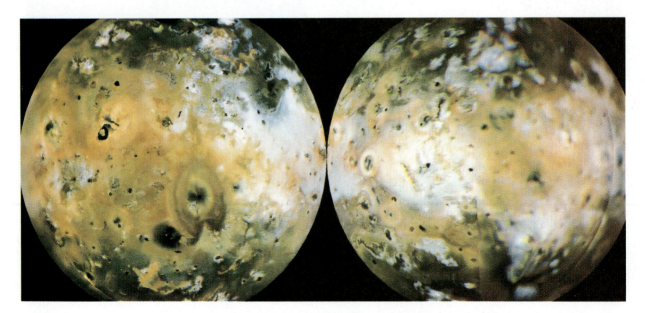

FIGURE 17.7   Io, showing the colorful volcanic features that dominate its highly active surface and make it unlike any other planet or satellite. (NASA/USGS)

In spite of its mainly rocky composition, Europa (Figure 17.6) has an ice-covered surface. In this way it is like the Earth, which also has global oceans of water, except that most of Europa's ocean may be frozen. There are very few impact craters, indicating that the surface of Europa has been capable of some degree of self-renewal. Additional indications of continuing internal activity are provided by an extensive network of cracks in its icy crust.

Io (Figure 17.7), the innermost of Jupiter's large satellites, might have been expected to be a twin of Europa. Instead, it displays a high level of volcanic activity, setting it off from the other objects in the planetary system.

(e) Volcanoes of Io

The discovery of active volcanism on Io was the most dramatic event of the Voyager flybys of Jupiter. Eight volcanoes were seen erupting when Voyager 1 passed in March 1979, and six of these were still active four months later when Voyager 2 passed. These eruptions consisted of graceful plumes that extended hundreds of kilometers into space (Figure 17.8). The material erupted is not lava or steam or carbon dioxide, all of which are vented by terrestrial volcanoes, but sulfur and sulfur dioxide (SO_2). Both of these can build up to high pressure in the crust of Io and then be ejected to tremendous heights. As the rising plume cools, the sulfur and SO_2 recondense as solid particles, which fall back to the surface in gentle "snowfalls" that extend as much as a thousand kilometers from the vent. The SO_2 snow is white, while sulfur forms red and orange deposits. Another sulfur compound detected on Io is H_2S. The surface of Io is slowly buried in these deposits, which accumulate at an average rate of a millimeter or so per year. Over millions of years, this is sufficient to cover any impact craters, so it is no surprise that no such craters have been seen on Io's surface.

Io displays other types of volcanic activity in addition to the spectacular plume eruptions. Images of its surface show numerous shield volcanoes and twisting lava flows hundreds of kilometers long. From their bright colors, these lava flows are thought to be sulfur. Further volcanic activity is indicated by hot spots, surface areas that are hundreds of degrees warmer than their frigid surroundings. (Note that on Io, where the average daytime temperature is only 130 K, even a 300-K area, the surface temperature of the Earth, would qualify as a hot spot.) The largest of these hot spots is a type of "lava lake" 200 km in diameter near the Loki eruption (Figure 17.9). The "lava" in this case is probably liquid sulfur. Telescopic observations made at Mauna Kea Observatory in Hawaii show that this Loki hot spot has been active for at least 15 years, and it accounts for about half of the total volcanic energy released by Io (Figure 17.10).

The SO_2 and other gases belched out by Io's volcanoes form a tenuous atmosphere. Io, however, orbits deep within the jovian magnetosphere, and its surface is subject to a tremendous bombardment by energetic ions of sulfur and oxygen (Section 16.4c). The molecules in Io's thin atmosphere are dissociated and ionized by these charged particles. Once ionized, they are swept up in Jupiter's magnetic field to form the Io plasma torus. Thus the volcanic eruptions on this satellite have a major influence on the huge magnetosphere of Jupiter.

(a)

(b)

FIGURE 17.8 Two views of erupting volcanoes on Io. (a) Crescent view with two eruptions near the edge of the image. (b) A large plume rising above the volcano called Pele. (NASA/JPL)

FIGURE 17.9   The Loki "lava lake" on Io, a hot spot in the form of a black horseshoe-shaped feature about 200 km across. The scale of this image is approximately the same as Figures 17.4b (Ganymede) and 17.6 (Europa). (NASA/JPL)

FIGURE 17.10   Image of the Loki hot spot obtained at 3.5 μm wavelength on Christmas night, 1989, with the NASA 3-m IRTF telescope in Hawaii. At this resolution (about 0.3 arcsec) the thermal glow of the volcano is easily visible against the sunlit surface of the satellite. (University of Hawaii, courtesy John Spencer)

How can Io maintain this remarkable level of volcanism, which exceeds that of much larger planets, such as Earth and Venus? The answer lies in tidal heating of the satellite by Jupiter. Io is about the same distance from Jupiter as is our Moon from the Earth, yet Jupiter is more than 300 times more massive than Earth, causing tremendous tides on Io. These tides pull the satellite into an elongated shape, with a bulge several kilometers high extending toward Jupiter. Now if Io always kept exactly the same face turned toward Jupiter, this tidal bulge would not generate heat. However, Io's orbit is not exactly circular, because of gravitational perturbations from Europa and Ganymede. In its slightly eccentric orbit, Io twists back and forth with respect to Jupiter, at the same time moving nearer and farther from the planet on each revolution. The twisting and flexing of the tidal bulge heats Io, much as repeated flexing of a wire coathanger heats the wire. In this way, the complex interaction of orbit and tides pumps energy into Io, melting its interior and providing power to drive its volcanic eruptions.

After billions of years, this tidal heating has taken its toll on Io, driving away H_2O and CO_2 and other gases, until now sulfur and sulfur compounds are the most volatile materials remaining. The inside is entirely melted, and the crust itself is constantly recycled by volcanic activity. Although Io was well mapped by Voyager, we expect that when re-imaged by the Galileo spacecraft, its surface will wear a partly unfamiliar face.

(f)   Triton and Its Volcanoes

Neptune's retrograde-orbiting satellite Triton is the smallest of the "large" satellites, with a diameter of only 2710 km. Its density is relatively large for such a small object—2.1 g/cm^3, the same as that of Pluto, which Triton also resembles in size. Like Pluto, it is probably composed of a mixture of approximately 75 percent rock and 25 percent H_2O ice.

The surface material of Triton is fresh ice or frost, with a very high average reflectivity—about 80 percent. As with Pluto, the top layer of frost is primarily nitrogen (N_2) and methane (CH_4). Because its bright surface reflects most of the incident solar energy, the surface temperature on Triton is the lowest to be found anywhere in the solar system: between 35 and 40 K. Most potential atmospheric gas is frozen at these temperatures, but a small quantity of N_2 vapor persists to form an atmosphere. The surface pressure of this atmosphere is only 16 millionths of a bar, yet this is sufficient to maintain a substantial ionosphere and to support haze or cloud layers, which were photographed by the Voyager 2 spacecraft in 1989.

Triton's surface, like that of many other satellites in the outer solar system, reveals a long history of geological evolution (Figure 17.11). While there are some impact craters, there are also many regions that have been flooded by "lava" (perhaps H_2O or H_2O/NH_3 mixtures) during past epochs of enhanced endogenic activity. The evidence for such activity includes a number of frozen "lava lakes" more than 100 km across

FIGURE 17.11 Global mosaic of Voyager images of Triton, showing a wide variety of surface features. The large polar cap dominates the lower part of this image. (NASA/JPL)

(Figure 17.12). There are also mysterious regions of jumbled or mountainous terrain that resemble the mountainous regions of Ganymede.

The Voyager flyby of Triton took place at a time when the satellite's southern pole was tipped toward the Sun and this part of the surface was enjoying a period of relative warmth. As shown in Figure 17.11, there appears to be a polar cap covering much of the southern hemisphere, apparently evaporating along its northern edge. This polar cap may consist of frozen nitrogen, deposited during the previous winter. Remarkably, the evaporation of this polar cap seems to generate geysers or volcanic plumes of N_2 that fountain to altitudes of about 10 km above the surface (Figure 17.13). These plumes differ from the volcanic plumes of Io in their composition, and also in that they derive their energy from sunlight warming the surface rather than from endogenic heat.

Triton has turned out to be a far more remarkable object than had been imagined before the Voyager encounter. Many planetary scientists would consider it, along with Io and Titan, among the most interesting objects in the outer solar system. These discoveries also make us wonder what Pluto would look like if we could observe it up close.

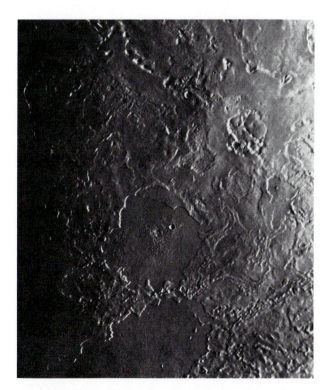

FIGURE 17.12 Old flooded "lava lakes" on Triton. These features, from 100 to 200 km in diameter, date from an earlier period of water volcanism on Triton.

FIGURE 17.13   Mysterious plume or geyser photographed on Triton at the limit of resolution of the Voyager cameras. The plume appears as a vertical dark line that rises 8 km into the atmosphere (center left), then abruptly is carried horizontally (to the right) by upper-atmosphere winds for a distance of about 150 km. The width of this frame is about 200 km. (NASA/JPL)

17.3   SMALL SATELLITES

(a)   Regular Satellites of Saturn

Saturn has six regular satellites with diameters between about 400 and 1600 km. Each of these six satellites has a surface that displays the spectral signature of water ice. Further, each has a density of about 1.3 g/cm³, close to the expected uncompressed density of an object composed half of water ice. (The uncompressed densities of Titan, Ganymede, and Callisto are also about 1.3 g/cm³.) Unlike the jovian system, there is no indication of a systematic variation in density and composition with distance from the planet. Evidently, Saturn was never hot enough to eliminate water ice from its inner satellites, as Jupiter seems to have done with Io and Europa.

The largest in this group of satellites is Rhea (Figure 17.14), with a diameter of 1530 km, just half as large as Europa. Rhea, and its smaller cousins Mimas, Dione, and Tethys, are all heavily cratered worlds with bright surfaces of relatively clean water ice (Figure 17.15). Although there are indications of some tectonic cracking and resurfacing early in their histories, all four of these objects seem to have stabilized geologically billions of years ago. In general, they behave as might be expected for icy objects of this size at this distance from the Sun, where any internal activity should have ceased early as the body cooled.

The other two of this group of six Saturn satellites are more unusual. Iapetus (Figure 17.16) is nearly as large as Rhea, and on one side it looks very much like Rhea, with a bright, heavily cratered surface of water ice. The other hemisphere is entirely different, however. The side of Iapetus that faces forward in its orbit—like all of these satellites, Iapetus always keeps the same side

FIGURE 17.14   Rhea, an icy satellite with heavily cratered surface. The diameter of this satellite is about 1500 km. (NASA/JPL)

FIGURE 17.16 Iapetus, perhaps the strangest satellite of Saturn, with one side as white as snow and the other as black as asphalt. Iapetus is the same size as Rhea, about 1500 km in diameter. (NASA/JPL)

FIGURE 17.15 Dione, another icy satellite with a cratered surface, also shows evidence of past endogenic activity. The diameter of Dione is about 1100 km. (NASA/JPL)

toward the planet—is covered with a very dark, carbon-rich material. The dark material is centered exactly on the forward hemisphere, in the form of a huge oval spot reflecting less than one-tenth as much light as the surface on the trailing (backward-facing) hemisphere. The contrast between the two kinds of surface material on this satellite is as great as that between a black asphalt pavement and freshly fallen snow.

Judging from its orientation, we conclude that this dark spot must have an external cause. Possibly impacts have vaporized enough ice on this side to concentrate near the surface the dark primitive dust that was originally mixed with the ice when Iapetus formed. Its surface, especially on the dark side, might be an extremely interesting place to search for evidence of the earliest chemistry of the solar system.

The other peculiar satellite is Enceladus (Figure 17.17). Although its diameter is only about 500 km, about half of its surface is nearly crater-free. There is evidence of powerful internal activity in the geologically recent past—within the past few hundred million years. In addition, the surface of Enceladus is among the most highly reflective of any planet or satellite, suggesting that it is covered with fine particles of fresh crystalline ice, like the glass beads on a projection screen. Finally, this satellite also seems to have associated with it a very tenuous ring, called the E Ring, circling Saturn, also made up predominantly of small (1 μm) ice particles. Something very strange is going on here! Has a recent volcanic eruption or impact sprayed out water droplets to freeze and form the E Ring, and

subsequently to coat the surface of Enceladus with bright material? Is such an event related to the crater-free areas on the surface? And most puzzling, what could maintain internal activity on a body as small as Enceladus, which should have cooled down very quickly after its formation? No one knows the answers to these questions.

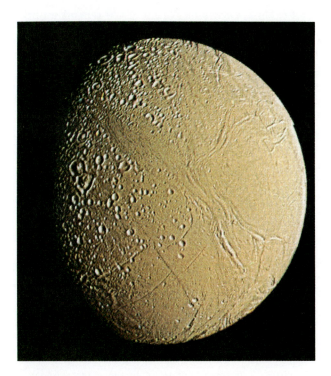

FIGURE 17.17 Enceladus, Saturn's most geologically interesting satellite, showing both heavily cratered and smooth regions of its surface. The diameter of Enceladus is only 500 km. (NASA/JPL)

(b) Small Satellites in Strange Orbits

There are 11 known small satellites of Saturn. In many ways, these small bodies are what one would expect to find: Heavily cratered, icy in composition, with irregular shapes, they appear to be fragments of once-larger parent bodies. What makes them interesting, however, is the variety of orbits they occupy, and the close interaction many of them have with the rings of Saturn (Section 17.5d).

One of these satellites, named Hyperion, has a very strange kind of rotation. Unlike the other satellites we have discussed, Hyperion is not in synchronous rotation—that is, it does not keep the same side toward its planet. In fact, it does not even have a well-defined period of rotation. It is in a state of *chaotic* rotation, in which gravitational interactions with Titan cause it to exchange angular momentum between its orbit about Saturn and its rotation.

Two satellites, Janus and Epimethius, are co-orbital—that is, they occupy nearly, but not quite, the same orbit around Saturn. If they had exactly the same period, they could avoid each other, but such a state is dynamically impossible. Instead, their orbits differ in radius by about 50 km, corresponding to a difference in orbital periods of 1 part in 2080. The inner co-orbital, following Kepler's laws, therefore catches up with the outer at a relative speed of about 9 m/s. Since these satellites are more than 100 km across, they cannot pass. Fortunately, they interact gravitationally before they get too close, exchanging orbits. Their relative motion therefore reverses, and they pull apart. This intricate orbital maneuver repeats about once every four years.

(c) Satellites of Uranus

The 15 known satellites of Uranus are conveniently divided into two groups: 5 larger bodies that had been discovered telescopically and 10 small darker moons, all relatively close to the planet, that were discovered by Voyager. We will focus our attention on the five larger satellites, all named after characters from Shakespeare or from Alexander Pope's poem "The Rape of the Lock."

The larger satellites of Uranus have diameters from about 1600 km down to 500 km, the same range as the regular satellites of Saturn. Their densities (1.4 to 1.6 g/cm³), however, are greater, indicating that there is a smaller proportion of ice, relative to rock, in the uranian satellites. Like the satellites of Saturn, their surfaces show the spectral signature of water ice, although their reflectivities are generally lower (ranging from 20 to 40 percent), suggesting that their surfaces are "dirtier." They are not nearly so dark as the small uranian satellites, however, which have reflectivities of only 5 percent.

As might be expected from our previous examination of the satellites of Jupiter and Saturn, these objects are all more or less heavily cratered. Presumably most of this cratering took place during the first billion years of solar-system history, at a time when there was more cometary debris present than there is today. If we apply the standard theories for cratering to these distant objects, we conclude that they are relatively inactive geologically, with surfaces that have been stable for billions of years. There is no young, active object like Io or Enceladus in the Uranus system.

Although all five of these satellites of Uranus are similar in being ice-rock mixtures and having heavily cratered surfaces, there are still striking differences in their individual geological histories. The two largest, Titania and Oberon, both have tectonic cracks or valleys rather like those of the Saturn satellites Dione and Tethys. However, the most geologically interesting are the two inner satellites, Ariel and Miranda.

Ariel (Figure 17.18), which has a diameter of 1160 km, is characterized by flat-floored tectonic valleys that seem to be the result of stretching of the crust—like the Valles Marineris of Mars. There are also relatively young flows that may represent a period of water volcanism in the history of this satellite. Since pure water seems unlikely to have been a fluid at the temperatures (well below 100 K) found in the uranian system, it is speculated that ammonia-water mixtures, or fluids involving carbon monoxide or methane, may have constituted the "lava" in this case.

Miranda, the smallest (484 km) and innermost of the five main satellites, is the most geologically diverse and mysterious of the uranian moons. Its surface (Figure 17.19), like that of Ganymede in the jovian system, consists of both older, heavily cratered terrain and widespread younger structures that nearly defy description. These include great valley systems with gorges as deep as 10 km and complex oval or trapezoidal ranges of mountains. These seem to represent a type of endogenic activity, with fluids erupted to the surface in response to the internal differentiation of the object. It is not known, however, why such activity should have been confined to this one rather small satellite.

17.4 PLANETARY RINGS

All four of the jovian planets have well-developed ring systems, consisting of billions of small particles or moonlets orbiting close to their planet. Each ring system displays a complex structure apparently related to interactions between the ring particles and the larger satellites. However, these rings are also very different from one another. Saturn's system, which is by far the largest, is made up primarily of small icy particles spread out into a vast flat ring with a great deal of fine

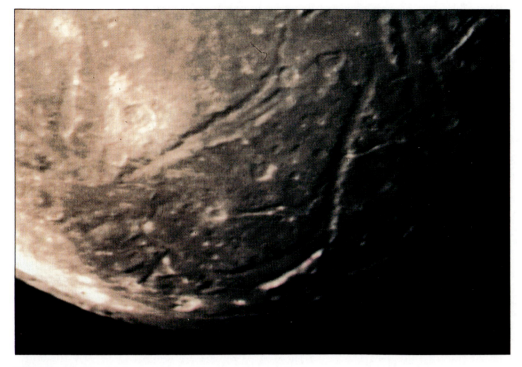

FIGURE 17.18   Ariel, a satellite of Uranus, displays flat-floored tectonic valleys that suggest an extensive early period of both tectonic and volcanic activity. Ariel is about the same size as Dione (Figure 17.15). (NASA/JPL)

structure. The uranian rings (with much smaller mass) are nearly the reverse, consisting of very dark particles confined to a few narrow rings, with broad gaps between. The neptunian rings are more tenuous but otherwise similar to those of Uranus, but with mysterious

thickening or condensations within them. Finally, the jovian ring is merely a faint transient dust band, constantly renewed by erosion of dust grains from its inner satellites. The main properties of these ring systems are summarized in Table 17.3.

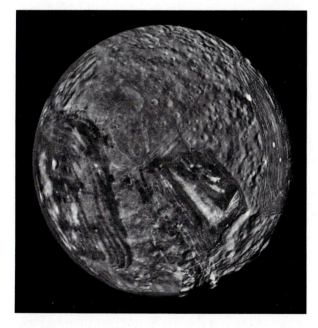

FIGURE 17.19   Miranda, the most geologically complex of the satellites of Uranus. It has experienced large scale endogenic modification, but the details are not well understood. Miranda is about the same size as Enceladus (Figure 17.17). (NASA/JPL)

(a)   Ring Origin and Dynamics

A ring is a collection of vast numbers of particles, each obeying Kepler's laws as it follows its own orbit around the planet. Thus the inner particles orbit faster than those farther out, and the ring as a whole does not rotate as a solid body. In fact, it is better not to think of a ring *rotating* at all, but rather to consider the *revolution* of its individual moonlets.

If the particles were widely spaced, they would move independently, like separate small satellites. However, in the rings of Saturn and Uranus the particles are close enough to one another to exert mutual gravitational influence, and occasionally even to rub together or

TABLE 17.3	PROPERTIES OF RING SYSTEMS			
	Outer Radius		Mass	Reflectivity
Planet	(km)	(R_{planet})	(kg)	(%)
Jupiter	128,000	1.8	10^{10} (?)	?
Saturn	140,000	2.3	10^{19}	60
Uranus	51,000	2.2	10^{14}	5
Neptune	63,000	2.5	10^{12}	5

bounce off of one another in low-speed collisions. Because of these interactions, phenomena such as waves can be produced that move across the rings, like water waves moving over the surface of the ocean.

There are two basic theories of ring origin. First is the breakup theory, which suggests that the rings are the remains of a shattered satellite. The second theory, which takes the reverse perspective, suggests that the rings are made of particles that were unable to come together to form a satellite in the first place.

In either theory, an important role is played by tidal forces. As we saw in Section 4.4f, tides are very sensitive to distance, with the tidal force varying as the inverse cube of the separation between two bodies. If objects approach too closely, their tidal bulges become so large that they are torn apart. The same effect takes place for some double stars, which can become so tidally distorted that one star leaks material over onto its companion.

Around each planet there exists a **tidal stability limit,** often called the *Roche limit* after the French mathematician who first calculated it. This is the distance within which a satellite with no internal strength (like a pile of gravel) would be disrupted by tides (differential gravitational forces). Alternatively, we may think of it as the distance within which the individual particles in a disk cannot attract one another to form a satellite. For most objects in the outer solar system, the tidal stability limit is at about 2.5 planetary radii from the center of a planet. All three known ring systems lie within the tidal stability limits for their respective planets (Figure 17.20).

This stability limit applies only to a satellite with no intrinsic strength. A solid object held together by its own strength will not necessarily break up inside the limit. This why we find small satellites (up to 100 km in diameter) orbiting within all four ring systems. If the satellite is large enough, however, its intrinsic strength becomes less important in comparison to the differential tidal forces, and breakup is more likely. Also, a satellite within the limit will break up if it is fractured by a large impact, while if it is outside the limit, it is likely to fall back together under its own gravitation after such an impact.

In the breakup theory of ring formation, we can imagine a satellite or a passing comet coming too close and being torn apart by tidal forces. A variant of this idea suggests that a small satellite near the stability limit might be broken apart in a collision, with the fragments then dispersing into a disk. The presence of many small inner satellites orbiting Saturn and Uranus near the stability limit suggests that this sort of breakup may have been fairly common over the history of the solar system.

Not all rings are made of substantial solid particles. The rings of Jupiter are composed of very small dust grains, apparently being eroded from the surfaces of the two known inner satellites, and perhaps from smaller unknown bodies within the rings. This dust is then gradually swept from the rings into the atmosphere of Jupiter. The observed amount of dust represents an equilibrium between production and loss. Similar tenuous sheets of dust are observed in other ring systems. These dust rings also apparently represent continuous erosion from inner satellites or larger ring particles.

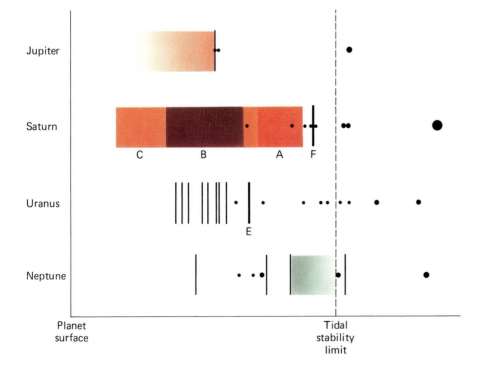

FIGURE 17.20 The ring systems of Jupiter, Saturn, Uranus, and Neptune, compared with the location of the tidal stability limit (Roche limit) for each planet. All four ring drawings are scaled to the diameters of their respective planets.

FIGURE 17.21 Voyager image of the rings of Saturn, photographed from below; note that except for the B Ring, these rings transmit enough sunlight to be clearly visible. (NASA/JPL)

(b)   The Rings of Saturn

The rings of Saturn circle the planet in its equatorial plane, which is tilted by 27° to the planet's orbital plane. As Saturn revolves about the Sun, we see one side of the rings for about 15 years, followed by the other side for the same period. The three brightest rings of Saturn, visible from Earth, are labeled (from outer to inner) the A, B, and C Rings. The outer radius of the A Ring is 136,780 km, while the inner edge of the C ring is just 12,900 km above the cloud tops of Saturn. The rings are illustrated in Figure 17.21, and Table 17.4 summarizes the dimensions of Saturn's main rings.

The B Ring is the brightest and has the most closely packed particles, while the A and C Rings are translucent. The total mass of the B Ring is about equal to that of an icy satellite 300 km in diameter. The B and A Rings are separated by a gap easily visible from the Earth, discovered in 1675 by the Italian-French astronomer J. D. Cassini and called the Cassini Division. Although it looks empty from the Earth, the Cassini Division actually contains many ring particles with considerable structure, including several true gaps, within it. The Cassini Division looks like a gap only in contrast with the denser A and B Rings on either side of it.

The rings of Saturn are very broad but very thin. The width of the main rings is 70,000 km, yet their thickness is only about 20 m. If we made a scale model of the rings out of paper the thickness of the sheets in this book, we would have to make the rings a kilometer across— about eight city blocks. On this scale, Saturn itself would loom as high as an 80-story building.

The ring particles are composed primarily of water ice, and they span a range of sizes from grains of sand up to house-sized boulders. An insider's view of the rings would probably resemble a bright cloud of floating snowflakes and hailstones, including a number of snowballs and larger objects, many of which are loose aggregates of smaller particles.

As revealed by Voyager, the rings of Saturn have a great deal of complex structure, including about a dozen gaps, each tens to hundreds of kilometers wide. Two of these gaps are in the C Ring, and two are in the A Ring. However, most of them are associated with the Cassini Division between the B and A Rings. Some of these gaps contain peculiar eccentric ringlets, that is, ribbons of particles that do not share the circular orbits of the other ring particles. For the ring as a whole to be eccentric, it is not sufficient that the individual particles have eccentric orbits; in addition, the major axes of these orbits must be aligned in space. Some of the gaps have wavy edges, and one of the gap ringlets is kinky.

The Pioneer and Voyager spacecraft revealed additional rings not visible from Earth. A faint D Ring lies inside the C Ring, and a very narrow F Ring, of radius 140,180 km, lies outside the A Ring. The F Ring is one of the most interesting features of the saturnian system, and it is the one Saturn ring that is similar in many ways to the rings of Uranus and Neptune. This ring (Figure 17.22) has a mass equivalent to an icy satellite a few kilometers in diameter. Within its 100-km width there are many ringlets, including a double bright ring with two components just a few hundred meters wide. In some places, the F Ring breaks up into two or three parallel strands, which sometimes show bends or kinks. Further, the F Ring as a whole is eccentric.

The major B Ring has no gaps, but it contains intricate structure, partly in the form of waves (Figure 17.23). Each wave corresponds to alternating ringlets where the ring particles are bunched together or spread more thinly. Photographed from the spacecraft, these waves, which are typically separated by 10 km or so, look like the grooves in a phonograph record. The A Ring has even more of this wave-like structure. However, the bulk of the structure in the A and B Rings is not wave-like, but apparently random and irregular. This structure has not been satisfactorily explained.

(c) The Rings of Uranus and Neptune

The rings of Uranus are narrow and black, making them almost invisible from the Earth. They were discovered accidentally on March 19, 1977, during observations of the occultation of a bright star by Uranus,

TABLE 17.4	RINGS OF SATURN		
	Outer Edge		Width
Ring Name	(R_s)	(km)	(km)
E	4	250,000	60,000
F	2.324	140,180	90
A	2.267	136,780	14,600
Cassini Division	2.025	122,170	4,590
B	1.949	117,580	25,580
C	1.525	92,000	17,490

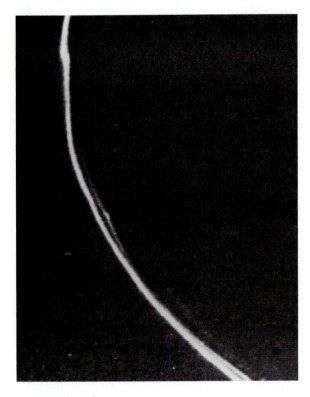

FIGURE 17.22 Voyager photograph of the narrow but complex F Ring of Saturn. (NASA/JPL)

FIGURE 17.23 Saturn's B Ring in detail. The structure seen here has scales of tens to hundreds of kilometers. Most of this structure has not been explained. (NASA/JPL)

observable only from the Indian Ocean and its sur-roundings. A team of Cornell University astronomers observed the occultation from above the middle of the Indian Ocean using the NASA 1-m airborne telescope aboard the Kuiper Airborne Observatory (Section

10.2d), while others operated telescopes in Australia, China, India, and South Africa.

About 20 minutes before its predicted occultation by the planet, the star briefly dimmed several times as it disappeared behind successive narrow rings. This pat-

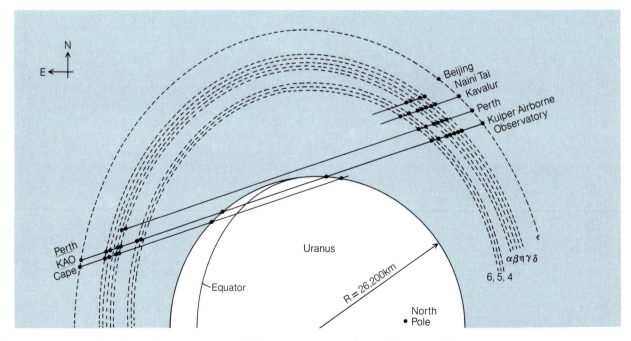

FIGURE 17.24 The rings of Uranus as revealed by measurements of occultations carried out from the Earth. Each dot corresponds to an occultation event observed at the observatory indicated. (Adapted from a summary prepared by J. Elliot of MIT)

TABLE 17.5	RINGS OF URANUS		
Name	Distance (km)	Width (km)	Eccentricity
U51 Epsilon	51,160	22–93	0.0079
U50 1986UR1	50,040	1–2	?
U48 Delta	48,310	3–9	0
U48 Gamma	47,630	1–4	0
U47 Eta	47,180	2	0
U46 Beta	45,670	7–11	0.0004
U45 Alpha	44,730	8–11	0.0008
U43 4 Ring	42,580	2	0.0011
U42 5 Ring	42,240	2–3	0.0019
U42 6 Ring	41,850	1–3	0.0010
U38 1986UR2	38,000	2500	?

tern of ring occultations was repeated later, as the opposite side of each arc passed in front of the star (Figure 17.24). From the symmetry of the occultation patterns on either side of Uranus, the observers quickly concluded that they had discovered continuous rings that circled the planet. Additional occultations led to the discovery of a total of nine narrow rings, and two more were added by Voyager in 1986 (Table 17.5).

Since the 1977 discovery, many more occultations have been observed to map out the uranian rings in detail, and in January 1986 the Voyager 2 spacecraft was able to study them at close range. Despite their low reflectivity—typically about 5 percent—they could be photographed by the spacecraft cameras (Figure 17.25). In addition, Voyager used observations, both of occultations of stars by the rings and an occultation of the spacecraft itself as it passed behind the planet, to probe the structure of the rings in greater detail than is possible from telescopes on the Earth.

The broadest and outermost of the rings of Uranus is called the Epsilon Ring (also called U51, since it is 51,000 km from Uranus). The main Epsilon ring is about the width of the Saturn F Ring, although it also has a much wider component of lower density. Its thickness is probably no more than 100 m, and from probes with the spacecraft radio system it appears that most of the particles are relatively large—several meters or more in diameter. The Epsilon Ring circles Uranus at a distance of 2.2 Uranus radii—near the position of the tidal stability limit. This ring probably contains as much mass as all of the other 10 rings combined. With one exception, all of the other rings are narrow ribbons less than 10 km in width—just the reverse of the broad rings of Saturn.

The rings of Neptune are also invisible from the Earth, and they too were discovered from their occultation of starlight. Beginning in 1985 several ring occultations were observed, but their meaning remained in dispute. Unlike the symmetric occultations observed at Uranus, the obscuration of a star by the "rings" on one side of Neptune was not repeated on the other side. At some occultation opportunities, the stars did not dim at

FIGURE 17.25 Voyager image of the narrow, dark rings of Uranus. (NASA/JPL)

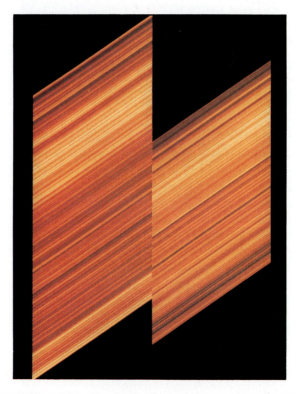

FIGURE 17.26 The Epsilon Ring of Uranus, shown in two high-resolution cross-sections derived from the Voyager occultation experiment. The colors are false; actually the rings are dark gray. (NASA/JPL)

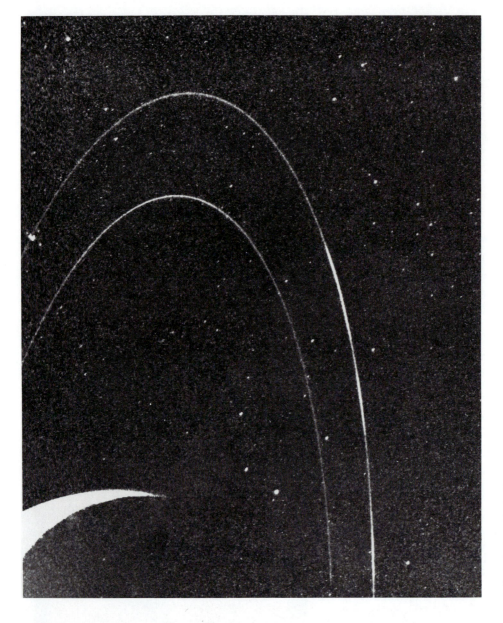

FIGURE 17.27 The rings of Neptune as photographed by Voyager in 1989. Note the three denser regions of the outer (N63) ring. The crescent image of Neptune is overexposed in this time exposure. (NASA/JPL)

all. Astronomers inferred that if there really were rings at Neptune, they must be discontinuous or clumpy.

As photographed by Voyager 2, the rings of Neptune revealed themselves as real, but much fainter than the rings of Uranus (Figure 17.26). They are composed of dark particles and appear to contain a larger proportion of fine material (dust) than the uranian rings. With one exception, they are too tenuous to block starlight and thus generate an observable occultation. That exception applies to three thicker condensations in the main ring (N63), each arc about 10° in length, which are the features that had been detected by occultations from the Earth. Table 17.6 lists the rings of Neptune as determined from Voyager data.

(d)   Satellite–Ring Interactions

After much study of the data from the Voyager flybys, astronomers have come to understand that most of the structure in the rings owes its existence to the gravitational effects of satellites. If there were no satellites, the rings would probably be flat and featureless. Indeed, if there were no satellites, there might be no rings at all, since, left to themselves, thin disks of matter gradually spread out and dissipate. The sharp edges as well as the fine structure of rings are due to the satellites.

In the Saturn system, the existence of some of the gaps and of the sharp outer edge of the A Ring result from gravitational resonances with Mimas and the two co-orbital inner satellites. A **resonance** takes place when two objects have orbital periods that are exact ratios of each other, such as 1:2, 1:3, etc. For example, any particle in the gap at the inner side of the Cassini Division would have a period exactly equal to one-half that of Mimas. Such a particle would be nearest Mimas in the same part of its orbit every second revolution. The repeated gravitational tugs of Mimas, acting always in the same direction, would perturb the particle, forc-

TABLE 17.6	RINGS OF NEPTUNE	
Name	Distance (km)	Width (km)
N63 1989NR1	62,900	15
N58 1989NR5	57,500	<1,000
N55 1989NR4	55,000	6,000
N53 1989NR2	52,300	<100
N42 1989NR3	41,900	

ing it into a new orbit that does not represent a resonance with a satellite.

Resonances can form gaps by ejecting material with periods that are an exact multiple of satellite periods, but they can also lead to circumstances where the boundary of a ring is stabilized by these gravitational effects. Such is the case for the sharp outer edge of the A Ring of Saturn, which is in a 7:6 resonance with the two co-orbital satellites, Janus and Epimetheus.

Much of the Saturn ring structure represents waves of higher and lower density that are propagating across the rings. Many of these features are tightly wound spirals, like the grooves in an old-fashioned phonograph record (Figure 17.27). These *spiral density waves* are produced at distances from Saturn corresponding to satellite resonances. The effect is very similar to that which is thought to generate the spiral arms of galaxies, with the role of individual stars being played here by the ring particles (Chapter 32).

Small satellites that orbit very close to rings can also control their shape and position. This is the case for Saturn's F Ring, which is bounded by the orbits of Pandora and Prometheus (Figure 17.28). These two small objects are referred to as **shepherd satellites,** since their gravitation serves to "herd" or "shepherd" the ring particles and keep them confined to a narrow ribbon. A similar situation applies to the Epsilon Ring of Uranus, which is shepherded by the satellites Cordelia and Ophelia. These two shepherd satellites, each about 50 km in diameter, orbit about 2000 km inside and outside the ring.

Theoretical calculations suggest that the other narrow rings in the uranian system should also be controlled by shepherd satellites, but none of these shepherds has been located. The calculated diameter for such shepherds—about 10 km—was just at the limit of deductibility for the Voyager cameras, so it is impossible to say if they are present or not. Presumably, similar mechanisms confine the rings of Neptune. However, while two small satellites were found near the inner edges of the main neptunian rings (N53 and N63), there are no corresponding outer shepherds. At best, the shepherd theory seems to explain only a small part of the structure observed in the rings.

FIGURE 17.28 Spiral density waves and bending waves in the A Ring of Saturn as photographed by Voyager. (NASA/JPL)

(e) Permanence of Rings

The dramatic differences among the four outer-planet ring systems, the problems with the shepherd theory, the presence of dusty rings that may not be stable over long time periods, and the existence of "arcs" in the neptunian rings are disturbing to many scientists. The rings have many characteristics that suggest random processes, as though each system was the product of

FIGURE 17.29 Voyager photograph of one of the shepherd satellites (Pandora) with Saturn's narrow F Ring. (NASA/JPL)

particular events that were not duplicated at other planets. Perhaps rings are transient phenomena with lifetimes that are short compared with the 4.5-billion-year age of the solar system.

There are several lines of evidence that point toward transient rings. Theoretical models indicate that rings should break up and dissipate within a few hundred million years, or about ¹⁄₁₀ of the age of the solar system. Constant bombardment by meteorites and other space debris also limits the lifetime of rings. At the same time, this bombardment is capable of creating new rings by breaking up preexisting inner satellites, which may be numerous. Rings as massive as Saturn's would require the destruction of an icy satellite at least 200 km in diameter, but the smaller rings of Uranus and Neptune might result from the destruction of a satellite only 10 km or so in diameter.

The study of rings is an exciting but immature field of planetary astronomy. Thanks to the Voyager encounters, supplemented by ground-based occultation measurements, we now have considerable data on more than 20 individual rings in four ring systems. Much has been revealed, and there is some understanding of the phenomena involved. But many aspects of the complex dynamics of these rings remain mysterious.

S U M M A R Y

17.1 The four jovian planets are accompanied by systems of satellites and rings. Jupiter has 16 satellites, Saturn 19, Uranus 15, and Neptune 8, many of which were discovered by Voyager. Most of these satellites are composed in part of H_2O ice (up to 60 percent by mass). Six of the satellites are of planetary dimensions. Saturn has the largest ring system (equivalent to a 300-km ice satellite), composed of H_2O ice. Uranus and Neptune have narrow rings of dark material, and Jupiter has an even more tenuous ring of dust.

17.2 The largest satellites are Ganymede, Callisto, and Titan (with diameters about 5000 km and masses about twice that of the Moon). They are all differentiated objects, composed half of H_2O. Callisto has an ancient cratered surface; Ganymede once experienced tectonic and volcanic activity; and Titan has a cloudy atmosphere (mostly N_2) with a surface pressure of 1.5 bars and interesting organic chemistry. Smaller and denser are Europa and Io, each about the size of the Moon. Io is the most volcanically active object in the solar system; its eruptions of sulfur and sulfur dioxide are powered by tidal interactions with Jupiter. Triton is the retrograde satellite of Neptune, smaller than Europa and made of about 75 percent rock and 25 percent ice. It has a thin atmosphere and eruptions of N_2 from within its evaporating polar cap.

17.3 The many smaller satellites are a varied group. At Saturn there are several intermediate-sized icy bodies, includ-

ing two-faced Iapetus and recently active Enceladus. Smaller objects include Hyperion, with its chaotic rotation, and the co-orbital satellites Janus and Epimethius. Uranus has five intermediate-sized satellites with varying degrees of past endogenic activity, the strangest of which is Miranda. In spite of their small sizes and low temperatures, many of these outer-planet satellites have experienced interesting geological histories, including volcanism based on water and other volatiles.

17.4 Rings are composed of vast numbers of individual particles orbiting a planet within its **tidal stability limit** (about 2.5 planetary radii). They may have formed from the breakup of one or more satellites or from material that never formed a satellite in the first place. The saturnian rings are broad, flat, and nearly continuous except for a handful of gaps. The particles are H_2O ice and are typically a few centimeters in dimension. The uranian rings are 100 times less massive and are narrow ribbons separated by wide gaps. The neptunian rings are similar but 100 times smaller yet. Both are made of dark particles; being nearly invisible from Earth, they were discovered and have been mapped out by occultation studies. Much of the complex structure of the rings is due to waves and **resonances** induced by the gravitational effects of inner satellites or by the effects of **shepherd satellites** embedded within the rings.

E X E R C I S E S

THOUGHT QUESTIONS

1. Why do you think the outer planets have such extensive systems of rings and satellites, while the inner planets do not?

2. Which would have the greater period, a satellite 1 million km from the center of Jupiter or a satellite 1 million km from the center of Earth? Why?

3. Ganymede and Callisto were the first icy objects to be studied from a geological point of view. Summarize the main differences between their geology and that of the rocky terrestrial planets.

4. Compare the properties of the atmosphere of Titan with those of the Earth's atmosphere.

5. Compare the properties of the volcanoes of Io with those of terrestrial volcanoes.

6. Would you expect to find more impact craters on Io or Callisto? Why?

7. Where did the nitrogen in Titan's atmosphere come from? Compare with the origin of the nitrogen in our atmosphere.

8. Do you think there are many impact craters on the surface of Titan? Why or why not?

9. Two of the differences between the Saturn and Jupiter systems are (a) the presence of icy rings at Saturn and (b) the absence of a dependence among the Saturn satellites of density on distance. Can you think of a common explanation for both of these differences?

10. Compare the satellites of Uranus with those of Saturn.

11. Compare the geology and volcanism of Triton with that of Io.

12. Explain why a large satellite is more likely to break up inside the tidal stability limit than a small satellite.

13. Three possibilities were suggested in the text for the origin of the rings of Saturn. List them and briefly summarize the arguments in favor of each.

14. Why do you suppose the rings of Saturn are made of bright particles, whereas the particles in the rings of Uranus and Neptune are black?

PROBLEMS

15. Saturn's A, B, and C Rings extend from a distance of about 75,000 to 137,000 km from the center of the planet. What is the approximate variation factor in the periods for various parts of the rings to revolve about the planet?

16. Occultations of stars by the rings of Uranus have yielded resolutions of 10 km in determining ring structure. What would be the angular resolution (in arcseconds) that a space telescope would have to achieve to obtain equal resolution from Earth orbit? How close to Uranus would a spacecraft have to come to obtain equal resolution with a camera having angular resolution of 2 arcsec?

*17. The main ring of Neptune (N63) has a radius of 63,000 km and a width of 15 km. Calculate the periods of revolution about Neptune for the inner and outer edges of the ring, and hence the relative velocities of the inner and outer edges. Now suppose a clump of material were placed at one location in the rings (like the three observed clumps in this ring). How long would it take for the differential speeds at the inner and outer edges to result in smearing the clump all of the way around the planet?

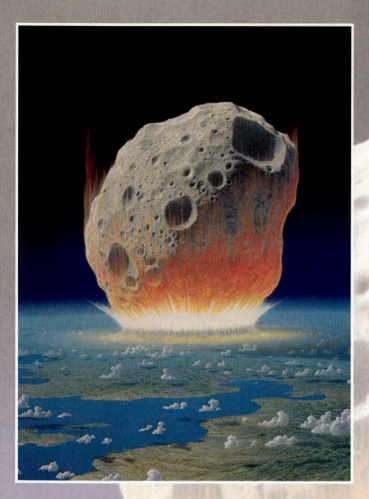

Artist's impression of the impact of an asteroid on the Earth.
(Painting by Don Davis)

In the last decade, several threads of evidence from geology, astronomy, and chemistry have led investigators to an astonishing but now widely shared conclusion: 65 million years ago, with no warning, an asteroid (or perhaps comet) about 10 km across slammed into what is today called the Yucatan region of Mexico, and exploded. After the impact, there was a crater some 200 km across—large enough to fit the entire Los Angeles metropolitan area and good number of suburban towns with it. More than 100,000 billion *tons* of dust were thrown up by the explosion, enough to blanket the Earth. Months of wintery darkness descended on our planet, and perhaps 70% of all living species perished from a variety of related causes. As we shall see in this and the next chapter, such collisions are a normal part of the Earth's cosmic environment. Smaller but still potent chunks of cosmic material have hit the Earth in the 20th century and recently pelted the planet Jupiter.

Asteroids are among the derelicts of the solar system. Astronomers are interested in them primarily because they are remnants of the formation of the planets 4.5 billion years ago. But the public is fascinated by asteroids because of their potential to destroy. In the past six chapters we discussed the craters formed on planets and satellites by the impacts of asteroids and comets. In this chapter and the next, we look at the kinds of objects that formed these impact craters in the past and will continue to collide with planets and satellites into the future.

This view of the solar system, with early and continuing acts of violence playing an important role, is a relatively new one among scientists. For much of its history, the central theme in planetary science was the action of long-range, gradual evolutionary forces. Now we know that sudden acts of cosmic violence have often punctuated the quiet evolutionary processes and made them veer on occasion in unexpected directions.

Eugene Shoemaker (b. 1928) of the U.S. Geological Survey began his career with a definitive study of the impact process at Meteor Crater in Arizona, and he played a leading role in the Apollo exploration of the Moon and the Voyager flybys of the satellites of the outer planets. Shoemaker is best known, however, for his studies of Earth-approaching asteroids and the role these bodies play in the cratering of the Earth and other planets. (Ramona Boudreau)

CHAPTER 18

ASTEROIDS AND IMPACTS

The asteroids, or minor planets, are small objects in orbits similar to those of the major planets. They differ from the planets primarily in size. Distinguishing them from their cousins the comets is a little more difficult. A comet is defined as a small object with a visible transient atmosphere—the coma or tail. If there is no atmosphere, a small body is called an asteroid. This practical definition reflects primarily a difference in composition between comets and asteroids: Comets contain water ice and other volatiles that vaporize when heated by the Sun, while asteroids are rocky objects with little volatile material.

All comets and most asteroids are *primitive* bodies, which have been relatively little altered chemically or physically since the formation of the solar system. They thus provide a window on our earliest history, the period before the planets formed. That is why asteroids are of such great interest to scientists, in spite of their small sizes and insignificant contribution to the total mass of the solar system. Indeed, it is precisely because they are small that the asteroids have been able to preserve a record of the early chemical history of the solar system.

18.1 DISCOVERY AND ORBITS

(a) Discovery of the Asteroids

Most of the asteroids are located between the orbits of Mars and Jupiter. From the time of Kepler, it was

recognized that this region of the solar system represented a gap in the spacing of planetary orbits. We have already seen that the Titius–Bode law predicts a planet at 2.8 AU from the Sun (Section 11.2). After the discovery of Uranus seemed to confirm this "law," many astronomers felt there should be a concerted effort to locate this missing planet.

The view that there was another planet was not held by everyone. The noted German philosopher Georg Wilhelm Hegel, for instance, argued that there could be only seven planets because, among other things, there are only seven openings in the human head. Nevertheless, in 1800 a methodical search for the missing planet was organized by the German Baron Francis Xavier von Zach. The plan was to divide the zodiac—that band around the sky centered on the ecliptic and through which the planets' orbits lie—into 24 sections and to assign the search of each section to a different astronomer. One of these was the Sicilian astronomer Giuseppe Piazzi.

Piazzi, however, had not yet been informed of his role in the search and had not received the charts of the region of the zodiac he was to survey. He was working independently on an entirely different project when, on January 1, 1801, he discovered a new object not on his star charts. The next night he observed it again and thought that it had moved; on the third night he was sure of it. Piazzi, like Herschel in his discovery of Uranus, thought he had found a comet. He observed the new object regularly until February 11, when his work was interrupted by a severe illness.

On January 24, Piazzi wrote to report his new comet to others, including Bode in Berlin. The news created a great deal of enthusiasm, and von Zach even published a report that the missing planet between Mars and Jupiter had been discovered! But by the time the astronomers in northern Europe were aware of the discovery, the new object was too close to the Sun to be observed, and it would be September before it became visible again.

By summer of 1801, Bode realized that the new planet was lost. Piazzi had observed it for only six weeks, too short a time to calculate its orbit by techniques then available. Fortunately, the discovery was saved by a 23-year-old genius from Brunswick, Germany: Karl Friedrich Gauss. Gauss was intrigued by an account of the discovery he had read in a newspaper. Putting his other work aside, he devised a new method of orbit calculation to deal with observations made over only a short arc of the total orbit. He finished his calculations in November 1801 and sent the results to von Zach.

On the last night of December, 1801, von Zach found the object almost precisely where Gauss had predicted it would be. At Piazzi's request, the object was named Ceres, for the Roman goddess of agriculture and protecting goddess of Sicily. The fame earned by Gauss for his efforts in the problem eventually led, in 1807, to his appointment as director of the Göttingen Observatory, where he remained the rest of his life.

Ceres was widely assumed to be the missing planet predicted by the Titius–Bode law. It came as a complete surprise, therefore, when in March 1802 Heinrich Olbers discovered a second moving star-like object—the asteroid to be named Pallas. It was a natural speculation that if there was room for two minor planets, there could be room for others as well, and a search for such objects began in earnest. The discovery of Juno followed in 1804 and of Vesta in 1807. But it was 1845 before Karl Hencke discovered the fifth minor planet, after 15 years of search. Subsequently, new ones were found with increasing frequency, until by 1890 more than 300 were known.

In 1891 Max Wolf of Heidelberg introduced the technique of astronomical photography as a means of searching for asteroids. The angular motion of a minor planet is large enough (especially if it is near opposition) so that during a long time exposure its image will form a trail. The object appears on the photograph, therefore, as a short dash rather than as a star-like point image (Figure 18.1). Brucia, the 323rd asteroid to be discovered, was the first to be found photographically.

Today discoveries of asteroids are usually accidental; they most often occur when the tiny objects leave their trails on photographs that are taken for other purposes. Literally thousands of minor planet trails appear on the photographs taken for systematic surveys of the sky. The majority of these trails are of objects that have never been catalogued. Most of them have been ig-

FIGURE 18.1   Time exposure showing trails left by asteroids (*arrows*). (Yerkes Observatory)

nored, however, and the searches carried out today usually focus on special groups of asteroids, such as those in orbits that bring them close to the Earth. More than 9000 asteroids have well-determined orbits.

By modern custom, after a newly found asteroid has had its orbit calculated, and it has been observed again after another circuit of the Sun since its first discovery, it is given a name and a number. The number is a running index that indicates the order of discovery among the minor planets. The discoverer is customarily given the honor of supplying the name. The full designation of the asteroid contains both the number and the name, thus: 1 Ceres, 2 Pallas, 433 Eros, 2410 Morrison, 4859 Fraknoi, and so on.

(b)   Statistical Studies

It would be a formidable task to discover, determine orbits for, and catalogue all the asteroids bright enough to be observed with modern telescopes. Nevertheless, the total number of such objects can be estimated by systematically sampling regions of the sky.

Several investigators have estimated the number of asteroids by photographing selected regions of the zodiac. The total number of asteroids bright enough to be photographed with the 18-in. Palomar Schmidt telescope, which is frequently used for such studies, is estimated to be about 100,000. Such a survey includes essentially all objects down to a diameter of about 5 km. It is estimated that our census of asteroids is 98 to 99 percent complete for objects down to 100 km diameter, and at least 50 percent complete for objects down to 10 km. It is therefore possible to estimate the total mass of the asteroids, which is about $\frac{1}{20}$ the mass of the Moon.

The largest asteroid is 1 Ceres, with a diameter of just under 1000 km. Two have diameters near 500 km, and about 30 are larger than 200 km (Table 18.1). The number of asteroids increases rapidly with decreasing size; the number of objects 10 km across, for example, is more than 100 times the number of objects 100 km across. We estimate that there are more than a million asteroids with diameters of 1 km or more. This size distribution, with many more small objects than large ones, is what we would expect to find if the asteroids have been battered by collisions for 4.5 billion years, so that all but the largest are collisonal fragments.

(c) Orbits of Asteroids

The asteroids all revolve about the Sun in the same direction as the planets (from west to east), and most of them have orbits that lie near the plane of the Earth's orbit. The mean inclination of their orbits to the plane of the ecliptic is 10°. A few, however, have orbits inclined more than 25°; the orbit of 2102 Tantalus is the most inclined (64°) to the ecliptic. The main **asteroid belt** contains minor planets with semimajor axes in the range of 2.2 to 3.3 AU, with corresponding periods of orbital revolution about the Sun from 3.3 to 6 years. The mean value of the eccentricities of the main belt asteroid orbits is 0.15, not much greater than the average for the orbits of the planets.

Some asteroids have orbits rather far outside the main asteroid belt. A few with semimajor axes around 1.9 to 2.0 AU are called the Hungarias (for the prototype 434 Hungaria), and a few with larger orbits, near 4 AU, are called the Hildas (for 153 Hilda). There are asteroids with even more extreme orbits, some of which cross the orbit of the Earth, and a few that cross the orbit of Jupiter. We shall return to these objects later.

An interesting characteristic in the distribution of asteroid orbits is the existence of several clear areas or gaps (Figure 18.2), corresponding to orbital periods that the asteroids seem to avoid. These unoccupied periods were interpreted in 1866 by the American astronomer Daniel Kirkwood as a resonance phenomenon caused by gravitational perturbations from Jupiter. For example, an asteroid at about five-eighths of Jupiter's distance from the Sun (3.3 AU) would have a period exactly half that of Jupiter. Every two times around the Sun it would find itself relatively near the planet. The repeated attractions toward Jupiter, always in the same direction, would eventually perturb the orbit of such an asteroid, just as resonances with satellites produce gaps in the rings of Saturn (Section 17.4d).

In 1917 the Japanese astronomer K. Hirayama found that a number of the asteroids fall into **families**—groups with similar orbital characteristics. He hypothesized that each family may have resulted from an explosion of a larger body or from the collision of two bodies. Slight differences in the initial velocities of the fragments account for the small spread in orbital characteristics now observed for the different asteroids in a given family. There are several dozen such families, and observations of the larger families (the Eos and Themis families in particular) show that their individual members are physically similar, as if they were fragments of a

TABLE 18.1   THE TWENTY LARGEST ASTEROIDS

Name	Discovery	Semimajor Axis (AU)	Diameter (km)	Class
1 Ceres	1801	2.77	940	C
2 Pallas	1802	2.77	540	C
4 Vesta	1807	2.36	510	*
10 Hygeia	1849	3.14	410	C
704 Interamnia	1910	3.06	310	C
511 Davida	1903	3.18	310	C
65 Cybele	1861	3.43	280	C
52 Europa	1868	3.10	280	C
87 Sylvia	1866	3.48	275	C
3 Juno	1804	2.67	265	S
16 Psyche	1852	2.92	265	M
451 Patientia	1899	3.07	260	C
31 Euphrosyne	1854	3.15	250	C
15 Eunomia	1851	2.64	245	S
324 Bamberga	1892	2.68	235	C
107 Camilla	1868	3.49	230	C
532 Herculina	1904	2.77	230	S
48 Doris	1857	3.11	225	C
29 Amphitrite	1854	2.55	225	S
19 Fortuna	1852	2.44	220	C

*Vesta has a very unusual (once thought unique) basaltic surface.
C = carbonaceous; S = silicaceous; M = metallic.

common parent. The existence of these families testifies to the frequency of asteroid collisions in the past.

(d) Spacing and Collisions

The asteroids in the main belt have the kind of size distribution that characterizes a population of fragments. With the exception of the largest objects (diameters greater than about 100 km), therefore, the asteroids are probably mostly

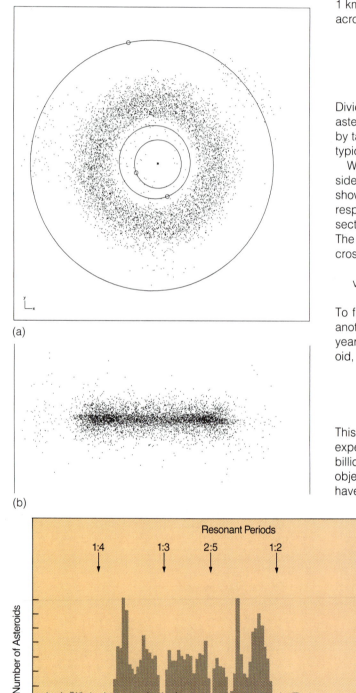

(a)

(b)

(c)

broken or shattered remnants of larger parent bodies. The Hirayama families are the most obvious result of such collisions, but the majority of the asteroids not in families may also have been involved in collisions further in the past. It has also been suggested that a very large collision or group of collisions in the asteroid belt about 4 billion years ago might have been responsible for the late heavy bombardment experienced by the Moon and the other terrestrial bodies.

It is easy to estimate the spacing of asteroids today and to calculate how often collisions take place. We have seen that there are about 100,000 objects in the belt with diameters of 1 km or more. These asteroids occupy a torus about 1 AU across and 0.5 AU thick, corresponding to a volume of:

$$v = \text{circumference} \times \text{width} \times \text{height}$$

$$v = 3 \text{ AU}^3 = 10^{25} \text{ km}^3$$

Dividing by the number of asteroids gives an average of one asteroid per 10^{20} km^3. The spacing between objects is found by taking the cube root, yielding about 5 million km as the typical asteroid separation in the main belt.

We can now calculate the frequency of collisions. Consider an asteroid with a radius of 5 km. Observations have shown that the typical relative speed of one asteroid with respect to others it approaches is 4 km/s. The cross-sectional area of our typical asteroid is πr^2, or about 75 km^2. The volume of space it moves through in a given time is its cross-section times its speed:

$$\text{volume swept out} = 300 \text{ km}^3/\text{s} = 1 \times 10^{10} \text{ km}^3/\text{yr}.$$

To find the frequency with which this object collides with another, we compare this volume of space swept out in a year with the volume of space associated with each asteroid, which we calculated above:

$$\frac{\text{volume per object}}{\text{volume swept out}} = \frac{10^{20}}{10^{10}} = 10^{10} \text{ yr}.$$

This result tells us that an individual 5-km asteroid can expect about one collision in 10 billion years. Thus in the 4.5-billion-year lifetime of the solar system, about half of the objects will have had at least one such collision, but few will have had a large number of them. If there were more aster-

FIGURE 18.2 Three views of the asteroid belt. (a) The positions of more than 6000 asteroids in February 1990 seen from above. Also shown are the planets Earth, Mars, and Jupiter. (b) The same distribution viewed from the plane of the solar system. (c) A plot of the number of asteroids with various semimajor axes (in AU). Some of the resonances are indicated at gaps (Kirkwood gaps) where the period of an asteroid would be a simple fraction of the period of Jupiter. (Parts a and b courtesy of Edward Bowell, Lowell Observatory)

oids in the past, collisions would have been even more frequent. This calculation supports our suggestion that many of the smaller asteroids are collisional fragments.

In a similar way we can estimate the probability that a spacecraft will impact an asteroid while crossing the belt. The spacecraft cross-section is about 10 m², and its speed is typically 20 km/s with respect to the asteroids. If the craft takes six months to cross the belt, the volume it sweeps out is about 3000 km³. Comparing this volume with the typical volume associated with each asteroid of 10²⁰ km³, we get the probability of collision:

$$\text{probability} = 3 \times \frac{10^3}{10^{20}} = 3 \times 10^{-17}.$$

Thus there is less than 1 chance in 10 million billion that a randomly aimed craft will collide with an asteroid of 1 km or larger diameter.

Finally, we can use this same kind of arithmetic to calculate how often a collision takes place somewhere in the belt. This is just the average time between collisions for any one asteroid divided by the number of asteroids, or:

$$\text{time between collisions} = \frac{10^{10}}{10^5} = 100,000 \text{ yr.}$$

18.2 THE ASTEROID BELT

(a) Geography of the Belt

The main asteroid belt stretches from its inner edge at 2.2 AU out to a rather sharp cutoff at 3.3 AU, the 2:1 resonance with Jupiter. Probably more than 90 percent of the asteroids are in the main belt. The belt is divided by the Kirkwood gaps into several subregions, separated by resonances with Jupiter.

Although there are nearly a million belt asteroids larger than 1 km in diameter, the asteroids are not closely spaced. The volume of the belt is actually very large, and the typical spacing between objects down to 1 km diameter is several million kilometers (Section 18.1e).

Not all asteroids are alike; in fact, their histories are surprisingly varied. As we shall see below, most of them are primitive objects, composed of silicates mixed with dark carbon-bearing compounds and metallic grains. These primitive asteroids appear to have formed directly from the original solar nebula and to have escaped major subsequent chemical modification by heating. Other asteroids, however, have undergone an extensive thermal evolution. Some of them are fully differentiated, like the planets and larger satellites. For the most part, the more primitive objects are in the outer part of the belt, while the rarer, differentiated objects are closer to the Sun.

(b) Asteroid Sizes

Let's look at how astronomers measure one of the most fundamental properties of asteroids—their size. Remember that they are all too small to be resolved with ground-based telescopes or even the Hubble Space Telescope. Since asteroid sizes cannot be measured directly with telescopes, we must use indirect methods.

The most accurate technique for measuring the size of an asteroid is by timing how long it takes to pass in front of (or **occult**) a star. All asteroids occult stars from time to time, and considerable effort has been made to coordinate observations over a wide geographical area to observe such occultations for the larger asteroids. From different places on Earth, different parts of an asteroid will appear to pass in front of the same remote star because the various observers see the asteroid in slightly different directions compared with the direction to the star. In effect, the asteroid casts its own moving shadow on the Earth. The combination of many observations of an occultation permits an accurate determination of the size and shape of the asteroid (Figure 18.3). About a dozen asteroid diameters have been measured by this technique, and the values so derived are accurate to within a few percent.

The method that works best to estimate the sizes of large numbers of asteroids is to compare their brightness in visible light (which is reflected sunlight) with the light they emit in the infrared—energy they have previously absorbed from the Sun. For a particular asteroid, we know its distance from the Sun and therefore how much sunlight falls on each area of its surface. The total sunlight it intercepts is equal to its cross-sectional area times the incident flux. Of that intercepted light, the

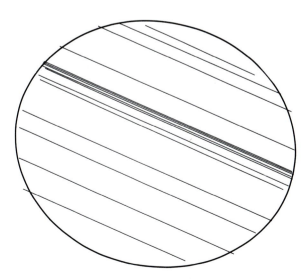

FIGURE 18.3 The diameter and shape of asteroid 3 Juno as determined from occultation measurements. The straight line segments represent the duration of the occultation timed from different observing sites. (Lowell Observatory, courtesy of Robert Millis)

asteroid reflects part, the fraction A (the albedo), and hence absorbs the rest, the fraction 1-A. The asteroid re-emits the energy it absorbs at infrared wavelengths. When we measure the infrared radiation coming from the object, we are recording how much energy it must have absorbed from the sunlight falling upon it. If we compare this measure with that of the light reflected from the object, we find what its reflectivity is. Then we can calculate what its size must be to account for the amount of light it reflects to us, that is, its observed brightness. Using observations made with the IRAS infrared satellite, astronomers have applied this technique to more than a thousand asteroids.

(c)   Composition and Classification

When the reflectivities of many asteroids are compared, a great diversity is seen. The majority are very dark, with reflectivities of only 3 to 5 percent, about as dark as a lump of coal. However, there is another large group with typical reflectivities of about 15 percent, a little brighter than the Moon, and still others with reflectivities as high as 60 percent. From these reflectivity measurements we can see that there are different kinds of asteroids, but further classification requires spectral information as well.

Spectroscopy is of limited use in the identification of the chemistry of solids and liquids. Almost all *gases* have their own unique pattern of sharp spectral lines that can be used for the analysis of starlight, for example. But solids tend to have only a few broad spectral bands, and their interpretation is often ambiguous. The common ices (H_2O, CO_2, NH_3, CH_4) are easy to distinguish by their infrared spectra, but the more complex silicate minerals that make up the surfaces of asteroids present the astronomer with a difficult challenge. In spite of these problems, however, astronomers have used spectra together with reflectivity measurements for a compositional classification of the asteroids (Figure 18.4).

The majority of the asteroids are revealed from spectral studies to be primitive bodies. The presence of dark organic materials reduces the asteroids' reflectivities to the 3 to 5 percent level observed. Many of these objects also include some water chemically bonded to the silicates. Two of the largest asteroids, Ceres and Pallas, are primitive, as are almost all of the objects in the outer third of the belt. We classify the dark primitive asteroids C asteroids, where the C stands for "carbonaceous," because they are similar to the carbonaceous class of meteorites (Chapter 20). Beyond 3 AU there are additional primitive forms with different spectra from the normal C asteroids. These appear to be composed of other mixtures of silicates and carbon compounds, not represented in our collection of meteorites.

The second most common type of asteroids are the S asteroids, where S stands for "silicate." In these asteroids the dark carbon compounds are missing, resulting in higher reflectivities and clearer spectral signatures of silicate minerals. The minerals present are similar to those that make up many meteorites, but the exact

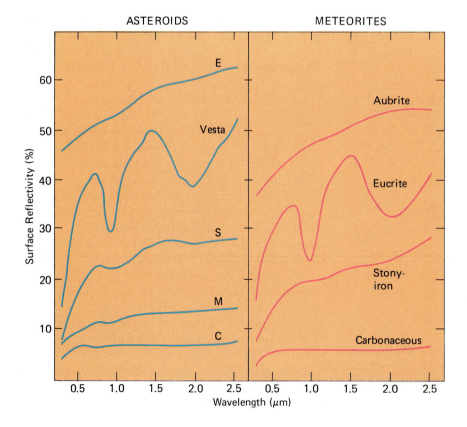

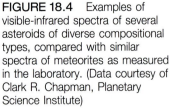

FIGURE 18.4 Examples of visible-infrared spectra of several asteroids of diverse compositional types, compared with similar spectra of meteorites as measured in the laboratory. (Data courtesy of Clark R. Chapman, Planetary Science Institute)

compositions of the S asteroids remain in dispute. In particular, we are unable to answer the basic question of whether these asteroids are primitive bodies or whether they are differentiated. Many scientists who study asteroids suspect that they are primitive, representing material that condensed in the inner part of the asteroid belt where dark carbonaceous materials were not present, but we really do not know if this is the case.

From spectral observations we have identified a few asteroids, not more than 5 percent of the total, that are clearly differentiated objects. These include the M class asteroids, which are thought to be made largely of metal (presumably the cores of differentiated parent bodies that were shattered in collisions). This supposition has been proved in the case of the largest M asteroid, 16 Psyche. When targeted by the Arecibo radar, Psyche was found to be much more reflective than would be a rocky object of the same size and distance. Metals are good radar reflectors, and these observations prove a sensitive test for metallic surfaces on asteroids.

Other differentiated asteroids have basaltic surfaces like the volcanic plains of the Moon and Mars. The large asteroid Vesta is in this category. Apparently some of the asteroids were heated early in the history of the solar system, but why these, and why only a small percentage of the total number, we do not know.

Detailed surveys of the population of the belt have revealed that the different compositional classes are at different distances from the Sun. Thus we may think of the belt as made up of overlapping rings of similar kinds of objects, with each ring having a width of about half an AU (Figure 18.5). The presence of this structure suggests to us that the asteroids must still be in approximately the positions in which they formed; the belt has not been totally mixed and homogenized. If we could sample the compositions of the different kinds of primitive asteroids, we could map out in some detail the nature of the solar nebula out of which they formed.

TABLE 18.2	MASSES AND DENSITIES OF ASTEROIDS	
Name	Mass (Moon = 1)	Density (g/cm³)
Ceres	1.4×10^{-2}	2.4
Pallas	2.9×10^{-3}	2.6
Vesta	3.7×10^{-3}	3.8
Hygiea	1.3×10^{-3}	2.6
Phobos	1.5×10^{-6}	2.0

(d) Vesta: A Volcanic Asteroid

Vesta is one of the most interesting of the asteroids. It orbits the Sun with a semimajor axis of 2.4 AU, and its relatively high reflectivity of almost 30 percent makes it the brightest of the main belt objects, clearly visible to the unaided eye if you know just where to look. But Vesta's real claim to fame is the fact that its surface is covered with basalt, indicating that Vesta was once volcanically active in spite of its small size (diameter about 500 km). It is probably completely differentiated, with a metal core. The density of Vesta, 3.8 g/cm³, is what we would expect for an asteroid this size that is differentiated rather than primitive. Vesta is the only large asteroid with a basaltic surface, although several very small objects (less than 10 km in diameter) composed of basalt have been discovered recently.

Greatly adding to the importance of Vesta is the fact that we apparently have samples of its surface to study directly in the laboratory. Meteorites (Chapter 20) have long been suspected to come from the asteroids, but there is generally no way to identify the particular source of a given meteorite that strikes the Earth. In Vesta's case, however, this identification seems fairly firm.

The meteorites that are believed to come from Vesta are called the **eucrites,** a group of about 30 basaltic meteorites of very similar composition (Figure 18.6). Chemical analysis of the eucrites has shown that they cannot have come from the Earth, Moon, or Mars. On

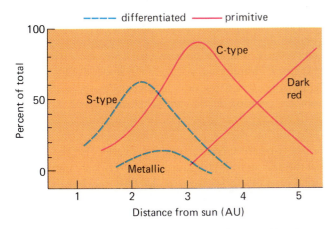

FIGURE 18.5 The distribution of asteroid compositional types with distance from the Sun. (Adapted from work carried out by D. Tholen and J. Gradie, University of Hawaii, and E. Tedesco, JPL)

FIGURE 18.6 Photograph of one of the eucrite meteorites, believed to be fragments from the crust of asteroid Vesta. (NASA)

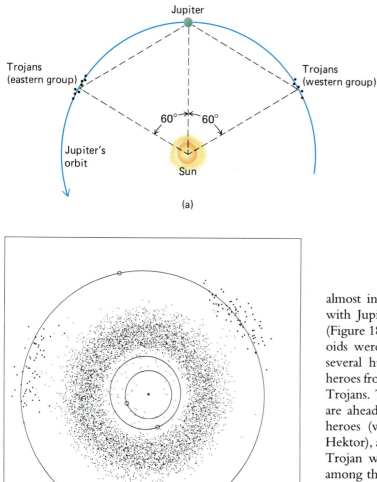

(a)

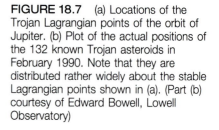

FIGURE 18.7   (a) Locations of the Trojan Lagrangian points of the orbit of Jupiter. (b) Plot of the actual positions of the 132 known Trojan asteroids in February 1990. Note that they are distributed rather widely about the stable Lagrangian points shown in (a). (Part (b) courtesy of Edward Bowell, Lowell Observatory)

(b)

the other hand, their spectra (measured in the laboratory) match perfectly the spectra of Vesta obtained telescopically. The age of the lava flows from which the eucrites derived has been measured at 4.5 billion years, less than 100 million years after the formation of the solar system. This age is consistent with what we might expect for Vesta; whatever process heated such a small object was probably intense and short-lived.

18.3   ASTEROIDS FAR AND NEAR

(a)   The Trojans

The Trojan asteroids are objects located far beyond the main belt, orbiting the Sun at about the same distance as Jupiter, 5.2 AU. As we saw in Section 4.2, the French mathematician Lagrange showed that there should be two points (called Lagrangian points) in the orbit of Jupiter near which objects like asteroids can remain

almost indefinitely. These are the two points which, with Jupiter and the Sun, make equilateral triangles (Figure 18.7). Between 1906 and 1908, four such asteroids were found; the number has now increased to several hundred. These asteroids are named for the heroes from Homer's *Iliad* and are collectively called the Trojans. Those that precede Jupiter (that is, those that are ahead of it in its orbit) are named for the Greek heroes (with the exception of the Trojan spy, 624 Hektor), and those that follow Jupiter are named for the Trojan warriors (with the Greek spy, 617 Patroclus, among them).

The Trojan asteroids circle the Sun with Jupiter's period of 12 years, one-sixth of a cycle ahead of or behind the planet. Their detailed motion, however, is very complicated; they slowly oscillate around the points of stability found by Lagrange, with some of their oscillations taking as long as 140 years.

Measurements of the reflectivities and spectra of the Trojans show that they are all very dark, primitive objects like those in the outer part of the asteroid belt. They appear faint because they are so dark and far away, but actually the larger Trojans are quite sizable. Four of them—Hektor, Diomedes, Agamemnon, and Patroclus—have diameters between 150 and 200 km. Hektor is about twice as long as it is wide, leading to the suggestion that it is a double asteroid, with two similar objects orbiting in contact with each other. Current estimates are that there must be more than 1000 Trojan asteroids in the region preceding Jupiter (the Greek camp) that are at least 15 km in diameter, and about 250 in the region following Jupiter (the Trojan camp). The total mass of the Trojan asteroids thus may be comparable to that of the main belt.

(b)   Hidalgo and the Centaurs

There may be many asteroids with orbits that carry them far beyond Jupiter, but they are difficult to detect and only a few have been discovered. Even among those that are known, there is some confusion whether they should be classified as asteroids or comets. Astronomers normally call a small object a comet only if it has a visible atmosphere, but at these distances from the Sun temperatures are too low for atmospheres to form. Thus there is no distinction far from the Sun between an inactive comet and a volatile-rich asteroid.

The first of these distant asteroids to be discovered was 944 Hidalgo; with its semimajor axis of 5.9 AU and a very large eccentricity of 0.66, Hidalgo has an aphelion outside the orbit of Saturn. Its diameter is at least 100 km. More distant is 2060 Chiron, with a semimajor axis of 13.7 AU; its orbit (eccentricity 0.38) carries it from just inside the orbit of Saturn at perihelion out almost to the orbit of Uranus. The diameter of Chiron is estimated to be about 200 km. In 1992, a still more distant object was discovered, named 5145 Pholus, with an orbit that takes it 33 AU from the Sun, beyond the orbit of Neptune. Pholus has the reddest surface of any object in the solar system, indicating a strange (and still unknown) composition. As more objects are discovered in these distant reaches, they will be given the names of Centaurs (half man, half horse); in ancient Greek mythology, Chiron and Pholus were the two "good" Centaurs.

If Chiron and the other Centaurs contain abundant volatiles (like H_2O or CO ice), they would probably develop atmospheres if they were heated by the Sun (in which case astronomers would call them comets). This is apparently what happened in 1988, when astronomers found that Chiron had brightened by about a factor of two, with the additional light reflected from gas or dust ejected from the surface. One year later, this new atmosphere was photographed (Figure 18.8). Chiron has been slowly approaching its perihelion for several years, and apparently the gradual increase in its temperature was sufficient by 1988 to initiate cometary activity.

Chiron and Pholus are much larger than the nuclei of any known comet. What a spectacular show they would put on if either were diverted into the inner solar system! Chiron has been shown to be in an unstable orbit, so sometime in the future this is exactly what could actually happen.

(c) Earth-Approaching Asteroids

There is an important class of asteroids with orbits that either come close to or cross the orbit of the Earth, called **Earth-approaching asteroids.** Some of these are the nearest approaching celestial objects, excepting the Moon. Ultimately many of these will collide with the Earth or Moon to produce major impact craters.

The Earth-approaching asteroids are divided into three groups on the basis of their orbits. The innermost are the Atens (named for 2062 Aten), which have orbits

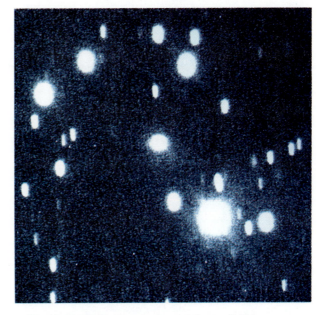

FIGURE 18.8 The atmosphere of "comet" Chiron, photographed in 1989 when Chiron had approached to within 12 AU of the Sun. (National Optical Astronomy Observatories, courtesy of Karen Meech, University of Hawaii)

with semimajor axes less than 1.0 AU. The best-known group are the Apollos (for 1862 Apollo, discovered in 1948), which cross the orbit of the Earth (or nearly do so) but have semimajor axes greater than 1.0 AU. The Apollos include 1566 Icarus, which has its perihelion inside the orbit of Mercury. The outer group are the Amors (for 1221 Amor), which are Mars-orbit-crossing asteroids with perihelion distances between 1.017 and 1.400 AU.

Only about 200 Earth-approaching asteroids have been located by astronomers so far, although more are being discovered every month. The largest, 433 Eros, is about 20 km across, similar in size to the satellites of Mars. Most of the known Earth-approaching asteroids, however, are only a few kilometers in diameter (Table 18.3). In composition, they seem similar of the main belt asteroids, with members of the C, S, and M classes, and even basaltic objects like Vesta.

Eugene Shoemaker of Lowell Observatory in Flagstaff, Arizona, leads one of the searches for new Earth-approaching asteroids and has carried out extensive calculations of the dynamics of the entire population. He estimates that there are about 2000 Earth-approaching asteroids down to a diameter of 1 km. About half of these are Apollos and half Amors, with only a few Atens.

Several of the Earth-approaching asteroids hold the distinction of being among the easiest objects in the solar system for round-trip travel from the Earth. In the best cases, the energy required for such a space flight is as small as the energy needed for a round trip to the Moon. Therefore, these little objects may become tar-

Name	Orbit Type	Discovery	Semimajor Axis (AU)	Eccentricity	Diameter (km)	Class
433 Eros[1]	Amor	1898	1.46	0.22	22.	S
1036 Ganymed[2]	Amor	1924	2.66	0.54	37.	S
1221 Amor	Amor	1932	1.92	0.44	—	—
1566 Icarus[3]	Apollo	1949	1.08	0.83	0.9	—
1862 Apollo	Apollo	1932	1.47	0.56	1.5	*
1866 Sisyphus[4]	Apollo	1972	1.89	0.54	8.2	—
2062 Aten	Aten	1976	0.97	0.18	0.9	S
3200 Phaethon[3]	Apollo	1983	1.27	0.89	6.9	*
4179 Toutatis[5]	Apollo	1989	2.51	0.64	4.0	—
4769 Castalia[5]	Apollo	1989	1.06	0.48	0.6	—

gets for human exploration, if and when we are prepared to venture once again beyond low-Earth orbit.

As Shoemaker and others have shown, the orbits of Earth-approaching asteroids are unstable. These objects will meet one of two fates: either they will impact one of the terrestrial planets, or they will be ejected from the inner solar system as the result of a near encounter with a planet. The probabilities for these two outcomes are about the same. The time scale for impact or ejection is only about 100 million years, very short in comparison with the age of the solar system.

If the current population of Earth-approaching asteroids will be removed by impact or ejection in 100 million years, there must be a continuing source of new objects. Some come from the main asteroid belt, where collisions can eject fragments into Earth-crossing orbits. Others may be dead comets that have exhausted their volatiles; as many as half of the Earth-approachers could be the solid remnants of former comets. We will discuss the fate of old comets in Chapter 19.

(d)   Radar Images

If an Earth-approaching asteroid comes close enough, it can be studied with radar as well as optical telescopes. It is even possible to use radar to produce an *image* of the asteroid. Steven Ostro and his colleagues at the NASA Jet Propulsion Laboratory have succeeded in imaging two such objects with sufficient resolution to establish their shape and size. The first of their targets was 4769 Castalia, which they studied in 1989 using the 300-m Arecibo radar in Puerto Rico. Although Castalia is only 600 m long, the radar images (Figure 18.9) show that it is a double object, consisting of two nearly equal lumps in a dumbbell configuration.

The second Earth-approaching asteroid to be suc-

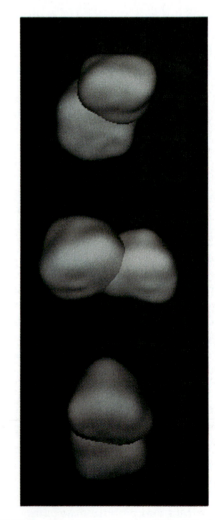

FIGURE 18.9   The near-Earth asteroid Castalla, imaged by the Arecibo planetary radar during a close pass by the Earth. The asteroid appears as two lumpy spheres, each about 500 m in diameter, rotating in contact. (Steve Ostro, NASA/JPL)

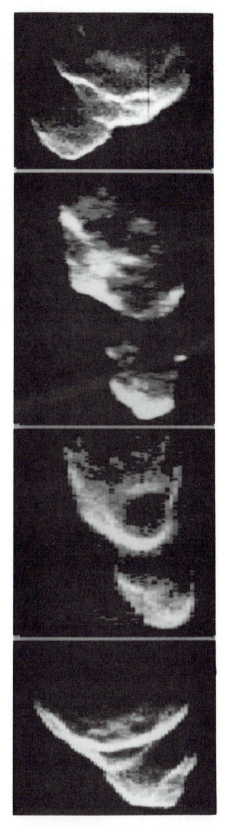

cessfully imaged was 4179 Toutatis, which approached to within 3 million km in 1992—less than 10 times the distance to the Moon. Several radar images of Toutatis are shown in Figure 18.10, obtained by Ostro using the NASA radar system in California. This too appears to be a double object, consisting of two irregular lumps with diameters of 3 km and 2 km, rotating in contact with each other. Judging from these two examples, double configurations may be common for such small asteroids.

18.4 ASTEROIDS UP CLOSE

Asteroids make up the last class of objects in the solar system to be visited by spacecraft; the first asteroid flyby did not take place until 1991. In this section we discuss the results of two spacecraft encounters with asteroids, together with spacecraft data on the two small moons of Mars.

(a) Phobos and Deimos

The two satellites of Mars (Section 14.2e) are probably captured asteroids. If they were captured, it must have been very early in solar system history. For more than 4 billion years they have been in orbit about Mars, and not out among the other asteroids. Still, they may be able to tell us something of what asteroids look like.

Phobos and Deimos were first studied at close range by the U.S. Viking orbiters in 1977. In 1989, Phobos was the target of the Russian spacecraft named Phobos, but unfortunately, the Phobos spacecraft failed before it could land experiments on the surface. Both martian satellites are rather irregular, somewhat elongated, and heavily cratered (Figure 18.11). The largest diameters of Phobos and Deimos are about 25 km and 13 km, respectively. Each is a dark brownish gray in color, and spectral analysis suggests that each is composed of dark materials similar to those out of which most asteroids are made. Apparently these two satellites are chemically primitive.

Some additional clues to the compositions of Phobos and Deimos can be derived from their densities, which have been calculated from the masses and volumes measured by the Viking spacecraft. Each has a density of only 2.0 g/cm^3, remarkably low for a rocky object—substantially lower than the densities of Ceres or Pallas, for example. In fact, this is about the same density as Pluto or Triton, which are thought to be made in part of water ice. At the temperatures of Phobos and Deimos, however, ice is not expected to be stable over the lifetime of the solar system, so their composition remains something of a mystery.

When studied in detail, Phobos is found to have some quite remarkable surface features. It is laced by long grooves or troughs associated with the large crater Stickney. Apparently the impact that produced Stickney very nearly ruptured the satellite.

FIGURE 18.10 Several radar images of near-Earth asteroid Toutatis, obtained in December 1992 with the NASA Goldstone radar imaging facility. These low resolution images show that Toutatis appears to consist of two parts somewhat squashed together, one of them about 3 km in diameter and the other 2 km in diameter. (Steve Ostro, NASA/JPL)

(a)

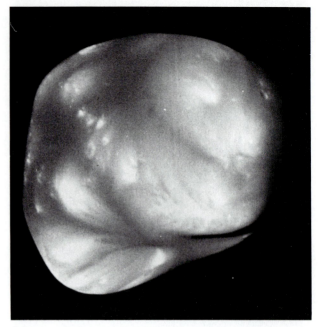

(b)

FIGURE 18.11(a) Phobos, the larger of the two martian moons, as photographed by Viking. (NASA/JPL, courtesy of Peter Thomas, Cornell University)

(b) Deimos, the smaller of the two satellites of Mars, as photographed by Viking. (NASA/JPL, courtesy of Peter Thomas, Cornell University)

(b) Encounter with Gaspra

The first close look at a main belt asteroid was provided by the Galileo spacecraft in October 1991, when it flew past 951 Gaspra at a distance of only 1600 km. Gaspra was known to be a member of the Flora family, suggesting that it is a fragment of a collision that broke up a preexisting asteroid to create Flora, Gaspra, and the other members of this family. It rotates with a period of 7 hours. Gaspra is a typical asteroid of the S (silicate) class, noteworthy only because its orbit carried it close to the Jupiter-bound Galileo spacecraft.

Spacecraft images (Figure 18.12) revealed Gaspra to be highly irregular, as befits a fragment from a catastrophic asteroid collision. It is 16 km long but only half as wide, similar in size to Deimos but more elongated. Large asteroids are expected to be approximately spherical in shape, but small ones can be almost any shape, especially if they were produced by the breakup of a larger parent object. Unlike Phobos and Deimos, Gaspra does not appear to be blanketed with a thick layer of dust. From the rather sparse number of craters on its surface, all of which must have been made by small impacts subsequent to the formation of the asteroid, the Galileo scientists estimate that Gaspra is only about 200 million years old—that is, the collision that formed Gaspra and other members of the Flora family took place about 200 million years ago.

(c) Ida and Its Moon

The Galileo spacecraft made a second pass through the asteroid belt in 1993, and this time it was targeted for a larger S-class asteroid named 243 Ida. Ida is 56 km long, and like Gaspra it has the irregular shape that might be expected for a collisional fragment (Figure 18.13). It is much more heavily cratered than Gaspra, suggesting that the surface has been exposed to small impacts for a longer time period, probably more than a billion years. Yet as a member of the Koronis asteroid family, Ida should be relatively young. Perhaps it acquired most of its craters in some special large impact event.

The greatest surprise of the Galileo flyby of Ida, and the most important result scientifically, was the discovery of a satellite (named Dactyl) in orbit about the asteroid (Figure 18.14). Although it is only 1.5 km in diameter, smaller than many college campuses, Dactyl provides scientists with something otherwise beyond their reach—a measurement of the mass of Ida using Kepler's laws. When the mass is combined with the volume measured from spacecraft images, we can also calculate the density of Ida. The satellite's distance of about 100 km and its orbital period of about 24 hours indicate that Ida is slightly denser than the moons of Mars, with a density that is similar to that of Ceres (roughly 2.5 g/cm^3). If this preliminary number is correct, then Ida is probably a chemically primitive object rather than a fragment of a differentiated parent body. By a logical extension of this reasoning, the entire Koronos family is probably made up of primitive bodies, and at least a part of the S class of asteroids is primitive rather than differentiated.

The origin of Dactyl is difficult for scientists to understand. No asteroidal satellite would be expected to survive for more than about 100 million years, due to

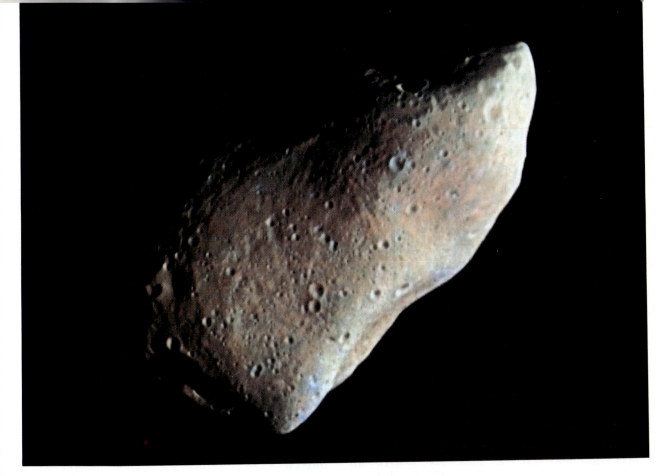

FIGURE 18.12   Galileo image of the small main-belt asteroid, Gaspra, from a distance of 1600 km, with a resolution of about 100 m. The color is highly exaggerated to bring out subtle differences in surface composition. The dimensions of Gaspra are approximately 19 × 12 × 11 km. (NASA/JPL)

FIGURE 18.13   Asteroid Ida photographed by the Galileo spacecraft in 1993. Features as small as 50 m across can be seen in this image of the 50 kilometer long asteroid. (NASA/JPL)

It is possible to calculate that the *gravitational sphere of influence* of a star—the distance within which it can exert sufficient gravitation to hold on to orbiting objects—is about ⅓ the distance to the nearest other stars. In the vicinity of the Sun, stars are spaced an average of 3 light years (200,000 AU) apart. Thus the Sun's sphere of influence extends only a little beyond 50,000 AU, and at such distances objects in orbit about the Sun will be perturbed by the gravitation of passing stars. Oort suggested, therefore, that the new comets were objects orbiting the Sun with aphelia near the edge of its sphere of influence, and that the perturbing effects of other nearby stars modified their orbits to bring them close to the Sun where humans could see them. This region from which the new comets are derived is now called the **Oort comet cloud.**

Several new comets are discovered each year, most of them probably approaching the Sun for the first time. If comets have been entering the inner solar system at this same rate for its entire history of 4.5×10^9 years, then the source region must have originally held at least 10^{10} comets. Using similar but more detailed arguments, Oort calculated there were at least 10^{11} comets in the cloud.

Just because most new comets have aphelia near 50,000 AU, we should be careful not to conclude that the Oort cloud consists of billions of comets in roughly circular orbits at this distance, like a shell around the Sun. There may be some comets in nearly circular orbits, but if so, we have no evidence of them. What we see are comets that are perturbed into the inner solar system, and these comets must have already had orbits of very large eccentricity. A passing star can cause only a slight change in eccentricity, so if a comet has an orbit with a perihelion of, say, 1 AU when we discover it, it must have had a perihelion of no more than a few AU on its previous unobserved passage near the Sun. For every comet that comes close enough to be seen, there must be many more skimming invisibly through the orbits of the outer planets.

(b) The Kuiper Belt

Not all comets originate in the Oort cloud. From detailed examination of their orbits, astronomers have concluded that many of the short-period and Jupiter-family comets originated in a much closer region of space than the 50,000 AU that is typical of new comets. This closer source is called the **Kuiper belt,** named after the Dutch-American planetary astronomer who suggested it as another source of comets, in addition to the Oort cloud.

Astronomers hypothesized that the Kuiper belt, like the asteroid belt, is flattened and lies in the plane of the solar system, somewhere beyond the orbit of Pluto. In this respect it is different from the Oort cloud, which is approximately spherical. Until recently, however, no one had actually observed any of the objects in the Kuiper belt, so it remained only a convenient concept, not a demonstrated reality.

The first actual residents of the Kuiper belt were discovered in 1992 by David Jewitt and Jane Luu of the University of Hawaii. Taking advantage of the excellent observing conditions on Mauna Kea, Jewitt and Luu obtained a number of long-exposure CCD images and carefully searched them for faint objects orbiting beyond Neptune. The first one is less than 100 km in diameter and follows a nearly circular orbit with a semimajor axis of 40 AU, very similar to that of Pluto. By March 1994, eight such objects had been discovered. Four of them have semimajor axes near 33 AU and may be Trojans of Neptune (Section 18.3a), but the other four are clearly at the distance of Pluto or beyond. Presumably these are just the largest and innermost of a population of millions of icy objects that orbit beyond the traditional limits of the planetary system. The comets in the Kuiper belt are thought to occupy a disk-shaped volume and to share the general east-to-west rotation of the planetary system.

(c) Total Mass of the Comets

By definition, the Oort comet cloud is the source region for most of the observed "new" comets. We see these comets because they come close to the Sun, and they come close to the Sun because at 50,000 AU their orbits can be perturbed by passing stars. We are also seeing a few of the objects at the inner edge of the Kuiper belt, just beyond the orbit of Pluto. We have no way, however, of detecting residents with stable orbits between 50 AU and 50,000 AU. We can only guess if this vast volume of space is filled with comets. We must rely on theory to estimate how many small, icy bodies might be present. Astronomers estimate there may be as many as 10^{13}, far more than the number in the observable Oort cloud.

What is the mass represented by 10^{13} comets? We can make an estimate if we assume something about typical comet sizes and masses. Let us suppose that the nucleus of Comet Halley is typical. Its estimated mass (Section 19.2) is about 10^{-10} Earth masses. The corresponding mass for all the comets is about 1000 Earth masses—greater than the mass of all the planets put together. Therefore, cometary material is the most important constituent of the solar system after the Sun itself. Various astronomers have made these calculations, and most think that the total mass of the comets is at least a few hundred times of the mass of the Earth.

(d) Formation of the Comet Cloud

There are two theories for the formation of the Oort cloud: either it is made of material that condensed in place, tens of thousands of AU from the Sun, or else the comets were formed closer to the Sun and subsequently ejected to these large distances.

It would be easy to imagine that the comets were condensates from the outer fringes of the solar nebula if there were not so many comets. As we have seen, the mass of the comets is a substantial fraction of the total mass of the solar system, yet models suggest that the solar nebula thinned out rapidly with increasing distance from the Sun. It is difficult to understand how any substantial amount of material could have condensed at such huge distances, or how small grains might have accumulated into bodies several kilometers across.

A more likely hypothesis is that the comets were ejected into the Oort cloud from initial orbits near the present orbits of Uranus and Neptune. Calculations indicate that if many icy planetesimals were still present after the giant planets formed, a substantial fraction of them would have been ejected by gravitational encounters with these two planets. If this theory is correct, then the comets are leftovers from the building blocks of the outer planets, preserved for 4.5 billion years in the deep freeze of space.

(e)   The Fate of Comets

Once a comet enters the inner solar system, its previously uneventful life history begins to accelerate. It may, of course, survive its initial passage near the Sun and return to the cold reaches of space where it spent the previous 4.5 billion years. At the other extreme, it may impact the Sun or pass so close that it is destroyed on its first perihelion passage. Observations from space indicate that at least one comet collides with the Sun every year, something that had never been expected from Earth-based observations. Frequently, however, the new comet does not come this close to the Sun, but instead interacts with one or more of the planets.

A comet coming within the gravitational influence of a planet has three possible fates: (1) it can impact that planet, ending the story at once; (2) it can be ejected on a hyperbolic trajectory, leaving the solar system forever; or (3) it can be perturbed into a shorter period. In the last case, its fate is sealed. Each time it approaches the Sun, it will lose part of its material, and it still has a significant chance of collision with a planet. Once in a short-period orbit, the comet's lifetime is measured in thousands, not billions, of years.

Measurements of the amount of gas and dust in the atmosphere of a comet permit an estimate of the total losses during one orbit. Typical loss rates are up to a million tons per day from an active comet near the Sun, adding up to some tens of millions of tons per orbit. This is equivalent to stripping off the top several meters from the nucleus. Comparing these loss rates with the total mass of the comet, we see that they amount to about 0.1 percent per orbit. At that rate, the comet will be gone after a thousand orbits.

Whether the comet evaporates away completely is not known. If the gas and dust are well mixed, we would expect the nucleus to shrink each time around the Sun until it has entirely disappeared. However, there remains the suggestion that many of the Earth-approaching asteroids are extinct comets. If there is a silicate core in the comet, or if the dirty snowball includes large blocks of nonvolatile material that are held gravitationally to the surface, then there could be a substantial solid residue after the ices are gone. We simply do not know which of these alternatives is correct.

(f)   Catastrophic Deaths of Comets

Some comets end catastrophically. Even if they avoid impacting a planet or being perturbed into an orbit that collides with the Sun, they can break apart for reasons that are not well understood. In 1846, for example, the nucleus of Comet Biela split into two parts, and on its next return in 1852 it appeared as two comets, separated by 2 million km. In 1866, the next perihelion year, nothing appeared; Comet Biela had simply and totally disappeared. A more recent example is Comet West. Shortly after its 1976 perihelion passage its nucleus split into four components that drifted apart at a rate of several hundred kilometers per day. The smallest fragment survived only a few days, but the other three retained their identities until the comet became invisible with increasing distance from the Sun.

The most spectacular example of cometary breakup is provided by a faint object discovered in 1993, called Shoemaker-Levy 9, the ninth comet found by the team of Carolyn and Gene Shoemaker and David Levy. Although it was much too faint to be seen without a telescope, photographs of the comet showed from its discovery that it had multiple nuclei stretched out in a row like beads on a string. The Hubble Space Telescope was able to image 22 separate nuclei, all gradually drifting apart (Figure 19.13). When astronomers calculated the orbit for this comet they found even more surprises. Shoemaker-Levy 9 was not actually in orbit around the Sun; instead it was captured into an elongated and loosely bound orbit around Jupiter. It had been disrupted into multiple fragments when it passed very close to Jupiter in June 1992, less than 50,000 km above the clouds.

The comet's orbit after this close encounter was not stable, and it soon became clear that all of the cometary fragments would crash into Jupiter in July 1994. As each object streaked in at 60 km/s, it disintegrated and exploded, forming a fireball that carried the comet dust as well as gases from the lower atmosphere to high altitudes. Unfortunately for astronomers, all of these impacts took place on the side of Jupiter that was turned away from the Earth, so none of the explosions could be seen directly. However, astronomers from all over the

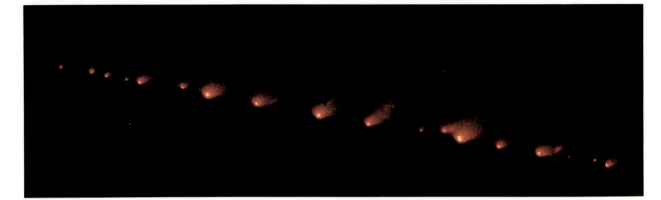

FIGURE 19.13   Comet Shoemaker-Levy 9 as photographed by the Hubble Space Telescope in May 1994, just two months before impact with Jupiter. Approximately 21 distinct objects can be seen, stretching over a distance of 1.1 million km. (NASA/STScI).

world trained their instruments on the planet to see the effects that these impacts had on the clouds and upper atmosphere as Jupiter rotated to bring the impact sites into view.

The larger fragments, which were each about 1 km in diameter, penetrated below the ammonia clouds but exploded before they reached the water clouds. The Galileo spacecraft provided the best view, since from its angle the impact sites could be seen directly. Many of the fireballs were so large, however, that they were visible even from Earth as they ascended over the horizon of Jupiter, reaching altitudes of several thousand kilometers. The individual explosions, releasing energies of tens of millions of megatons, sprayed out dark clouds of ejecta that settled into the stratosphere of Jupiter, producing long-lived "bruises" that were larger

than the Earth and could easily be seen even in small telescopes (Figure 19.14). Millions of people all over the world peered at Jupiter though telescopes or followed the event via television, and astronomers will be analyzing their data for many years to come.

19.4   METEORS

Whatever the fate of the remnants of a cometary nucleus after the volatiles are exhausted, we do know what happens to the dust that is carried away from a comet by the evaporation of the nucleus. This dust fills the inner part of the solar system. The Earth is surrounded by it, and each of the larger dust particles that reaches the Earth creates a **meteor.**

FIGURE 19.14   Scars on Jupiter produced by the comet impacts. The large dark feature was formed by dust ejected in the explosion; since the comet entered the atmosphere from the south at a 45° angle, much of the resulting ejecta was thrown back along the direction of approach. At the time this image was taken, 105 minutes after the impact, there was a central dark spot 1500 km in diameter, a larger thin ring 7500 km in diameter, and a diffuse outer ring about 12000 km in diameter, or about the same size as the Earth. Later, the winds on Jupiter blended these features into a broad spot that remained visible for more than a month. (NASA/STScI)

(a) The Phenomenon of a Meteor

Although the layperson often confuses comets and meteors, these two phenomena are very different. Comets can be seen when they are many millions of miles away from the Earth and may be visible in the sky for weeks or even months, slowly shifting their positions from day to day. They rise and set with the stars, and during a single night they appear motionless to the casual glance. Meteors, on the other hand, are small, solid particles that enter the Earth's atmosphere from interplanetary space. Since they move at speeds of many kilometers per second, the high friction they encounter in the air vaporizes them. The light caused by the luminous vapors formed in such an encounter appears like a star moving rapidly across the sky, fading out within a few seconds. This is why meteors are commonly called "shooting stars" or "falling stars." In Afghan myth, each meteor marks the destruction of a devil by an angel in the eternal cosmic struggle between good and evil.

On a dark, moonless night an alert observer can see half a dozen meteors per hour. To be visible, a meteor must be within 200 km of the observer. Over the entire Earth, the total number of meteors bright enough to be visible must total about 25 million per day. More meteors can generally be seen in the hours after midnight than in the hours before, since we are then on the leading edge of the Earth as it moves through space.

Meteors become visible at an average height of 95 km. The highest meteors form at heights of 130 km. Nearly all meteors completely disintegrate, and their luminous paths end, by the time they reach altitudes of 80 km. A few meteors that strike the atmosphere at grazing incidence doubtless skip out again, returning to space before they are completely burned up.

The typical bright meteor is produced by a particle with a mass less than 1g—no larger than a pea. Of course, the light you see comes from the much larger region of glowing gas surrounding this little grain of interplanetary material. A particle the size of a marble produces a bright fireball when it plunges through the atmosphere, and one as big as a fist has a fair chance of surviving its fiery entry to become a meteorite, if its approach speed is not too high. The total mass of meteoritic material entering the Earth's atmosphere is estimated to be about 100 tons per day.

(b) Meteor Showers

Many—perhaps most—of the meteors that strike the Earth can be associated with specific comets. These interplanetary dust particles retain approximately the orbit of their parent comet, and the particles travel together through space. When the Earth crosses such a dust stream, we see a sudden burst of meteor activity, usually lasting several hours. These events are called **meteor showers.**

The meteors in a meteor shower all seem to radiate or diverge from a single point on the celestial sphere. That point is called the **radiant** of the shower. Recurrent showers are named for the constellation within which the radiant lies or for a bright star near the radiant.

The apparent divergence of shower meteors from a common point is easily explained. The meteors in a shower are all traveling in the same direction through space. When the Earth passes through such a dust stream, it is struck by many particles, all approaching it from the same direction. On the ground, if we look toward the direction from which the particles are coming, their paths all seem to diverge from a point. Similarly, if we look along railway tracks, those tracks, although parallel to each other, seem to diverge from a point in the distance (Figure 19.15).

Meteoric dust is not always evenly distributed along the orbit of the comet, so that sometimes more meteors are seen when the Earth intersects the dust stream and sometimes fewer. A very clumpy distribution is associated with the Leonid meteors, which in 1833 and again in 1866 (after an interval of 33 years—the period of the comet) yielded the most spectacular showers ever recorded. The last good Leonid shower was on November 17, 1966, when in some southwestern states in the U.S., up to 100 meteors could be observed per second.

The best meteor shower that can be depended on at present is the Perseid shower, which appears for about three nights near August 11 each year. In the absence of bright moonlight, meteors can be seen with a frequency of about one per minute during a typical Perseid shower. It is estimated that the total combined mass of the particles in the Perseid swarm is nearly 5×10^{11} kg. This gives at least a lower limit for the original mass of its associated comet, called Comet 1862 III. The mass of Comet Halley, however, is about a thousand times greater, suggesting that only a very small fraction of the original material survives in the meteor stream.

No shower meteor has ever survived its flight through the atmosphere and been recovered for laboratory analysis. However, there are other ways to investigate the nature of these particles and thereby to gain additional insight into the comets from which they are derived. Analysis of the photographic tracks of meteors shows that most of them are very light or porous, with densities typically less than $1.0 \, g/cm^3$. Apparently a fist-sized lump, if you placed it on a table, would fall apart under its own weight. Such particles break up very easily in the atmosphere, accounting for the failure of even relatively large shower meteors to produce meteorites. Comet dust, apparently, is fluffy, rather inconsequential stuff. This fluff, by its very nature, does not reach the Earth's surface intact.

The characteristics of some of the more famous meteor showers are summarized in Table 19.3. Other spectacular meteor showers can occur, however, at

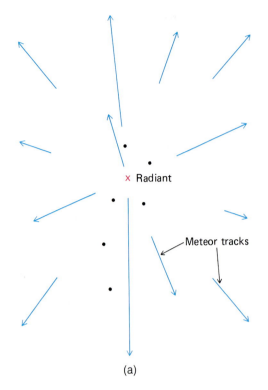

× Radiant

Meteor tracks

(a)

(b)

FIGURE 19.15 Radiant of a meteor shower.

almost any time, just as some bright comets appear unexpectedly.

(c) Zodiacal Dust

Comet dust can be seen directly if you have access to a really dark, clear sky. The faint glow of sunlight reflected from this dust is called the **zodiacal light.** It is brightest along those parts of the zodiac nearest the Sun and is best seen in the west in the few hours after sunset or in the east before sunrise. Under the most favorable circumstances, the zodiacal light rivals the Milky Way in brilliance. It is sometimes called the "false dawn" because of its visibility in the morning hours before twilight.

The total mass of the material responsible for the zodiacal light is estimated at about 10^{16} kg. Calculations show that these dust particles are gradually spiraling into the Sun, but they are replaced from the comets and from collisional fragmentation of objects in the asteroid belt, in the amount of about 10 tons (10^4 kg) per second.

The zodiacal dust has also been detected in the infrared region. In 1983 the IRAS satellite carried out an all-sky survey at wavelengths from 10 to 100 μm, in which the presence of this material is obvious. IRAS also detected previously unrecognized structure in the emission, indicating the presence of three distinct dust bands. These bands apparently represent the debris from asteroid collisions, and they have been identified with two of the largest asteroid families, the Eos and Themis families.

TABLE 19.3 MAJOR ANNUAL METEOR SHOWERS			
Shower Name	Date of Maximum	Associated Comet	Comet Period (yr)
Quadrantid	Jan 3	Unknown	—
Lyrid	Apr 21	Thatcher	415
Eta Aquarid	May 4	Halley	76
Delta Aquarid	July 30	Unknown	—
Perseid	Aug 11	Swift-Tuttle	105
Draconid	Oct 9	Giacobini-Zinner	7
Orionid	Oct 20	Halley	76
Taurid	Oct 31	Encke	3
Leonid	Nov 16	Tempel-Tuttle	33
Geminid	Dec 13	Phaethon*	1.4

* An Earth-approaching asteroid, not a comet.

Spacecraft that have explored beyond the asteroid belt confirm that most of the interplanetary dust is located within a few astronomical units of the Sun. This is to be expected, whether the dust originates mostly in comets or whether the asteroids also make an important contribution.

S U M M A R Y

19.1   Most comets orbit the Sun on very elongated ellipses—nearly parabolas. Edmund Halley was the first to show that some comets are on closed orbits and return periodically to the Sun. Today several new comets are discovered each year. Those with periods of less than 200 years are called short-period comets. About 50 comets with periods between five and ten years are associated with Jupiter.

19.2   The solid heart of a comet is its **nucleus,** a few kilometers in diameter and composed of volatiles (primarily H_2O) and solids (including both silicates and carbonaceous materials). Whipple first suggested this "dirty snowball" model in 1950, and it has been fully confirmed by spacecraft studies of Comet Halley (by the Soviet VEGA and European Giotto probes). As the nucleus approaches the Sun, its volatiles evaporate (perhaps in localized jets or explosions) to form the comet's **head,** or atmosphere, which escapes at about 1 km/s. This jet action produces nongravitational forces, which modify the comet's orbit. From spectroscopic observations of molecules and **radicals** in the comet's atmosphere, we deduce the **parent molecules** released from the nucleus, such as H_2O, CO_2, CO, CH_4, and NH_3. Driven by the solar wind, the atmosphere streams away from the Sun to form a long **tail**—often both a dust tail and a plasma tail.

19.3   As first proposed in 1950 by Jan Oort, new comets are derived from the **Oort comet cloud,** which surrounds the Sun out to about 50,000 AU, or one-third of the distance to the nearest stars. There are about 10^{12} comets in this cloud, and perhaps ten times more in the **Kuiper belt** nearer to the Sun. The total cometary mass is similar to that of the planets. Comets are ancient, primitive bodies, presumably left-over debris from the formation of the outer solar system. They gradually leak inward owing to the gravitational perturbations of passing stars. Once a comet is diverted into the inner solar system, it survives no more than about 1000 perihelion passages before it loses all its volatiles.

19.4   When a fragment of interplanetary dust strikes the Earth's atmosphere, it burns up to create a **meteor.** Streams of dust particles traveling through space together produce **meteor showers,** in which we see the meteors diverging from a spot in the sky called the **radiant** of the shower. Many meteor showers recur each year and are associated with particular comets. Reflected sunlight from interplanetary dust can be seen in the form of the **zodiacal light.** Part of this dust is derived from asteroid collisions, as well as comets.

E X E R C I S E S

THOUGHT QUESTIONS

1. Describe the nucleus of a typical comet, and compare it with an asteroid of similar size.

2. Describe the origin and eventual fate of comets.

3. Comets that have been associated with meteor showers are all periodic comets. Why do you suppose that showers have not been identified with comets having near-parabolic orbits?

4. How did the 1986 spacecraft encounters with Comet Halley change our ideas about the nature of comets?

5. Comets are considered to be relics from the origin of the solar system. Why do astronomers think they are older and more primitive than the planets and their satellites?

6. Suppose that next year a comet is discovered approaching the Sun on a distinctly hyperbolic orbit. List the possibilities for its previous history, and evaluate how likely it is that this comet is truly a visitor from interstellar space rather than a member of the solar system.

7. If meteors are really as small as a pea, how can they be visible at distances of 100 km or more?

8. Two particles of the same mass enter the Earth's atmosphere at the same instant. Both produce observable meteors. One is moving at 70 km/s, the other at 35 km/s. Which meteor gives out more light? Estimate how many times as much light it produces as the other meteor.

9. Suggest reasons why comets have been associated in Western culture with disaster and evil.

PROBLEMS

10. What is the period of revolution for a comet with aphelion at 5 AU and perihelion at the orbit of the Earth?

11. Find the period of a comet that at perihelion just grazes the Sun, and whose aphelion distance from the Sun is (a) 200 AU; (b) 2000 AU; (c) 20,000 AU; (d) 200,000 AU.

12. On the assumption that a periodic comet can survive 1000 perihelion passages, find the lifetimes of the first two comets in the previous exercise.

13. Suppose that the Oort comet cloud contains 10^{12} comets, with an average comet diameter of 10 km. Calculate the mass of a comet 10 km in diameter, assuming that is it composed mostly of water ice (density of 1 g/cm³). Next calculate the total mass of the comet cloud. Finally, compare this mass with that of the Earth and Jupiter.

Collisions between planetesimals during the formative stages of the solar system. Many meteorites are fragments of these planetesimals, dating back to the origin of our planetary system.

(Painting by Don Dixon)

Although most of what we know about the universe (and what we discuss throughout this text) comes from analyzing cosmic radiation, occasionally astronomers can get their hands on some laboratory-size samples from our immediate neighborhood. In fact, the Moon rocks returned to Earth by the Apollo astronauts are among the most valuable legacies of lunar exploration. Each Apollo sample is treated with the respect of a priceless gem, and none are privately owned. Yet the Apollo rocks are not the only cosmic samples that we have here on Earth. Every year dozens of *meteorites* fall and can be collected for free. These rocks from space include pieces from both the Moon and Mars, as well as dozens of different asteroids. Because of their variety, the meteorites are even more valuable scientifically than the Apollo samples. Yet private individuals can and do own them; one of the authors of this book regularly keeps a small (1-g) piece of Mars in his pocket to produce when needed to make a point to his scientific colleagues.

Astronomers can use the data from the analysis of meteorites, together with other information (much of it from primitive asteroids and comets), to check and improve their theories of the formation of the planets. Although the genesis of the solar system took place 4.5 billion years ago, astronomers and planetary scientists have developed detailed models of the formation process. Much of this chapter focuses on testing these models by comparing them with the wealth of information we have acquired about the solar system.

METEORITES AND THE ORIGIN OF THE SOLAR SYSTEM

Harold C. Urey (1893-1981) won the Nobel Prize for chemistry in 1934 for the discovery of deuterium. Later in his career he became interested in the origin of the Earth and planets, and his recognition of the role of primitive meteorites as remnants from the birth of the solar system laid the foundation for much of the modern interest in both the meteorites and the broader study of cosmochemistry.

Where did we come from? This is one of the most basic questions asked by astronomers (or anyone else). In previous chapters we have traced much of the history of the Earth and the other planets since they were formed. But we have not addressed the more basic problem of the origin of the solar system itself.

Much of our current understanding of the formation process is derived from the study of the meteorites, a scientific field known as *meteoritics*. During the past two decades, this branch of research has steadily gained in importance with the increasing sophistication of its laboratory techniques and the impact of its conclusions. In addition to the traditional meteorites, which are fragments ejected from the asteroids, we now have samples from the Moon and Mars available for study. The power of laboratory analysis is so great that the primary objectives of future space missions to the planets are likely to involve the return of additional samples for study using the techniques developed for the investigation of meteorites.

20.1 METEORITES: STONES FROM HEAVEN

A fragment of interplanetary debris in space is called a **meteoroid.** Occasionally, a meteoroid survives its flight through the atmosphere and lands on the ground as a **meteorite.** This happens with extreme rarity in any one locality, but over the entire Earth hundreds of meteorites fall each year. These rocks from the sky carry a remarkable record of the formation and early history of the solar system.

(a)   Extraterrestrial Origin of Meteorites

While occasional meteorites have been recovered throughout history, their extraterrestrial origin was not accepted by scientists until the beginning of the 19th century. Before that, these strange stones were either ignored or treated with supernatural respect.

The earliest recovered meteorites are lost in the fog of mythology. A number of religious texts speak of stones from heaven, which sometimes arrive at opportune moments to smite the enemies of the authors of the texts. At least one sacred meteorite has survived in the form of the Ka'aba, the holy black stone in Mecca that is revered by Islam as a relic from the time of the Patriarchs.

Ancient people apparently found practical uses for iron meteorites at a time when this metal was difficult or impossible for them to refine from available terrestrial ores. The legendary sword Excalibur of the Arthurian legend in ancient Britain may have been made of extraterrestrial iron. So probably is the iron dagger found in the 3500-year-old burial tomb of the Egyptian Pharaoh Tutankhamen in the Valley of the Kings. The Greek myth of the gift of iron from the gods to Prometheus and similar stories in Japan and in other cultures may all have their roots in the fall of meteoritic iron.

The modern scientific history of the meteorites begins in the late 18th century, when a few scientists suggested that some peculiar stones that had been found around the world were of such odd composition and structure that they were probably not of terrestrial origin. Their ideas revived interest in the stories of falling stones. The general acceptance that indeed "stones fall from the sky" occurred after the French physicist Jean Baptiste Biot described the circumstances of a fall in the Orne village of l'Aigle on April 26, 1803, in which many witnesses observed the explosion of a brilliant meteor, after which many meteoritic stones were found, reportedly still warm, on the ground.

A fall of meteorites may represent a group of meteoroids that were moving together in space before they collided with the Earth, but more likely the different stones are fragments of an original meteoroid that broke up during its violent passage through the atmosphere. It is important to remember that such a *shower of meteorites*

has nothing to do with a *meteor shower*. No meteorites have ever been recovered in association with meteor showers. Whatever the ultimate source of the meteorites, they do not appear to come from the comets or their associated dust streams.

(b) Orbits

One way to investigate the source of meteorites is to determine the orbit of the meteoroid while it is still in space, before it encounters the Earth's atmosphere. Once the meteorite has fallen, of course, its former orbit cannot be reconstructed. However, there have been three cases in which photographic patrols have yielded the path of the fireball from which a meteorite was later recovered.

Successful orbits have been calculated for the Pribram (Czechoslovakia, 1959), Lost City (U.S., 1970), and Innisfree (Canada, 1977) meteorites. All three of these meteoroids were on eccentric orbits that carried them from the main asteroid belt to the Earth, similar to the orbits of many Apollo asteroids (Figure 20.1). While not conclusive, these results suggest the asteroids as the source for at least some of the meteorites.

(c) Meteorite Falls and Finds

Meteorites are found in two ways. First, they may be seen to fall, or the freshly fallen rocks may be identified following sightings of a brilliant fireball. These are called meteorite **falls.**

Second, unusual-looking rocks are occasionally found that turn out to be meteoritic. These are termed **finds.** Members of the public frequently send suspected meteorites to experts for identification. Some of these turn out to be real meteorites; others are "meteorwrongs." Typically, a dozen new meteorites are found each year, part of them falls and part of them random finds.

In addition, an especially rich source of meteorites has been identified in the Antarctic, where the ice cap collects and concentrates them. Meteorites that fall in regions where ice accumulates are buried and carried, with the motion of the ice, to other areas where the ice is gradually worn away. After tens of thousands of years, the rock finds itself again on the surface, along with other meteorites carried to the same location. Since there are few other exposed rocks on the Antarctic ice, it is relatively easy to collect the meteorites (Figure 20.2).

Several thousand Antarctic meteorites have been collected during the past decade. Most of them are small stones, weighing less than a kilogram. Often, later analysis indicates that many of these small pieces are fragments from a single fall, but the number of new falls represented by the Antarctic meteorites still exceeds a thousand. The Antarctic meteorites have more than

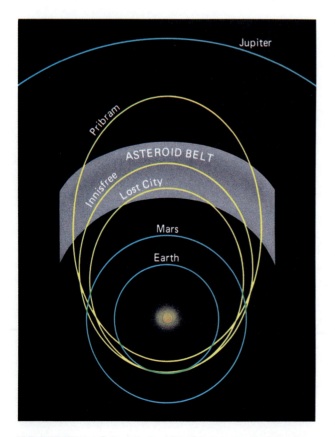

FIGURE 20.1 Calculated orbits of 3 recovered meteorites: Pribram (1959), Lost City (1970), and Innisfree (1977).

FIGURE 20.2 An iron meteorite lying on the Antarctic ice just before it was added to our collections. (NASA)

doubled the total size of our meteorite collections, and in addition these samples are being treated with much more care to avoid terrestrial contamination than was common in the past.

(d) Classification and Nomenclature

The meteorites in our collections include a wide range of compositions and histories, but traditionally they have been placed into three broad classes. First, there are the **irons,** which are composed of nearly pure metallic nickel-iron. Second are the **stones,** which is the term used for any silicate or rocky meteorite. Third are the much rarer **stony-irons,** which are (as the name implies) made of mixtures of stony and metallic materials.

Of these three types, the irons and stony-irons are the most easily recognized because of their metallic content. Native, or unoxidized, iron almost never occurs naturally on Earth. This metal is always found here as an oxide or other mineral ore. Therefore, if you ever come across a chunk of metallic iron, it is sure to be either a product of human industry or a meteorite.

The stones are much more common than the irons, but harder to recognize (Figure 20.3). Often a laboratory analysis is required to demonstrate that a particular sample is really of extraterrestrial origin, especially if it has lain on the ground for some time and has been subject to weathering. The most scientifically valuable stones are those that are collected immediately after they fall or the Antarctic samples that have been preserved in a nearly pristine state by the ice.

Meteorites have traditionally been named for the town nearest to the place where they are found. For example, the large fall that took place in 1969 near the village of Pueblito de Allende in northern Mexico has given us the Allende meteorite. Since there are no towns in the Antarctic, the meteorites collected there are designated by a combination of letters and numbers. An example is the first meteorite of lunar origin, found near the Allan Hills and known as ALHA 81005. The numbers indicate that this meteorite was found in 1981, and that it was the fifth sample collected in that year at the Allan Hills site.

(e) Some Meteorite Trivia

The largest meteorite ever found on the Earth is Hoba West, near Grootfontein, Namibia. It has a volume of about 7 m^3 and an estimated mass of about 60 tons. The

FIGURE 20.3 Stony meteorite of the type called ordinary chondrites. To the layperson, such a meteorite looks very much like a terrestrial rock. (NASA/JSC)

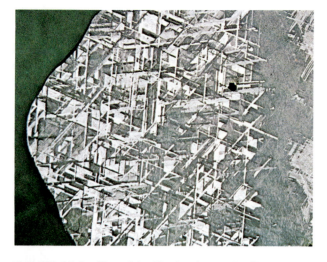

FIGURE 20.5 Slice of the Kamkas iron meteorite, polished and etched to show the crystal pattern in the metal. (Ivan Dryer)

FIGURE 20.6 Polished slice of the Albin stony-iron meteorite. This type of meteorite, called a pallasite, consists of nickel iron mixed with crystals of the green mineral olivine. (Ivan Dryer)

which we use the word) after it lands on the Earth. Most, but not all, stone meteorites are primitive.

Differentiated meteorites are fragments of differentiated parent bodies. Like the igneous terrestrial rocks, they have been heated above their melting points and subjected to a degree of chemical reshuffling. The irons (Figure 20.5), which are derived from the metallic cores of their parent bodies, and the stony-irons (Figure 20.6), which probably represent the interface between iron core and silicate mantle, are all differentiated meteorites. Some stones are also differentiated, including several groups composed of basalt that must have originated on volcanically active parent bodies.

The great majority of the meteorites that reach the Earth are primitive stones. The primitive meteorites are also called *chondrites*, because most of them contain small rounded granules or *chondrules* of unknown origin (Figure 20.7). The ordinary chondrites are mostly composed of light-colored gray silicates with some metallic grains mixed in, but there is also an important group of darker stones called **carbonaceous meteorites** (or car-

FIGURE 20.7 Cross-section of a chondrite stony meteorite showing the small spherical chondrites. (NASA)

bonaceous chondrites). As their name suggests, these meteorites contain carbon, various complex organic compounds, and often chemically bound water; they are also depleted in metallic iron. The carbonaceous meteorites are presumably related to the dark, carbonaceous asteroids, which we saw (Figure 18.6) were concentrated in the outer part of the asteroid belt. Carbonaceous meteorites probably come from a source region more distant from the Sun than that of the lighter colored ordinary primitive stones—most likely the outer half of the main asteroid belt.

Many meteorites that are classed as primitive have actually experienced some modifications by heat or water. Particularly intriguing is the evidence for liquid water alteration of the minerals in some carbonaceous meteorites. Was there a time, early in solar system history, when water in the liquid form was present on the asteroidal parent bodies of these meteorites?

Differentiated meteorites are rare, amounting to less than 10 percent of observed falls. In this respect the meteorites parallel the asteroids, which are predominantly primitive objects.

Table 20.2 summarizes the frequencies of the different classes of meteorites among falls, finds, and the Antarctic meteorites.

(c) Basaltic Meteorites

The basaltic meteorites are samples from the surfaces of differentiated parent bodies that have experienced active volcanism. Because most asteroids are too small to have retained the internal heat necessary for volcanic eruptions, any asteroid that has experienced volcanism is automatically interesting. But equally intriguing is that some of these meteorites are derived from the major, not the minor, planets.

The first basaltic meteorite to yield a definitive identification of its parent body was ALHA 81005, which was found at the Allan Hills Antarctic site in 1981 (Figure 20.8). This meteorite is lunar in origin, similar in many ways to the samples returned in the Apollo program. Subsequently, several other lunar samples have been discovered in the Antarctic. The presence of these lunar fragments demonstrates that cratering impacts can eject

FIGURE 20.8 Antarctic meteorite ALHA 81005, which is a fragment of lunar material ejected from the Moon in an ancient impact. (NASA)

material with high enough velocity to escape from the Moon and impact the Earth.

Another closely related group of basaltic meteorites is the SNC meteorites, mentioned in Chapter 14, which have solidification ages of 1.3 billion years (see Figure 14.16). These stones represent samples of the martian surface, presumably from the Tharsis area.

The third and largest group of basaltic meteorites with an identified parent body comprises the eucrites, mentioned in Chapter 18 (Figure 18.7). Largely by a process of elimination of alternatives, these approximately 30 stones are thought to represent samples of the surface of the asteroid 4 Vesta. With solidification ages of nearly 4.5 billion years, the eucrites date from a period of volcanism at the very beginning of solar-system history, probably within 20 million years of the collapse of the solar nebula.

(d) The Allende and Murchison Primitive Meteorites

The carbonaceous meteorites are the most primitive materials available for laboratory study, excepting the tiny micrometeorites from comets (Section 19.2). Two large carbonaceous meteorites that fell within a few months of each other have proved particularly valuable in probing the birth of the solar system.

The Allende meteorite fell in Mexico, and the Murchison meteorite fell in Australia. Arriving in 1969 at the same time that many laboratories were preparing for analyses of the first Apollo lunar samples, these two meteorites were widely studied from the beginning. Indeed, Allende served as a "dry run" for the Apollo 11 samples (Figure 20.9).

TABLE 20.2 FREQUENCY OF OCCURRENCE OF DIFFERENT METEORITE CLASSES			
	Finds (%)	Falls (%)	Antarctic (%)
Primitive stones	51	87	85
Differentiated stones	1	8	12
Irons	42	3	2
Stony-irons	5	1	1

FIGURE 20.9 The Allende carbonaceous meteorite that fell in Mexico in 1969. (NASA/JSC)

Murchison is best known for the variety of organic, or carbon-bearing, chemicals that it has yielded. Most of the carbon compounds in these meteorites are complex, tar-like substances that defy exact analysis. However, Murchison also contains 16 amino acids, 11 of which are rare on Earth. Unlike terrestrial amino acids, which are formed by living things, the Murchison chemicals include equal numbers with right-handed and left-handed molecular symmetry. The presence of these amino acids and other complex organic compounds in Murchison demonstrates that a great deal of interesting chemistry must have taken place in the solar nebula. Perhaps some of the molecular building blocks of life on Earth were actually derived from the primitive meteorites and comets.

The Allende meteorite is a rich source of information on the formation of the solar system because it contains many individual grains with varied chemical histories. As much as 10 percent of the material in Allende has been estimated to be of pre–solar system origin— interstellar dust grains that were not destroyed in the processes that gave rise to our own system. Allende also yeilds the oldest reliable radioactive ages: 4.559 billion years.

(e) Technical Classification of Meteorites

Over many years meteoriticists (scientists who study meteorites) have developed a rather complex nomenclature to classify the meteorites. We summarize that classification below, with emphasis on the more important types of meteorites.

The *iron meteorites* all have approximately the same composition, primarily metallic iron with from 5 to 15 percent, by weight, of nickel and trace quantities of other metals. They are divided into three groups on the basis of their nickel content, which shows up in crystal patterns (called Widmannstätten patterns) that can be observed when the meteorite is polished and etched (Figure 20.5). Most iron meteorites are called *octahedrites*. Those with more than 13 percent nickel are the *ataxites,* and those with less than 6 percent are the *hexahedrites.*

There are two main groups of *stony-iron meteorites.* The *pallasites* are the most beautiful of all meteorites, consisting of crystals of the transparent green mineral olivine embedded within a metallic nickel-iron matrix (Figure 20.6). They are thought to be samples from the core-mantle interface of their parent body. The second main type of stony-irons is the *mesosiderites,* which are breccias with mixed iron-silicate composition.

We turn now to the differentiated stony meteorites, called *achondrites,* since they do not contain chondrules. One group of achondrites consists primarily of the mineral enstatite; these are called *enstatite achondrites,* or *aubrites.* They are presumably fragments from the mantles of their differentiated parent bodies. The second major group of achondrites comprises the *basaltic achondrites,* which are lavas from the crusts of their parents. As we have already discussed, the basaltic achondrites include the *eucrites,* which appear to be fragments from asteroid Vesta, and the *SNC meteorites* (*shergottites, nakhlites,* and *chassignites*), which appear to come from Mars.

The great majority of meteorites are primitive, and most of these contain chondrules; collectively they are known as *chondrites.* In spite of their name, the chondrites are defined by their general chemical nature and not by the presence (or absence) of chondrules.

One small group of chondrites is composed primarily of the mineral enstatite and is known as *enstatite chondrites;* these may be condensates from a hot, oxygen-depleted region of the solar nebula. Most primitive meteorites, however, fall into the category of *ordinary chondrites,* characterized by the presence of silicates formed in a moderately oxidizing environment together with some tens of percent metallic iron. They are classified primarily on the basis of their iron content, which in turn reflects their degree of oxidation. These classes are called high-iron (*H* chondrites), low-iron (*L* chondrites), and very-low-iron (*LL* chondrites). Finally, there are the *carbonaceous chondrites,* which contain significant quantities of carbon and water but relatively little iron. They are also more highly oxidized than the ordinary chondrites. All of these distinctions refer to the basic chemical composition of the chondrites, and presumably they reflect conditions in the solar nebula, where they formed.

An additional classification of the ordinary and carbonaceous chondrites can be made on the basis of their degree of thermal or chemical alteration since formation—in other words, their degree of metamorphism. The most common system uses a number from 1 to 6 to express the degree of metamorphic alteration. Thus, for example, an L5 chondrite is a rather highly modified low-iron meteorite. In the case of the carbonaceous chondrites, the least modified types (equivalent to class 1 for the ordinary chondrites) are usually called *CI* chondrites, while more modified types are called *CM, CO,* and *CV* chondrites. Generally, the degree of thermal alteration is less for carbonaceous chondrites than for other chondrites, but their aqueous modification may be significant.

20.3   FORMATION OF THE SOLAR SYSTEM

The comets, asteroids, and meteorites are surviving remnants from the origin of the solar system. The planets and the Sun, of course, also are the products of the formation process. We are now ready to put together the information from the past eight chapters to discuss what is known of the origin of the solar system.

(a)   Observational Constraints

There are certain basic properties of the planetary system that any theory of formation should explain. These may be summarized under three categories: dynamical constraints, chemical constraints, and age constraints.

Dynamical Constraints.   The planets all move around the Sun in the same direction and approximately in the plane of the Sun's own rotation. In addition, most of the planets share this same sense of rotation, and most of the satellites also move in counterclockwise orbits. With the exception of the comets, the members of the system define a disk shape. On the other hand, exceptions are possible in the form of retrograde rotation, like that of Venus.

Chemical Constraints.   The planets Jupiter and Saturn have approximately the same composition as the Sun and stars, dominated by hydrogen and helium. Each of the other members is, to some degree, lacking in the light elements. A careful examination of the composition of solar system objects shows a striking progression from the metal-rich inner planets through predominantly rocky materials out to ice-dominated composition in the outer solar system. The comets are also icy objects, whereas the asteroids represent a transitional rocky composition with abundant dark, carbon-rich material. Many specific constraints are also imposed by detailed analyses of the meteorites and samples from Mars and the Moon. There are also the problems of understanding the peculiar chemical compositions of the Moon and Mercury, as well as the presence of large quantities of water on Earth and Mars, surprising if these planets formed in a region where the temperature was too hot for ice to condense.

Age Constraints.   Radioactive dating demonstrates that there are rocks on the surface of the Earth that have been present for at least 3.8 billion years and lunar samples that are 4.4 billion years old. In addition, the primitive meteorites all have radioactive ages near 4.5 billion years. The age of these unaltered building blocks is considered the age of the planetary system. The similarity of the measured ages tells us that planets formed and their crusts cooled within the first hundred million years of the solar system. Further, detailed examination of primitive meteorites indicates that they are made primarily from material that condensed or coagulated out of a hot gas; few identifiable fragments or grains survived from before this hot vapor stage 4.5 billion years ago.

(b)   The Solar Nebula

All of the above constraints lead to the conclusion that the solar system formed 4.5 billion years ago out of a rotating cloud of hot vapor and dust of approximately cosmic composition. This cloud is called the *solar nebula*. The general idea of such an origin appears to have been first suggested by the German philosopher Immanuel Kant (1724–1804), and it was developed into a specific model by the French astronomer Marquis Pierre Simon de Laplace in 1796. The Kant-Laplace idea is known as the *nebular hypothesis*.

The modern concept of the solar nebula, with the detailed constraints outlined previously, dates from work carried out in the 1940s and 1950s by the German theoretical physicist Carl von Weizsacker, the Dutch-American astronomer Gerard P. Kuiper, and the American Nobel prize–winning chemist Harold Urey. In Chapter 28 we will return to the question of how such a nebula might form in the first place. Here we begin our discussion with such a rotating nebula in place, surrounding a central condensation that would evolve into the Sun (Figure 20.10).

Because of its rotation, the nebula was unable to fall entirely into the central star. Instead, it collapsed into a disk that rotated about the central condensation. As material fell toward the plane of the disk, it was heated by its own gravitational energy as well as by radiation from the protosun. As temperatures rose near the plane of the disk, any solid material that was originally present was vaporized. Some fraction of the interstellar dust probably survived the high temperatures, but there is insufficient evidence to be sure how much interstellar material might have thus been preserved. The existence of this disk-shaped, rotating nebula explains the primary dynamical properties of the solar system as described in the previous section.

There is considerable uncertainty today regarding the total mass of the solar nebula. Some calculations suggest that the material in the disk might have been only perhaps five to ten times more massive than the planets today; other scientists calculate that the total amount of material present in the disk was comparable to the mass of the Sun. In either case, the mass in the disk was much larger than the total mass of the planetary system today, so that the material we see now is no more than a remnant of the material originally present.

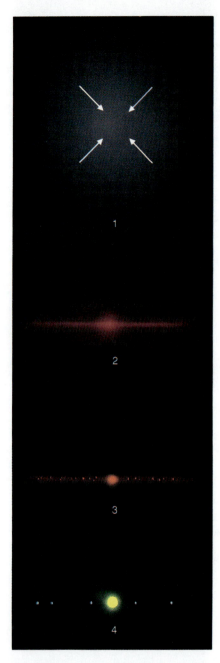

FIGURE 20.10 Schematic representation of the formation of the solar system. (1) The solar nebula contracts. (2) As the nebula shrinks, its rotation causes it to flatten until (3) the nebula is a disk of matter with a concentration near the center, which (4) becomes the protosun. Meanwhile, solid particles condense as the nebula cools, giving rise to the planetesimals, which are the building blocks of the planets.

(c)   Condensation and Accretion

The solar nebula stabilized when the supply of infalling gas and dust was exhausted. With no more gravitational energy to heat it, the disk began to cool, except near its center, where the newly formed Sun kept the temperatures up. As the nebula cooled, the gas atoms interacted

chemically to produce compounds (molecules), and eventually these compounds condensed into liquid droplets or solid grains.

The sequence of compounds that should form in such a cooling nebula is illustrated in Figure 20.11. As the temperature dropped, the first materials to form grains were the metals and various rock-forming silicates and other minerals. Once the gas cooled below about 600 K, these were joined by sulfur compounds and by carbon- and water-rich silicates such as those now found abundantly among the asteroids. However, in the inner parts of the nebula the temperature never dropped low enough for these materials to condense, so they are lacking on the innermost planets. While temperatures stabilized in the inner nebula, at greater distances the gas continued to cool. Beyond about 4 AU it soon cooled to below 300 K, and the oxygen combined with hydrogen to condense in the form of water ice. Beyond the orbit of Saturn, temperatures fell below 100 K, allowing carbon and nitrogen to combine with hydrogen and condense as additional ices such as methane (CH_4) and ammonia (NH_3).

This chemical condensation sequence explains how the observed pattern of planetary compositions originated. In particular, note the condensation point for H_2O at about 300 K. Since hydrogen and oxygen were both abundant in the solar nebula, H_2O was a major constituent of the condensing material in the outer, cooler parts of the disk. But closer to the Sun than about

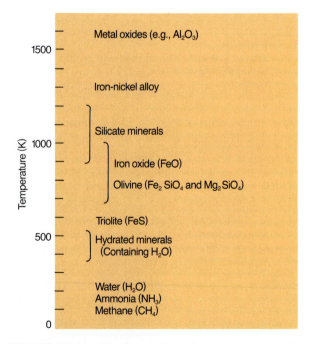

FIGURE 20.11 The chemical condensation sequence in the solar nebula, showing the primary chemical species that would be expected to form in a cooling gas cloud of solar composition under equilibrium conditions. (Adapted from diagrams published by John Lewis, University of Arizona)

4 AU, the temperature remained above this level, and H_2O could not condense. Thus the original composition of the Earth and other terrestrial plants did not include any water or ice.

Eventually the cooling nebula was filled with solid grains, which were sorted chemically by distance from the Sun and mixed with the still abundant hydrogen and helium gas (and an uncertain component of original interstellar dust). It is believed that these grains rather quickly formed into larger and larger aggregates, until most of the solid material was in the form of chunks astronomers called **planetesimals**, each a few kilometers to a few tens of kilometers in diameter. The Russian theorist Victor Safronov studied the formation and properties of planetesimals in the late 1960s, and his theory of formation of the planets is sometimes called the planetesimal hypothesis. The number of planetesimals must have exceeded 100 trillion (10^{14}).

Some planetesimals were large enough to attract their neighbors gravitationally and thus to grow by the process called **accretion.** Calculations indicate that objects of at least lunar size were built up rather quickly—perhaps within a few hundred thousand years. George Wetherill of the Carnegie Institution of Washington calls these "planetary embryos." In the inner part of the solar system, these embryos became the building blocks of the four terrestrial planets. In the outer solar system, where the nebular condensates included ices as well as silicates, the embryos combined to form much larger bodies, with masses of 10 to 20 times the mass of the Earth. These larger bodies became the cores of the giant planets.

(d)   Formation of the Giant Planets

Astronomers calculate that these planetary cores of the outer solar system became so large that they were able to attract the surrounding nebular gas. As hydrogen and helium rapidly collapsed onto their cores, the giant planets were heated by the energy of contraction, just as the contraction of the solar nebula heated the Sun and triggered its nuclear source of power. (The generation of the Sun's energy is described in Chapter 27.) But these giant planets were far too small to achieve the central temperatures and pressures necessary to become miniature suns. Calculations show that 70 times the mass of Jupiter is required for even the smallest self-sustaining star. After glowing dull red for a few thousand years, the giant planets gradually cooled to their present state.

The collapse of nebular gas onto the cores of Jupiter and Saturn explains how these objects came to have about the same hydrogen-rich composition as the Sun itself. For some reason much less nebular gas was captured by Uranus and Neptune, which is why these two planets have compositions dominated by the icy build-ing blocks that made up their large cores, rather than by hydrogen and helium.

Some time after the formation of the giant planets, the newly formed Sun, like other very young stars, went through a stage in which it developed a very strong solar wind. Blasts of hot plasma flowed away from its atmosphere, sweeping through the remains of the solar nebula. Although this intense solar wind had little effect on the planets and other solid material, it interacted strongly with the gas still present, driving it out of the system. The solar nebula was dissipated, leaving a new star surrounded by a few planets and a disk of smaller bodies. As we will see in Chapter 28, the dissipation of nebulae by strong stellar winds and the presence of disks of solid material have been observed in association with other young stars, so the processes we have described are not unique to our own system.

20.4   DYNAMICAL EVOLUTION OF THE PLANETARY SYSTEM

All of the processes described in Section 20.3, from the collapse of the solar nebula to its dissipation by the solar wind, took place within at most a few million years, and possibly even faster. However, the story of the origin of the solar system is not complete at this stage. We have not yet discussed the formation of the terrestrial planets or the fate of the planetesimals and other debris that did not accumulate to form the planets.

(a)   Formation of the Inner Planets

The formation of the terrestrial planets was a violent and chaotic process. Theoretical simulations carried out by George Wetherill suggest that many collisions must have taken place between the planetary embryos during the first few million years, when as many as 100 embryos larger than the Moon were loose in the inner solar system, as well as many more smaller surviving planetesimals. While many of these collisions led to the growth of larger objects, sometimes the result was the fragmentation of larger objects into smaller ones. Eventually, however, these essentially random processes led to the formation of the four terrestrial planets we have today. According to these ideas, all four of these planets should have nearly similar bulk compositions, since all were formed from a variety of smaller building blocks, drawn from about the same source region. As we have seen, this is approximately true for Earth, Venus, and Mars, but not for the Moon or Mercury.

As the four inner planets stabilized, the possibilities of further fragmentation declined. However, the remaining solid matter in the solar system—the planetesimals and embryos—continued to interact gravitationally with the planets. Most close encounters led either to a

FIGURE 20.13 The ancient lunar highlands, which record impacts dating back to 4.4 billion years ago. (NASA)

impacts. The retrograde rotation of Venus would not be expected for a planet forming by the aggregation of planetesimals in a spinning solar nebula. Also unexpected are the rotational orientations of Uranus and Pluto, both of which are tipped on their sides. All of these departures from the general rotational symmetry of the solar system can be understood if the final stages of planetary accretion were marked by giant impacts. In effect, these large, essentially random events struck three of the nine planets so hard that their rotation axes were knocked on their sides or reversed.

(f) Conclusion

All of the larger violent events terminated by about 4.4 billion years ago, which is when the oldest rocks in the lunar crust solidified (Figure 20.13). Since then there has been a continuing sweep-up by the planets of remaining debris, but at a much slower rate than during the accretionary period. Planetary orbits have remained stable, and no errant large objects remain to threaten planetary collisions. There was a particularly intense period of bombardment about 4 billion years ago (described in Chapter 13), but even this was minor in comparison with the events of the early history of the planetary system, much to the gratification of the life-forms that eventually evolved on the third planet of the system.

S U M M A R Y

20.1 A **meteoroid** that survives to reach the ground is a **meteorite.** The extraterrestrial origin of meteorites was established at the beginning of the 19th century, but their great value to science was not realized until the mid-20th century. A meteorite seen to fall is a **fall;** otherwise it is a **find.** The simplest classification is the **stones,** the **irons,** and the **stony-irons.** Analysis of meteorites reveals that their **parent bodies** were a large number of relatively small bodies. Orbits determined for three meteorites indicate that they originated in the asteroid belt, and it is generally thought that the parent bodies are asteroids.

20.2 All meteorites (except those from the Moon or Mars) were formed 4.5 billion years ago—the figure that represents the age of the solar system. Most meteorites are primitive stones, derived from primitive parent bodies. These include the dark **carbonaceous meteorites,** such as Murchison and Allende. Differentiated meteorites (from differentiated parent bodies) account for only about 10 percent of falls. These include the irons and stony-irons and the basaltic meteorites, such as the eucrites, which are samples of asteroid Vesta.

20.3 Theories of the origin of the solar system must account for the observed dynamical, chemical, and age properties of the system. Today we are confident that the planets formed with the Sun from a rotating solar nebula. Rapid condensation led to the formation of **planetesimals** (their composition depending on distance from the Sun), and further **accretion** built up larger objects (planetary cores). In the outer solar system these cores grew to masses ten times that of Earth and attracted large quantities of hydrogen and helium from the solar nebula, thus forming the jovian planets. A strong solar wind then dissipated the remaining gas.

20.4 In the inner solar system dozens (perhaps hundreds) of objects of roughly lunar dimensions formed. Dynamical interactions eventually formed the terrestrial planets. The asteroids are a remnant population of rocky planetesimals between Mars and Jupiter, and the comets are remnants of icy planetesimals from the outer solar system. Giant impacts during the late stages of planet formation formed the Moon from ejected terrestrial mantle material, and Mercury lost most of its silicate mantle by a similar process. All of these violent events terminated by about 4.4 billion years ago.

E X E R C I S E S

THOUGHT QUESTIONS

1. Meteors apparently come primarily from comets, while the meteorites are thought to be fragments of asteroids. This may seem contradictory. Explain why we do not believe meteorites come from comets, or meteors from asteroids.

2. Explain why iron meteorites represent a much higher percentage of finds than of falls.

3. Why is it more useful to classify meteorites according to whether they are primitive or differentiated, rather than into stones, irons, and stony-irons?

4. Which meteorites are the most useful for defining the age of the solar system? Why?

5. Suppose a new primitive meteorite is discovered and analysis shows that it contains a trace of amino acids, all of which show the same rotational symmetry (unlike the Murchison meteorite). What might you conclude from this finding?

6. What is the main difference between the giant planets and the terrestrial planets? How can this difference be understood in terms of the theory of the formation of the solar system described in this chapter?

7. How do we know when the solar system formed? Usually we say that the solar system is about 4.5 billion years old. What does this age correspond to? Are there parts of the solar system that might be substantially older than this?

8. Give some everyday examples of the conservation of angular momentum.

9. Describe the chemical building blocks that are thought to have been available in the grains that condensed from the solar nebula. If each planet formed in place from these grains, what would be the chemical composition of objects at 0.4 AU, 1.0 AU, 5.0 AU, and 25 AU from the Sun?

10. We have suggested that the SNC meteorites are fragments from Mars. Suppose we are wrong. What other body might be the parent to these meteorites?

11. The suggestion that a giant impact that nearly destroyed the Earth was necessary for the formation of the Moon seems like a drastic hypothesis. Indicate why planetary scientists are invoking such a seemingly implausible idea for the origin of the Moon.

PROBLEMS

12. Consider the differentiated meteorites. We think the irons are from the cores, the stony-irons from the interface between mantle and core, and the stones from the mantles of their differentiated parent bodies. If these parent bodies were like the Earth, what fraction of these meteorites would you expect to consist of irons, stony-irons, and stones? Is this consistent with the observed numbers of each?

13. The angular momentum of an object is proportional to the square of its size divided by the period of rotation (D^2/P). If angular momentum is conserved, then any change in size must be compensated for by a proportional change in period, so as to keep D^2 divided by P a constant. Suppose that the solar nebula began with a diameter of 10,000 AU and a rotation period of 1 million years. What would be its rotation period when it had shrunk to the size of the orbit of Pluto? To the orbit of Jupiter? To the orbit of the Earth?

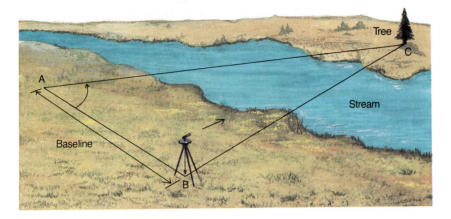

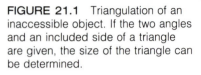

FIGURE 21.1 Triangulation of an inaccessible object. If the two angles and an included side of a triangle are given, the size of the triangle can be determined.

Our eyes are separated by a baseline of a few inches, so our two eyes see an object in front of us from slightly different directions. The brain, like an electronic computer, solves the triangle and gives us an impression of the distance of the object. The greater the parallax, the nearer the object. Hold a pencil a few inches in front of your face and look at it first with one eye and then with the other. Note the large apparent shift in direction against the more distant wall across the room. Now hold the pencil at arm's length and note how the parallax is less. If an object is fairly distant, the shift—the parallax—is too small to notice with the eyes. Depth perception fails for objects more than a few tens of meters away. It would take a larger baseline than the distance between the eyes to see the parallax of an object 500 m distant.

Since astronomical objects are very far away, a very large baseline must be found or highly precise angular measurements must be made, or both. The Moon is the only object near enough that its distance can be found fairly accurately with measurements made without a telescope. Hipparchus, a Greek astronomer who worked in the second century B.C., determined the distance to the Moon to within about 20 percent of the correct value. He used measurements of the shift of position or parallax of the Moon relative to the Sun from two different spots on the Earth at the time of a total solar eclipse. Hipparchus assumed that the Sun is so far away that its parallax is zero.

With the aid of telescopes, later astronomers were able to measure the distances to the nearer planets using the Earth's diameter as a baseline. (Modern determinations of the distances to planets depend on the use of radar rather than triangulation—Section 21.2.) To use triangulation to reach for the stars, however, requires a much longer baseline.

(b) Distances to Nearby Stars

We do have such a baseline available to us as a result of the motion of the Earth around the Sun. As we view the sky first from one side of the Sun and then six months later from the opposite side, nearby stars should be seen to shift their positions relative to very distant stars (Figure 21.2). The amount of the parallax provides a quanti-

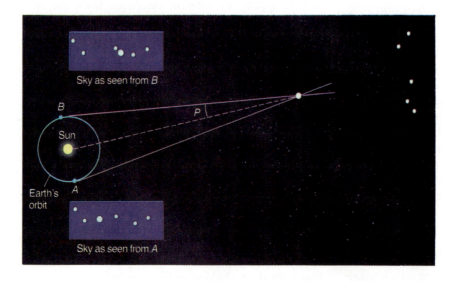

FIGURE 21.2 As the Earth revolves around the Sun, the direction in which we see a nearby star varies with respect to distant stars. The parallax of the nearby star is defined to be one-half of the total change in direction and is measured in arcsec.

A
(LY
yea
(abl
LY.
emj
are
froi
Wh
the
we
and

(b)

Thε
acci
orb
17tl
syst
nin
dist
calc
beg
nev
Sur
15 ͺ
T
prε
plaι
pro
vati
reπ
plaι
req
plaι
nor
dist
$c =$
uni
km

tative measurement of distance. We define the parallax of a star as one-half of the total change in its apparent direction as viewed from opposite sides of the Earth's orbit.

It is important to realize that Figure 21.2 is deceptive. In reality, the distance to even the nearest stars is much greater than we can show on such a diagram (or else the Earth's orbit would be too small for you to see). As a result, the angles we have to measure are terribly small and dauntingly difficult to pin down accurately. This makes the distances we obtain more and more uncertain as we probe farther and farther out.

Even for the nearest stars, measured parallaxes are only small fractions of a second of arc (arcsec.) A second of arc is ⅟₆₀ of a minute of arc, which, in turn, is ⅟₆₀ of a degree. You will recall that the dome of the sky, from horizon to horizon, spans 180°. How small is an arsec? If a friend 5 km away held up a U.S. quarter, the diameter of that coin as seen by you would be an arsec.

Today, the measurement of parallaxes is being revolutionized by the use of instruments in space. The European Space Agency's Hipparcos satellite, launched in 1989, has an ambitious program to measure the parallaxes of 120,000 stars. We anticipate that the repaired Hubble Space Telescope, with its capability of measuring much smaller images unblurred by the atmosphere, will achieve parallaxes of one-thousandth of an arcsecond or better.

(c)   The "Skinny" Triangle

In astronomy we frequently have to measure distances that are very large in proportion to the length of the available baseline, and the triangle to be solved is thus long and "skinny." Suppose it is found that the displacement in the direction of an object (at O, in Figure 21.3) viewed from opposite sides of the Earth is the angle p; p, then, is the angle at O subtended by the diameter of the Earth. Imagine a circle, centered on O, that passes through points A and B on opposite sides of the Earth. If the distance of O is very large

compared with the size of the Earth, then the length of the chord AB is very nearly the same as the distance along the arc of the circle from A to B. This arc is in the same ratio to the circumference of the entire circle as the angle p is to 360°. Since the circumference of a circle of radius r is $2\pi r$, we have

$$\frac{AB}{2\pi r} = \frac{p}{360°}.$$

By solving for r, the distance to O, we find

$$r = \frac{360°}{2\pi} \cdot \frac{AB}{p}.$$

If p is measured in arcseconds, rather than in degrees, it must be divided by 3600 (the number of seconds in 1°) before its value is inserted in the above equation. After such arithmetic, the formula for r becomes

$$r = 206{,}265 \frac{AB}{p \text{ (in arcseconds)}}.$$

As an example, suppose p is 18 arcsec (about what would be observed for the Sun). Since AB, the Earth's diameter, is 12,756 km,

$$r = 206{,}265 \frac{12{,}756}{18} = 1.46 \times 10^8 \text{ km}.$$

(d)   The Parallactic Ellipse

As is so often true in science, the actual problem of measuring parallax is slightly more complicated than suggested by Figure 21.2. As the Earth moves about its orbit, the place from which we observe the stars continually changes. Consequently, the positions of the comparatively near stars, projected against the more remote ones, are also continually changing. If a star is in the direction of the ecliptic, it seems merely to shift back and forth in a straight line as the Earth passes from one side of the Sun to the other. A star that is at the pole of the ecliptic (90° from the ecliptic) seems to move about in a small circle against the background of more

(c)

To
par
app
its
Grε
to f
ary
dist
too
to .

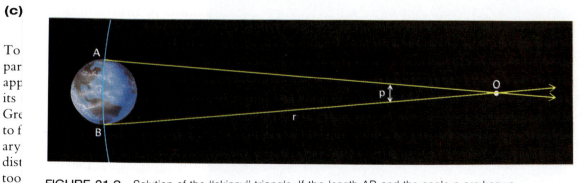

FIGURE 21.3   Solution of the "skinny" triangle. If the length AB and the angle p are known, the distance to O can be calculated.

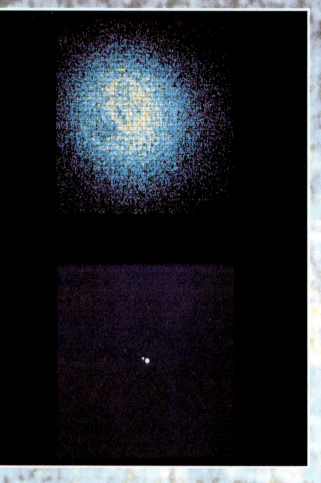

The determination of stellar masses depends on measurement of the gravitational effects of one star on a companion. New techniques make it possible to measure very close binary stars. The top image shows a long time exposure of the 4th magnitude spectroscopic binary star Sigma Herculis. Atmospheric turbulence produces a blurred image of both stars that is about 2 arcsec in diameter. New techniques can remove this blurring and reveal two stars separated by 0.07 arcsec.

(Anthony Readhead/Palomar Observatory,
California Institute of Technology)

How much mass is there in a star like the Sun? If we told you that the Sun contains more than 10^{56} atoms, you would probably feel that this number, while impressive, is so large, it is not really possible to get a good feel for its magnitude. But when we tell you that the Sun has enough mass to make 333,000 Earths, chances are that will make a more profound impression. As we move from the study of the planetary system to the stars and groups of stars, more and more we will have to consider numbers that dwarf everything in the realm of human experience.

How do astronomers measure masses like this? For testing our theories, we need mass measurements for planets, for stars, and even for the galaxies of stars, far beyond our own Milky Way. In this chapter we discuss how some of the principles of gravity and motion first derived in the 17th century allow us to determine the masses of cosmic objects. (Again, we will find a remarkable universality to these basic principles. The rule Kepler found centuries ago seems to apply in a distant cluster of galaxies as effectively as it does in the solar system.)

Textbooks like ours, which must summarize an entire field so that you can read about it in a single course, can never do full justice to how difficult the kinds of measurements you will read about in this chapter really are. While you are reading, bear in mind that many generations of astronomers around the world (and their unsung assistants) have contributed painstaking, careful work to building up our data bank of stellar measurements. As we have seen before, such data improve as our instruments improve; today we can make remarkably good estimates of a wide range of masses in the universe.

BINARY SYSTEMS: WEIGHING AND MEASURING THE STARS

Edward Charles Pickering (1846–1919), American astronomer, was a pioneer in the study of stellar spectra, and especially in the study of the spectra of binary stars. He was the first to demonstrate the existence of spectroscopic binaries, and he also carried out fundamental work in the investigation of eclipsing binary stars. (Yerkes Observatory)

At least half the stars around the Sun are found in pairs (binary stars) or in systems of three or more, ranging up to clusters of thousands—each star moving under the combined gravitational influence of the other stars in the same system. This situation is fortunate, because analyses of these systems provide us with our best means of learning stellar masses and sizes—fundamental data we need to piece together the story of how stars live and die.

23.1 DETERMINATION OF THE SUN'S MASS

The masses of stars must be inferred from their gravitational influences on other objects. To understand how masses can be derived, we will begin by showing how astronomers determined the mass of the Sun, a mass that can be measured more reliably than that of any other star.

The most direct way to calculate the mass of the Sun is from the orbit of the Earth. For the sake of illustration, let us assume that the orbit of the Earth is circular. Then the force required to keep the Earth in its orbit is the centripetal force (Section 3.2):

$$F = \frac{v^2 M_e}{R},$$

where v is the Earth's orbital speed, M_e is the mass of the Earth, and R is the radius of its orbit. This acceleration must be provided by the gravitational attractive force of the Sun on the Earth; that is,

$$F = \frac{v^2 M_e}{R} = \frac{G M_s M_e}{R^2},$$

where G is the universal gravitational constant and M_s is the mass of the Sun. Solving the above equation for the Sun's mass, we obtain

$$M_s = \frac{v^2 R}{G}.$$

Both v and R are known from observation, and G is determined from laboratory measurements (Section 3.2). In metric units, $v = 3 \times 10^4$ m/s, $R = 1.49 \times 10^{11}$ m, and $G = 6.67 \times 10^{-11}$ m³/kg·s². Substituting these values into the above formula, we find for the mass of the Sun,

$$M_s = 2 \times 10^{30}\, \text{kg} = 2 \times 10^{27}\, \text{tons}.$$

Because the Earth's orbit is nearly circular, the value thus found for the mass of the Sun is very nearly the

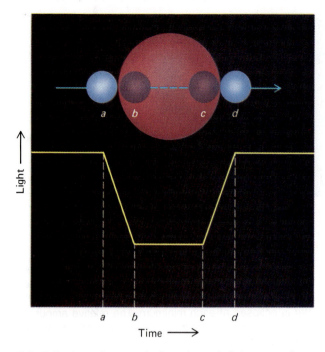

FIGURE 23.8 Contacts in the schematic light curve of a hypothetical eclipsing binary with central eclipses.

are visible in the composite spectrum of the binary, both radial-velocity curves can be observed. Then the size of the relative orbit can be found, and we can determine the actual radii of the stars in kilometers. In other words, the velocity of the small star with respect to the large one is known and, when multiplied by the time intervals from first to second contacts and from first to third contacts, gives, respectively, the diameters of the small and large stars.

In general, the orbits are not exactly edge on, and the eclipses are not central. However, it is a relatively simple geometry problem, at least in principle, to use measurements of the depths of the minima and the exact instants of the various contacts to calculate both the inclination of the orbit and the sizes of the stars relative to their separation.

The foregoing discussion applies only to eclipses that are total and annular. If they are partial, the analysis is far more difficult, although it can still be accomplished.

To summarize: From the analysis of the light curve of an eclipsing binary, we can find the inclination of the orbit and the sizes of the stars relative to their separation. If, in addition, we can measure the Doppler shifts of the spectral lines of both stars during their period of revolution, we can obtain their velocity curves. The analysis of the velocity curves, as described in the preceding section, leads to a determination of lower limits to the masses of the stars. Knowledge of the inclination of the orbit allows us to convert these minimum values for the masses to actual masses for the individual stars. We can also convert the lower limit to the separation of the stars (that is, the semimajor axis of their relative orbit) to the actual value when the inclination is known. Since the sizes of the stars relative to this separation are found from the light curve, we find their actual diameters. Finally, from the relative depths of the primary and secondary minima, we can calculate the relative surface brightnesses of the stars and hence their effective temperatures

(the surface brightness of a star is proportional to the fourth power of its effective temperature—Sections 7.2 and 22.4d). Note that we do not need to know the distance to an eclipsing-spectroscopic binary to determine its mass, as we do in the case of a visual binary. Among the thousands of known eclipsing binaries, however, only a few dozen are so favorably disposed for observation that all the necessary data can be obtained.

(f) The Range in Stellar Masses

What is the largest mass that a star can have? The limit is not known for sure, but searches for massive stars indicate that very few stars have masses greater than about 60 times the mass of the Sun. There is no convincing evidence that there are any stars with masses that significantly exceed about 100 times the mass of the Sun. The rarity of stars with large masses is illustrated by the fact that there are no stars within 30 LY of the Sun that have masses greater than four times the mass of the Sun. According to theoretical calculations no star can have a mass less than about $1/12$ the mass of the Sun.

(g) The Mass–Luminosity Relation

Studies of binary stars have provided measurements of the masses of over a hundred individual stars. When the masses and luminosities of these stars are compared, it is found that, in general, the more massive stars are also the more luminous. This relation, known as the **mass–luminosity relation,** is shown graphically in Figure 23.9. It is estimated that about 90 percent of all stars

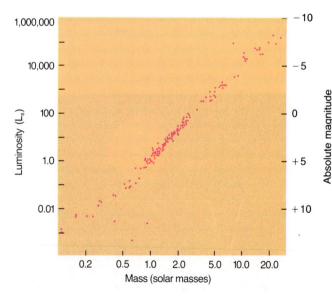

FIGURE 23.9 Mass–luminosity relation. The plotted points show the masses (abscissas) and luminosities (ordinates) of stars for which both of these quantities are known to an accuracy of 15 to 20 percent. The three points lying below the sequence of points are all white dwarf stars (*see* Chapter 24).

obey the mass-luminosity relation illustrated in Figure 23.9.

The range of stellar luminosities is much greater than the range of stellar masses. Luminosities of main-sequence stars are roughly proportional to their masses raised to a power of 3.5 to 4.0. Most stars have masses between 0.1 and 100 times that of the Sun. According to the mass-luminosity relation, the corresponding luminosities of stars at either end of the range are, respectively, less than 0.001 and greater than a million times the Sun's brightness.

The mass-luminosity relation provides a useful means of estimating the masses of stars of known luminosity that do not happen to be members of visual or eclipsing binary systems.

(h) Mass Exchange in Close Binaries

If two stars revolve about each other in circular orbits, there is an equilibrium point between them at which the gravitational force exerted by each is exactly balanced by that of the other. A low-mass body located exactly at this point can remain stationary with respect to the two stars. Should the small body be shoved closer to one of the revolving stars, however, it is drawn to it by gravitational attraction. We shall also see that as stars age, the generation of nuclear energy in their interiors causes them to distend their outer layers greatly, so that those stars become giants (Chapter 29). If such a star is a member of a close binary system, the atoms in its expanding outer layers may reach and pass through this balance point. Once it is beyond this point, the matter from the expanding star can flow to the other star.

Mass exchange is believed to occur between many stars in close binary systems. This exchange of mass can have profound effects on the evolution of the stars in a system. Not only may matter stream from one star to another in a close binary system, but also it apparently causes such explosive phenomena as novae and supernovae (Chapter 30).

23.3 DIAMETERS OF STARS

Eclipsing binary systems allow us to determine the diameters of stars, but this technique works only for those stars that happen to be members of eclipsing systems for which the necessary analysis can be carried out. It would be convenient if angular sizes could be measured directly for many stars of known distances. Then their linear diameters in kilometers could be calculated just as they are for the Moon or planets. But the stars are so far away, they look like points of light. The Sun is the only star whose angular size can be resolved visually and whose diameter can be calculated

simply. There are many other stars, however, whose angular sizes are only slightly beyond the limit of resolution of the largest telescopes and can be measured with special techniques. These include high-speed observations of the dimming of light from stars being occulted by the Moon or planets.

In all, several hundred angular stellar diameters have been measured by one or more of these methods. There are some difficulties with each technique (for example, in connection with limb darkening—see Section 24.4b). Still, these data give us some confidence in the correctness of the less direct determinations of stellar diameters, to which we now turn.

(a) Finding Stellar Radii from Radiation Laws

For most stars we must calculate their diameters from theory. We can use the Stefan-Boltzmann law (Section 7.2) to calculate the radius of a spherical blackbody that has the same luminosity and effective temperature that a star does. The diameter, then, is twice the radius. Applications of this technique to stars whose sizes are already known from other measurements show that the diameter of this hypothetical blackbody is a fairly good estimate of the diameter of the star.

The luminosity of a star can be obtained by the procedure discussed in Chapter 22, and the effective temperature of a star can be obtained from its color or its spectrum (Section 22.4d). The energy emitted per unit area of a star (given by the Stefan-Boltzmann law), multiplied by its entire surface area, gives the star's total output of radiant energy, that is, its luminosity. Since the surface area of a sphere of radius R is $4\pi R^2$, the luminosity of a star is given by

$$L = 4\pi R^2 \times \sigma T^4.$$

The above equation can be solved for the radius of the star.

Note that the temperature appearing in the above formula is raised to the fourth power. If it is in error, therefore, the computed value of the star's radius can be substantially incorrect. In particular, because stars are *not* true blackbodies, values of stellar temperatures as determined by different methods do not all agree precisely. Different kinds of stellar temperatures were discussed more fully in Section 22.4d. Despite this uncertainty, observations indicate that we can use the Stefan-Boltzmann law to find the sizes of most stars with an accuracy of 10 to 20 percent.

The results of the measurements of stellar size confirm that most nearby stars have roughly the size of the Sun—typically a million kilometers or so in diameter. Faint stars, as might be expected, are generally smaller than more luminous stars. There are, however, dramatic exceptions. Some stars have diameters as small as that of the Earth, while others are so large that they would fill the entire inner solar system almost as far as Jupiter.

As an application of the radiation laws to the calculations of stellar diameters, consider, first, a star whose red color indicates that it has a temperature of about 3000 K, roughly half the temperature of the Sun. Each square meter of the star, therefore, emits only $\frac{1}{16}$ as much light as the Sun (because the light emitted is proportional to the fourth power of the temperature). Suppose, however, that the star is 400 times as luminous as the Sun. It must be many times larger than the Sun to emit more light despite its much lower surface brightness. We can find its radius, in terms of the Sun's, by noting that L is proportional to $R^2 T^4$, and thus (since the constants of proportionality cancel in each of the ratios),

$$\frac{R_\star}{R_s} = \left(\frac{L_\star}{L_s}\right)^{1/2} \left(\frac{T_s}{T_\star}\right)^2 = (400)^{1/2} \times 4 = 80.$$

(The subscripts $\star$ and s refer to the star and Sun, respectively.) This star has 80 times the Sun's radius; if the Sun were placed at its center, the star's surface would reach past the orbit of Mercury.

Next, consider a star whose blue color indicates a temperature of about 12,000 K—twice the Sun's tem-perature. Suppose, however, that this star has a lu-minosity of only $\frac{1}{100}$ that of the Sun. We find for the star's radius,

$$\frac{R_\star}{R_s} = \left(\frac{L_\star}{L_s}\right)^{1/2} \left(\frac{T_s}{T_\star}\right)^2 = \left(\frac{1}{100}\right)^{1/2} \left(\frac{1}{2}\right)^2 = \frac{1}{40}.$$

The star has only $\frac{1}{40}$ the Sun's radius—less than three times the radius of the Earth. These two examples are not extreme cases but are more or less typical of two types of stars that are called *red giants* and *white dwarfs*, respectively (see Chapter 24).

(b)   Summary of Stellar Diameters

The few dozen good geometrical determinations of stellar radii come from (1) direct measure of the Sun's angular diameter, (2) measures of the angular diameters of several hundred stars by special techniques, including lunar occultations, and (3) analyses of the light curves and radial-velocity curves of eclipsing binary systems. All other determinations of stellar radii make use of the radiation laws. The validity of this indirect method is verified by noting that it gives the correct radii for those stars whose sizes can also be measured by geometrical means.

S U M M A R Y

23.1   Application of Kepler's third law as modified by New-ton to the Earth-Sun system and to the Earth-Moon system yields the ratio of the mass of the Sun to the mass of the Earth. Since the mass of the Earth is known, the mass of the Sun can be calculated. The mass of the Sun is 333,000 times greater than the mass of the Earth.

23.2   Visual binaries are double stars in which both com-ponents can be seen directly. The orbital motions of **spectroscopic binaries** are detected by measurements of the Doppler shifts of lines in the spectra of the member stars. **Eclipsing binaries** are double stars in which one star passes directly in front of or behind the other, as viewed from the Earth, thereby causing variations in brightness. Such binary systems are important because they can be used to determine the masses of the stars. Such masses range from $\frac{1}{12}$ the mass of the Sun up to about 100 times the mass of the Sun. Stars with the highest masses are exceedingly rare. Most stars obey the **mass–luminosity relation:** The more massive a star is, the greater the total energy it radiates.

23.3   The diameters of a small number of stars can be derived because they are in eclipsing binaries or because they are occulted by the Moon or planets. For most stars, diameters must be estimated by use of the Stefan-Boltzmann law ($L = 4\pi R^2 \sigma T^4$), which permits the calculation of the radius R if the luminosity L and the temperature T are known. Most stars have diameters similar to that of the Sun, but some have diameters as small as the Earth's, while others are large enough to fill the entire solar system almost as far as the orbit of Jupiter.

E X E R C I S E S

THOUGHT QUESTIONS

1.   A few stars are both visual binaries and spectroscopic binaries (their variations in radial velocity can be de-tected). Why do you suppose such stars are rare?

2.   There are fewer eclipsing binaries than spectroscopic binaries. Explain why. Within 50 LY of the Sun, visual binaries outnumber eclipsing binaries. Why? Which is easier to observe at large distances—a spectroscopic bi-nary or a visual binary?

3. Why do most visual binaries have relatively long periods and most spectroscopic binaries relatively short periods? Under what circumstances could a binary with a relatively long period (over a year) be observed as a spectroscopic binary?

4. Although the periods of known eclipsing binaries range from 4^h39^m to 27 years, the average of their periods is less than the average period of all known spectroscopic binaries. Can you suggest an explanation?

5. Describe the apparent relative orbit of a visual binary whose true orbital plane is edge on to the line of sight. Describe the apparent motions of the individual stars of the system among the background stars in the sky.

PROBLEMS

6. What, approximately, would be the periods of revolution of binary-star systems in which each star had the same mass as the Sun, and in which the semimajor axes of the relative orbits had these values? **(a)** 1 AU; **(b)** 2 AU; **(c)** 6 AU; **(d)** 20 AU; **(e)** 60 AU; **(f)** 100 AU.

★ 7. In each of the binary systems in Problem 6, at what distance would the two stars appear to have an angular separation of 1 arcsec? (Assume circular orbits.)

8. Show that the semimajor axis of the true relative orbit of a visual binary system, in astronomical units, is equal to its angular value, in seconds of arc, times the distance of the system, in parsecs.

9. The true relative orbit of Xi Ursae Majoris has a semimajor axis of 2.5 arcsec, and the parallax of the system is 0.127 arcsec. The period is 60 years. What is the sum of the masses of the two stars in units of the solar mass?
Answer: 2.1 solar masses

10. In a particular visual-spectroscopic binary, the maximum value of the radial velocity of one star with respect to the other is 60 km/s, the inclination of the orbital plane to the plane of the sky is 30°, and the period is 22 days. If the stars in the system have circular orbits, what is the sum of their masses?
Answer: 3.8 solar masses

★11. The observed component of a hypothetical astrometric binary system is found to move in an elliptical orbit of semimajor axis 1 AU about the barycenter of the system in a period of 30 years. If the visible star is assumed to have a mass equal to that of the Sun, what is the mass of its unseen companion?
Answer: About 0.11 solar mass

12. A hypothetical spectroscopic-eclipsing binary star is observed. The period of the system is three years. The maximum radial velocities, with respect to the center of mass of the system, are as follows: Star A, $(4/3)\pi$ AU per year; Star B, $(2/3)\pi$ AU per year.
  a. What is the ratio of the masses of the stars?
  b. Find the mass of each star (in solar units). Assume that the eclipses are central.

13. In an eclipsing binary in which the eclipses are exactly central, and in which a small star revolves about a considerably larger one, the interval from first to second contacts is 1 hr and from first to third contacts is 4 hr. The entire period is three days. The centers of the stars are separated by 11,460,000 km. What are the diameters of the stars?
Answer: 1 million and 4 million km

14. How many times more massive than the Sun would you expect a star to be that is 1000 times more luminous? What if it were 10,000 times more luminous? (Assume that the mass-luminosity relation holds for these stars.)

★15. Measured angular diameters of several stars are given in the following table:

Star	Angular Diameter	Distance (pc)	Linear Diameter (In Terms of Sun's)
Betelgeuse (α Orionis)	0".034* 0.042	150	500 750
Aldebaran (α Tauri)	0.020	21	45
Arcturus (α Bootis)	0.020	11	23
Antares (α Scorpii)	0.040	150	640
Scheat (β Pegasi)	0.021	50	110
Ras Algethi (α Herculis)	0.030	150	500
Mira (o Ceti)	0.056*	70	420

* Variable in size

Which of the stars are larger than the orbit of the Earth? Are any larger than the orbit of Mars? Of Jupiter?

16. Show how the measured angular diameters and observed energy fluxes of stars can be used to measure their effective temperatures. (*Hint*: Use the inverse-square law of light, and recall the Stefan-Boltzmann law.)

17. What is the radius of a star (in terms of the Sun's radius) with the following characteristics:
  a. Twice the Sun's temperature and four times its luminosity?
  b. Eighty-one times the Sun's luminosity and three times its temperature?

★18. Assume the wavelength of maximum light of the Sun to be exactly 500 nm, its temperature exactly 6000 K, and its absolute bolometric magnitude exactly 5.0. Another star has its wavelength of maximum light at 10,000 nm. Its apparent visual magnitude is 15.5, its bolometric correction is 0.5, and its parallax is 0.01 arcsec. What is its radius in terms of the Sun's?
Answer: $R/R_s = 0.4$

This small region in the constellation of Scorpius includes many stars, together with a complex cloud of gas and dust that are the raw materials of the stars. Note that stars of several different colors can be seen in this image. The "arm" of dust can help you get a sense of depth here: In the region of the dust, only the foreground stars (the ones closer to us than the dust) can be seen. Outside the dust cloud, you can see far deeper into the galaxy and count many more stars.

(Courtesy David Malin and the Anglo-Australian Telescope Board)

In 1984, a Gallup Poll revealed that 55 percent of American teenagers (ages 13–18) believed that astrology works. How did the pollsters come up with this depressing result? Clearly, they did not interview every one of the millions of teenagers in the U.S. They did what all pollsters do: They spoke with a much smaller but representative cross-section of such teenagers and then extrapolated the results to the entire population.

In the same way, astronomers cannot possibly measure the characteristics of every one of the billions of stars in our Milky Way Galaxy. But, like pollsters, they can take a thorough survey of a cross-section of stars and draw conclusions from that work. The trick is to figure out how to make such a poll representative: to be sure that we've got a fair share of all the different sorts of stars in the sample. For example, as you will see, the stars that happen to be the brightest and thus easiest to detect in our sky are by no means representative of the vast majority of stars.

When we do a good survey of the stars with which we share our galaxy, we can begin to divide them into convenient types. In this chapter we will introduce some of the main categories that astronomers have found especially useful; in later chapters you will discover that these categories represent stages in the life cycle of a star and begin to understand their significance in stellar evolution.

But even this introductory chapter should be enough to fire your imagination. We will meet stars so bloated that if we replaced the Sun with them, the Earth would be *inside* the star. We'll examine stars that, at the end of their lives, become so compact that a pea-sized piece would weigh more than a fully loaded truck. As we found in the chapters on the outer planets, nature delights in variety and has found a remarkable number of ways to arrange and display the atoms of which everything, including the stars, is made.

THE STARS: A CELESTIAL CENSUS

Henry Norris Russell (1877–1957), American astronomer, was a professor at Princeton University and one of the most respected astronomers of our century. His many interests included the study of stellar evolution. Russell and Ejnar Hertzsprung independently discovered the main sequence of stars, best illustrated on the famous diagram now known as the Hertzsprung–Russell diagram. (Princeton University Archives)

Having learned in the previous chapters how astronomers measure the various characteristics of the stars—distance, motion, brightness, color, mass, radius, and composition—we are now ready to organize and classify them.

We begin by exploring what we know about our nearest stellar neighbors.

24.1 THE NEAREST AND THE BRIGHTEST STARS

Let us consider first our most conspicuous stellar neighbors, the brightest-appearing stars in the sky. Appendix 14 lists some of the properties of the brightest 20 stars. Many of these are double- or triple-star systems. Data are given for each component of these systems.

(a) The Brightest Stars

Recall from Chapter 22 that, since not all stars are alike in brightness, a star could look bright in our sky *either* because it is nearby *or* because it is intrinsically so luminous that it looks bright even at a significant distance. In fact, the most striking thing about most of the brightest-appearing stars is that they are bright because they are actually of high intrinsic luminosity. Of the 20

brightest stars listed in Appendix 14, only 6 are within 10 pc of the Sun. Recall that the absolute magnitude of a star (Section 22.1b) is the apparent magnitude that it would have if it were at a distance of 10 pc. Since the brightest 20 stars are of apparent magnitude 1.5 or brighter, the 14 of them that are more distant than 10 pc must have absolute magnitudes less (that is, brighter) than 1.5. This means that their intrinsic luminosities must exceed that of the Sun by at least a factor of 20. (For comparison, the absolute magnitude of the Sun is +4.8.) Even among the approximately 3000 stars with apparent magnitudes less than 6.0, only about 60 are within 10 pc. Most stars seen with the unaided eye are tens or even hundreds of parsecs away and are many times more luminous than the Sun. Indeed, among the 6000 stars visible to the unaided eye, at most 50 are intrinsically fainter than the Sun. Figure 24.1 is a histogram showing the distribution in luminosity of the 30 brightest-appearing stars.

From Appendix 14 or Figure 24.1, we might gain the impression that the Sun is far below average among stars in luminosity. This is not so. Most stars, as we shall see in the next subsection, are much less luminous than the Sun is. They are too faint, in fact, to be conspicuous unless they are nearby. Stars of high luminosity are rare—so rare that the chance of finding one within a small volume of space, say, within 10 pc of the Sun, is

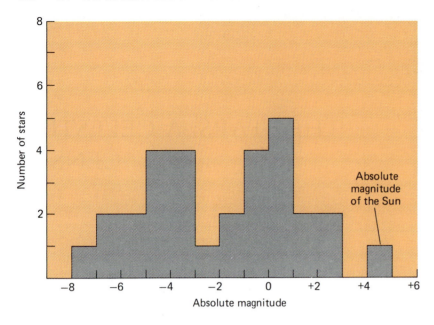

FIGURE 24.1. The absolute magnitudes of the 30 brightest-appearing stars. The units are the numbers of stars per unit of absolute magnitude.

very slight. Why, then, are most of the stars that we see with the unaided eye the very rare stars of high luminosity? And why are the most common, intrinsically faint stars not among those that we see when we look up at the heavens without a telescope?

The question is best answered with the help of some numerical examples. The Sun, whose absolute visual magnitude is +4.8, would appear as a very faint star to the naked eye if it were 10 pc away. Stars much less luminous than the Sun would not be visible at all at that distance. Stars with intrinsic luminosities from 10^{-2} to 10^{-4} L_s (where L_s stands for the luminosity of the Sun) are very common, but a star with a luminosity of 10^{-2} L_s would have to be within 1.6 pc to be visible to the naked eye. Only Alpha Centauri and its companions are closer than this. The intrinsically faintest star observed has an absolute visual magnitude of about +20 (10^{-6} L_s; since this star is very cool, most of its energy is actually emitted in the infrared region of the spectrum and its bolometric luminosity is 10^{-4} L_s). For this star to be visible to the naked eye, it would have to be within 0.025 pc, or 5200 AU. No star is that close to us!

It is clear, then, that the vast majority of nearby stars, those less luminous than the Sun, do not send enough light across interstellar distances to be seen without optical aid. For example, the star closest to the Sun is Proxima Centauri, a companion of Alpha Centauri. Proxima Centauri has an absolute visual magnitude of +16, and it is invisible to the naked eye.

In contrast, consider the highly luminous stars. Stars with absolute magnitudes of 0 have luminosities of about 100 L_s. They are far less common than stars less luminous than the Sun, but they are visible to the naked eye even out to a distance of 160 pc. A star with an

intrinsic luminosity of 10,000 L_s can be seen without a telescope to a distance of 1600 pc (if there is no dimming of light by interstellar dust—see Chapter 25). Such stars are very rare, and we would not expect to find one within a distance of only 10 pc. The volume of space included within a distance of 1600 pc, however, is about 4 million times that included within a distance of only 10 pc. Hence many stars of high luminosity are visible to the unaided eye.

(b) The Nearest Stars

Evidently, the brightest stars seen with the unaided eye do not provide a representative sample of the stellar population in the neighborhood around the Sun. Let us turn then to the nearest known stars. Appendix 13 lists 44 stars within 5 pc (some are double or multiple systems) from data provided by the U. S. Naval Observatory. (Additional nearby stars are discovered from time to time; the total number of stars within 5 pc may be double the number listed. Moreover, the measurements of distances, luminosities, and so on for nearby stars are being continually refined. Still the table does indicate the general characteristics of the Sun's nearest stellar neighbors.)

The table shows that only 3 of the 43 nearest stars (other than the Sun) are among the 20 brightest stars: Sirius, Alpha Centauri, and Procyon. This fact is further confirmation that the nearest stars are not the brightest-appearing stars. The nearby stars also tend to have large proper motions, as would be expected (Section 21.5). In fact, the large proper motions of many of these stars led to the discovery that they are located nearby. Another interesting observation is that 13 of the 44 stars are really binary- or multiple-star systems; the table thus contains, actually, 59 rather than 44 stars. Twenty-eight of these 59 stars, or nearly half, are members of systems containing more than one star. Two or three other stars in the list are suspected of having companions.

The most important datum concerning the nearest stars is that most of them are intrinsically faint. Only ten of the nearest stars are individually visible to the unaided eye. Only 3 are as intrinsically luminous as the Sun; 43 have luminosities less than 0.01 L_s. If the stars in our immediate stellar neighborhood are representative of the stellar population in general, and we think they are, we must conclude that the most numerous stars are those of low luminosity. In this sample of stars in the solar neighborhood, only about 1 star in 20 is intrinsically as luminous as the Sun.

An estimated lower limit can be established for the mean density of stars, i.e., the number of stars per cubic parsec, in the solar neighborhood. There are at least 59 stars within 5 pc (counting the members of binary- and multiple-star systems and the Sun). A sphere of radius 5 pc has a volume of $\frac{4\pi(5)^3}{3}$, or about 524 pc³. Since this volume of space contains at least 59 stars, the density of stars in space is at least 1 star for every 9 pc³. The actual stellar density, of course, can be greater than this figure if there are undiscovered stars within 5 pc. The mean separation between stars is the cube root of 9, or about 2.1 pc. If the matter contained in stars could be spread out evenly over space, and if a typical star has a mass 0.4 times that of the Sun, the mean density of matter in the solar neighborhood would be about 3×10^{-24} g/cm³. That's a very small number, showing how much more of our neighborhood consists of space than stars!

The nearest stars constitute a much more nearly representative sample of the stellar population in the vicinity of the Sun than do the brightest stars. We are still not sure, however, that we have identified all of the faintest stars in the solar neighborhood. Moreover, there do not happen to be any stars of high luminosity in this "tiny" volume of space. Yet we can identify all the luminous stars, with a reasonable degree of completeness, out to a much greater distance. If we make allowance for the different volumes of space that we must survey to catalogue large samples of stars of different intrinsic luminosities, we can gain some indication of their relative abundances. For example, within 10 pc there are about 12 known stars brighter than approximately 2 L_s (absolute magnitude +4), while within 5 pc there are about 57 known stars fainter than about 2 L_s. We would expect eight times as many of these faint stars within 10 pc (for that distance includes a volume of space eight times as large). Since $8 \times 57 = 456$ stars fainter than 2 L_s within 10 pc, the ratio of stars with intrinsic luminosities less than 2 L_s to the number of more luminous stars is about 456 to 12, or 38:1.

(c) The Luminosity Function

Once the numbers of stars of various absolute magnitudes or intrinsic luminosities have been found, the relative numbers of stars in successive intervals of luminosity within any given volume of space can be established. This relationship is called the **luminosity function.** Figure 24.2 shows the luminosity function for stars in the solar neighborhood. Compare Figure 24.2 with Figure 24.1.

The Sun, we see, is more luminous than the vast majority of stars. On the other hand, the relatively few stars of higher luminosity than the Sun compensate for their small numbers by their high rate of energy output.

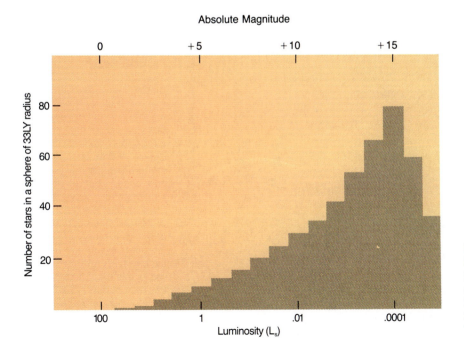

Absolute Magnitude

FIGURE 24.2. Luminosity function of stars in the solar neighborhood. Note that faint stars are much more common than bright ones. (Note that absolute magnitudes are shown at the top, for comparison with Figure 24.1.)

It takes only 10 stars of absolute magnitude 0, which is 100 times the luminosity of the Sun, to outshine 1000 stars fainter than the Sun, and only 1 star of absolute magnitude −5 to outshine 10,000 stars fainter than the Sun. Most of the starlight from our part of space, it turns out, comes from the relatively few stars that are more luminous than the Sun.

24.2 THE HERTZSPRUNG–RUSSELL DIAGRAM

In this and the three preceding chapters, we have described something about the characteristics of stars and how we measure those characteristics. You may feel somewhat overwhelmed by all of this new information. It will be easier to understand and remember what stars are like if we can now find some patterns that describe the relationships between size, mass, luminosity, and temperature. Fortunately, such patterns do exist.

In 1911 the Danish astronomer Ejnar Hertzsprung compared the colors and luminosities of stars within several clusters by plotting their magnitudes against their colors. In 1913 the American astronomer Henry Norris Russell undertook a similar investigation of stars in the solar neighborhood by plotting the absolute magnitudes of stars of known distance against their spectral classes. These investigations by Hertzsprung and by Russell led to an extremely important discovery concerning the relation between the luminosities and surface temperatures of stars. The discovery is exhibited graphically with a diagram named in honor of the two astronomers—the **Hertzsprung–Russell** or **H–R diagram.**

(a) Features of the H–R Diagram

Two easily derived characteristics of stars of known distances are their absolute magnitudes (or luminosities) and their surface temperatures. The absolute magnitudes can be found from the known distances and the observed apparent magnitudes. The surface temperature of a star is indicated by either its color or its spectral class.

An H–R diagram for selected nearby stars is shown in Figure 24.3. In plotting these diagrams, astronomers always adopt the convention that temperature increases toward the left and luminosity toward the top. Figure 24.4, a schematic H–R diagram for a large sample of stars, is also shown to make the various features more apparent.

The most significant feature of the H–R diagram is that the stars are not distributed over it at random, exhibiting all combinations of absolute magnitude and

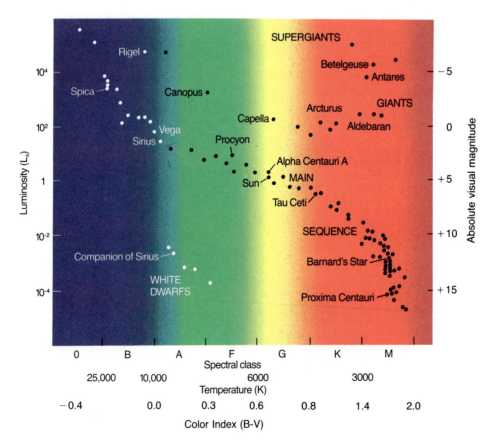

FIGURE 24.3.
Hertzsprung–Russell diagram for a selected sample of stars. Note that the stars do not have all possible values of temperature and luminosity, but rather that most are found along the main sequence, a band that stretches from upper left to lower right in the diagram, or from high temperature and high luminosity to low temperature and low luminosity. A few stars are red giants (high luminosity, low temperature) or white dwarfs (high temperature, low luminosity).

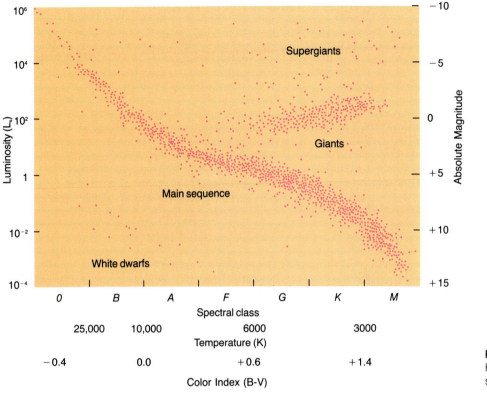

Figure 24.4. Schematic H–R diagram for many stars.

temperature. Rather they cluster into certain parts of the diagram. The majority of stars are aligned along a narrow sequence running from the upper left (hot, highly luminous) part of the diagram to the lower right (cool, less luminous) part. This band of points is called the **main sequence.** The characteristics of main-sequence stars of different spectral types are summarized in Table 24.1. As we shall see in Chapter 28, the factor that determines where a star falls along the main sequence is its *mass.* The more massive stars are the hotter and more luminous stars.

A substantial number of stars, however, lie above the main sequence on the H–R diagram, in the upper right (cool, high luminosity) region. These are called **giants.** At the top part of the diagram are stars of even higher luminosity, called **supergiants.** Finally, there are stars

in the lower left (hot, low luminosity) corner known as **white dwarfs.** To say that a star lies "on" or "off" the main sequence does not refer to its position in space, but only to the point that represents its luminosity and temperature on the H–R diagram.

An H–R diagram, such as Figure 24.3, that is plotted for stars of known distance does not show the relative proportions of various kinds of stars, because only the nearest of the intrinsically faint stars can be observed. To be truly representative of the stellar population, an H–R diagram should be plotted for all stars within a certain distance. Unfortunately, our knowledge is reasonably complete only for stars within a few parsecs of the Sun, among which there are no giants or supergiants. It is estimated that about 90 percent of the stars in our part of space are main-sequence stars and

TABLE 24.1	CHARACTERISTICS OF MAIN-SEQUENCE STARS			
Spectral Type	Mass (Sun = 1)	Luminosity (Sun = 1)	Temperature	Radius (Sun = 1)
O5	40	5×10^5	40,000 K	18
B0	16	2×10^4	28,000 K	7
A0	3.3	80	10,000 K	2.5
F0	1.7	6	7,500 K	1.4
G0	1.1	1.3	6,000 K	1.1
K0	0.8	0.4	5,000 K	0.8
M0	0.4	0.03	3,500 K	0.6

about 10 percent are white dwarfs. Fewer than 1 percent are giants or supergiants.

(b)   Distances from Stellar Spectra

One of the most important applications of the H–R diagram is in the determination of stellar distances. Suppose, for example, that a star is known to be a spectral-class G2 star on the main sequence. Its intrinsic luminosity could then be read off the H–R diagram at once. It would be about 1 L_s (absolute magnitude $+5$). From this and the star's apparent magnitude, its distance can be calculated (Section 22.1c).

In general, however, the spectral class alone is not enough to fix, unambiguously, the intrinsic luminosity of a star. The G2 star described in the last paragraph could have been, for example, a main-sequence star of absolute magnitude $+5$, a giant of absolute magnitude 0 (100 L_s), or a supergiant of still higher luminosity. We recall, however (Section 22.3b), that pressure differences in the atmospheres of stars of different sizes result in slightly different degrees of ionization for a given temperature. It will be seen in the next subsection that giant stars are larger than main-sequence stars of the same spectral class and that supergiants are larger still. It is thus possible to classify a star by its spectrum, not only according to its temperature (spectral class), but also according to whether it is a main-sequence star, a giant, or a supergiant.

The most widely used system of classifying stars according to their luminosities divides stars of a given spectral class into as many as six categories, called **luminosity classes.** These luminosity classes are as follows:

Ia Brightest supergiants

Ib Less luminous supergiants

II Bright giants

III Giants

IV Subgiants (intermediate between giants and main-sequence stars)

V Main-sequence stars

The full specification of a star, including its luminosity class, would be, for example, for a spectral-class F3 main-sequence star, F3 V. For a spectral-class M2 giant, the specification would be M2 III.

Figure 24.5 illustrates the approximate mean positions of stars of various luminosity classes on the H–R diagram. The dashed portions of the lines represent those spectral classes (for a given luminosity class) for which there are very few or no stars.

With both its spectral class and luminosity class known, a star's position on the H–R diagram is uniquely determined. Its absolute magnitude, therefore, is also known, and its distance can be calculated. Distances determined this way, from the spectral and luminosity classes, are said to be obtained from the method of **spectroscopic parallaxes.**

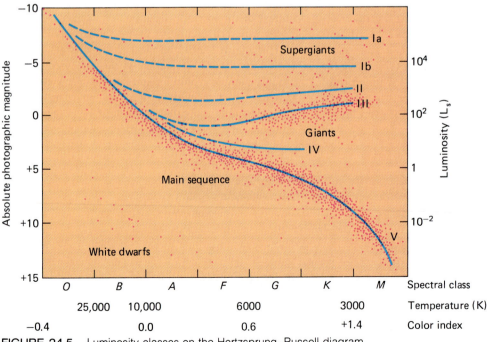

FIGURE 24.5.   Luminosity classes on the Hertzsprung–Russell diagram.

24.3 EXTREMES OF STELLAR CHARACTERISTICS

(a) The Most Luminous Stars

The H−R diagram is useful for investigating the extremes in stellar size, luminosity, and density. The most massive blue stars have masses up to about 100 solar masses. These are also the most luminous main-sequence stars. They have absolute magnitudes of −6 to −8. A few stars are known that have absolute bolometric magnitudes of −10 (10^6 L_s). These superluminous stars, most of which are at the upper left on the H−R diagram, are very hot spectral-type O and B stars and are very blue. These are the stars that would be the most conspicuous at very great distances in space.

The cool giants and supergiants are located at the upper right corner of the H−R diagram. These stars are at least a few hundred times more luminous than the Sun (if they are giants) or some thousands of times more luminous (if they are supergiants). They also have very much larger diameters than the Sun. Consider, for example, a typical red, cool supergiant that has a surface temperature of 3000 K and an intrinsic luminosity of 10^4 L_s. Since each unit area of a star with half the temperature of the Sun emits only $\frac{1}{16}$ as much light as a unit area of the Sun (Section 7.2), its total surface area must be greater than the Sun's by 160,000 times. Since the surface area is equal to $4\pi R^2$, where R is the radius, the radius of this red supergiant must be 400 times the Sun's radius. If the Sun could be placed in the center of such a star, the star's surface would lie beyond the orbit of Mars and the Earth would be inside the star. We see, then, that supergiants represent extremes in radius as well as luminosity.

Such red giant stars have extremely low mean densities. The volume of the star described in the last paragraph would be 64 million times the volume of the Sun. The masses of such giant stars, however, are probably at most only 50 solar masses, and very likely much less. If we assume that a supergiant star with 64 million times the Sun's volume has only 10 times its mass, we find that it has just over 10^{-7} times the Sun's mean density, or only about 2×10^{-7} times the density of water. The outer parts of such a star would constitute an excellent laboratory vacuum.

(b) The Least Luminous Stars

The very common red, cool stars of low luminosity at the lower end of the main sequence are much smaller and more compact than the Sun. As an example, consider such a red dwarf, the star Ross 614B, which has a surface temperature of 2700 K and an absolute bolometric magnitude of about +13 (5×10^{-4} L_s). Each unit area of this star emits only $\frac{1}{20}$ as much light as a unit area of the Sun, but to have only $\frac{1}{2000}$ the Sun's luminosity, the star must have only about $\frac{1}{100}$ the Sun's surface area, or $\frac{1}{10}$ its radius. The faintest red dwarfs have luminosities only about 10^{-4} times as great as that of the Sun. A star with such a low luminosity also has a low mass (Ross 614B has a mass about $\frac{1}{12}$ that of the Sun) but still would have a mean density about 80 times that of the Sun. Its density must be higher, in fact, than that of any known solid found on the surface of the Earth.

The faint red main-sequence stars are not the stars of the most extreme densities. The white dwarfs, at the lower left corner of the H−R diagram, have ever higher densities.

(c) The White Dwarfs

The first white dwarf to be discovered was the companion to Sirius, the brightest-appearing star in the sky. The companion was first seen telescopically in 1862. Sirius is the brightest star in the constellation Canis Major, Orion's big dog. (Incidentally, Procyon, the brightest star in Orion's other dog, Canis Minor, also has a white dwarf companion.)

The companion of Sirius has a mass about 5 percent greater than that of the Sun. From its temperature and luminosity, however, we find its diameter to be only about 1 percent of the Sun's, or about twice that of the Earth. Thus the white dwarf must have a mean density more than 100,000 times that of the Sun and 1.6×10^5 times that of water. Some white dwarfs have much higher mean densities, and many have central densities in excess of 10^7 times that of water. A teaspoonful of such material would weigh nearly 50 tons on Earth!

The British astrophysicist Sir Arthur Eddington described the first known white dwarf this way:

> The message of the companion of Sirius, when decoded, ran: I am composed of material three thousand times denser than anything you've come across. A ton of my material would be a little nugget you could put in a matchbox. What reply could one make to something like that? Well, the reply most of us made in 1914 was, "Shut up; don't talk nonsense."

Today, we know that at the densities reached by white dwarfs, matter cannot exist in its usual state. Although it is still gaseous, its atoms are completely stripped of their electrons. Most stars are believed to become white dwarfs near the end of their lives. Eventually, after many billions of years, white dwarfs radiate away their internal heat, cooling off to become black dwarfs—cold, dense stars no longer shining. We shall describe the theory of white dwarfs in Chapter 30, where we shall also learn that there are stars millions of times denser yet!

24.4 THE DISTRIBUTION OF STARS IN SPACE

In the immediate neighborhood of the Sun, the stars seem to be distributed more or less at random (except for their tendency to form small clusters). The larger the volume of space we survey, the more stars we find, and if allowance is made for the fact that the faintest stars become invisible at larger distances, the number of stars we can count is roughly proportional to the cube of the distance to which we look. Eventually, however, the stars do thin out more rapidly in some directions than in others. The way they thin out is a clue to the nature of the stellar system to which the Sun belongs.

(a) Herschel's Star Gauging

The idea that the Sun is a part of a large system of stars was suggested as early as 1750 by Thomas Wright in his *Theory of the Universe*. Immanuel Kant, the great German philosopher, suggested the same hypothesis five years later. It was the German-English astronomer William Herschel, the discoverer of Uranus, who first demonstrated the nature of the stellar system.

Herschel sampled the distribution of stars about the sky by a procedure he called *star gauging*. He observed that in some directions he could count more stars through his telescope than in other directions. In 1784 and 1785 he presented two papers giving the results of gauges or counts of stars that he was able to observe in 683 selected regions scattered over the sky. While in some of these fields he could see only a single star, in others he was able to count nearly 600. Herschel reasoned that in those directions in which he saw the greatest numbers of faint stars, the stars extended the farthest, and in other directions they thinned out at relatively shorter distances. As a result of his star gauging, Herschel arrived at the conclusion (only partially correct, as we shall see) that the Sun is inside a great sidereal system, and that the system is shaped like a wheel with the Sun near the center.

(b) The Phenomenon of the Milky Way

If you have looked at the sky on a moonless night away from the glare of city lights, you may have noticed the Milky Way (Figure 24.6), a faint, luminous band of light that completely encircles the sky. Galileo turned his telescope on the Milky Way and saw that it really consists of a myriad of faint stars. Herschel's picture that the stars are distributed in a wheel-shaped pattern explains why the Milky Way appears as a band all the way around the sky.

It must be recalled that we view our sidereal system from the inside. Figure 24.7 shows an idealized portion of the "wheel" of stars, viewed edge on. The Sun's position is at *O*. If we look from *O* toward either face of

FIGURE 24.6. This wide-angle picture covers over 50° of the sky in the direction of the center of the Milky Way. The galactic center is totally obscured at optical wavelengths. (Anglo-Australian Telescope Board)

the wheel, that is, in directions *a* or *b*, we see only those stars that lie between us and the nearest boundary of the stellar system. In these directions in the sky, therefore, we see only scattered stars. On the other hand, if we look edge on through the wheel, say, in directions *c* or *d*,

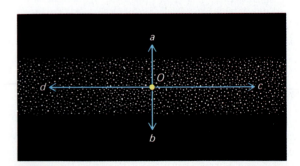

FIGURE 24.7. We view our own Galaxy from inside. If we imagine that the Sun is located at *O*, then we see the band of stars that forms the Milky Way when we look through the Galaxy edge-on (directions *c* and *d*). We see many fewer stars when we look in directions *a* and *b*.

FIGURE 24.8.   A copy of a diagram by Herschel, showing a cross section of the Milky Way system of stars. The yellow circle shows the location of the Sun.

we encounter so many stars along our line of sight that we get the illusion of a continuous band of light. Since the greatest dimensions of the wheel extend in all directions along its flat plane, the band of light extends completely around the sky. This band of light is the Milky Way; it is simply the light from the many distant stars that appear lined up in projection when we look edge on through our own flattened stellar system. Figure 24.8 is a copy of one of Herschel's diagrams, showing a cross-section of the stellar system, as he derived its shape.

We call our stellar system the **Galaxy,** or sometimes the **Milky Way Galaxy,** or even just the Milky Way.

The Galaxy is far more complicated than Herschel's image of a wheel (Figure 24.9). We now know that the Galaxy is a vast pinwheel with billions of stars and clouds of glowing gas and dark dust arranged in a series of spiral arms. Indeed it is the dust, which dims the light of distant stars, that masks the true extent of the Galaxy and misled Herschel into thinking that the Sun is at the center of our stellar system.

In the next several chapters, we will explore the properties of the gas and dust in the Galaxy, then trace the evolution of individual stars, and finally turn to an analysis of the structure of the Galaxy and to speculation about its origin.

FIGURE 24.9   A photograph showing a rich star field in the direction toward the center of the Milky Way Galaxy. In addition to stars, the Galaxy contains clouds of glowing gas and dark clouds of dust, which obscure the light of distant stars. (Anglo-Australian Observatory Board)

SUMMARY

24.1   Most of the stars that appear brightest in the nighttime sky are bright not because they are nearby but because they are intrinsically very luminous. Most of the nearest stars are invisible to the unaided eye because they are intrinsically very faint. The **luminosity function** indicates how the numbers of stars depend on intrinsic brightness. On average, only 1 of every 20 stars in the immediate solar neighborhood emits more energy than the Sun. The typical distance between stars in the solar neighborhood is about 2 pc.

24.2   The **H–R diagram** is a plot of stellar luminosity as a function of temperature. Most stars lie on the **main sequence,** which extends diagonally across the H–R diagram from high temperature and high luminosity to low temperature and low luminosity. The position of a star along the main sequence is determined by its mass. About 90 percent of the stars near the Sun lie on the main sequence. About 10 percent of the stars are **white dwarfs.** Fewer than 1 percent of the stars are **giants** or **supergiants.** From examination of stellar spectra, it is possible to determine the **luminosity class** of a star and hence its **spectroscopic parallax,** which gives its distance.

24.3   The luminosities of stars range from (approximately) 10^{-4} to $10^6 L_s$. The densest stars are the white dwarfs. In these stars, which are nearing the end of their lives, a mass equal to that of the Sun occupies a volume comparable to that of the Earth.

24.4   All stars visible to the unaided eye are members of the **Milky Way Galaxy,** which is a flattened, wheel-shaped system of billions of stars, gas, and dust.

EXERCISES

THOUGHT QUESTIONS

1. Describe an everyday situation that is analogous to the fact that most stars seen with the unaided eye are of far more than average stellar luminosity.

2. If stars of all kinds were uniformly distributed through space, what would the approximate luminosity function have to be in order for intrinsically faint stars to be the most common among stars seen with the unaided eye?

3. Given the luminosity function, how would stars have to be distributed in space in order for the intrinsically faint ones to be most common among stars seen with the unaided eye?

4. Plot the luminosity functions of the nearest stars (Appendix 13) and of the 20 brightest stars (Appendix 14). Explain how and why these two luminosity functions differ.

5. Why are most faint-appearing stars blue?

6. It is possible to construct the equivalent of an H–R diagram for human beings by plotting height against weight. Try doing so for your classmates; obtain additional information for children and babies. Do all combinations of height and weight occur? Can you think of special examples of human beings that deviate from normal relationships?

7. Several very bright stars are identified in Figure 24.3. Select some that are visible at this time of year. What color would you expect each to be? Go outside at night and find these stars. Do they actually have approximately the color that you expected?

8. Suppose you wanted to use a space telescope to search for main-sequence stars with very low mass. Would you design your telescope to detect light in the ultraviolet or in the infrared part of the spectrum? Why?

9. Consider data on five stars with the following apparent magnitudes and spectral types:

$$m = 15; \text{ G2 V}$$
$$m = 20; \text{ M3 Ia}$$
$$m = 10; \text{ M3 V}$$
$$m = 15; \text{ B9 V}$$
$$m = 15; \text{ M5 V}$$

(a) Which is hottest? **(b)** Coolest? **(c)** Most luminous? **(d)** Least luminous? **(e)** Nearest? **(f)** Most distant? In each case, give your reasoning.

10. Suppose you had data on the apparent magnitudes and colors of several hundred stars in a cluster. Explain how you could use these data to determine the distance to the cluster.

11. For normal stars what is the approximate range of **(a)** effective temperature; **(b)** mass; **(c)** radius; **(d)** luminosity?

12. Why do you suppose that most visual binaries are stars of low luminosity?

13. Suppose the Milky Way were a band of light extending only halfway around the sky (that is, in a semicircle). What then would you conclude about the Sun's location in the Galaxy?

PROBLEMS

14. (a) At what distance would a star of absolute magnitude +15 appear as a fifth-magnitude star? (b) At what distances would a star of absolute magnitude −10 and one of absolute magnitude +15 appear brighter than the fifth apparent magnitude?

15. If the brightest-appearing star, Sirius, were three times its present distance, would it still make the list of "Twenty Brightest Stars" (Appendix 14)? What about the second brightest star, Canopus?

16. Would any of the stars within 5 pc (Appendix 13) be able to be seen with the naked eye at a distance of 100 pc? If so, which ones?

*17. Verify that the mean density of stellar matter in the solar neighborhood is about 3×10^{-24} g/cm³.

18. Suppose that within 10 pc there were 11 stars of the same intrinsic luminosity as the Sun, and that within 30 pc there were 11 stars with luminosities of 100 L_s. Estimate the true ratio of the number of stars of luminosity 1 L_s to the number with luminosities of 100 L_s in a given volume of space.

19. Find the distances to the following stars (see Figure 24.5):
    a. $m = +10$, spectral designation A0 Ib
    b. $m = +5$, spectral designation K5 III
    c. $m = 0$, spectral designation G2 V

20. Suppose you weigh 70 kg on the Earth. How much would you weigh on the surface of a white dwarf star that has the same size as the Earth, but a mass 300,000 times the Earth's (nearly the mass of the Sun)?

This region contains some of the most colorful clouds of interstellar gas and dust ever photographed. The blue region at the top is a cloud of dust surrounding the star Rho Ophiuchi; the blue color is starlight reflected from grains of dust. At the lower right, the bright star Antares (Alpha Scorpii), a cool red supergiant, is embedded in a thick dust cloud of its own making. The reddish nebulae glow with light emitted by hydrogen atoms. Immediately to the left of Antares, is M4, a much more distant cluster of stars.

(David Malin; copyright Royal Observatory, Edinburgh)

We now turn to the exploration of the raw material from which stars are born and into which they recycle some of their material when they die. The full understanding of this material has only come in recent decades, for much of the gas and dust between the stars is invisible to our eyes. But in a few places, where starlight illuminates or is absorbed by this raw material, it is revealed to us in spectacular clouds of light and dark.

The first evidence that interstellar space was not empty came in the 18th century, when increasingly powerful telescopes revealed some faint patches of glowing material that turned out to be regions of hot gases or dust clouds reflecting the light of nearby stars. Since those first tentative hints, our instruments have shown the interstellar medium to be a far more complex and active realm than early astronomers could have imagined.

Among the "ingredients" of these gas and dust clouds are complicated molecules—including some of the key building blocks of life—that appear to form where the dust is thickest and can afford the most protection. Vast reservoirs of various molecules have been found out there in recent years, among them, water, ammonia, formaldehyde (embalming fluid), and even alcohol. Some scientists now speculate that some of the molecules in the cloud that give birth to our solar system may have been the very earliest chemical ancestors of the readers of this book.

BETWEEN THE STARS: GAS AND DUST IN SPACE

Bengt Georg Daniel Strömgren (1908–1987), Danish astronomer, spent much of his productive career in the United States, especially as Professor and Director of the Yerkes Observatory (University of Chicago). He received many honors for his fundamental work in the study of the structure and evolution of stars, and especially for his pioneering investigation of the physics of the interstellar medium. (John B. Irwin)

By earthly standards the space between the stars is empty, for in no laboratory on Earth can so complete a vacuum be produced. Yet throughout large regions of space this "emptiness" consists of vast clouds of gas and tiny solid particles. Some of these tenuous clouds are visible, or partially so, in the form of **nebulae** (Latin for "clouds"). Other clouds can be detected by the energy they emit at infrared or radio wavelengths. Still other clouds make their existence known through their effect on the light that passes through them.

The conditions in this tenuous matter between the stars, which astronomers refer to as the *interstellar medium*, vary widely. There are dense clouds with temperatures as low as 10 K and low-density regions with temperatures of a million degrees. The interstellar medium is dynamic. Clouds form, collide, coalesce, and fragment to form stars. Knowledge of the origin of interstellar matter and of the physical processes that control its evolution is critical to understanding where and how stars form.

25.1 THE INTERSTELLAR MEDIUM

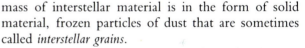

The primary components of the interstellar medium are gas and dust (Figure 25.1), and the gas is composed mainly of hydrogen and helium. About 1 percent by mass of interstellar material is in the form of solid material, frozen particles of dust that are sometimes called *interstellar grains*.

Interstellar material is concentrated between the stars in the spiral arms of our own and other galaxies. The density of the interstellar matter in the arms of our Galaxy in the neighborhood of the Sun, for example, is estimated to be 3 to 20 times that of the interarm regions. The gas and dust are not distributed smoothly throughout the spiral arms, however, but have a patchy, irregular distribution, being denser in some areas than in others, hence forming "clouds." In the spiral arms, on average, there is about one atom of gas per cubic centimeter in interstellar space. There are, in addition, a few hundred to a few thousand tiny particles or dust grains, each less than a micrometer (a thousandth of a millimeter) in diameter, per cubic kilometer. In some of the denser clouds, the density of gas and dust may exceed the average by as much as a thousand times or even more, but even this density is more nearly a vacuum than any attainable on Earth. In air, for contrast, the number of molecules per cubic centimeter at sea level is on the order of 10^{19}. Indeed, there is more gas in the Earth's atmosphere in a hypothetical vertical tube with a cross-section of 1 m^2 than would be encountered by extending that same tube from the top of the atmosphere all the way to the edge of the observable universe, which is 10 to 15 billion LY away.

FIGURE 25:1 Clouds of luminous gas and opaque dust, such as the nebulosity NGC 3603, are found between the stars. (Anglo-Australian Telescope Board)

FIGURE 25.2 NGC 3576 and NGC 3603. Clouds of luminous gas surround compact clusters of very hot stars. These hot stars ionize the hydrogen in the gas clouds. When electrons then recombine with protons and move back down to the lowest energy orbit, emission lines are produced. The strongest line in the visible region is in the red part of the spectrum and is responsible for the color of the photograph. (Anglo-Australian Telescope Board)

While the density of interstellar matter may be very low, its total mass is substantial. Stars occupy only a small fraction of the volume of the Milky Way Galaxy. For example, it takes only about 2 s for light to travel a distance equal to the radius of the Sun but more than 4 years to travel from the Sun to the nearest star. Even if the density of gas and dust surrounding the Sun and lying between it and the nearest stars is small, the volume of space filled by this low-density material is so large that the total mass of gas and dust in the Milky Way Galaxy is equal to about 5 percent of the mass contained in stars. The total mass of interstellar gas and dust in our Galaxy therefore amounts to several billion times the mass of the Sun.

25.2 INTERSTELLAR GAS

Some of the most spectacular astronomical photographs (Figure 25.2) show interstellar gas located near hot stars. This gas is heated to temperatures close to 10,000 K by the nearby stars and glows because it emits strong lines of hydrogen. Hydrogen makes up about three-quarters of the interstellar gas, and hydrogen and helium together compose 96 to 99 percent of it by mass.

Most interstellar gas, however, is located far from hot stars and is cold and nonluminous. This gas can be detected because it absorbs starlight passing through it. In the radio region of the spectrum, the cold gas produces the hydrogen line at 21 cm, and this line is one of the most important tools for tracing the distribution of interstellar gas. Most surprisingly, observations at ultraviolet and x-ray wavelengths show that some interstellar gas—even gas at large distances from stars—is very hot indeed. Temperatures of 10^6 K have been found, and one challenge for theorists is to explain what has heated this gas when it lies far from any sources of energy.

In this section, we describe the properties of the interstellar gas.

(a) H II Regions—Gas Near Hot Stars

Interstellar gas near very hot stars is ionized by the ultraviolet radiation from those stars. Since hydrogen is the main constituent of the gas, we often characterize a region of interstellar space according to whether its hydrogen is neutral, in which case we call it an **H I region,** or ionized. A region of ionized hydrogen is called an **H II region.**

All ultraviolet radiation of wavelength 91.2 nm (912 Å) or less can be absorbed by neutral hydrogen, and in the process the hydrogen is ionized (Section 7.4). An appreciable fraction of the energy emitted by the hottest stars lies at wavelengths shorter than 91.2 nm. If such a star is embedded in a cloud of interstellar gas, the ultraviolet radiation from that star ionizes the hydrogen in the gas, converting it into a plasma of positive hydrogen ions (protons) and free electrons. The detached protons in the gas are continually colliding with free electrons and capturing them, becoming neutral hydrogen again. As the electrons cascade down through the various energy levels of the hydrogen atoms on their way to the lowest energy levels or ground states, they emit light in the form of emission lines (Figure 25.2). Lines belonging to all the series of hydrogen (Section 7.4) are emitted—the Lyman series, Balmer series, Paschen series, and so on—but the lines of the Balmer series are most easily observed from the surface of the Earth because our atmosphere blocks the light from the Lyman and most other hydrogen series lines. Part of the ultraviolet light from the star is thus transformed into light in the Balmer emission lines of hydrogen. Color photographs of H II regions appear red because the strongest of the Balmer lines falls in the part of the electromagnetic spectrum that the eye sees as red.

When a proton in an H II region captures an electron, light is emitted as the electron cascades down to lower energy levels. The proton then loses that electron again almost immediately by the subsequent absorption of another ultraviolet photon from the star. Thus, al-though neutral hydrogen is responsible for absorbing and emitting light in H II regions, almost all the hydrogen, at any given time, is in the ionized state.

The interstellar gas, of course, contains other elements besides hydrogen. Many of them are also ionized in the vicinity of hot stars and are capturing electrons and emitting light, just as hydrogen does.

This process for producing clouds of glowing gas near hot stars is called **fluorescence.** The light emitted from regions of ionized gas consists largely of emission lines, so they are also called **emission nebulae.** Those emission nebulae in which the gas happens to be much denser than average (it occasionally reaches densities of 10^3 or 10^4 atoms per cubic centimeter—still an extremely high vacuum on Earth) are especially conspicuous. The best known example is the Orion nebula, which is barely visible to the unaided eye, but easily seen with binoculars, in the middle of the sword of the hunter (Figure 25.3).

Besides ionizing hydrogen and other elements, the radiation from hot stars also heats the gas in the surrounding nebula. When atoms absorb enough energy to become ionized, the electrons normally carry away some kinetic energy. When they subsequently collide with other particles in the H II region, the electrons share their energy, increase the velocities of those other particles, and so heat the nebula. The gas is cooled by the radiation that escapes from the nebula, with oxygen emission being a particularly important source of cooling. The balance of heating and cooling leads to a steady-state temperature of about 10^4 K for H II regions in our Galaxy.

FIGURE 25.3 The central "star" of the three forming the sword of Orion is in fact a group of four stars known as the Trapezium cluster. These stars are easily visible with binoculars and are the brightest members of a substantial cluster hidden by the dust in the nebula. These stars are at a distance of about 450 pc (Anglo-Australian Telescope Board, 1981)

(b)   Forbidden Radiation

It was explained in Section 7.4 that an atom or ion can be excited in either of two ways: by absorbing radiation or by collision with another particle. Atoms of singly ionized nitrogen and singly and doubly ionized oxygen all contain energy levels that correspond to low energies above their ground states. The ions are easily excited to these "low-energy" levels by collisions with free electrons in an H II region (most of these electrons have been freed from hydrogen atoms by ionization). Ordinarily, observed emission lines originate from atoms or ions that have remained excited for only a very brief period—on the order of a hundred millionth to a ten millionth of a second—before becoming de-excited by the emission of radiation. The levels to which oxygen and nitrogen ions are excited by collision, however, are said to be *metastable levels*. They are metastable because all of the routes by which the atom might spontaneously lose energy and return to the ground state are extremely unlikely. The situation is somewhat analogous to that of a mountain climber who has reached a summit and cannot find a way to get back down.

Since there are no likely ways in which interstellar oxygen and nitrogen can radiate energy, the ions will normally remain in the metastable levels for periods of hours before they emit energy in some improbable way and drop to their ground states. The lines produced by these improbable transitions to lower energy are not seen in the laboratory. Even the lowest densities achievable on Earth are much higher than those in interstellar space. In the laboratory, therefore, an ion in a metastable level is normally de-excited by a collision that transfers energy to another atom before it can spontaneously radiate energy and return to the ground state. Since the lines seen in interstellar space that involve metastable levels are not seen in the laboratory, they are called *forbidden lines*.

In the interstellar gas, however, the cards are stacked in favor of forbidden radiation. At the temperatures of H II regions, many of the free electrons have just the right kinetic energy to excite oxygen and nitrogen ions to their metastable levels. Although transitions from these levels are slow to occur, so many ions are excited to them at any time that many such "forbidden" transitions or de-excitations occur in an H II region. Since H II regions are transparent to visible light, the photons emitted through the entire depth of an H II region contribute to visible emission lines. Indeed, the forbidden radiation often constitutes half or more of the observable light from H II regions. The most important forbidden lines in the spectra of emission nebulae are two green lines (500.7 and 495.9 nm) due to doubly ionized oxygen, and these lines play an important role in cooling H II regions. Other important forbidden lines are two ultraviolet lines (near 372.7 nm) due to singly ionized oxygen, two red lines (658.4 and 654.8 nm) due to singly ionized nitrogen, and two ultraviolet lines (386.7 and 396.8 nm) due to doubly ionized neon.

When the green forbidden oxygen lines were first observed in the spectra of emission nebulae, their origin was a mystery. For a time, they were ascribed to an unknown element, *nebulium*, named for its apparent prevalence in gaseous nebulae. The correct explanation of the "nebulium lines" was provided by the American physicist I. S. Bowen in 1927.

(c)   Size and Brightness of H II Regions

If a cloud of gas surrounding a hot star is not very extensive, some of the ultraviolet radiation emitted by the star that is capable of ionizing hydrogen may leak out through the gas, and the apparent boundary of the emission nebula will be the actual edge of the gas cloud. Such a nebula is said to be *optically thin*, and the H II region is said to be *gas- or density-bounded*. Usually, however, an H II region is *optically thick*, which means that all of the star's ultraviolet radiation (of wavelengths less than 91.2 nm) is absorbed within the gas, and the H II region is said to be *radiation-bounded*. In this case, the boundary of the H II region is merely the limiting distance through the gas to which the star's ultraviolet radiation penetrates.

If the interstellar gas were distributed with absolute uniformity, and if it and the stars were all at rest, every emission nebula would be a spherical H II region exactly centered on a hot star. Because the distribution of the gas is patchy and irregular, and because the ionizing stars are often moving through it, actual H II regions are only approximately spherical, with irregular boundaries corresponding to the irregularities in the gas density. Further irregularities may result

when two or more stars are responsible for the radiation and their H II regions overlap.

The theory of H I and H II regions was worked out in detail by the Danish astronomer B. Strömgren in 1939. The more or less spherical emission regions are therefore sometimes called *Strömgren spheres*. The extent of an H II region depends on two things: (1) the ultraviolet luminosity of the central star, that is, how much energy it emits per second in wavelengths less than 91.2 nm; and (2) the density of the gas in the nebula. The higher the density, the more circumstellar hydrogen per unit volume there is to ionize, and the shorter is the distance through the gas that the ultraviolet energy can penetrate before it is completely absorbed. If the gas density is very low, and the star is very hot and luminous, the H II region can be very large. If the density is one atom per cubic centimeter, a main-sequence spectral-type O6 star can ionize a region more than 100 pc (about 300 LY) in diameter. Main-sequence stars of types B0 and A0 would produce in the same gas H II regions having diameters of 40 pc and 1 pc, respectively.

(d)   Temperatures of H I and H II Regions

When we speak of the temperature of a gas, we usually mean its *kinetic temperature*, which is a measure of the energy with which typical particles in the gas are moving. To understand the temperature of the interstellar medium, we must first consider the processes that heat the gas and then those that cool it. Finally, we calculate the equilibrium temperature that exists when these two processes—the heating and the cooling of the gas—exactly balance each other.

In H II regions the heating is principally by ionization of hydrogen, which is by far the most abundant element in the gas. The neutral hydrogen atom, in becoming ionized, can absorb *any* photon of wavelength shorter than 91.2 nm. The energy absorbed by the atom, in excess of that required for its ionization, is converted to kinetic energy of the freed electron. By collisions, the free electrons gradually share their energy with the other particles of an H II region. Excess energy absorbed in the process of ionization of hydrogen, therefore, is slowly converted into heat in the gas.

Electrons can lose energy by collisions with ions of oxygen and nitrogen, because these ions can be excited by such collisions and can eventually radiate the energy away in the form of "forbidden" emission lines. This mechanism, in other words, takes energy *from* the gas and *cools* it. Calculations show that the heating (by ionization) and cooling (by the emission of forbidden lines) should balance each other at an equilibrium temperature in the range of 7000 K to 20,000 K. If the temperature should drop much below 10,000 K, the rate of collisional excitations of oxygen and nitrogen would decrease (because the electrons would be moving more slowly) and the gas would heat up. If the temperature rose much above 20,000 K, the collisional excitations would become so numerous that the gas would cool rapidly. The heating and cooling mechanisms, therefore, act like a thermostat keeping the gas in the H II region at a relatively even temperature. Measurements indicate that most H II regions have temperatures between 8000 and 10,000 K—just in the range predicted by theoretical calculations.

The temperature of H I regions is also dictated by a balance of heating and cooling mechanisms, but the processes are more complicated, and they vary in effectiveness from place to place. Heating can occur by ionization of atoms of such heavier elements as carbon and silicon and by cosmic rays. Cooling mechanisms include collisional excitation of carbon and other atoms, and collisions between atoms and solid grains. Measures mostly derived from radio observations indicate that cold clouds of H I have temperatures that range from 20 K to 125 K.

(e) Neutral Hydrogen Clouds

While ionized hydrogen gas is the type of interstellar matter that is most often photographed, observations show that it is not the most abundant form of interstellar matter. The very hot stars required to produce H II regions are rare, and only a small fraction of interstellar matter is close enough to such hot stars to be ionized by them. To discover and study the interstellar matter that is spread throughout the Galaxy, astronomers had to devise new observational techniques to measure it.

Interstellar matter located at large distances from stars does not produce the strong emission lines that make H II regions visible. A cold cloud of gas will, however, produce dark absorption lines in the spectrum of light from a star that lies behind it. First evidence for absorption by interstellar clouds came from analysis of a spectroscopic binary star. While most of the lines in the spectrum shifted alternately from longer to shorter wavelengths and back again, as one would expect from the Doppler effect for one star in orbit around another, a few lines in the spectrum did not vary in wavelength. Subsequent work showed that these lines were formed in a cold cloud of gas located between us and the binary star.

There are other ways of knowing that the lines seen in the spectrum of a star do not originate in the star itself. Since the interstellar gas is cold, most of its atoms are neutral and in the state of lowest energy. The lines produced by these atoms are, therefore, generally not the same as the ones that are produced by the atoms of the hot gases in stellar photospheres. Moreover, the lines of the cold interstellar gas are very narrow, while those formed in stellar photospheres show the characteristic broadening associated with the spectra of hot gases at relatively high pressure (Section 22.4).

Interstellar lines have been found of most of those elements for which observable lines would be expected. The most conspicuous optical interstellar lines are produced by sodium and calcium. Lines of some of the common elements are also observed. Molecular bands produced by CN, CH, and CH^+ are seen. Ultraviolet observations made with orbiting telescopes have detected lines of carbon, hydrogen, oxygen, nitrogen, and other elements, of molecular hydrogen, and of CO (carbon monoxide).

The strengths of interstellar lines lead to estimates of the relative abundances of the elements that produce them. For some elements such estimates do not differ markedly from their relative abundances in the Sun and other stellar photospheres. For others the relative abundance is noticeably lower, especially for elements that most easily condense into solids at relatively high temperatures (notably aluminum, calcium, and titanium, as well as iron, silicon, and magnesium). As we shall see in Section 25.4, it is likely that these elements have indeed formed tiny solid grains of interstellar dust.

Sometimes the strength of an interstellar line seen in the spectrum of a star provides an indication of the distance to that star. A very strong interstellar line, for example, indicates that the starlight has traversed a considerable amount of interstellar gas. Because the density of gas in space is very low, the starlight would have to travel a long distance to encounter that much material; that is, the star would have to be very far away. Sometimes the interstellar lines are double or even multiple, which indicates that the starlight has traversed two or more gas clouds, moving with respect to each other, whose different radial velocities produce different Doppler shifts.

(f) Radio Observations of Cold Clouds: The 21-cm Line

Radio observations of the spectral line of hydrogen at a wavelength of 21.11 cm have provided critical information about the sizes and locations of cold interstellar gas. A hydrogen atom possesses a tiny amount of angular momentum by virtue of the axial spin of its electron and the electron's orbital motion about the nucleus (proton). In addition, the proton has an axial spin of its own. If the spins of the two particles are in opposite directions, the atom as a whole has a very slightly lower energy than if the two spins are aligned (Figure 25.4). If an atom in the lower energy state (spins opposed) acquires a small amount of energy, the spins of the proton and electron can be aligned, leaving the atom in a slightly *excited state*. If the atom then loses that same amount of energy again, it returns to its ground state. The amount of energy involved is that associated with a photon of 21-cm wavelength.

Neutral hydrogen atoms can be excited by collisions with electrons and other atoms. Such collisions are extremely rare in the sparse gases of interstellar space. An individual atom may wait many years before such an encounter aligns the spin of its proton and electron. Nevertheless, over many millions of years a good fraction of the hydrogen atoms are so excited. An excited atom can then lose its excess energy either by a subsequent collision or by radiating a photon of 21-cm wavelength. An excited atom will wait, on the average, about 10 million years before emitting a photon and returning to its state of lowest energy. Despite this long

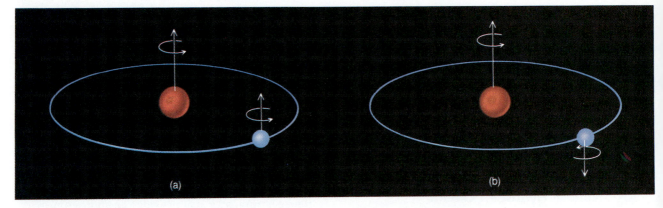

FIGURE 25.4 Formation of the 21-cm line. When the electron in a hydrogen atom is in the orbit closest to the nucleus, the proton and the electron may be spinning either in the same direction (*left*) or in opposite directions (*right*). When the electron flips over, the atom either gains or loses a tiny bit of energy and either absorbs or emits electromagnetic energy with a wavelength of 21 cm.

wait, there is a definite chance that the atom will radiate away its energy before a second collision can carry away its energy of excitation.

Equipment sensitive enough to detect the 21-cm line of neutral hydrogen became available in 1951, and since that time many other radio lines produced by both atoms and molecules have been discovered. Observations at 21 cm show that the neutral hydrogen in the Galaxy is confined to an extremely flat layer, most of it in a sheet less than 100 pc thick, extending throughout the plane of the Milky Way. Individual cold hydrogen clouds turn out to have temperatures of about 100 K and densities of 50 atoms per cubic centimeter. The diameters of cold clouds range from about 1 to 10 pc, and a light beam that travels 1000 pc through space in the plane of the Galaxy will encounter, on average, about six of these clouds. About 2 percent of interstellar space is filled with cold clouds. The masses of cold clouds are typically in the range 1 to 1000 times the mass of the Sun, with clouds of low mass being the most common.

Not all of interstellar hydrogen is found in such cold clouds. A comparable amount of mass is found in warm clouds with temperatures of 3000 to 6000 K and typical densities of 0.3 hydrogen atoms per cubic centimeter. The hydrogen in these clouds is not ionized. At least 20 percent of the space between stars in the plane of the Milky Way is filled with warm clouds of neutral hydrogen.

(g)   Hot Interstellar Gas

Before the launch of astronomical observatories into space, models of the interstellar medium assumed that most of the region between stars was filled with cool hydrogen. Astronomers were therefore surprised, when ultraviolet observations were made above the Earth's atmosphere, to discover interstellar lines at wavelengths of 103.2 and 103.8 nm. These lines are produced by oxygen atoms in the interstellar medium that have been ionized five times. To strip five electrons

from their orbits around an oxygen nucleus requires a lot of energy. In fact, these observations imply that the temperature of the interstellar medium where these atoms occur must be approximately 1 million degrees. The density of hydrogen nuclei in these regions is typically a few times 10^{-3} per cubic centimeter, with large variations in density from place to place.

25.3   A MODEL OF THE INTERSTELLAR GAS

(a)   Structure and Distribution of Interstellar Clouds

Table 25.1 summarizes the characteristics of the various types of clouds that populate interstellar space. There are cold, dense clouds in which hydrogen is not ionized. There are clouds so hot that molecules cannot survive and in which atoms are mainly ionized. The challenge for the theoretician is to assemble from the observations a model of the interstellar medium that tells us where we might expect to find the various types of clouds, what the structure of an individual cloud is like, and how the clouds change as time passes.

The observations described in this chapter offer several important clues. First, the interstellar material is distributed in a patchy way. There are regions where the interstellar material is concentrated into clouds and regions of very low density between the clouds. We can detect the individual gas clouds by looking at the interstellar absorption lines that appear in the spectra of distant stars. In many cases, we can see several absorption lines, each formed in a separate cloud moving with its own distinct radial velocity, superposed on the spectrum of a single distant star.

A second vital clue to the structure of the interstellar medium was the discovery that much of interstellar space is filled with low-density gas at a temperature of 10^6 K. These high temperatures are almost certainly produced by the explosive force of supernovae. Stars nearing the ends of their lives (Chapter 30) explode and send high-temperature gas, moving at velocities of thousands of kilometers per second, out into interstellar space. This high-speed gas will sweep away any low-density gas, compressing it in the process and so perhaps creating some of the raw material required for the formation of cold clouds.

Astronomers estimate that there is about one supernova explosion every 25 years somewhere in the Galaxy. On the average, the hot gas from a supernova will sweep through any given point in the Galaxy about once every 2 million years. At this rate, the sweeping action is continuous enough to keep most of the space between clouds filled with gas at a temperature of a million degrees.

The final constraint in developing a model of the structure of the interstellar medium is that the clouds and the gas between the clouds must be at approximately the same pressure. Suppose they were not. If the cloud pressure were higher, the cloud would expand until its pressure matched that of its environment. If, on the other hand, the pressure of the hot gas were greater than that of a cloud embedded in it, the hot gas would compress the cloud and force it to shrink until its pressure became high enough to resist further compression.

The pressure in a gas is proportional to the temperature T and to the number of particles per cubic meter in the gas n. Specifically, the pressure P of a gas can be calculated according to the formula

$$P = nkT,$$

where k is a constant (and is equal to 1.38×10^{-23} J deg^{-1}).

If the pressure of a gas cloud is to be equal to the pressure of the intercloud gas that surrounds it, then

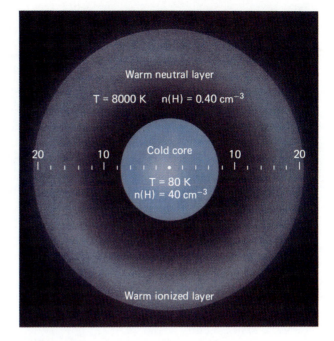

FIGURE 25.5   A typical interstellar cloud consists of a dense, cold core surrounded by a warm envelope. The horizontal scale shows the radius in light years.

$$\frac{n(\text{cloud})}{n(\text{intercloud})} = \frac{T(\text{intercloud})}{T(\text{cloud})}.$$

In words, this equation says that if the gas pressures in these two regions are equal, then the region at higher temperature must have fewer particles per cubic centimeter, that is, it must have a lower density.

Figures 25.5 and 25.6 show in a schematic way the most widely accepted model of what interstellar clouds look like based on the requirement that gas pressures must be nearly the same everywhere. Individual clouds are scattered at random throughout the Galaxy. The typical cloud may be a few tens of light years in diame-

TABLE 25.1	INTERSTELLAR GAS		
Type of Region	Temperature (K)	Density (number/cm³)	Description
H I: cold clouds	10^2	50	Hydrogen atoms; distributed in clouds with typical diameter of 1–10 pc; fills 2 percent of interstellar space
H I: warm clouds	3–6×10^3	0.3	Hydrogen not ionized; fills 20 percent of interstellar space
Hot gas	10^5–10^6	10^{-3}	Found well above and below as well as in galactic plane; hydrogen ionized; probably heated by supernovae explosions
H II regions	10^4	10^3–10^4	Found near hot stars; hydrogen mostly ionized

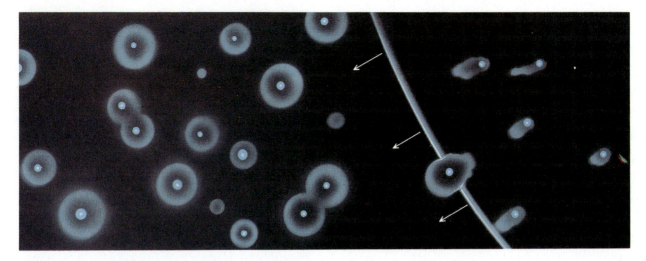

FIGURE 25.6 Interstellar clouds are embedded in hot low-density gas, which is heated to temperatures as high as 10^6 K by supernova explosions. In the upper right, a supernova remnant is shown sweeping through interstellar space. (Diagram is taken from work published in *The Astrophysical Journal* by C. McKee and J. Ostriker.)

ter. Since this cloud is embedded in gas with a temperature of a million degrees or so, the outer portions of the cloud are heated by conduction. The temperature of the outer portion of the cloud is typically about 8000 K. If the cloud is large enough, it can shield its innermost core from being heated, and the core may have a temperature that is 100 times lower and a density correspondingly 100 times higher. Typical values for the cloud core are a temperature of 80 K and a density of 40 hydrogen atoms per cubic centimeter.

The most controversial part of this particular model is the assumption that 70 to 80 percent of interstellar space is filled with gas at a temperature of 10^6 K. In fact, observations do not yet tell us just how much hot gas there really is, and some observers have estimated that perhaps only 20 percent of space contains hot gas. It is also not certain that individual clouds are approximately spherical. Interstellar gas may instead be distributed in sheets and long, thin filaments. Despite these uncertainties, the model presented here is the one most often used by astronomers in describing the overall characteristics of the interstellar gas.

(b)   Interstellar Matter Around the Sun

The Sun is located in a region where the density of interstellar matter is unusually low. The temperature of interstellar matter in the immediate vicinity of the Sun is about 10^6 K, and its density is only 5×10^{-3} hydrogen atoms per cubic centimeter. This region of low-density gas is called the Local Bubble and extends to a distance of about 100 pc from the Sun.

If interstellar space near the Sun contained the normal number of clouds, we would expect to have detected approximately 2000 of them within the Local Bubble. We have not. Clouds of the type shown in Figure 25.6 are conspicuously absent.

We do not know whether conditions within the Local Bubble are typical of large regions of the Galaxy or whether some mechanism blew away all high-density interstellar gas. One possibility is that a supernova explosion in the past 10^5 to 10^7 years swept the region we now see as the Local Bubble nearly clean of interstellar matter and at the same time heated the small amount of remaining gas to very high temperatures.

While typical cold hydrogen clouds are very rare within the Local Bubble, some clouds of lower density do exist. The Sun itself seems to be inside a cloud with a density of 0.1 hydrogen atoms cm^{-3} and a temperature of 10^4 K. This cloud is so tenuous that it is referred to as Local Fluff.

There is one sizable warm cloud in the direction toward the galactic center but within 20 pc of the Sun. It may be that the Local Fluff is the warm, partially ionized edge of a denser, cooler cloud (see Figure 25.5) and that the Sun is just entering this cloud.

(c)   Evolution of Interstellar Clouds

The model of interstellar gas described here presents a picture of the interstellar medium as it appears, on the average, at any given time. The individual clouds do, however, change with time. Clouds collide with one another, and such collisions may cause clouds to coalesce or, if the collision is a violent one, to fragment into many smaller clouds.

Schematically, we think that clouds are formed initially from the expanding gas around supernovae or hot stars. These clouds are relatively small and do not exceed 100 times the mass of the Sun. The clouds then grow through collisions with other clouds. Ultimately, this process may lead to the formation of giant clouds with diameters as large as 200 LY and masses that exceed 100,000 times the mass of the Sun. It is in these giant clouds that the most vigorous star formation occurs. The newly formed stars then evolve and become supernovae and in the process eject gaseous material, thus starting the cycle over again.

25.4 COSMIC DUST

Figure 25.7 shows a striking example of what is actually a common phenomenon—a dark region on the sky that appears nearly empty of stars. For a long time astronomers debated whether these dark regions were "tunnels" through which we looked beyond the stars of the Milky Way Galaxy into intergalactic space or, alternatively, were dark clouds that obscured the light of the stars beyond. We now know that the latter explanation is correct. Indeed, there are so many clues in support of this interpretation that, with perfect hindsight, it is difficult to understand why astronomers debated the issue for so long.

(a) Dark Nebulae

The obscuration of starlight illustrated in Figure 25.7 is produced by a relatively dense cloud or *dark nebula* of tiny solid grains, which are commonly called interstellar dust. Opaque clouds are conspicuous on any photograph of the Milky Way (Figures 24.6 and 32.1). The "dark rift," which runs lengthwise down a long part of the Milky Way and appears to split it in two, is produced by a collection of such obscuring clouds.

While dust clouds are invisible in the optical region of the spectrum, they glow brightly in the infrared. Small dust grains absorb optical and ultraviolet radiation very efficiently. The grains are heated by the absorbed radiation, typically to temperatures between 20 and about 500 K, and reradiate this heat at infrared wavelengths. We can use Wien's law (Section 7.2) to estimate in what part of the spectrum this radiation will fall. For a temperature of 100 K, this maximum is at about 30 μm (1 μm = 1 micrometer = 10^3 nm), while grains as cold as 20 K will radiate most strongly near 150 μm. The

FIGURE 25.7 A star cluster (NGC 6520) next to a dark cloud of interstellar matter. Old stars in our Galaxy are yellowish in color and form the brightest part of the Milky Way. Superimposed on these background stars in this picture is a dark cloud (Barnard 86), which is visible only because it blocks out the light from the stars beyond. Also in this picture, and possibly associated with the dark cloud, is a small cluster of young blue stars. (Anglo-Australian Observatory)

Earth's atmosphere is opaque to radiation at these wavelengths, and so emission by interstellar dust is best measured from space.

Observations from above the Earth's atmosphere by IRAS (the Infrared Astronomical Satellite) show that thermal emission from dust is seen throughout the plane of the Milky Way (Figure 25.8). The bright patches of emission have been given the name **infrared cirrus.** The closest infrared cirrus clouds are about 100 pc away. Dust is typically found in the same regions as hydrogen gas (Figure 25.1).

FIGURE 25.8 The galactic plane as viewed in the infrared region by IRAS. The color coding is such that the warmest dust appears blue and the coldest dust red. The galactic center is located in the bright region just to the right of center. The wispy emission extending above and below the plane is produced by infrared cirrus. (NASA)

(b) Reflection Nebulae

The tiny interstellar grains absorb only a portion of the starlight they intercept. At least half of the starlight that interacts with a grain is merely scattered—that is, it is redirected helter skelter in all directions (Figure 25.9). Since neither the absorbed nor the scattered starlight reaches us directly, both absorption and scattering make stars look dimmer. The effects of both processes are termed **interstellar extinction.**

Some dense clouds of dust contain luminous stars within them and scatter enough starlight to become visible. Such a cloud of dust, illuminated by starlight, is called a **reflection nebula.** One of the best known examples is the nebulosity around each of the brightest stars in the Pleiades cluster (Figure 25.10).

Blue light is scattered more than red by the dust. A reflection nebula, therefore, usually appears bluer than its illuminating star (Figure 25.11). A reflection nebula could be red only if the star that is the source of its light were very red. However, it takes a bright star to illuminate a dust cloud sufficiently for it to be visible to us, and in the regions of the Galaxy where dust clouds are found, the brightest stars are usually blue main-sequence stars. Most of the reflection nebulae, therefore, are very blue, since they are illuminated by blue stars.

Gas and dust are generally intermixed in space, although the proportions are not everywhere exactly the same. The presence of dust is apparent on many photographs of emission nebulae (Figure 25.1). Spectra of H II regions often reveal the faint continuous spectrum (with absorption lines) of the central star, whose light is reflected to us by the dust associated with the gas.

Both the emission component (due to the gas) and the reflection component (due to the dust) are brighter the hotter and more luminous the central star. Stars cooler than about 25,000 K have so little ultraviolet radiation of wavelengths shorter than 91.2 nm (that is, which can ionize hydrogen) that the reflection nebulae around such stars outshine the emission nebulae. Stars hotter than 25,000 K emit enough ultraviolet energy so that the emission nebulae produced around them generally outshine the reflection nebulae.

(c) Interstellar Reddening

It is fortunate that interstellar extinction affects blue light more than red. The fact that interstellar obscuration is *selective*, which means that light of short wavelengths is obscured more readily than that of long wavelengths, provides a means of estimating the *amount* by which starlight is dimmed.

Seventy years ago, astronomers were puzzled by the existence of stars whose spectral lines indicate that they are intrinsically hot and blue, of spectral-type B, although they actually appear as red as cool stars of spectral-type G. We know today that the light from these stars has been *reddened* by interstellar material. Most of their violet, blue, and green light has been obscured, but some of their orange and red light, of longer wavelengths, penetrates the obscuring dust and reaches Earth-based telescopes.

We have all seen an example of reddening. The setting Sun appears much redder than it does at noon. Sunlight is scattered by molecules in the Earth's atmosphere.

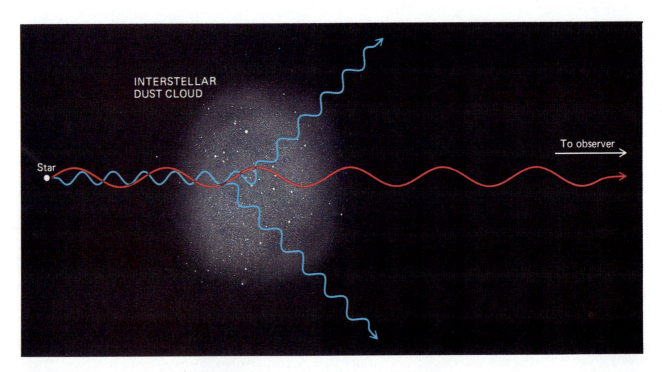

FIGURE 25.9 Interstellar dust scatters blue light more efficiently than red light, thereby making distant stars appear redder.

FIGURE 25.10 The Pleiades open star cluster. This cluster contains hundreds of stars and is located about 120 pc from the Sun. The nebulosity is starlight reflected by interstellar dust in a cloud that happens to be passing through the cluster at the present time. (Anglo-Australian Telescope Board)

When the Sun is low in the sky, its light must traverse a greater distance through the atmosphere than when the Sun is high in the sky. Over this greater distance there is a higher probability that sunlight will be scattered. Since red light is less likely to be scattered than blue light, the Sun appears more and more red as it approaches the horizon (Figure 25.12).

There is another practical consequence of the fact that light of short wavelength is absorbed more efficiently than light of long wavelength. It is much easier to get sunburned at noon than in the late afternoon, even though at 4:00 P.M., for example, the Sun feels nearly as hot as at noon. The reason for this is that tans and sunburns are caused primarily by sunlight with wavelengths between 280 and 320 nm. Sunlight at these short wavelengths is so efficiently absorbed by ozone in the Earth's atmosphere that very little of it penetrates the long distance that it must travel to reach the ground early in the morning and late in the afternoon. The heat that we feel, however, is produced mainly by infrared radiation, and these long wavelengths can reach the surface of the Earth even when the Sun is low in the sky.

The manner in which the dimming or extinction of starlight depends on wavelength can be evaluated by comparing, at various wavelengths, the relative brightnesses of two appropriate stars. Suppose there are two stars whose spectral lines show them to be approximately identical. Suppose that one is dimmed and reddened by interstellar dust, while the other lies in a direction in the sky that is free of interstellar obscuration. Suppose further that the two stars are at the same distance so that, in the absence of interstellar obscuration, they would appear to be the same brightness. Of course, the star that lies behind the dust cloud will actually appear to be fainter. Since blue light is dimmed more than red, this star will appear to be even fainter at short wavelengths than at long wavelengths. At a wavelength of 1000 nm, for example, the obscured star might appear to be 0.5 magnitude fainter than the unobscured star. At 500 nm it would then appear about 1 magnitude fainter, and at 330 nm it would be about 1.5 magnitudes fainter than the unobscured star.

From studies of many pairs of obscured and unobscured stars, the extinction of the interstellar material at various wavelengths has been determined. Over the visible spectral region, it turns out that the extinction, expressed in magnitudes, is roughly inversely proportional to wavelength, as in the example in the previous paragraph.

We can estimate the total amount by which a star is dimmed from the amount that it is reddened. The extinction of the light from a star increases its apparent color index (the redder the star, the greater the color index—see Section 22.2). The difference between the *observed* color index and the color index that the star *would have* in the absence of obscuration and reddening is called the **color excess.** The $B - V$ color excess, for example, is the amount by which the difference between the blue and visual magnitudes of a star is increased by reddening. In most directions in the Galaxy, the total absorption, in visual magnitudes, is found empirically to be about three times the $B - V$ color excess.

In addition to the overall trend of increasing extinction with decreasing wavelength, interstellar absorption is especially strong at wavelengths where there are spectral features characteristic of silicates (9.7 μm). A few clouds show absorption at 3.1 μm because of water ice. Ultraviolet observations from satellites show an absorption feature at 220 nm characteristic of small particles, which are probably made of carbon. These absorption features, therefore, provide critical clues to the composition of interstellar material.

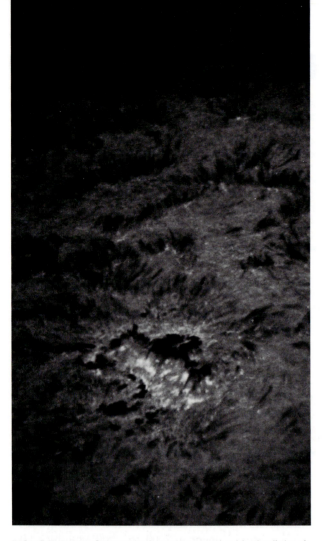

FIGURE 26.2 Solar spicules, photographed in the light of Hα. (National Solar Observatory/National Optical Astronomy Observatories)

they are viewed near the limb of the Sun, so many are seen in projection that they give the effect of a forest. They consist of gas jets moving upward at about 30 km/s and rising to heights of 5000 to 20,000 km above the photosphere. Individual spicules last only 10 minutes or so. Through the spicules matter continually flows into the corona. It is believed that the transition region may be wrapped around the spicules like a cloak.

Figure 26.3 shows the temperature structure of the solar atmosphere from the photosphere to the corona.

(e) The Corona

The chromosphere merges into the outermost part of the Sun's atmosphere, the **corona.** Like the chromosphere, the corona was first observed only during total eclipses (Figure 26.4), but unlike the chromosphere, the corona has been known for many centuries. It was referred to by Plutarch and was discussed in some detail by Kepler. The corona extends millions of miles above the photosphere and gradually thins to a sparse wind of ions and electrons flowing outward through the entire solar system. The corona emits half as much light as the full moon; its invisibility, under ordinary circumstances, is due to the overpowering brilliance of the photosphere. Like the chromosphere, the corona can now be photographed, with the coronagraph and other instruments, even when there is no eclipse.

The corona is also observed at radio wavelengths. In Great Britain in 1942, unexpected noise was picked up on radar receivers. It was subsequently learned that the source of this noise was the Sun. Not only do radio waves originate in the corona, but the corona produces

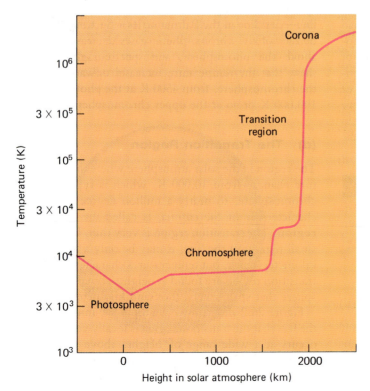

FIGURE 26.3 Temperatures in the solar atmosphere as a function of height above the photosphere. Note the very rapid increase in temperature over a very short distance in the transition region between the chromosphere and the corona.

FIGURE 26.4   An image of the Sun taken at the time of the solar eclipse on February 16, 1980. Since the light from the brilliant surface (photosphere) of the Sun is blocked by the Moon, it is possible to see the tenuous outer atmosphere of the Sun, which is called the corona. (High Altitude Observatory/NCAR)

Most of the coronal light seen in photographs is reflected sunlight. Spectra show that bright emission lines also are produced in the corona. In 1942 the Swedish physicist B. Edlen identified these lines as forbidden lines of calcium, iron, and nickel (see Section 25.2b for a discussion of forbidden radiation).

Analysis of the coronal spectral lines shows that the corona is very hot—millions of Kelvins. The atoms are all very highly ionized. For example, there are ultraviolet lines of iron ionized 16 times. The density, however, is very low. At the base of the corona there are about 10^9 atoms per cubic centimeter, compared with 10^{16} per cubic centimeter in the upper photosphere and 10^{19} per cubic centimeter at sea level in the Earth's atmosphere. Thus, despite the high temperature of the corona (a measure of how fast the particles are moving), its density is so low that the actual heat (energy content per cubic centimeter) is very low. In that near-vacuum it would take a long time for the hot coronal gases to warm up a cup of coffee (of course, the radiant energy from the nearby photosphere would do the job in a hurry).

Because of its high temperature, the solar corona is a source of x rays. Images of the Sun taken with an x-ray telescope launched on a rocket show that the corona has loops, plumes, and both bright and dark regions (Figure 26.5). Observations from Skylab in 1973 were the first to reveal that there are sometimes large regions of the corona that are relatively cool and quiet. These **coronal holes** are places of extremely low density and are usually (but not always) found in the polar regions of the Sun. They cause the empty spaces that can be seen on

variations in the signals from distant radio sources when they are observed through its outer part. The phenomenon is somewhat analogous to the twinkling of the light from stars caused by the Earth's atmosphere. Variations in the radiation of remote radio sources observed 90° away from the Sun in the sky show that the corona actually reaches out beyond the Earth. Coronal atoms are also regularly detected in the vicinity of the Earth by space vehicles.

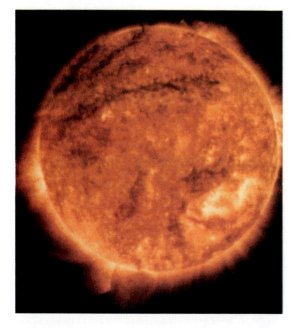

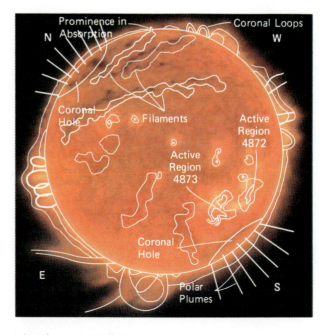

FIGURE 26.5   An x-ray/extreme ultraviolet image of the Sun taken from a sounding rocket 100 miles above White Sands Missile Range on October 23, 1987. The features shown in the photograph are identified in the drawing. (Art Walker/Stanford University and NASA)

some of the eclipse photographs of the solar corona (see also Figure 26.6).

The high temperature of the chromosphere and corona should seem surprising. The surface (photospheric) temperature of the Sun is only about 6000 K. How is it possible to heat the outer layers of the Sun's atmosphere to much higher temperatures? Observations indicate that magnetic fields play a major role. Magnetic fields apparently store energy and carry it to the chromosphere and corona, where it heats the gases of the Sun's outer atmosphere. The precise way in which magnetic energy is converted to heat energy is not yet understood. Explaining this process is one of the challenges facing solar astronomers.

(f) The Solar Wind

The **solar wind** is a stream of charged particles, mainly protons, flowing outward from the Sun at a speed of about 400 km/s. This solar wind exists because the gases in the corona are too hot to be confined by solar gravity. In just the same way, an atmosphere of light gas would quickly escape from the Moon.

In optical photographs, the solar corona appears to be fairly uniform and smooth. In x-ray and radio pictures, it is very patchy (Figures 26.5 and 26.6). Hot gas is present mainly where magnetic fields have trapped and concentrated it. The coronal holes lie between these concentrations of gas. The solar wind comes predominantly from these coronal holes, streaming through them into space, unhindered by magnetic fields.

The speed of the solar wind near the Earth's orbit averages about 400 km/s, and its density is usually two to ten ions per cubic centimeter. Both the speed and the density of the solar wind, however, are highly variable. Some of the atomic particles streaming away from the Sun have energies in the low-energy cosmic ray range. Thus the Sun can be an occasional source of weak cosmic rays. These, however, constitute only a tiny fraction of the total cosmic ray influx to the Earth.

At the surface of the Earth, we are protected from the solar wind by the atmosphere and by the Earth's magnetic field. The solar wind does, however, disrupt the ionized layers of gas in the ionosphere. This rain of particles is responsible for the aurora (Figure 26.7). As the particles in the solar wind strike atoms and molecules in the upper atmosphere, they excite them. When the electrons then rejoin the atoms and return to lower energy states, characteristic emission lines are produced. It is these emission lines that give rise to the aurora. The most spectacular auroras occur at altitudes of 75 to 150 km.

(g) Solar Rotation

By recording the apparent motions of the sunspots as the turning Sun carried them across its disk (Figure 26.8), Galileo first demonstrated that the Sun rotates on its axis. He found that the rotation period of the Sun is a little less than one month. The Sun's rotation rate can also be determined from the difference in the Doppler shifts of the light coming from the receding and approaching limbs (Section 22.4d).

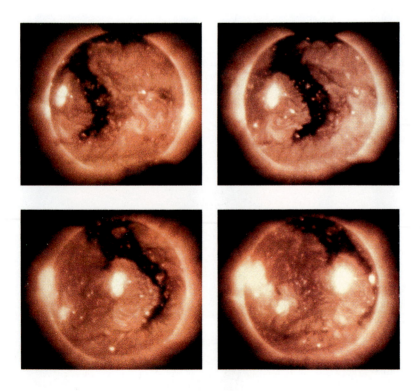

FIGURE 26.6 X-ray images of the Sun taken from the Skylab satellite show hot coronal gas. The corona is patchy, with bright spots indicating regions where hot gas is concentrated. The long, dark area where there is no x-ray emission is called a coronal hole. In these regions hot gas streams away from the solar surface out through the solar system. This stream of particles is called the solar wind. The four successive images of the Sun clearly show how the positions of solar features change as the Sun rotates. (Harvard College Observatory/NASA)

FIGURE 26.7 An auroral arc in the Earth's atmosphere, as photographed from space. The frequency of occurrence of the aurora depends on the level of solar activity. (NASA)

Measurements show that the rotation period of the Sun is about 25 days at the equator, 28 days at latitude 40°, and 36 days at latitude 80°, in the direction west to east (like the orbital motions of the planets). The Sun, being a gas, need not rotate like a solid body.

The apparent motions of sunspots are not usually straight lines across the Sun's disk but, rather, slight arcs, because the axis of rotation of the Sun is not exactly perpendicular to the plane of the ecliptic. The angle of inclination of the solar equator to the ecliptic is about 7°.

26.2 THE ACTIVE SUN

Overall, the Sun is quite stable, and we rely on that stability to maintain a life-sustaining environment on Earth. The surface of the Sun is not at all quiet, how- ever. It is a seething, bubbling cauldron of hot gas. Sunspots come and go. Gas is ejected into the chromosphere and corona. Occasionally, there are giant explosions on the Sun that have major effects on the Earth.

(a) Photospheric Granulation

Observations show that the photosphere has a mottled appearance resembling grains of rice spilled on a dark tablecloth. This structure of the photosphere is now generally called **granulation** (Figure 26.9). Typically, granules are 700 to 1000 km in diameter. They appear as bright areas surrounded by narrow darker regions.

The motions of the granules can be studied by the Doppler shifts in the spectra of gases just above them. It is found that the granules themselves are columns of hotter gases rising from below the photosphere. As the rising gas reaches the photosphere, it spreads out and sinks down again. The darker intergranular regions are the cooler gases sinking back. The centers of the granules are hotter than the intergranular regions by 50 to 100 K. The vertical motions of gases in the granules have speeds of 2 or 3 km/s. Individual granules persist for about 8 min. The granules, then, are the tops of convection currents of gases rising through the photosphere.

(b) Sunspots

The most conspicuous of the photospheric features are the **sunspots** (Figure 26.10). Occasionally, spots on the Sun are large enough to be visible to the naked eye. Galileo first showed that sunspots are actually on the surface of the Sun.

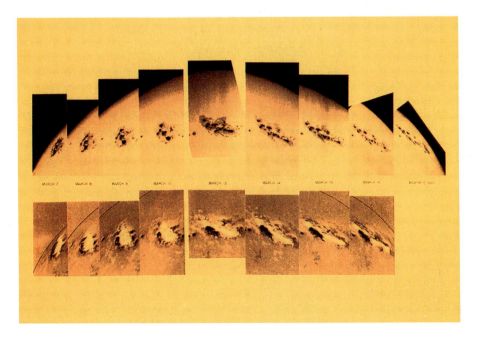

FIGURE 26.8 Photographs of the surface of the Sun showing a large group of sunspots. The series of exposures follows the rotation of sunspots across the visible hemisphere of the Sun. The top sequence shows the sunspots; the bottom sequence shows chromospheric emission. (National Solar Observatory/National Optical Astronomy Observatories)

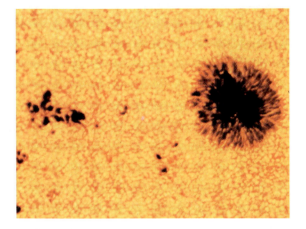

FIGURE 26.9   Solar granulation in the vicinity of sunspots. Each small, bright region is a rising column of hotter gas about 1000 km across. Cooler gas descends in the darker regions between granules. (National Solar Observatory/ National Optical Astronomy Observatories)

Sunspots are regions of the photosphere where the gases are up to 1500 K cooler than those of the surrounding photosphere. Sunspots are nevertheless hotter than the surfaces of many stars. If they could be removed from the Sun, they would be seen to shine brightly. They appear dark only by contrast with the hotter, brighter surrounding photosphere.

Individual sunspots have lifetimes that range from a few hours to a few months. If a spot lasts and develops, it is usually seen to consist of two parts: an inner darker core, the *umbra*, and a surrounding less dark region, the *penumbra*. Many spots become much larger than the Earth, and a few have reached diameters of 50,000 km. Frequently spots occur in groups of 2 to 20 or more. If a group contains many spots, it is likely to include two large ones, one approximately east of the other, with many smaller spots clustered around the two principal ones (Figure 26.10). The largest groups are very complex and may have over a hundred spots. Like storms on the Earth, sunspots may move slowly on the surface of the Sun, but their individual motions are slow when compared with the solar rotation, which carries them across the disk of the Sun.

(c)   The Sunspot Cycle

In 1851 Heinrich Schwabe, a German apothecary and amateur astronomer, published a paper in which he concluded that the number of sunspots visible, on the average, varied with a period of about ten years. Since Schwabe's work, the **sunspot cycle** has been clearly established. Although individual spots are short-lived, the total number of spots visible on the Sun at any one time is likely to be very much greater during certain periods, the periods of sunspot maximum, than at other times, the periods of sunspot minimum (Figure 26.11). Sunspot maxima have occurred at an average interval of 11.1 years, but the intervals between successive maxima have ranged from as little as 8 years (from 1830 to 1838)

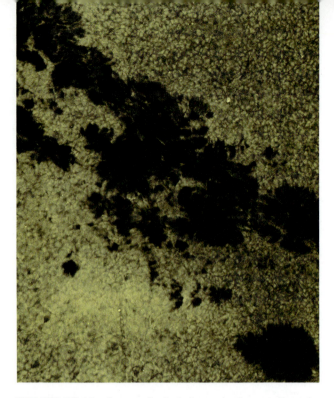

FIGURE 26.10   An excellent photograph of sunspots and solar granulation in the surrounding photosphere. (National Solar Observatory/National Optical Astronomy Observatories)

to as long as 16 years (from 1888 to 1904). During sunspot maxima, more than 100 spots can often be seen on the Sun at once. During sunspot minima, the Sun sometimes has no visible spots. Activity was near maximum in 1990 and 1991.

At the beginning of a cycle, just after a minimum, a few spots or groups of spots appear at latitudes of about 30° on the Sun. As the cycle progresses, the successive spots occur at lower and lower latitudes, until, at the maximum of the cycle, their average latitude is about 15°. Near minimum, the last few spots of a cycle appear at about 8° latitude. About the same time, the next cycle begins with a few spots occurring simultaneously at higher latitudes. Sunspots almost never appear at latitudes greater than 40° or less than 5°.

(d)   Magnetic Fields on the Sun

A spectral line is usually split up into several components in the presence of a magnetic field, the phenomenon known as the Zeeman effect (Section 22.4d). In 1908 the American astronomer George E. Hale observed the Zeeman effect in the spectrum of sunspots and found evidence for strong magnetic fields. The magnetic fields observed in sunspots range from 100 to nearly 4000 gauss. This field is over a thousand times greater than the Earth's magnetic field. The general solar magnetic field in regions where there are no sunspots is only a few gauss.

Whenever sunspots are observed in pairs or in groups containing two principal spots, one of the spots usually has the magnetic polarity of a north-seeking magnetic pole, and the other has the opposite polarity. Moreover, during a given cycle, the leading spots of pairs (or leading principal spots of groups) in the Northern Hemisphere all tend to have the same polarity, while those in the Southern Hemisphere all tend to have the opposite polarity. During the next sunspot cycle, however, the polarity of the leading spots is reversed in each hemisphere. For example, if during one cycle the leading spots in the Northern Hemisphere all had the polarity of a north-seeking pole, the leading spots in the Southern Hemisphere would have the polarity of a south-seeking pole. During the next cycle, the leading spots in the Northern Hemisphere would have south-seeking polarity, and those of the Southern Hemisphere would have north-seeking polarity. We see, therefore, that the sunspot cycle does not repeat itself with regard to magnetic polarity until two 11-year maxima have passed. The **magnetic cycle** of the Sun is therefore sometimes said to last 22 years, rather than 11.

Magnetic fields hold the key to explaining why sunspots are cooler and darker than the regions without strong magnetic fields. The forces produced by the magnetic field resist the motions of the bubbling columns of rising hot gases. Since these rising columns of hot gas carry most of the heat from inside the Sun to the surface, there is less heating where there are strong magnetic fields. As a result, darker, cooler sunspots appear in regions where magnetic forces are strong.

Magnetism is not strong enough in the photospheric layers to control the convective bubbling motions of the gas. In the chromosphere and corona, however, the gas density and pressure are enormously less. In these regions the magnetic fields are relatively strong and play an important role in influencing the motions of ionized gases. Far out in the corona, magnetic lines of force manifest themselves by organizing ionized coronal gases into streamers, which are easily seen and photographed during total solar eclipses. Low in the corona, magnetic fields guide the motions of ions in solar prominences (see Figure 26.14). The solar magnetic field extends into interplanetary space and is measured in the vicinity of the Earth and other planets with magnetometers carried on space probes. It can even be deduced from changes in the Earth's field.

(e) Plages

To see portions of the Sun that lie directly above the photosphere, we may observe in spectral regions where the photospheric gases are especially opaque—at the centers of strong absorption lines such as those of hydrogen and calcium. It is most common to isolate one of the absorption lines of ionized calcium (the K line) in the ultraviolet region or the Hα line of hydrogen in the red. These spectral lines appear dark when viewed against the rest of the solar spectrum, but they are not completely dark. What light remains in the centers of the lines is emitted from atoms in the chromosphere. There are special filters that transmit light only in narrow spectral regions, and now astronomers routinely observe the Sun through such monochromatic filters (Figure 26.12). Images taken in this way are called **spectroheliograms.**

Spectroheliograms of the Sun in the light of calcium and hydrogen show bright "clouds" in the chromosphere in the magnetic field regions around sunspots. These bright regions (formerly called flocculi—"tufts of wool") are known as **plages.** Calcium and hydrogen plages are also sometimes seen in regions where there are no visible sunspots, but these regions are generally those of higher than average magnetic fields.

The plages are not, of course, concentrations of calcium or hydrogen, but are regions where calcium and hydrogen happen to be emitting more light at the observed wavelengths. These elements are partially ionized throughout most of the visible chromosphere, and some of the atoms emit light as they capture electrons and become neutral (or less ionized) or as those atoms (or ions) cascade down through the various ex-

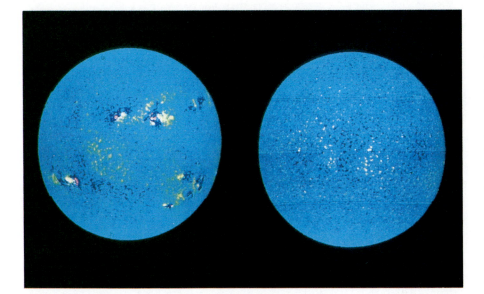

FIGURE 26.11   A comparison of the number of sunspots and magnetic activity on the active (*left-hand image*) and quiet Sun. The computer-generated images use yellow to indicate positive or north polarity, and dark blue for negative or south polarity. In the image of the active Sun, note that pairs of sunspots have opposite polarity. Note also that the polarity of the leading spot is different in the upper and lower hemispheres. At solar minimum (*right-hand image*), there are no large sunspots and the magnetic fields are weak. (National Solar Observatory/National Optical Astronomy Observatories)

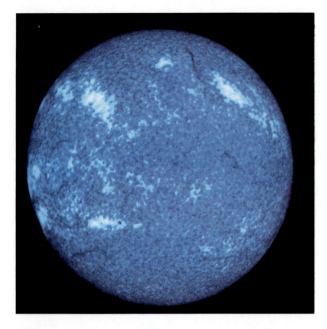

FIGURE 26.12   A spectroheliogram showing the Sun in the light of singly ionized calcium. The picture was taken on March 18, 1990. (National Solar Observatory/National Optical Astronomy Observatories)

cited energy levels. The plages, then, are regions where some of the atoms of the element observed are changing their states of ionization or excitation and are emitting more light than in the surrounding areas. Plages of hydrogen and calcium usually occur in approximately the same projected regions at the same time.

Plages sometimes emit light at many wavelengths and can be seen in the direct image of the Sun. These "white-light" plages are called *faculae* ("little torches") and were first described by Galileo's contemporary Christopher Scheiner. Faculae are seen best near the limb of the Sun, where the photosphere is not so bright and the contrast is more favorable for their visibility.

(f)   Prominences

Among the more spectacular of coronal phenomena are the **prominences.** They appear as red, flame-like protuberances rising high above the limb of the Sun. The features of some quiescent prominences may remain nearly stable for many hours, or even days, and may extend to heights of tens of thousands of kilometers above the solar surface (Figure 26.13). Others, the more active prominences, move upward or have loops or arches that surge slowly back and forth (Figure 26.14). The relatively rare eruptive prominences appear to send matter upward into the corona at speeds up to 700 km/s, and the most active surge prominences may move upward at speeds up to 1300 km/s. Some eruptive prominences have reached heights of over 1 million km above the photosphere. When seen silhouetted against the disk of the Sun, prominences have the appearance of irregular, dark filaments.

Superficially, prominences appear to be material ejected upward from the Sun, but motion pictures show that whereas a prominence may grow in size and rise higher and higher above the photosphere, the actual material in the prominence most often appears to move downward in graceful arcs, evidently along lines of magnetic force. Apparently, most prominences form from coronal material that cools and moves downward, even though the disturbance that characterizes the prominence may move upward.

Prominences are cool and dense regions in the corona where atoms and ions are capturing electrons and emitting light. Their origin is unknown, but it is significant that they usually originate near regions of sunspot activity and lie on the boundary between regions of opposite magnetic polarity. Quiescent prominences are supported by coronal magnetic fields, and eruptive prominences evidently result from sudden changes in the magnetic fields. Prominences seem to be further symptoms of the same general disturbances that produce spots and plages, that is, local magnetic fields.

(g)   Flares

The most awesome event on the surface of the Sun is a **solar flare.** A typical flare lasts for 5 to 10 min and releases a total amount of energy equivalent to that of about a million hydrogen bombs. The largest flares last for several hours and emit enough energy to power the entire U.S. at its current rate of electric consumption for 100,000 years. Near sunspot maximum, small flares occur several times per day, and major ones may occur every few weeks.

Flares are often observed in the red light of hydrogen (Figure 26.15), but the visible emission is only a tiny fraction of what happens when a solar flare explodes. At the moment of the explosion, the matter associated with the flare is heated to temperatures as high as 10^7 K. At such high temperatures, a flood of x-ray and extreme ultraviolet radiation is emitted.

There is evidence that flares occur when magnetic fields pointing in opposite directions interact and destroy each other, releasing energy. The gases at the solar surface are in constant motion, and this motion frequently brings oppositely directed fields together. If the magnetic interactions cover a large volume in the solar corona, then immense quantities of coronal material—mainly protons and electrons—may be ejected at high speeds (500–1000 km/s) into interplanetary space; such *coronal mass ejections* can affect the Earth in several ways.

The most obvious effect of coronal mass ejections on the Earth is the appearance of the aurora borealis. The high-speed particles interact with the Earth's magnetic field, weakening it so that Earth's atmosphere is no longer shielded from the solar wind. Auroras occur preferentially near the magnetic poles of the Earth because the charged particles from the Sun tend to flow down into the Earth's atmosphere along the magnetic

Figure 26.13 A nonerupting prominence. These gases reach as much as 50,000 km above the photosphere. (National Solar Observatory/National Optical Astronomy Observatories)

field and to penetrate to lowest altitudes near the poles. Only unusually bright auroras are seen in the southern U.S.

The changes in the Earth's magnetic field caused by interactions with the charged particles in the coronal mass ejections in turn generate changing electric currents. The effects are most noticeable in long power lines and can even cause components to burn out in power stations. As a result of the coronal mass ejection and flare in March 1989, parts of Montreal and Quebec Province were without power for up to 9 hr. Other effects occurred as well. For example, people found their automatic garage doors opening and closing for no apparent reason.

The ultraviolet and x-ray emission from flares can affect the ability of the atmosphere (specifically the ionosphere) to reflect radio waves and can disrupt short-wave radio transmissions. The March 1989 event affected shortwave radio communications for 24 hr.

The short-wavelength radiation produced during so-lar flares heats the outer atmosphere of the Earth. In 1981, a very large solar flare occurred while the space

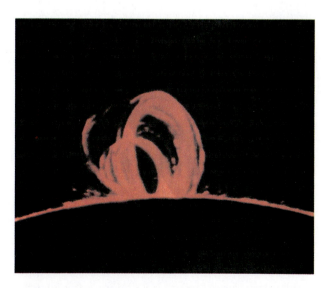

FIGURE 26.14 A loop prominence. The distinctive shape of the prominence results from strong magnetic fields in the region bending the hot plasma into a loop. (National Solar Observatory/National Optical Astronomy Observatories)

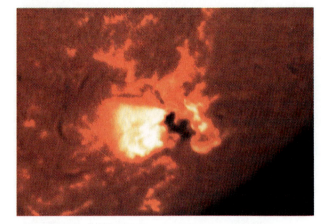

FIGURE 26.15 A giant solar flare is shown at the upper left edge of the Sun. This flare occurred on March 6, 1989, and produced a brilliant aurora that was seen as far south as Tucson, Arizona. (National Solar Observatory/National Optical Astronomy Observatories)

shuttle Columbia was in orbit. The astronauts aboard made measurements that showed that the flare, which lasted for 3 hr, increased the temperature of the atmosphere at an altitude of 260 km from its normal value of 1200 K to 2200 K. When the outer atmosphere is heated, it also expands. As a consequence, friction between the atmosphere and spacecraft increases and drags satellites to lower altitude. At the time of the flare in March 1989, the system responsible for tracking some 19,000 objects orbiting the Earth temporarily lost track of 11,000 of them because their orbits were changed by the expansion of the Earth's atmosphere. During solar maximum, a great many satellites are brought to such a low altitude that they are destroyed by friction with the atmosphere.

The level of solar activity is a critical factor in calculating the lifetimes and orbits of the shuttle and satellites in near-Earth orbit. Obviously, it would be extremely valuable to be able to predict both the overall level of solar activity and the occurrence of individual flares. Solar astronomers are working very hard to try to learn how to make reliable predictions, but accurate forecasts of solar "weather" are proving to be as elusive a goal as reliable forecasts of the weather on Earth.

Sunspots, flares, and bright regions in the chromosphere and corona tend to occur together on the Sun. That is, they all tend to have similar longitudes and latitudes but, of course, to be located at different heights in the atmosphere. For example, the flare in March 1989 occurred in a region where there was a large and long-lived group of sunspots. A place on the Sun where these phenomena are seen is called an **active region.**

Active regions possess strong magnetic fields. Plages, prominences, and solar flares all occur more frequently at times of sunspot maximum, when magnetic fields are strong, and all vary together in the semiregular activity cycle of 22 years.

The solar cycle is thus closely related to magnetism in the Sun, and it is the changing magnetic field of the Sun that provides the driving force for many aspects of solar activity. Theoretical models indicate that rotation and convection just below the solar surface can distort magnetic fields, causing them to grow and then decay, being regenerated with opposite polarity approximately every 11 years. As the fields become stronger, they float to the surface in the form of loops. When the loop penetrates the surface it creates an active region. Where the loop emerges, it has one polarity; where it reenters, it has the opposite polarity. Remember that the leading and trailing sunspots in an active region have opposite polarity.

26.3 IS THE SUN A VARIABLE STAR?

The Sun is one of the few truly constant objects in our daily lives. It rises faithfully at a time that can be precisely calculated. Each day it deposits energy on the Earth, warming it and sustaining life. But we know that the Sun varies on time scales ranging from minutes for flares to 22 years for solar activity. Does the Sun vary on longer time scales? And do any of these variations affect the Earth's climate?

Variations in the total amount of energy emitted by the Sun, if any do exist, must be subtle. The existence of life on Earth demonstrates that there have been no major recent changes in the climate of the Earth. We do know, however, that there are ice ages, separated by relatively warm interglacial periods. Do these changes have anything to do with the Sun?

(a) Variations in the Number of Sunspots

The most obvious variation of the Sun is the number of sunspots. There is considerable evidence that the number of sunspots was much lower from 1645 to 1715 than it is now. This interval of low activity was first noted by Gustav Spörer in 1887 and by E.W. Maunder in 1890 and is now called the **Maunder Minimum.** The incidence of sunspots over the past four centuries is shown in Figure 26.16. Sunspot numbers were also somewhat lower than they are now during the first part of the 19th century, and this period is called the Little Maunder Minimum.

When the number of sunspots is high, the Sun is active in a number of other ways (Figure 26.17), and this activity affects the Earth directly. For example, auroras (see Figure 26.7) are caused by the impact of charged particles from the Sun on the Earth's magnetosphere. Energetic charged particles are much more likely to be ejected by the Sun when the Sun is active and when the sunspot number is high. There is a strong correlation between sunspot number and the frequency of auroral

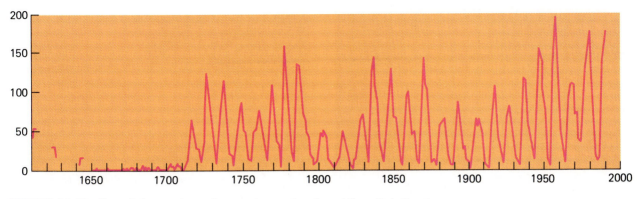

FIGURE 26.16 The relative numbers of sunspots as a function of time. Note the absence of sunspots from 1645 to 1715.

displays. Historical accounts indicate that auroral activity was low during the Maunder Minimum.

The best quantitative evidence of variations in the level of solar activity comes from studies of the radioactive isotope carbon-14. The Earth is constantly bombarded by cosmic rays, which are high-energy charged particles, including protons and nuclei of heavier elements. The rate at which cosmic rays from sources outside the solar system reach the upper atmosphere depends on the level of solar activity. When the Sun is active, charged particles streaming away from the Sun out into the solar system carry the Sun's strong magnetic field with them. This magnetic field shields the Earth from incoming cosmic rays. At times of low activity, when the Sun's magnetic field is weak, cosmic rays reach the Earth in larger numbers.

When the energetic cosmic ray particles impact the upper atmosphere, they produce several different radioactive isotopes. One such isotope is carbon-14, which is produced when nitrogen is struck by high-energy cosmic rays. The rate of production of carbon-14 is higher when the activity of the Sun is lower and the solar magnetic field does not shield the Earth from bombardment by cosmic rays.

Some of the radioactive carbon is contained in carbon dioxide molecules, which are ultimately incorporated into trees through photosynthesis. By measuring the amount of radioactive carbon in tree rings, we can estimate the historical levels of solar activity. Correlations with visual estimates of sunspot numbers over the past 300 years indicate that the carbon-14 estimates of solar activity are indeed valid. Because it takes about 10 years, on the average, for a carbon dioxide molecule to be absorbed from the atmosphere or ocean into plants, this technique cannot provide data on the 11-year solar cycle. It can be used, however, to look for long-term (over several decades) changes in the level of solar activity.

Estimates of the amount of carbon-14 in tree rings now extend continuously back about 8000 years into the past. Variations in solar activity levels have occurred throughout this period, and the Sun has been at times both more and less active than it is now. The measurements confirm that the amount of carbon-14 was unusually high, and solar activity correspondingly low, during both the Maunder Minimum and the Little Maunder Minimum. During the past thousand years, activity was also low in the years 1410 to 1530 and 1280 to 1340. Between about 1100 and 1250, the level of solar activity may have been even higher than it is now.

(b)   Solar Variability and the Earth's Climate

Variations in the overall level of the Sun's activity seem to be well established. Do these variations have any direct impact on the Earth or its climate? It has long been known that the period of the Maunder Minimum was a time of exceptionally low temperatures in Europe—so low that this period is described as a Little Ice Age. The global climate also appears to have been unusually cool from 1400 to 1510, another period of low solar activity.

The relationship between solar luminosity and activity level was determined only recently from measurements made by a satellite orbiting the Earth. The changes in solar luminosity are too small to be measured reliably from the ground because of uncertainties in estimating how much of the Sun's energy is transmitted by the Earth's atmosphere. Precise measurements from space show that the luminosity of the Sun varies as a result of its rotation by about 0.1 percent. The size of this short-term variation in luminosity depends on what fraction of the surface is covered by sunspots. The Sun is fainter when more sunspots are present, as one might expect given that sunspots are dark. However, a change of 0.1 percent in the energy received from the Sun is probably not enough to affect the Earth's climate.

Surprisingly, measurements over the entire solar cycle, rather than just over one rotation period, indicate that we receive the maximum amount of radiation from the Sun at times of solar maximum. The source of the extra radiation at sunspot maximum is not known. Nevertheless, the total energy emitted from the Sun is greater when it is more active.

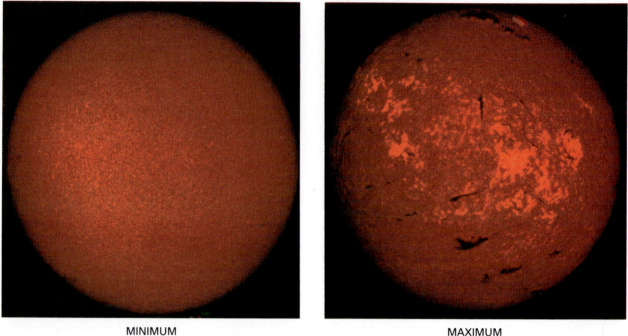

MINIMUM   MAXIMUM

FIGURE 26.17 These images show the appearance of the Sun in the red light of hydrogen at the times of a minimum (*left*) and a maximum (*right*) in solar activity. Hydrogen emission, the number of plages and prominences, and the number of flares all vary in the same cycle as the number of sunspots. (National Solar Observatory/National Optical Astronomy Observatories)

These observations appear to support the idea that the Maunder Minimum was indeed associated with the Little Ice Age. The unusually cold temperatures at that time imply a drop in the solar luminosity of about half a percent. It seems possible that such a drop in luminosity might have accompanied a long period of reduced solar activity.

We can get additional clues about whether large variations in activity are likely to occur over several centuries by looking at activity cycles in stars that are similar to the Sun. It turns out that the Sun's variation (which amounts to just 0.1 percent of the total energy emitted) is unusually small. Most stars vary by 0.3 percent, and some by as much as 1 percent. Therefore, it seems reasonable to assume that the energy output of the Sun at the time of the Maunder Minimum might indeed have been low enough to cause unusually cold temperatures on Earth. The observations of higher variability in the energy output of most other solar types stars also suggest that the stability of solar behavior over the past three centuries—and the stability of the Earth's climate—may be unusual.

In addition to variations in the Sun's luminosity, a wide variety of other phenomena affect the global climate (Chapter 12), including variations in the shape of the Earth's orbit, the amount of carbon dioxide in the atmosphere, and changes in the transparency of the atmosphere because of injections of dust by volcanic explosions. Because of the complex circulation patterns of the Earth's atmosphere, local effects may differ from global effects. There can also be strong variations from one year to the next in the severity of either summers or winters that may mask long-term trends. Nevertheless, it appears likely that changes in the Sun do affect the climate of the Earth on time scales of several hundred years.

S U M M A R Y

26.1 The outer layers of the solar atmosphere are, in order of increasing distance from the center of the Sun, the **photosphere,** with temperatures in the range of 4500 K to 6800 K; the **chromosphere,** with temperatures of 10^4 to 10^5 K; the **transition region,** a zone that may be only a few kilometers thick, where the temperature increases rapidly from 10^4 K to 10^6 K; and the **corona,** with temperatures of a few million degrees Kelvin. **Solar wind** particles stream out into the solar system through **coronal holes.** Hydrogen and helium together make up 98 percent of the mass of the Sun. The Sun rotates more rapidly at its equator, where the rotation period is about 25 days, than near the poles, where the period is slightly greater than 36 days.

26.2 The Sun's surface is mottled with upwelling currents of hot, bright **granules. Sunspots** are dark regions where the temperature is up to 1500 K cooler than in the surrounding photosphere. The number of visible sunspots varies on an 11-year time scale. Spots frequently occur in pairs. During a given 11-year cycle, all leading spots in the northern hemisphere have the same magnetic polarity; all leading spots in the southern hemisphere have the opposite polarity. In the subsequent 11-year cycle, the polarity reverses. For this reason the **magnetic cycle** of the Sun is often said to last for 22 years. Sunspots, **solar flares,** and bright regions, including **plages,**

tend to occur in **active regions**—that is, in places on the Sun with the same latitude and longitude but at different heights in the atmosphere.

26.3 Evidence exists that there are long-term (100 years or more) variations in the level of solar activity and in the number of sunspots. It appears that there is a tendency for the Earth to be cooler when the number of sunspots is unusually low for several decades.

E X E R C I S E S

THOUGHT QUESTIONS

1. The astronomer William Herschel (1738–1822) proposed that the Sun has a cool interior and is inhabited. Give at least three good arguments against this idea.

2. How might you convince an ignorant friend that the Sun is not hollow?

3. What are the main differences between the composition of the Earth and the composition of the Sun?

4. Suppose you were to take two photographs of the Sun, one in light at a wavelength centered on a strong absorption line, and the other at a wavelength region in the continuum away from strong lines. In which photograph would you be observing deeper, hotter layers? Why?

5. If the rotation period of the Sun is determined by observing the apparent motions of sunspots, must any correction be made for the orbital motion of the Earth? If so, explain what the correction is and how it arises. If not, explain why the Earth's orbital revolution does not affect the observations.

6. Suppose an (extremely hypothetical) elongated sunspot formed that extended from a latitude of 30° to a latitude of 40° along a fixed line of longitude. How would the appearance of that sunspot change as the Sun rotates?

7. If the corona, which is outside the photosphere, has a temperature of 1 million K, why do we measure a temperature of 6000 K for the surface of the Sun?

8. Why is it difficult to determine whether or not small changes in the amount of energy radiated by the Sun have an effect on the Earth's climate?

PROBLEMS

9. **(a)** What is the distance to the Sun in parsecs? **(b)** What is the distance modulus of the Sun? **(c)** How many times farther away would the Sun be if it were removed to a distance of 10 pc? **(d)** How many times fainter would the Sun appear at 10 pc?

10. Use Kepler's third law in the form $(M_1 + M_2)P^2 = D^3$ to calculate the ratio of the mass of the Sun plus the mass of the Earth to the mass of the Earth plus the mass of the Moon. Show that this ratio is nearly the same as the ratio of the mass of the Sun to the mass of the Earth.

11. Use the data in Table 26.1 to confirm the result that the density of the Sun is 1.4 g/cm³. What kinds of materials have similar densities? One such material is ice. How do you know that the Sun is not made of ice?

12. From the data in Section 26.1, find how long it takes solar wind particles, on the average, to reach the Earth from the Sun.

13. Suppose an eruptive prominence rises at 150 km/s. If it did not change speed, how far from the photosphere would it extend in 3 hr?

14. Would the material in the prominence in Problem 13 escape the Sun? Why? (See Section 4.1.)

15. From the Doppler shifts of the spectral lines in the light coming from the east and west edges of the Sun, it is found that the radial velocities of the two edges differ by about 4 km/s. Find the approximate period of rotation of the Sun.

The McMath-Pierce Solar Telescope is the largest telescope in the world dedicated to observing the Sun. The building is white but appears red in this photograph because of high-altitude haze following the 1991 eruption of Mount Pinatubo in the Philippines. This is a 9-hour exposure, and the three equally spaced star trails on the right mark the belt of Orion. Left of these is a band of star trails formed by red Betelgeuse, blue Bellatrix, and white Procyon.

(W. Livingston/National Solar Observatory)

In this chapter, we examine the energy source and internal structure of the Sun, not just because the Sun is our star, but because it can serve as a prototype for what happens inside stars in general. Our understanding of the interior of the Sun, like so many other astronomical discoveries, did not come about from the work of astronomers alone. In science, discoveries in one field of research often set the stage for major advances in another.

During the 1930s, physicists discovered the subatomic particles called the positron and the neutron, and also measured the rates at which nuclei of atoms interact with one another. In April 1938, Hans Bethe, a young scientist who was familiar with these developments in nuclear physics, attended a conference that discussed the problem of trying to determine the source of energy for the Sun and stars. Within six months, Bethe had written a paper that showed that the energy production of stars is due "entirely to the combination of four protons and two electrons into an alpha particle [that is, into a helium nucleus]." He identified the specific interactions of atomic nuclei that resulted in the formation of helium, and he showed that the temperature required for these interactions to occur was in just the range that models of the interiors of stars predicted. Bethe's paper is one of the most important ever written in astrophysics, and laid the basis for quantitative studies of how stars change as they age.

THE SUN: A NUCLEAR POWERHOUSE

Sir James Chadwick (1891–1974), British physicist. In addition to his knighthood in 1945, he received the Nobel Prize in physics for his basic contributions to our understanding of the atomic nucleus. He proved the existence of the neutron in 1932 by bombarding beryllium nuclei with alpha particles (helium nuclei) and also worked on the generation of chain reactions and nuclear fission. (American Institute of Physics)

What makes the Sun shine? The power output of the Sun has been, according to geological evidence, not very different since the formation of the Earth billions of years ago. Moreover, the amount of energy that the Sun has poured forth over these billions of years is enormous. The rate at which the Sun emits electromagnetic radiation into space, and thus the rate at which energy must be generated within it, is about 4×10^{26} watts. The challenge for scientists was to find what source of power can provide the gigantic amounts of energy required to keep stars like the Sun shining for so long.

27.1 THERMAL AND GRAVITATIONAL ENERGY

Two large stores of energy in a star are its internal heat, or thermal energy, and its gravitational energy. The heat stored in a gas is simply the energy of motion (kinetic energy) of the particles that compose it. If the speeds of these particles decrease, the loss in kinetic energy is radiated away as heat and light. This is how a hot iron cools after it has been unplugged (except that the atoms in a solid vibrate within a crystalline structure, rather than moving freely, as in a gas).

(a) Conservation of Energy

The source of heat energy that is most familiar to us here on Earth is burning (the chemical term is "oxida-tion") of wood, coal, gasoline, or other fuel. However, even if the immense mass of the Sun consisted of a burnable material like coal or wood, it could not produce energy at its present rate for more than a few thousand years. Geologists have found fossils in rocks that are 3.5 billion years old. We know, therefore, that the Sun must have been heating the Earth to nearly its current temperature for at least that long.

In the 19th century, scientists used the *law of conservation of energy* to look for a source of energy for the Sun. The law of conservation of energy simply says that energy cannot be created or destroyed, but it can be transformed. The steam engine, which was the key to industrial development during the 19th century, relies on the transformation of heat energy to mechanical energy. The steam from a boiler drives the motion of a piston.

The reverse is also true. Mechanical motion can be transformed into heat. If you clap your hands vigorously, your palms will become hotter. If you rub ice on the surface of a table, the heat produced by friction will melt the ice. In the 19th century, scientists considered the possibility that the motion of meteorites falling into the Sun might provide an adequate source of heat. Calculations show, however, that to produce the total amount of energy—heat and light—emitted by the Sun, the mass in meteorites that would have to fall into it every 100 years would equal the mass of the Earth. The increase in the mass of the Sun would, according to

Kepler's third law, change the period of the Earth's orbit by 2 s per year. Such a change would be easily measurable and has not been detected.

(b) Gravitational Contraction As a Source of Energy

As an alternative, the German scientist Hermann von Helmholtz and the British physicist Lord Kelvin in about the middle of the 19th century proposed that the outer layers of the Sun might "fall" inward, and thereby produce heat energy. The outer layer of the Sun is a gas made up of individual atoms, all moving about in random directions. Temperature is simply a measure of the speed of their motion. Now imagine that this outer layer starts to fall inward. The atoms acquire an additional velocity because of this falling motion. As the outer layer falls inward, it also contracts and the atoms move closer together. Collisions become more likely. Some collisions serve to transfer the velocity associated with the falling motion to other atoms, and increase their velocities, and so increase the temperature of the Sun. Other collisions may actually excite electrons within the atoms to higher energy orbits. When these electrons return to their normal orbits, they emit photons, which can then escape from the Sun as heat or light.

Kelvin and Helmholtz calculated that a contraction of the Sun at a rate of only about 40 m per year would be enough to provide for its total energy output. Over the time span of human history, the decrease in the Sun's size from such a slow contraction would be undetectable. The amount of energy that has been released since the presolar cloud began to contract is of the order of 10^{42} joules. This is the amount, according to the Helmholtz and Kelvin theory, that the Sun could have converted to thermal energy and luminosity. Since the present luminosity of the Sun is 4×10^{26} joules/s, or about 10^{34} joules per year, its contraction could have kept it shining at its present rate for a period of the order of 100 million years.

In the 19th century, this length of time seemed adequate. But in the 20th century geologists have shown that the Earth (and hence the Sun) has an age of several billion years. Contraction of the Sun therefore cannot account for the luminosity it has generated over its lifetime.

Even as geologists were ruling out one hypothesis about the source of the Sun's energy, physicists were developing a new one. The key lies in the nucleus of the atom and in Einstein's theory of special relativity.

27.2 MASS, ENERGY, AND THE THEORY OF SPECIAL RELATIVITY

According to the law of conservation of energy, energy cannot be created or destroyed but only converted from one form to another. One of the remarkable results of Einstein's theory of special relativity is the recognition that mass and energy are equivalent (Section 8.3) and are related by the famous equation:

$$E = mc^2.$$

Remember that in this equation, E is the symbol for energy, m is the symbol for mass, and c, the constant that relates the two in a precise mathematical way, is the speed of light. What this equation says is that mass can be converted to energy, and energy can be converted to mass. Because c^2, the speed of light squared, is a very large quantity, the conversion of even a small amount of mass results in a very great amount of energy.

We can apply this idea to the Sun. If we can find a set of interactions of atoms that lead to the destruction of some of the Sun's most abundant element (hydrogen) and the conversion of that lost mass into energy, then we will have identified a source of energy for the Sun that can last for billions of years. With Einstein's equation $E = mc^2$, it is possible to calculate that the amount of energy radiated by the Sun could be produced by the complete conversion of about 4 million tons of matter to energy each second. This sounds like a lot of matter, but in fact the Sun contains enough mass to continue shining at its present rate for about 10 billion years before it exhausts its supply of fuel.

To understand how the conversion of mass to energy actually occurs, it is necessary to explore the structure of the atom.

(a) Elementary Particles

The fundamental components of matter are called **elementary particles.** The most familiar of the elementary particles are the **proton, neutron,** and **electron,** which are the constituent particles of ordinary atoms (Section 7.3).

We have learned in the 20th century that protons, neutrons, and electrons are by no means all the particles that exist. First, for each kind of particle, there is a corresponding **antiparticle.** If the particle carries a charge, its anti has the opposite charge. The antielectron is the **positron,** of the same mass as the electron but positively charged. The antiproton has a negative charge. The antineutron, like the neutron, has no charge, but interacts with other matter opposite to the way the neutron does.

Whole atoms of positrons, antiprotons, and antineutrons could exist. They constitute what is called *antimatter.* Such atoms do not exist on Earth because when a particle comes in contact with its antiparticle, the two annihilate each other, turning into energy. Antimatter in our world of ordinary matter, therefore, is highly unstable (in large doses, it would be mighty dangerous!), but individual antiparticles are found in cosmic rays and can be formed in the laboratory.

The existence of another type of particle, the **neutrino,** was originally postulated in 1933 by physicist Wolfgang Pauli to account for small amounts of energy that appeared to be missing in certain nuclear reactions. Neutrinos were presumed to be massless and to move with the speed of light. They interact very weakly with other matter and so are very difficult to detect; most of them pass completely through a star or a planet without being absorbed. Yet, neutrinos have actually been detected.

Experiments are not sufficiently precise to prove that neutrinos have *exactly* zero mass. If neutrinos turn out to have even a tiny mass, it could have interesting consequences for cosmology (Chapters 34 and 35) and for models of the interior of the Sun (Section 27.5).

The properties of the proton, electron, neutron, and neutrino are summarized in Table 27.1.

(b) Binding Energy

Just as gases give up gravitational potential energy when they come together to form a star, particles release energy in uniting to form an atomic nucleus. Energy given up is called the *binding energy* of the nucleus.

The binding energy is greatest for atoms with a mass near that of the iron nucleus, and it is less for both the lighter and the heavier atoms. In general, therefore, if light atomic nuclei come together to form a heavier one (up to iron), energy is released. This joining together of atomic nuclei is called nuclear **fusion.** On the other hand, if heavy atomic nuclei can be broken up into lighter ones (down to iron), energy is also released; this process is called nuclear **fission.** Nuclear fission sometimes occurs spontaneously, as in natural radioactivity.

Since mass and energy are equivalent, the binding energy released by either fission or fusion must correspond to a decrease in the mass of the atomic nuclei that are produced. Indeed, we find that the mass of every nucleus (other than the simple proton nucleus of hydrogen) is less than the sum of the masses of the nuclear particles that are required to build it. This slight deficiency in mass, or **mass defect,** is always only a small fraction of a mass unit (we define the unit of mass, u, as $\frac{1}{12}$ the mass of an atom of carbon or approximately the mass of a proton).

A nuclear transformation is a buildup of a heavier nucleus from lighter ones or a breakup of a heavier nucleus into lighter ones. In any such nuclear transformation, if the mass defect increases, the equivalent amount of energy is released. That energy, of course, is the difference in mass defect times the square of the speed of light. On the other hand, if, in the nuclear transformation, the mass defect decreases, a corresponding amount of energy must be put into the system.

(c) Nuclear Reactions in the Sun's Interior

The Sun taps the energy contained in the nuclei of atoms through nuclear fusion. Deep in the Sun's core, four hydrogen atoms combine or fuse together to form a helium atom. The helium atom is slightly less massive than the four hydrogen atoms that combine to form it, and that lost mass is converted to energy.

The steps required to form one helium nucleus from four hydrogen nuclei are shown in Figure 27.1. First, two protons combine to make a deuterium nucleus, which by definition contains one proton and one neutron. In effect, one of the original protons has been converted to a neutron. Electric charge is conserved in nuclear reactions, and so the positive charge originally associated with one of the protons is carried away by a positron.

This positron will instantly collide with an electron, and both will be annihilated, producing pure electromagnetic energy in the form of gamma rays. After about 10^7 years, this electromagnetic energy makes its way to the surface of the Sun, being constantly absorbed and re-emitted by atoms along the way and converted to photons of longer wavelength and lower energy in the process. The photons that we observe directly are only those that are emitted so close to the surface of the Sun that they can escape without being absorbed again.

In addition to the positron, the fusion of two hydrogen atoms to form deuterium results in the emission of a neutrino. Neutrinos produced by fusion reactions near the center of the Sun travel directly to the Sun's surface and then on toward the Earth without interacting with other atoms along the way.

The next step in forming helium from hydrogen is to add a proton to the deuterium nucleus and form a helium nucleus that contains two protons and one neutron. In the process, more gamma radiation is emitted. Finally, this helium nucleus combines with another just like it to form normal helium, which has two protons and two neutrons in its nucleus. The two protons that are left over can participate in still more fusion reactions.

This series of reactions can be described succinctly through the following equations:

$$^1H + {}^1H \rightleftarrows H + e^+ + \nu$$

$$^2H + {}^1H \rightarrow {}^3He + \gamma$$

$$^3He + {}^3He \rightarrow {}^4He + 2\,{}^1H,$$

TABLE 27.1	PROPERTIES OF SOME ELEMENTARY PARTICLES	
Particle	Mass (kg)	Charge
Proton	1.67262×10^{-27}	$+1$
Neutron	1.67493×10^{-27}	0
Electron	9.11×10^{-31}	-1
Neutrino	0	0

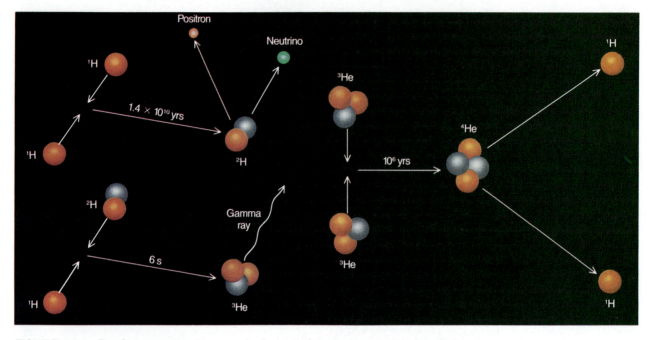

FIGURE 27.1   The Sun generates its energy by fusing four hydrogen nuclei to form helium. The steps involved in the process are shown.

where the superscripts indicate the total number of neutrons and protons in the nucleus, e^+ is the symbol for the positron, ν is the symbol for neutrino, and γ indicates that gamma rays are emitted.

Why do these reactions take place? Protons are positively charged, and positive charges repel each other. These reactions can occur only in regions of very high temperature, where the velocities of the protons are high enough to overcome the electrical forces that try to keep protons apart. In the Sun, hydrogen fusion takes place only in regions where the temperature is greater than about 10 million K and the velocities of the protons average 1000 km/s or more. Such extreme temperatures are reached only in the regions surrounding the center of the Sun, which has a temperature of 15 million K. Calculations show that nearly all of the Sun's energy is generated within about 150,000 km of its core, or within about one-quarter of its total radius.

Even at these high temperatures, it is exceedingly difficult to force two protons to combine. On average, a proton will rebound from other protons for about 14 billion years, at the rate of 100 million collisions per second, before it fuses with a second proton. Of course, some protons are lucky and take only a few collisions to achieve a fusion reaction. It is those protons that are responsible for producing the energy radiated by the Sun. Since the Sun is only about 4.5 billion years old, most of its protons have not yet been involved in fusion reactions. The low probability of the interaction of protons is fortunate for us, since it means that the Sun's fuel lasts for a long time—long enough to permit the slow biological processes on Earth to produce complex forms of life.

After the deuterium nucleus is formed, the remaining reactions happen very quickly. After about 6 s on average, the deuterium nucleus will be converted to ^{3}He. About a million years after that, two ^{3}He nuclei each formed in the same way, will combine to form ^{4}He.

We can compute the amount of energy generated by these reactions by calculating the difference between initial and final mass. The masses of hydrogen and helium atoms are 1.007825 u and 4.00268 u, respectively. Here we include the mass of the entire atoms, not just the nuclei, because the electrons are involved as well. When hydrogen is converted to helium, two positrons are created, and these are annihilated with two free electrons, adding to the energy produced.

4×1.007825
$= 4.03130$ u (mass of initial hydrogen atoms)
$- 4.00268$ u (mass of final helium atom)
$ 0.02862$ u (mass lost in the transformation)

The mass lost, 0.02862 u, is 0.71 percent of the mass of the initial hydrogen. Thus if 1 kg of hydrogen turns into helium, 0.0071 kg of material is converted into energy. The velocity of light is 3×10^8 m/s, so the energy released by the conversion of 1 kg of hydrogen to helium is

$$E = 0.0071 \times (3 \times 10^8)^2$$
$$= 6.4 \times 10^{14} \text{ joules.}$$

This amount of energy is a thousand times greater than the amount of energy required to raise the 5-m Palomar telescope 150 km above the ground.

To produce the Sun's luminosity of 4×10^{26} joules, some 6×10^{11} kg of hydrogen must be converted to helium each second, with the simultaneous conversion of about 4×10^9 kg of matter into energy. As large as these numbers are, the store of nuclear energy in the Sun is still enormous. Suppose 10 percent of the Sun's mass of 2×10^{30} kg is hydrogen that can ultimately be converted into helium; then the total store of nuclear energy would be 10^{44} joules. Even at the Sun's current rate of energy expenditure, 10^{34} joules per year, the Sun could survive for more than 10^{10} years.

At temperatures that prevail in the Sun and in less massive stars, most of the energy is produced by the reactions that we have just described, and this set of reactions is called the **proton-proton cycle.** Protons collide directly to build into helium nuclei. In hotter stars, another set of reactions, called the **carbon-nitrogen-oxygen (CNO) cycle,** accomplishes the same net result. In the CNO cycle, carbon, nitrogen, and oxygen nuclei are involved in collisions with hydrogen nuclei (protons), eventually ending with carbon again and a new helium nucleus. The CNO cycle is important at temperatures above 15×10^6 K. The details of the CNO cycle are given in Appendix 8.

27.3 THE INTERIOR OF THE SUN: THEORY

Fusion of protons will occur in the center of the Sun only if the temperature exceeds 10^7 K. How do we know whether the Sun is actually this hot? To determine what the interior of the Sun is like, it is necessary to resort to mathematical calculations. In effect, astronomers teach a computer everything they know about the physical processes that are going on in the interior of the Sun. The computer then calculates the temperature and pressure at every point inside the Sun, and determines what nuclear reactions, if any, are going on. The computer can also calculate how the Sun will change with time.

The Sun must change. In its center the Sun is slowly depleting its supply of hydrogen and creating helium instead. Will this change in composition have measurable effects? Will the Sun get hotter? Cooler? Larger? Smaller? Brighter? Fainter? Ultimately, the changes must be catastrophic, since the hydrogen fuel will eventually be exhausted. Either a new source of energy must be found, or the Sun will cease to shine. What will happen to the Sun will be described in Chapters 29 and 30. For now, let's look at what we need to teach the computer about the Sun in order to carry out the calculations.

(a) The Sun Is a Gas

The Sun is so hot that the material in it is gaseous throughout. The particles that constitute a gas are in rapid motion, frequently colliding with one another. This constant bombardment is the **pressure** of the gas (Figure 27.2). The greater the number of particles within a given volume of the gas, the greater the pressure is because the combined impact of the moving particles increases with their number. The pressure is also greater when the molecules or atoms are moving faster. Since their rate of motion is determined by the temperature of the gas, the pressure is greater when the temperature is higher.

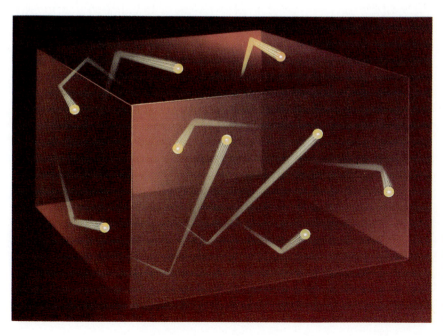

FIGURE 27.2 Gas pressure. The particles in a gas are in rapid motion and produce pressure through collisions with the surrounding material. Here particles are shown bombarding the sides of a container.

Most students have run across these concepts in high school, in the form of Boyle's law, which states that the pressure of a gas at constant temperature is proportional to its density, and Charles' law, which states that the pressure (at constant volume) is proportional to the temperature of the gas. These two ideas combine to give us the **perfect gas law,** which can be written in the form

$$P = nkT,$$

where P is the pressure, n is the number of molecules per unit volume, and T is the temperature. The constant k is equal to 1.38×10^{-23} joules/K.

The perfect gas law thus provides a mathematical relation between the pressure, density, and temperature of a perfect, or ideal, gas (one in which intermolecular or interatomic forces can be ignored). The gases in most stars closely approximate an ideal gas; thus, they must obey this law. The exceptions are very massive stars, where radiation pressure (Section 19.2e) can play an important role, and collapsed stars or the collapsed cores of stars, where the matter is degenerate (Chapter 30).

(b)   The Sun Is Stable

Apart from some very low-amplitude pulsations (Section 27.5), the Sun, like the majority of other stars, is stable. It is neither expanding nor contracting. Such a star is said to be in a condition of equilibrium. All the forces within it are balanced, so that at each point within the star the temperature, pressure, density, and so on, are maintained at constant values. We shall see (Chapters 29 and 30) that even these stable stars, including the Sun, are changing as they evolve, but such evolutionary changes are so gradual that to all intents and purposes the stars are still in a state of equilibrium.

The mutual gravitational attraction between the masses of various regions within the Sun produces tremendous forces that tend to collapse the Sun toward its center. Yet the Sun has been emitting approximately the same amount of energy for billions of years and so has managed to resist collapse for a very long time. The gravitational forces must therefore be counterbalanced by some other force, and that force is the pressure of the gases within the Sun (Figure 27.3). To exert enough pressure to prevent the collapse of the Sun due to the force of gravity, the gases at the center of the Sun must be at a temperature of 15 million K. So we see that temperatures high enough to fuse protons are *required* by the fact that the Sun is not contracting.

If the internal pressure in a star were not great enough to balance the weight of its outer parts, the star would collapse somewhat, contracting and building up the pressure inside. If the pressure were greater than the weight of the overlying layers, the star would expand,

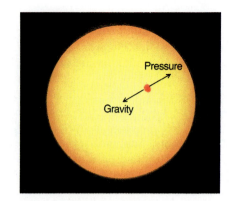

FIGURE 27.3 Hydrostatic equilibrium. In the interior of a star, the inward force of gravity is exactly balanced at each point by the outward force of gas pressure.

thus decreasing the internal pressure. Expansion would stop, and equilibrium would be reached, when the pressure at every internal point again equaled the weight of the stellar layers above that point. An analogy is an inflated balloon, which will expand or contract until an equilibrium is reached between the excess pressure of the air inside over that of the air outside and the tension of the rubber. This condition is called **hydrostatic equilibrium.** Stable stars are all in hydrostatic equilibrium; so are the oceans of the Earth, as well as the Earth's atmosphere. The pressure of the air keeps the air from falling to the ground.

(c)   The Temperature of the Sun Is Not Changing

From observations we know that electromagnetic energy flows from the surfaces of the Sun and stars. According to the second law of thermodynamics, heat always tries to flow from hotter to cooler regions. Therefore, as energy filters outward toward the surface of a star, it must be flowing from inner hotter regions. The temperature cannot ordinarily decrease inward in a star, or energy would flow in and heat up those regions until they were at least as hot as the outer ones. We conclude that the highest temperature occurs at the center of a star and that temperatures drop to successively lower values toward the stellar surface. (The high temperature of the Sun's chromosphere and corona may therefore appear to be a paradox. These high temperatures are believed to be maintained by magnetic heating, shock waves, or some other process that would not exist for a gas in which heat is simply flowing outward by means of radiation or convection, processes described in the next section.) The outward flow of energy through a star, however, robs it of its internal heat and would result in a cooling of the interior gases, were that energy not replaced. There must therefore be a source of energy within each star. In the case of the Sun that source of energy is the fusion of hydrogen to form helium.

If a star is in a steady state (that is, in hydrostatic equilibrium and shining with a steady luminosity), the temperature and pressure at each point within it must remain approximately constant. If the temperature were to change suddenly at some point, the pressure would similarly change, causing the star to contract suddenly, or to expand, or otherwise to deviate from hydrostatic equilibrium. Energy must be supplied, therefore, to each layer in the star at just the right rate to balance the loss of heat in that layer as it passes energy outward toward the surface. Moreover, the rate at which energy is supplied to the star as a whole must, at least on the average, exactly balance the rate at which the whole star loses energy by radiating it into space. That is, the rate of energy production in a star is equal to the luminosity. We call this balance of heat gain and heat loss for the star as a whole and at each point within it the condition of **thermal equilibrium.**

(d) Heat Transfer in a Star

Since the nuclear reactions that generate the Sun's energy occur deep within it, we must find a way to transport heat from the center of the Sun to its surface. There are three ways in which heat can be transported: by *conduction*, by *convection*, and by *radiation*. Conduction and convection are both important in planetary interiors, while radiation is unimportant because of the low temperatures and great opacity of planetary materials. In stars, which are so much more transparent, radiation and convection are important, while conduction can be ignored unless the gas is degenerate (Chapter 30) or is very hot (as in solar flares, the solar corona, and the interstellar medium).

Stellar **convection** occurs as currents of gas flow in and out through the star (Figure 27.4). While these convection currents travel at moderate speeds and do not upset the condition of hydrostatic equilibrium or result in a net transfer of mass either inward or outward, they nevertheless carry heat very efficiently outward through a star. However, convection currents cannot be maintained unless the temperatures of successively deeper layers in a star increase rapidly in relation to the rate at which the pressures increase inward. In a similar way, convection in planetary atmospheres is important only in the troposphere, which also has a temperature that decreases rapidly with height. Convection does occur, nevertheless, in certain parts of many stars, and convection currents may travel completely through some of the least luminous M stars.

Unless convection occurs, the only significant mode of energy transport through a star is by electromagnetic **radiation,** which gradually filters outward as it is passed from atom to atom. However, radiative transfer is not an efficient means of energy transport, because gases in the interiors of stars are very opaque—that is, a photon does not go far before it is absorbed by an atom (typically, in the Sun, about 0.01 m). The energy absorbed by atoms is always re-emitted, but it can be re-emitted in *any* direction. A photon that is traveling outward in a star when it is absorbed has almost as good a chance of being reradiated back toward the center of the star as toward its surface. A particular quantity of energy being passed from atom to atom, therefore, zigzags around in an almost random manner and takes a long time to work its way from the center of the star to the surface (Figure 27.5). In the Sun, the time required is of the order of 1 million years. If the photons were not absorbed and re-emitted along the way, then they would travel at the speed of light and could reach the surface in a little over 2 s, just as the neutrinos do (Figure 27.5).

The measure of the ability of matter to absorb radiation is called its **opacity.** It should be no surprise that

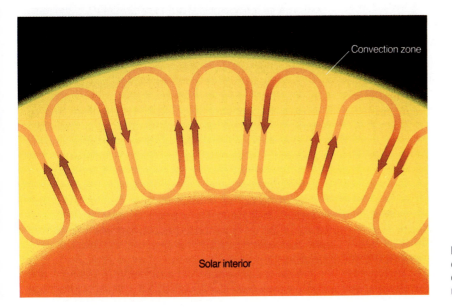

FIGURE 27.4 Rising convection currents carry heat from the interior of the Sun to its surface. Cooler material sinks downward.

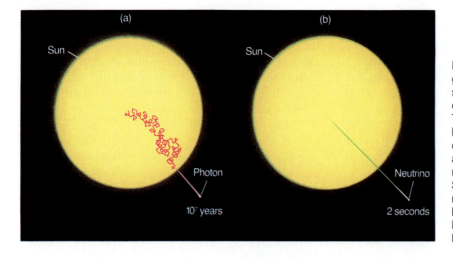

FIGURE 27.5   (a) Photons generated by fusion reactions in the solar interior travel only a short distance before they are absorbed. The re-emitted photons usually have lower energy and may travel in any direction. As a consequence, it takes about 11 million years for energy to make its way from the center of the Sun to its surface. (b) In contrast, neutrinos do not interact with matter but traverse the Sun at the speed of light, reaching the surface in only a little more than 2 seconds.

the gases in the Sun are opaque. If they were completely transparent, we would be able to see all the way through the Sun. We have discussed earlier (Section 7.4) the processes by which atoms and ions can interrupt the flow of energy—such as by becoming ionized. In addition, individual electrons can scatter radiation helter-skelter. For a given temperature, density, and composition of a gas, all of these processes can be taken into account, and the opacity can be calculated. The computations are very complicated and thus require large computers.

Once the opacity is known, we can find how each layer or shell of the Sun or a star impedes the outward flow of radiation. Of course, there is a net outward flow of the energy generated by thermonuclear reactions in the interior, or the star would have no luminosity. Thus from the opacity we calculate how the temperature must increase inward through the shell to force the observed radiation out, and thereby learn the temperature distribution throughout the interior.

If the temperature difference across some region of a star is high enough, then convection currents, rather than radiation, carry most of the energy. Within those regions the variation of temperature with depth is determined by the expansion of outward-moving masses of gas and the contraction of inward-moving ones. Here again, knowledge of the energy transport mechanism within a star makes possible the calculation of the temperature distribution.

27.4   MODEL STARS

We now have enough theory to determine the internal structure of a star. We must combine the principles we have described: hydrostatic equilibrium, the perfect gas law, thermal equilibrium, energy transport, the opacity

of gases, and the rate of energy generation from nuclear processes. These physical ideas are formulated into mathematical equations that are solved to determine the march of temperature, pressure, density, and other physical quantities throughout the stellar interior. The set of solutions so obtained, based upon a specific set of physical assumptions, is called a *theoretical model* for the interior of the star in question.

(a)   Computation of a Stellar Model

It is relatively easy to write down the equations that describe the structure of a star. However, it is relatively difficult to solve those equations, and large, fast computers are required to calculate stellar models.

A stellar model lists the values of four quantities at any distance, r, from the star's center. The distance r varies from zero at the center to R, the radius of the star, at the stellar surface. The four quantities are the pressure, $P(r)$, the temperature, $T(r)$, the mass, $M(r)$, contained within a sphere of radius r concentric with the star's center, and the contribution to the star's total luminosity, $L(r)$, that is generated within this sphere. As input to the models, we must also specify the composition of the star and its total mass, both of which can be estimated from observations.

After we have computed a stellar model that gives us values of pressure, temperature, mass, and luminosity through the star, we can calculate other physical properties. For example, if we know the pressure and temperature at a position r, then we can use the perfect gas law to calculate the density at that same position. The rate of energy generation depends on the pressure, temperature, and chemical composition of the stellar material. The opacity, that is, the ability of the star to impede the outward flow of radiation, also depends on pressure, temperature, and chemical composition.

In order to calculate the four quantities $P(r)$, $T(r)$, $M(r)$, and $L(r)$, we must have four equations. Each equation describes

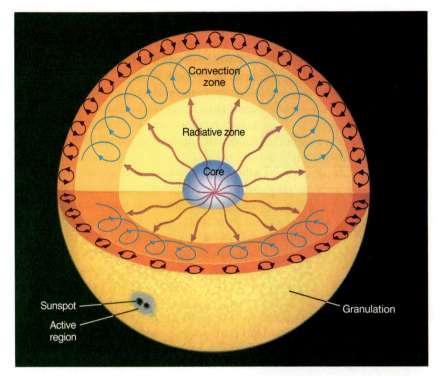

FIGURE 27.6 The interior structure of the Sun. Energy is generated in the core by the fusion of hydrogen to form helium. This energy is transmitted outward by radiation, that is, by the absorption and re-emission of photons. In the outermost layers, energy is transported mainly by convection.

how one of the quantities changes through a small spherical shell within the star. The first equation describes the change in pressure across such a shell, which is given by the condition of hydrostatic equilibrium. The second equation relates the mass of the shell to its density, which in turn is given by the pressure and temperature in the shell and the perfect gas law. The third equation specifies the change in luminosity across the shell, which is determined by the rate at which energy is generated within it. The fourth equation describes the change in temperature, which is governed by the way in which energy is transported through the shell. If energy is transported by radiation, the opacity determines the temperature variation. If a region of the star is in convection, the expansion and contraction of the gases determine the change in temperature. The simultaneous solution of these four equations for a shell at any position r in the star then gives us the four desired quantities at that position.

The calculation of the stellar model can begin at the center of the star ($r = 0$), where $M(r)$ and $L(r)$ are known to be zero. The four equations are then used to calculate how the values of pressure, temperature, mass, and luminosity change over a short distance, thus yielding the values of these quantities at that new position. The equations are then used again to calculate the changes over the next short distance. Step by step, the pressure, temperature, mass, and luminosity are calculated throughout the star from its center to its surface.

The mass and initial chemical composition of a star determine its structure and evolution. Specifically, the luminosity and radius can be derived from the calculation of a stellar model. The size of a star of a given mass is just what it needs to be so that the energy radiated at the surface is just balanced by the rate at which energy is generated in the stellar core. In the computer, we can also let a short time pass, and then calculate how the luminosity, radius, and other quantities will change with time, thereby determining how the star evolves.

(b)   A Model for the Sun

Figure 27.6 illustrates schematically what the interior of the Sun is like. Energy is generated through fusion in the core of the Sun, which extends only about one-quarter of the way to the surface. This core contains about one-third of the total mass of the Sun. At the center the temperature reaches a maximum of about 15 million K, and the density is nearly 150 times the density of water. The energy generated is transported toward the surface by radiation until it reaches a point about 70 percent of the distance from the center to the surface. At this point convection begins, and energy is transported the rest of the way primarily by rising columns of hot gas.

Figure 27.7 shows how the temperature, density, composition, and rate of energy generation vary from the center of the Sun to its surface.

(c)   The Russell–Vogt Theorem and the Interpretation of the Main Sequence

If a star is in hydrostatic and thermal equilibrium, and if it derives all its energy from nuclear reactions, then its structure is completely and uniquely determined by its total mass and by the distribution of the various chemical elements throughout its interior. In other words,

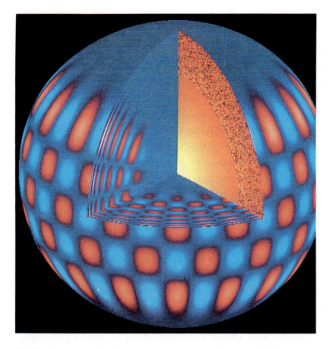

FIGURE 27.8 New observational techniques permit astronomers to measure small differences in velocity at the surface of the Sun to infer what the deep interior of the Sun is like. In this computer simulation, red denotes regions of the surface that are moving away from the observer; blue marks regions moving toward the observer. Note that the velocity changes penetrate deep into the interior of the Sun. (National Optical Astronomy Observatories)

neutrinos. About 3 percent of the total energy generated by nuclear fusion in the Sun is carried away by neutrinos. Neutrinos only rarely interact with matter, and the neutrinos created in the center of the Sun make their way directly out of the Sun and to the Earth at the speed of light. So far as neutrinos are concerned, the

Sun is transparent. If we can devise a way to detect some of the 300 billion billion (3×10^{20}) solar neutrinos that pass through each square meter of the Earth's surface every second, then we can obtain information directly about what is going on in the center of the Sun.

Unfortunately, the very property that makes neutrinos an interesting source of information about the interior of the Sun makes them very difficult to detect. Neutrinos pass through material on the Earth as readily as they escape from the Sun. On very, very rare occasions, however, a neutrino will interact with another atom. Several experiments have been devised to detect these interactions.

The first of these experiments uses chlorine. Raymond Davis, Jr., and his colleagues at Brookhaven National Laboratory placed a tank containing nearly 400,000 liters of cleaning fluid (C_2Cl_4) 1.5 km beneath the surface of the Earth in a gold mine at Lead, South Dakota. Calculations show that solar neutrinos should produce about one atom of argon-37 daily in this tank.

These individual argon-37 atoms can be detected, and the results are that only about one-third as many neutrinos reach the Earth as is predicted by standard models of the solar interior. A second experiment carried out by Japanese astronomers, who looked for the interaction of neutrinos with water, confirmed this deficiency of solar neutrinos.

Calculations show that fewer than 1 percent of the neutrinos produced in the Sun have energies high enough to be detected by the chlorine, and the water experiment is also sensitive only to the neutrinos with very high energies. Therefore, a small decrease in the central temperature of the Sun relative to what is assumed in standard solar models might be able to account for the deficiency of high-energy neutrinos.

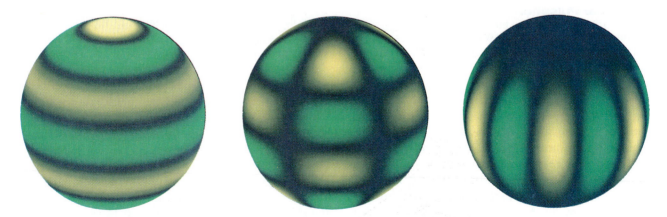

FIGURE 27.9 Waves resonating in the interior of the Sun cause the Sun's surface to oscillate in complex patterns. These images show some of the possible patterns. Green regions are approaching the observer; dark regions are receding. Gray regions have zero radial velocity. All of the patterns of oscillation shown, plus several million others, are present at the same time. The velocities measured at the solar surface are produced by the superposition or summation of all of these sound waves. (National Optical Astronomy Observatories)

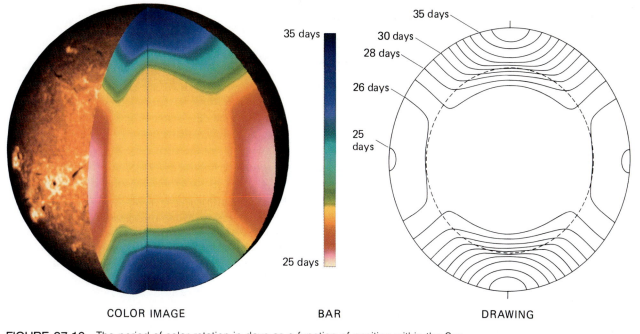

COLOR IMAGE   BAR   DRAWING

FIGURE 27.10   The period of solar rotation in days as a function of position within the Sun. Note that the rotation period increases from 25 days at the equator to 36 days at the pole. These same periods persist inward throughout the Sun's convective zone to a depth of 200,000 km. Deeper inside, the Sun appears to rotate as a solid body with a period of 27 days. (Kenneth Libbrecht/California Institute of Technology)

A small change in central temperature could not, however, account for a change in the number of low-energy neutrinos. Nearly all of the solar energy, as well as nearly all of the solar neutrinos, are produced when two protons combine to form deuterium (Section 27.2). Since we know how much total energy is emitted by the Sun, we know very accurately how many times each second two protons combine to form deuterium. We therefore also know how many times each second the associated neutrinos are produced by this interaction. Once it became clear that standard solar models predicted far more high-energy neutrinos than are detected on Earth, scientists began to design experiments to look for the low-energy neutrinos associated with the proton-proton reaction.

These low energy neutrinos interact with the element gallium. In the early 1990s, the first results of two gallium experiments, one in Russia and the other in Italy, were reported. Those observations indicate that the number of low-energy neutrinos is about two-thirds the value predicted by standard solar models.

Are the solar models wrong? Or, as many scientists are beginning to think is more likely, are we wrong about the properties of neutrinos? There are three types of neutrinos, and standard theory assumes that all three types have no mass. There is actually no laboratory evidence that the mass of a neutrino is exactly zero. If neutrinos have even a tiny mass, then it is possible for

one type of neutrino to change to another type on its journey from the center of the Sun to the photosphere and on to the Earth. The Sun produces only one type of neutrino, the so-called electron neutrino, and the chlorine, water, and gallium experiments are sensitive only to this one type of neutrino. If some of the electron neutrinos change to another type of neutrino on their way from the center of the Sun to the Earth, then the experiments performed so far would have missed those neutrinos.

To test this idea, we need a new experiment that can measure all types of neutrinos. If the total number of neutrinos emitted by the Sun were found to be greater than the number of electron neutrinos, then we would have evidence that the mass of the neutrino is not zero. Such an experiment, which involves heavy water (water with the hydrogen atoms replaced by deuterium), is now being built in Canada, and the results should be available in the late 1990s.

It will be another 10 years before we have the results of the new experiments on solar pulsations and neutrinos. Until then, we will not know whether our models of the Sun are wrong or whether the assumption that the neutrino has no mass is wrong. While it may seem discouraging that there are questions for which definitive answers are not yet available, science often works this way. Observations lead to the development of a model. This model then suggests a number of other measurements that can be made and predicts the outcome of those measurements. Frequently, the predictions are incorrect, and the models must be modified to take into account the new measurements. And so science moves forward by successive approximations, each step providing a better and more complete description of what is actually occurring. Rather than finding the lack of final answers discouraging or frustrating, scientists find such situations a never-ending challenge. The possibility of learning something never before known by anyone is what attracts many scientists to research.

S U M M A R Y

27.1 The Earth is 4.5 billion years old, so the Sun must have been shining for at least this long. Neither chemical burning nor gravitational contraction can account for the energy radiated by the Sun during this time.

27.2 The source of the Sun's energy is the **fusion** of hydrogen to form helium. A helium atom is about 0.71 percent less massive than the four hydrogen atoms that combine to form it, and that lost mass is converted to energy.

27.3 Even though we cannot see inside the Sun, it is possible to calculate what the solar interior must be like. As input for these calculations, we use what we know about the Sun. It is a gas. Apart from some very tiny changes, the Sun is neither expanding nor contracting (it is in **hydrostatic equilibrium**) and puts out energy at a constant rate. Fusion of hydrogen occurs in the center of the Sun, and the energy generated is carried to the surface by **radiation** and **convection.**

27.4 A stellar model describes the structure of the interior of a star. Specifically, it describes how the pressure, temperature, mass, and luminosity depend on the distance from the center of the star. If a star is in hydrostatic and **thermal equilibrium** and derives its energy from nuclear reactions, then its structure is determined by its total mass and chemical composition. This principle is known as the **Russell–Vogt theorem.** Stars on the main sequence differ in mass but have similar compositions. Stars not on the main sequence must have a structure that is different from main-sequence stars. Their chemical composition is different either because of nuclear reactions that have depleted hydrogen and produced heavier elements in their interiors (red giants) or because they do not derive their energy from nuclear reactions (white dwarfs).

27.5 The calculated solar model explains most of the observations. It predicts, however, that we should be able to detect more **neutrinos** than we do. The model also does not accurately reproduce the pattern of small oscillations that are observed in the photosphere. More observations are required to determine exactly why the model is wrong about these details.

E X E R C I S E S

THOUGHT QUESTIONS

1. In what way is a neutrino very different from a neutron?

2. Which of the following transformations is fusion and which is fission: the transformation of **(a)** helium to carbon; **(b)** carbon to iron; **(c)** uranium to lead; **(d)** boron to carbon; **(e)** oxygen to neon?

3. What makes the Sun shine?

4. Stars that are hotter than the Sun also derive their energy by fusing hydrogen to form helium, but a different set of reactions is involved. This set of reactions is called the carbon-nitrogen-oxygen (CNO) cycle, and the individual steps in the process are given in Appendix 8. Draw a picture like Figure 27.1 that shows what happens in each step of the CNO cycle.

5. In the CNO cycle, carbon and nitrogen are referred to as *catalysts*, since the carbon and nitrogen nuclei are required to make the reactions proceed, but the total number of carbon and nitrogen atoms does not change when all of the steps are completed. Show from your drawing for the previous problem that this statement is true.

6. Why is a higher temperature required to fuse hydrogen to helium by means of the CNO cycle than by the process that occurs in the Sun, which involves only isotopes of hydrogen and helium?

7. After a star converts its hydrogen to helium, it must then use helium as a fuel. The specific reaction that is involved is the conversion of three helium nuclei to a carbon nucleus. The steps are given in Appendix 8. Again, draw a picture like Figure 27.1 that shows what happens in this two-step process. How do you think the interior structure of the star must change to make the conversion of helium to carbon possible?

8. The Earth's atmosphere is in hydrostatic equilibrium. Explain what this means. Would you expect the pressure in the Earth's atmosphere to increase or decrease as you climb from the bottom of a mountain to its summit. Why?

9. Give some everyday examples of the transport of heat by convection and by radiation.

10. Why do you suppose so great a fraction of the Sun's energy comes from its central regions? Within what fraction of the Sun's radius does practically all of the Sun's luminosity originate? (See Figure 27.7.) Within what radius of the Sun has its original hydrogen been partially used up? Discuss what relation the answers to these questions bear to one another.

11. Why do we not expect nuclear fusion to occur in the surface layers of stars?

12. The Sun obtains its energy by fusing four hydrogen nuclei to form a helium nucleus. It is also possible to obtain energy by breaking up atomic nuclei of such heavy elements as uranium and plutonium to form lighter nuclei. This process is called nuclear *fission*. Why is fission not an important source of energy in the Sun?

PROBLEMS

13. Verify that some 600 million tons of hydrogen are converted to helium in the Sun each second.

14. Stars exist that are as much as a million times more luminous than the Sun. Consider a star of mass 2×10^{32} kg and luminosity 4×10^{32} watts. Assume that the star is 100 percent hydrogen, all of which can be converted to helium, and calculate how long it can shine at its present luminosity. There are about 3×10^7 seconds in a year.

15. Perform a similar computation for a typical star less massive than the Sun, such as one whose mass is 1×10^{30} kg and whose luminosity is 4×10^{25} watts.

16. If the atmospheric pressure were the same on two different days, but if one day were much hotter than the other, what could you say about the relative density of the air on the two days?

17. If, in a vacuum chamber, the pressure is only one-millionth of sea-level pressure, how does the density of the gas in the chamber compare with the average density of air at sea level?

18. If an observed oscillation of the solar surface has a period of 10 min and the average radial velocity is 1 m/s in and out, calculate the total displacement of the surface that is involved in this particular oscillation mode. What percent is this of the total radius of the Sun?

★19. Let ΔP denote the pressure increase inward through a spherical shell of inner radius r and thickness Δr. Multiply ΔP by the area of the inner surface of the shell to find the total outward force on the shell. Now equate this outward force to the weight of the shell pulling it inward, and show that

$$\Delta P = \frac{GM(r)}{r^2} \rho \Delta r,$$

where $M(r)$ is the mass of the star interior to r and ρ is the density of the shell.

★20. Use the equation derived in Exercise 19 to make a very rough estimate of the pressure at the center of the Sun. For this estimate suppose Δr to be the entire radius of the Sun (that is, the entire Sun is taken as one shell) and ΔP to be the increase in pressure from the surface to the center. For ρ use the mean density of the Sun, and for $M(r)$ the entire solar mass. See Appendix 5 for needed data. Your estimate cannot be expected to be better than a few orders of magnitude, but it should give some indication of the amount of pressure.

A group of very small, dense, dark clouds of dust and gas (called Bok globules after the astronomer who first studied them) are seen silhouetted against the glowing hydrogen gas in the nebula called IC 2944. At least some Bok globules appear to contain stars in the earliest stages of formation.
(Anglo-Australian Telescope Board)

With this chapter, we begin our in-depth exploration of the life story of the stars, from their birth out of the raw material we studied in the last chapter to their collapse and death when their story is done. No scenario is more important to astronomers who seek to chronicle the history of the universe, and none is more relevant to understanding our own cosmic antecedents.

As part of the story of star birth, we must address the question of the formation of planets. Are they a rare and accidental addition to the retinue of a star (as astronomers believed a century ago) or are they a regular accompaniment to the process of making stars throughout the Galaxy (as evidence is beginning to indicate today)? Clearly, if we want to find other Earths in the depths of space, this is a question of major consequence.

The quest for planets elsewhere then leads us naturally to ask the question many students find especially interesting: Are we the only example of intelligent life in the cosmos, or do we have counterparts among the stars? Today, you can find astronomers who will argue vigorously for both points of view. Such arguments are possible because in science it is always the experimental evidence that is the ultimate judge among competing theories. And, despite the often misguided claims of UFO believers, we do not yet have any direct evidence of extraterrestrial life.

From a philosophical point of view, one could argue that we have always been wrong in the past when we have claimed that the Earth is somehow unique. Since the time of Copernicus, we have known that we do not occupy the center of the universe. The Sun itself is a rather undistinguished star, and we shall see later that our position in the Galaxy is also not especially significant. It would be surprising if the beginnings or the evolution of life, on the other hand, were absolutely limited to the surface of our planet and happened nowhere else. But such arguments are not as good as actually finding sentient life elsewhere; and so we will discuss the demanding experiments that some astronomers are carrying out to see if we can actually find intelligence out there and settle the debate once and for all.

Sir Fred Hoyle (b. 1915), the well-known British astrophysicist and cosmologist, is also well known for his science fiction and even for an opera libretto. Hoyle was one of the pioneers in the modern study of stellar evolution, and in the 1950s, his brilliant deduction about the nature of the carbon nucleus enabled us to understand where the atoms of our bodies originated. (Floyd Clark/California Institute of Technology)

THE BIRTH OF STARS

Astronomers estimate that there are more than 10,000 very luminous young stars in our Galaxy—stars with lifetimes that are measured in only millions of years. If, as seems likely, such highly luminous stars have been present throughout the billions of years that our Galaxy has existed, then as these stars die, they must be replaced by new ones. On average, in fact, one new bright star must be formed somewhere in our Galaxy every 500 to 1000 years if the total number of highly luminous stars is to remain approximately constant. And for every such luminous star that is formed, there are many others of more modest mass and luminosity.

As we shall see in this chapter, star formation is a continuous process that is going on *right now*. Stars of all masses, low as well as high, are being formed, and that formation process is taking place in the interiors of clouds of dust and gas, which provide the necessary raw material.

example, deep in the interior of the Sun, 600 million tons of hydrogen are converted to helium every second, with the simultaneous conversion of about 4 million of these tons to energy. At this rate the Sun can continue to shine for more than 10 billion years, so massive is its fuel supply. But what about stars that are consuming their nuclear fuel more rapidly?

Stars more massive and more luminous than the Sun exhaust their fuel supply much more rapidly than does the Sun. The most massive stars have only 50 to 100 times the mass of the Sun, yet their luminosities—and correspondingly the rate at which they consume their supply of hydrogen—are a million times greater. Accordingly, these massive stars must exhaust their fuel supply, burn themselves out, and become unobservable in no more than a few million years. The brightest hot star in Orion—Rigel—cannot have been shining when the first human-like creatures walked the Earth.

28.1 STELLAR EVOLUTION

(a) The Lifetimes of Stars

It is natural to think of the stars as fixed, permanent, and unchanging. Yet stars are radiating energy at a prodigious rate, and no source of energy can last forever. For

(b) The H–R Diagram and Stellar Evolution

How is the astronomer to study the evolution of stars when the changes we would like to understand require tens of thousands, millions, or even billions of years to take place, depending on the mass of the star? In our brief lifetimes, we have only a single snapshot of the

Galaxy. Our problem is similar to that of scientists who wish to study the growth and life cycle of humans but are allowed only a few seconds to observe a group of people. Within that few seconds we might see one or two deaths, but nothing else would change; even the birth of a baby requires longer than a few seconds. So what do we do?

One approach would be to tabulate the properties of a large number of humans. We would see, for example, humans of many different sizes, from 40 cm or so up to about 2 m in length. Some of the variation may reflect differences from one individual to another, but careful examination would probably indicate that there is also an evolutionary process. Small humans become large humans. By noting the numbers of humans in each size range, we could calculate the rate of growth and the fraction of a human lifetime required for growth to full size, even though we never actually measured any change in an individual during our brief interval of observation.

Astronomers use a similar technique to study stellar evolution. We see a snapshot of the stellar population, consisting of individual stars at all stages of their life cycles. One of the best ways to represent this snapshot is by plotting the properties of stars on the H−R diagram, which relates stellar temperatures to their luminosities.

Some of the differences between stars are due, as we have seen, to their different masses. Others represent changes as a star ages. Even though we do not see dramatic changes in any one star (except for a few violent deaths, as discussed in Chapter 30), we can use the distribution of stars in the H−R diagram, together with stellar models, to trace out the evolution of a star.

As a star uses up its nuclear fuel, its luminosity and temperature change. Thus, the position where it is plotted in the H−R diagram changes as the star ages. Astronomers often speak of a star *moving* on the diagram or of its evolution *tracing out a path* in the diagram. Of course, the star does not really move at all in a spatial sense. This is just a shorthand way of saying that its temperature and luminosity change as it evolves.

28.2 STAR FORMATION

(a) Molecular Clouds: Stellar Nurseries

If we want to find the very youngest stars—stars still in the process of formation—we must look in places where there is plenty of the raw material required to make stars. Stars are made of gas, and so we must look in dense clouds of gas (Figure 28.1). Giant molecular clouds (Section 25.6) are prime places to look, and

FIGURE 28.1
Stars form in clouds of gas and dust. M16 is a cluster of stars that formed about 2 million years ago. The dark areas visible across the face of the nebula are thought to be condensations of material that might one day collapse into yet more stars. (Anglo-Australian Observatory Board)

observations confirm that most star formation does take place within them.

Molecular clouds are not smooth but contain clumps or *dense cores* of material that have very low temperature (10 to 50 K) and densities much higher (10^4 to 10^5 atoms per cubic centimeter) than is typical of most of the rest of the interstellar medium. Both of these conditions—low temperature and high density—are favorable to the star formation process. If a star is to form, it is necessary that a region with a mass comparable to that of a star shrink in radius and increase its density by nearly a factor of 10^{20} from that typical of a dense core to that of a star. This collapse is brought about by the force of gravity.

The story of stellar evolution is the story of the competition between two forces—*gravity* and *pressure*. As described in Chapter 27, the force of gravity tries to make a star collapse. Internal pressure produced by the motions of the gas atoms tries to force the star to expand. When these two forces are in balance, the star is stable. Major changes in the structure of a star occur when one or the other of these two forces gains the

FIGURE 28.3   An infrared picture of the region of the Trapezium stars. Because infrared radiation penetrates the dust, observations at 2.2 μm reveal the cluster of young stars within the molecular cloud. Compare this image with the optical photograph in Figure 25.3; note how many fewer stars are visible in that photograph. (National Optical Astronomy Observatories)

upper hand. Low temperature and high density both work to give gravity the advantage and so facilitate the star formation process.

(b)   The Orion Molecular Cloud

The best studied of the stellar nurseries is the Orion region (Figure 28.2). A luminous cloud of dust and gas can be seen with binoculars in the middle of the sword in the constellation of Orion. Associated with the luminous material is a much larger molecular cloud, which is invisible in the optical region of the spectrum. Near the center of the optically bright region of the nebula is the Trapezium cluster of very luminous O-type stars (Figure 25.3). An infrared picture of this same region shows hundreds of stars, invisible on photographs taken at optical wavelengths (Figure 28.3).

The long dimension of the Orion molecular cloud stretches over a distance of about 30 pc. The total quantity of molecular gas is about 200,000 times the mass of the Sun. Star formation began about 12 million years ago at one edge of this molecular cloud near the right-hand (western) shoulder of Orion and has slowly moved through the molecular cloud, leaving behind groups of newly formed stars. The oldest of these groups is the one farthest from the site of current star formation, and the ages become progressively younger the closer the groups are to the Trapezium cluster. The stars in the belt are about 8 million years old, and the stars near the Trapezium are less than 2 million years old. Even younger stars are still hidden in the molecular cloud and can be observed only in the infrared region.

Star formation is not a very efficient process and uses typically only a few percent—at most perhaps 25 percent—of the gas in a molecular cloud. The leftover material is heated either by stellar radiation or by su-

FIGURE 28.2   The nearest molecular cloud is in Orion. The extensive network of very faint filaments, which are traceable over most of the constellation of Orion, is optical evidence of a substantial dark cloud of molecular gas and dust. At radio wavelengths, this cloud fills most of the field pictured here. The bright nebula in the lower right contains the Trapezium stars and is a region of active star formation. The Horsehead Nebula appears in the upper left. (Anglo-Australian Observatory Board)

pernova explosions and blown away into interstellar space. The oldest groups of stars can therefore be easily observed optically because they are no longer shrouded in dust and gas (Figure 28.4).

Because of the correlation between stellar ages and position in the Orion region, we know that star formation has moved progressively through this molecular cloud. While we do not know what caused stars to begin forming in Orion, there is good evidence that the first generation of stars triggered the formation of additional stars, which led to the formation of still more stars (Figure 28.5). The basic idea is as follows. When a massive star is formed, it emits copious amounts of ultraviolet radiation, which heats the surrounding gas in the molecular cloud. This heating increases the pressure in the gas and causes it to expand. When massive stars exhaust their supply of fuel, they explode, and the energy of the explosion also heats the gas. The hot gases burst into the surrounding cold cloud, compressing the material in it until the cold gas is at the same pressure as the expanding hot gas. In the conditions typical of molecular clouds, the compression is enough to increase the gas density by a factor of 100. At densities this high, stars can begin to form in the compressed gas. This process seems to have occurred not only in Orion but also in many other molecular clouds.

(c)   The Birth of a Star

The earliest stages of star formation are still shrouded in mystery. There is almost a factor of 10^{20} difference between the density of a molecular cloud core and that of the youngest stars that can be detected. So far we have been unable to observe directly what happens within a cloud as material comes together and collapses gravitationally through this range of densities to form a star. We do not know what causes a single large cloud to fragment and form individual stars, nor do we know why multiple-star systems and planets form.

FIGURE 28.4   The Rosette Nebula. The cluster of blue stars at the center of the nebula formed less than a million years ago. The gas and dust have been driven away from the bright stars by radiation and intense stellar winds. (Anglo-Australian Telescope Board)

Observations of this stage of stellar evolution are nearly impossible for several reasons. First, the interiors of molecular clouds where stellar births take place cannot be observed with visible light. The dust in these clouds acts like a thick blanket of interstellar smog, which cannot be penetrated by visible radiation (Figure 28.6). It is only with the new techniques of infrared and millimeter radio astronomy that we are able to make any measurements at all. Even so, the time scale for the initial collapse, which is measured in thousands of years, is very short, astronomically speaking. Furthermore, the collapse occurs in a region so small (0.1 pc) that in most cases we cannot resolve it with existing techniques. Accordingly, we have yet to catch a star in the act, so to speak, of its initial collapse. Nevertheless, through a combination of theoretical calculations and the limited observations that are available, astronomers

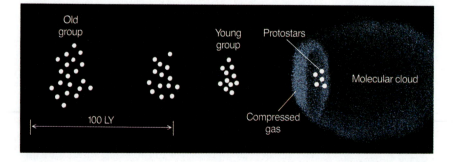

FIGURE 28.5   Schematic diagram showing how star formation can move progressively through a molecular cloud. The oldest group of stars lies to the left of the diagram and has expanded because of the motions of individual stars. Eventually the stars in the group will disperse and will no longer be recognizable as a cluster. The youngest group of stars lies to the right, next to the molecular cloud. This group of stars is only 1 to 2 million years old. The pressure of the hot, ionized gas surrounding these stars compresses the material in the nearby edge of the molecular cloud and initiates the gravitational collapse that will lead to the formation of more stars.

The[...]
in t[...]
For[...]
mu[...]
ind[...]
pla[...]

[...]
ind[...]
que[...]
evi[...]
dis[...]
rpr[...]
par[...]
tra[...]
the[...]
par[...]
wo[...]
An[...]
at t[...]
du[...]

[...]
the[...]
Th[...]
is r[...]
tha[...]
the[...]
cer[...]

mi[...]
reg[...]
sin[...]
the[...]
tin[...]
ste[...]
bil[...]
ye[...]

28

O[...]
se[...]
tu[...]
ou[...]
ou[...]

(a

If[...]
ev[...]
w[...]
te[...]
cc[...]
w[...]
lig[...]

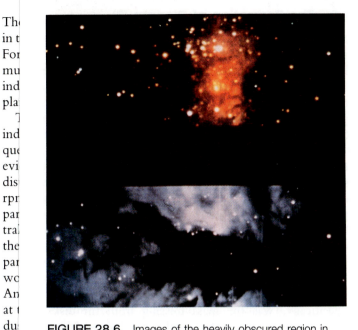

FIGURE 28.6 Images of the heavily obscured region in the Orion molecular cloud. The lower image was taken in visible light and shows what appears to be an empty sky in the central region of the picture. In fact, this region is not empty at all but, rather, contains a substantial amount of dust that obscures visible light. The upper image was made with an infrared detector at a wavelength of about 2.2 μm. Electromagnetic radiation at this long wavelength can penetrate the obscuring dust easily. Infrared observations, therefore, provide a way to detect the recently formed stars within the cloud. (National Optical Astronomy Observatories)

have pieced together a picture of what the earliest stages of stellar evolution are likely to be.

During the period when a condensation of matter in a molecular cloud is contracting to become a true star, we call the object a **protostar.** In their theoretical studies of stellar evolution, astronomers compute a series of *models* for stars, each successive model representing a later point in time. Given one model, it is possible to calculate how a star should change (in the case of the young protostars under discussion, due to gravitational contraction) and hence what the star will be like at a slightly later time. At each step, calculations give the luminosity and radius of the protostar, and from these its surface temperature.

Let us follow the evolution of a stellar condensation within a molecular cloud (Figure 28.7a). Since a molecular cloud is more massive than a typical star by a factor of 100,000 or more, many stars must form within each cloud. The first step in the process of creating stars is the formation within the cloud—through a process that we do not yet understand—of the dense cores of material. These cores then attract additional matter because of the gravitational force that they exert on the cloud material that surrounds them. Eventually, the gravitational force becomes strong enough to overwhelm the pressure exerted by the cold material that forms the dense cores. The material undergoes a rapid collapse, the density of the core increases greatly, and a protostar is formed.

Theory indicates that if the collapsing core is rotating and if its density is strongly peaked toward its center, then a flattened disk will form (Figure 28.7b; Section 20.3). The protostar and disk are embedded in an en-

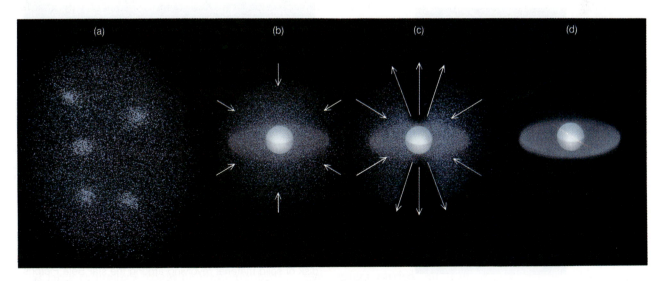

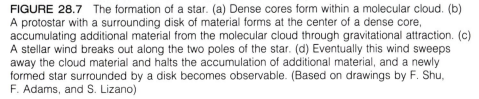

FIGURE 28.7 The formation of a star. (a) Dense cores form within a molecular cloud. (b) A protostar with a surrounding disk of material forms at the center of a dense core, accumulating additional material from the molecular cloud through gravitational attraction. (c) A stellar wind breaks out along the two poles of the star. (d) Eventually this wind sweeps away the cloud material and halts the accumulation of additional material, and a newly formed star surrounded by a disk becomes observable. (Based on drawings by F. Shu, F. Adams, and S. Lizano)

period of 12 years, equal to that of Jupiter. From the observed motion and the period, they could deduce the mass and distance of Jupiter using Kepler's laws.

Contemporary astronomical measurements from the Earth's surface have a precision of a few milli-arcseconds, sufficient to detect jovian-mass companions around the nearer low-mass stars. Many decades of positional measurements with small refracting telescopes yielded claims in the 1960s of planetary detection, but all of these claims have since been found to be in error. More recent applications of this technique with higher precision ground-based instruments have turned up nothing. A specially built telescope in space, however, could easily detect a planet with only one-tenth of Jupiter's mass around any of several hundred nearby stars, if such planets are present.

Another approach, which involves the search for the tiny changes in radial velocity of the star as it moves around its orbit, has had better luck. To understand how this technique works, consider the Sun. Its radial velocity changes by about 10 m/s with a period of 12 years owing to the gravitational force of Jupiter. A jovian-mass planet at a distance of 1 AU would induce velocity changes of 50 m/s with a period of just 1 year. Velocity differences of this magnitude in other stars can be detected by very high-resolution spectroscopy to measure the Doppler effect. Several such surveys are currently under way, and there are indications that they are proving successful.

In 1988, astronomers discovered an object with a mass that may be as small as 11 times the mass of Jupiter in an 84-day orbit about the nearby star HD 114762. While this object is probably a brown dwarf (Section 28.6), its mass is getting close to that of a true planet. Meanwhile, a Canadian project carried out in Hawaii searched for radial-velocity variations in about 16 stars. Although variations were found, none repeated in the regular way that would be expected if a planet were causing the velocity changes. These observations suggest that Jupiter-size planets may be rare. Many more observations of other stars will be required to establish whether this conclusion is correct.

The search for planets around other stars seems to be a project whose time has come. If there are numerous large planets out there, it is within the capability of modern astronomy to find them. If successful, the search will allow astronomers for the first time to consider the formation of our own planetary system within a broader cosmic context. If the search is carried out with negative results, however, the implication will be almost equally interesting. An absence of planets accompanying single, solar-type stars would challenge our current understanding of the process of star and planet formation and would send astronomers "back to the drawing boards" to develop better theories. A negative result would also raise the profound question of the uniqueness of our own solar system and hence of life in the universe.

28.4 LIFE IN THE UNIVERSE

It appears likely that many stars will prove to have planets circling them and that we will soon be able to detect some of the more massive of these planets. But is there life on these planets? That is a much more speculative question and one that is much more difficult to answer.

(a) Life in the Solar System

In the quest for life in the universe, it is appropriate to begin with the planets we know best, those in our own solar system. By examining the nature of life on Earth and its possible adaptability to other planets, we can gain a perspective on the environments elsewhere that might also support living things.

The life that we are familiar with developed with our planet, influencing the Earth as it evolved (Chapter 12). While we do not know the details of its origin, life clearly seems to have been the product of the chemical evolution of the early atmosphere and oceans, at a time when there was no free oxygen and conditions were at least mildly reducing (that is, dominated chemically by hydrogen rather than oxygen). Temperatures on Earth must have been roughly similar to those today, with at least part of the oceans consisting of liquid water at all seasons, and with plentiful sunlight to provide the energy needed for life to grow and diversify.

Laboratory experiments have reproduced many of the early chemical steps along the road to life. We know that naturally occurring reactions would have resulted in plentiful organic chemicals, enough to make the early oceans a sort of organic "soup." (Similar reactions are taking place today in the atmosphere of Saturn's satellite Titan.) Among the prebiological compounds that must have been present are amino acids, sugars, and proteins. Within the first billion years of its formation, the Earth had developed self-replicating molecules that could use this organic soup to create copies of themselves, beginning the long evolutionary course leading to plants, pigs, and even people.

Could a similar sequence of events have taken place on other planets in the solar system? In the outer planets and their larger satellites, we see reducing atmospheres and a nonbiological organic chemistry that mimic in many ways the conditions on the primitive Earth. Certainly the atmospheres of Titan and the jovian planets contain many of what are believed to be the building blocks of life. Yet there is no evidence that life actually developed. Perhaps liquid water is essential, and the outer satellites were simply too cold to have this vital substance. Or perhaps the lack of solid surfaces on the giant planets denied them a stable enough environment for life to prosper. In any case, the evidence suggests that life is probably not present, since we do not see any unusual chemistry in the atmospheres of the outer planets that indicates the activity of large numbers of living organisms.

Among the inner planets, Mars has always seemed the most likely abode of life. It is interesting that early in the 20th century, educated persons all over the world believed that the scientific evidence (as promoted by Percival Lowell and others) indicated the existence of not just life, but intelligent life, on Mars. But with increasing knowledge of the red planet, this hope has faded. Following the Viking landings in 1976, it seems clear that conditions on Mars are too harsh for the development of life such as we know it on Earth. The absence of liquid water, and particularly the lethal ultraviolet radiation that reaches the surface unimpeded by ozone or other absorbers in the thin martian atmosphere, are not conducive to biological activity.

We find, then, several planets in our solar system that seem to offer possibilities for the development of life, but only one—the Earth—where living things have proliferated and survived to the present. Perhaps not coincidentally, Earth is also the only planet with liquid water on its surface. If Mars were a little larger and a little closer to the Sun, it might also have a climate like our own. Similarly, Venus could have developed differently if it were farther from the Sun, escaping the runaway greenhouse effect that led to the loss of its water and the scorching temperatures it now experiences. But only on the Earth were conditions just right for life to prosper.

One of the fascinating properties of life is that its chemical building blocks are composed primarily of some of the most common elements in the universe: hydrogen, oxygen, carbon, and nitrogen. These elements should be present in any planetary system that forms anywhere in the Galaxy. Organic compounds are even prevalent in interstellar space, as we saw in Chapter 25. There is nothing special or unusual about the chemistry of life. If self-replicating organic molecules once form, the universe is full of the chemical food that they require for their growth.

But under what circumstances will such complex, self-replicating molecules form? That is a very difficult question. Part of the problem arises because we have only a single form of life to study, chemically speaking. All the tremendous diversity of living things on Earth represents the same genetic material. Only the packaging differs: from paramecium to parakeet, from rose to rhinoceros. Therefore we do not know what other kinds of self-replicating molecules, and hence what other chemical variants, are possible. If alternative forms once existed on Earth, they were destroyed by the dominant DNA-based life that we know today.

(b) The Possibility of Intelligent Life in the Galaxy

The first step in mounting a search for intelligent life is to estimate how likely it is to occur. Although the numerical results are extremely uncertain, the exercise is instructive because it leads to a strategy for possible discovery and communication with other intelligent creatures.

University of California astronomer Frank Drake has pioneered the attempt to estimate the number of potentially communicative civilizations in the Galaxy. Drake's famous equation expresses the number, N, of currently extant civilizations in the Galaxy as the product of seven factors:

$$N = R_s f_p n_p f_b f_i f_c L_c,$$

where R_s is the rate of star formation in the Galaxy, f_p is the fraction of those stars with planetary systems, n_p is the mean number of planets suitable for life per planetary system, f_b is the fraction of those planets suitable for life on which life has actually developed, f_i is the fraction of those planets with life on which intelligent organisms have evolved, f_c is the fraction of those intelligent species that have developed communicative civilizations, and L_c is the mean lifetime of those civilizations.

An analogy may help explain why the Drake equation works. Consider the chance of finding a good partner at a college party. The probability that you will find one at any given time depends on the rate at which new partygoers enter the room times the probability that each of them has the characteristics you are looking for (right sex, age, major, personality, interests, etc.) times the length of time each candidate remains in the room.

In the Drake equation, the first three factors are astronomical ones, the next two are biological, and the last two are sociological. We can make some educated estimates regarding the astronomical factors, we are on shaky ground with the biological ones, and the last two are *very* uncertain. Yet some interesting estimates can be made.

The number of new stars formed per year in our Galaxy is about 10. This is a reasonable guess for the average star formation rate over the past few billion years. Thus we adopt $R_s = 10$ stars per year.

The Sun originated from a cloud of gas and dust whose rotation caused it to flatten into a disk from which the planets formed. We expect a similar process to be commonplace around single stars. Let's be very optimistic and assume that all stars form planetary systems, adopting $f_p = 1$.

In our own solar system, at least one planet had suitable conditions for the development of life. Other systems may have none, but if we assume that on the average, a star of the right sort with a planetary system has one suitable planet, we can adopt $n_p = 1$.

Many biologists believe that given the right kind of planet and enough time, the development of life is inevitable. Let us assume optimistically that they are right. The corresponding value of f_b is thus 1.

Similarly, given the emergence of life, there is a widespread view that with enough time and natural selection a highly intelligent species will certainly evolve. Even were it inevitable that an intelligent species evolve on every planet with life, however, how long should it take? On Earth, it took 4.5×10^9 years. What if we happened to be quick about it, but that the average intelligent species takes, say, 20×10^9 years? Moreover, of the many parallel lines of evolution on Earth, only one (so far) has produced a being with enough intelligence to build a technology. Certainly, one could not rule out a probability as low as 1 percent. For the sake of argument, let us adopt $f_i = 0.01$.

Not all intelligent societies would necessarily develop a technology capable of interstellar communication. We are on the threshold of that capability and possess a natural curiosity about the rest of the universe. It is not certain, however, that this human trait is fundamental to intelligence. Insects, while sometimes highly organized, do not appear to have any curiosity at all. Even if a society were curious, it might have good reason for wishing to have nothing to do with any other civilization. Let us assume that one-tenth of intelligent species form communicative societies; that is, $f_c = 0.1$.

It is generally agreed that the final factor, L_c, is the most uncertain. It is useful, therefore, to leave L_c as an unknown and see what the rest of the equation yields with the numbers we have suggested. If we substitute the optimistic estimates made above into the equation, we find

$$N = 10 \times 1 \times 1 \times 1 \times 0.01 \times 0.1 \times L_c$$
$$= 0.01 \times L_c.$$

We can obtain the actual number of communicative civilizations present at any time in the Galaxy by replacing L_c with our estimate of the average lifetime in years.

The only known technology, of course, is our own, and we have only just reached the capability of interstellar communication. Some pessimists have argued that our technology might well end in a few decades. If so, and if we are typical, then L_c might be about 100 years. In rebuttal, some contend that if a communicative society can manage to survive for 100 years, it might well maintain itself for a billion years. We simply do not know.

If $L_c = 100$, then even with our fairly optimistic estimates, the number of communicative civilizations in the Galaxy at any time is only just one—at the present, ourselves! One can make equally plausible arguments for adopting much more pessimistic estimates of the various quantities in the Drake equation, in which case the average number of communicative civilizations drops to 0.0001 or even less, suggesting that there are long intervals in which no one is out there to communicate with. In either case, the search for other civilizations will be fruitless.

On the other hand, if $L_c = 1$ billion years, the optimistic value for N is 10 million—a Galaxy teeming with civilizations with which to communicate. The distance to the nearest such civilization would be expected to be less than 30 pc. While we cannot choose among optimistic and pessimistic estimates, we see the rationale by which estimates are made. We also come to one very important conclusion that will influence any strategy to search for other communicative civilizations.

28.5 SETI: THE SEARCH FOR EXTRATERRESTRIAL INTELLIGENCE

(a) The Basic Problem

The Drake equation tells us that if the lifetime of communicative civilizations is short, then basically we are alone in the Galaxy. This conclusion is valid whatever the particular choice you make of numbers for the various factors in the calculations. Putting it the other way around, a search for extraterrestrial civilizations will succeed only if the average such civilization has a lifetime much longer than our own—perhaps many millions of years longer.

The important conclusion that follows is that any civilization we contact is likely to be very much older, and very much more advanced, than our own. The chance of coming across another civilization like ours, just a few decades after its discovery of radio astronomy, is vanishingly small. Thus if we are going to search, it must be with the expectation of discovering a civilization far in advance of our own. It also follows that the best approach is to let them do the talking and to assume that they will be ahead of us in considering the problems of interstellar communication. The proper strategy for SETI, the search for extraterrestrial intelligence, is to wait or, better, to listen.

(b) Direct Contact

One alternative to listening as a means of detecting other civilizations in our own Galaxy might be interstellar travel. We have seen, however, that the nearest neighboring civilization is expected to be at least a few hundred, and probably a thousand, or even tens of thousands of light years away. Because nothing can travel faster than light, a visit to another civilization would involve at least hundreds, and more likely thousands, of years.

On the other hand, even now we can send, and have sent, material messages into interstellar space at speeds of a few kilometers per second. Both the Pioneer and Voyager spacecraft, for example, have entered orbits on which they will eventually escape the solar system. It is unlikely that they will ever be seen or recovered by another intelligent species, but it is remotely possible. Partly for this reason, each Pioneer carries a plaque bearing line drawings of human beings and messages describing the world from which it came. The Voyagers contain phonograph recordings with messages from and descriptions of Earth. It is doubtful that the message on the plaque will ever be decoded or the recording heard. More important by far is the message carried by each spacecraft itself. Its discovery would convey a great deal of information about the species that launched it and the state of our technology.

Another possibility for direct communication is a visit to the Earth by extraterrestrial visitors. If the nearest civilizations are hundreds of light years away, they cannot have come to see us as a result of learning about us, for even radio waves that we have inadvertently been emitting into space—our radio and television programs—have been on their way for only a few decades. They could have reached only the very nearest stars, and it is highly unlikely that anyone there has received them and dispatched spaceships to look us over. If we have been visited, it must have been by random selection by interstellar travelers.

The popular literature is full of accounts of sightings of UFOs, presumably operated by some intelligence, of abductions by aliens, and even of alleged evidence for highly intelligent beings that have visited the Earth and taught people to build such magnificent structures as the pyramids, Easter Island statues, and other marvels. Most scientists are highly skeptical of the extraterrestrial interpretation of reports of lights or erratically accelerating shiny objects in the sky. Hard evidence of objects from space is lacking. Not one of the many purported visitors has ever left a piece of extraterrestrial material. Scientists, more than anyone, would delight in finding concrete evidence of alien life—there is so much we have to learn from it! But we still need evidence that can be analyzed by scientists qualified to judge its extraterrestrial origin. Rumors, hearsay, secondhand reports, and eyewitness accounts all require positive verification before being taken as convincing evidence for life in the universe beyond the Earth.

(c) Contact by Radio

While direct contact with other galactic civilizations seems unlikely, there is a very real possibility that we may be able to communicate with them by radio. We do not necessarily mean two-way communication, for the radio waves would probably require hundreds of years at the very least for their round-trip travel between each question and answer. (Interstellar communication would be between civilizations, not individuals.) On the other hand, if there are communicative civilizations in the Galaxy, they may already be trying to communicate or at least to send one-way messages to other possible civilizations. There is a good chance that we would recognize an intelligent message—for example, a binary-coded broadcast of the number π repeated over and over. With even our present technology we could send such messages ourselves to other stars in the Galaxy.

With existing radio-astronomy facilities, a number of limited searches for intelligently coded signals have already been made by radio astronomers in the U.S. and the Soviet Union. To date, there has been no success, but no success would have been expected from such meager efforts. Just what would it take to learn if messages are being beamed in our direction from other civilizations in the Galaxy?

The problem is difficult because we do not know in advance either the location or the nature of any possible broadcast. The entire electromagnetic spectrum is available for potential communication. Most of those who have considered this problem have concluded that radio waves—more specifically, microwaves—offer the most promising part of the electromagnetic spectrum. At these wavelengths, the absorption of energy by the interstellar medium and the emission of competing background radiation from natural sources are both near their minimum values.

All of the early searches for extraterrestrial radio signals were limited to one band or very few frequency bands, just as we tune a radio to a single station at a time. The fact that so few frequency bands were searched is the primary reason that no success was expected. The key to an effective SETI program is the development of sensitive receivers that can listen at many bands simultaneously.

The first modern SETI program began in the early 1980s. Financed in part by public contributions to the Planetary Society, this search uses an old 60-ft radio telescope that Harvard University was planning to decommission (Figure 28.10). The strategy is to point the telescope sequentially at each part of the sky and to measure any radiation received in each of its frequency channels, analyzing the results with sophisticated computer codes that can identify an artificial signal amid the natural babble of the Galaxy. The powerful million-channel receiver was paid for by movie-maker Steven Spielberg and placed into operation by the Planetary Society in 1986.

Another approach that is being tried is to search for very narrow-band signals, such as might be produced by an interstellar navigation beacon, in data that are regularly collected with radio telescopes for astronomi-

FIGURE 28.10 This dish in Massachusetts is being used regularly to search for radio signals from extraterrestrial civilizations. (D. Morrison)

cal purposes. The trick in this technique is to eliminate all of the signals from nearby intelligent life—ourselves! Man-made electronic devices produce several thousand events each day of the type that might also be broadcast by extraterrestrial intelligence. This program has just begun, and it is far too early to estimate its probability of success.

In the early 1990s, NASA started the development of very sensitive receivers and data processing systems to carry out a decade-long SETI survey of unprecedented sensitivity. Unfortunately, funding for the project was canceled by the U.S. Congress just as the project was scheduled to begin.

A portion of the planned NASA program will be carried out with private donations. Existing radio telescopes will be pressed into service. The primary program will be to examine a few hundred nearby solar-type stars at more than a billion separate frequencies. If enough observing time can be scheduled, then this targeted search will be billions of times more comprehensive than all previous surveys combined. The original NASA program also included a search of the entire sky, but with somewhat less sensitivity than planned for the observations of the nearby stars. There are now no funds for this part of the program.

The consequences of the detection of a message from an advanced civilization are hard to imagine. In part, the significance would depend on our success in deciphering the message. We can hope that any civilization

that wishes to be found by the relatively primitive techniques we know of will also have worked out a way to make its message intelligible to us. For one plausible scenario of the detection and decryption of an interstellar signal, read Carl Sagan's novel *Contact*.

28.6 TRACING EVOLUTION IN THE H–R DIAGRAM

Let's now turn back from speculation about life in the universe to the story of protostars. Now that we know how stars form, we can use the H–R diagram to summarize how they evolve. Theoretical calculations give us quantitative information about how the luminosity, radius, and temperature of protostars change with time. We can thus find where any star (or its embryo) should be represented on the H–R diagram. By plotting the temperature and luminosity of the collapsing protostar as it changes with time, we can trace the track that the star follows on the H–R diagram.

(a) Evolutionary Tracks

In its early contraction phases, a star transports its internal energy not by radiation but by leisurely convection currents (Section 27.3d). The Japanese astrophysicist C. Hayashi first showed that such stars must lie in a zone on the H–R diagram extending nearly vertically from the lower main sequence to the regions occupied by red giants and red supergiants. (Although protostars fall in the same part of the H–R diagram as red giants and supergiants, we reserve the terms "red giants" and "red supergiants" for older stars, which we will discuss in Chapter 29.)

According to calculations, stars in the initial stages of their evolution contract and move (on the H–R diagram) downward along lines we now call *Hayashi tracks*. Representative tracks for stellar embryos of several masses and of chemical composition like that of the Sun are shown in Figure 28.11. During this phase of evolution, protostars generate energy by gravitational contraction by the process described by Kelvin and Helmholtz (Section 27.1). It is not yet hot enough in the stellar core for fusion of hydrogen to occur. Stars first become visible only after the stellar wind clears away the surrounding dust and gas, and this occurs when they near the bottom of the Hayashi tracks.

After some thousands or millions of years of contraction and heating, the convection currents cease at the center of the protostar, and energy must be transported by radiation in those regions. The central zone in radiative equilibrium gradually grows in size, while the convection currents extend less and less deeply beneath the stellar surface. In this stage of its evolution, the

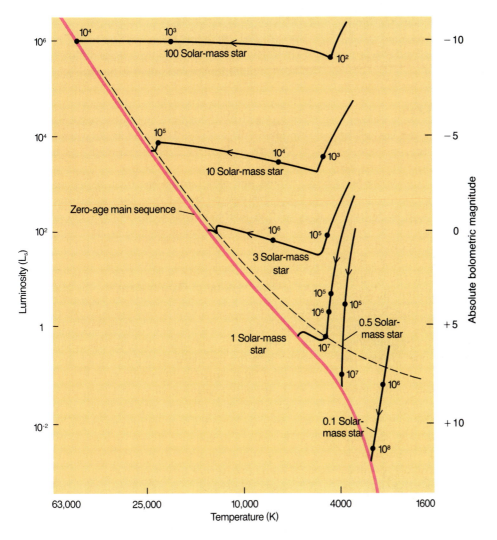

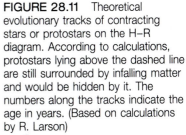

FIGURE 28.11 Theoretical evolutionary tracks of contracting stars or protostars on the H–R diagram. According to calculations, protostars lying above the dashed line are still surrounded by infalling matter and would be hidden by it. The numbers along the tracks indicate the age in years. (Based on calculations by R. Larson)

protostar, still slowly shrinking and deriving its energy from gravitational contraction, turns sharply on the H–R diagram and moves left, almost horizontally, toward the main sequence. Only stars with very low masses remain convective all the way to the core and evolve directly from the vertical Hayashi track onto the main sequence.

Protostars with masses similar to that of the Sun that are on the horizontal portion of their pre-main-sequence evolutionary tracks are called *T Tauri stars*. This class of stars was named after the prototypical example, which is seen in the direction of the constellation Taurus. T Tauri stars are typically between 500,000 and 3 million years old, have circumstellar disks, derive their energy from gravitational contraction, and often vary in brightness.

Eventually, as the release of gravitational energy continues to heat up the star's interior, its central temperature becomes high enough to support nuclear reactions. Soon this new source of energy supplies heat to the interior of the star as fast as energy is radiated away. The central pressures and temperatures are thus maintained,

and the contraction of the star ceases. It is now on the main sequence. By this time, the infall of matter is complete and the star is fully formed. The small hooks in the evolutionary tracks of the stars shown in Figure 28.11, just before they reach the main sequence, are the points (according to theory) where the onset of nuclear energy release occurs.

Note that, as shown in Figure 28.11, stars of different mass arrive at different places along the main sequence and take significantly different times to get there. The numbers labeling the points on each evolution track in Figure 28.11 are the times, in years, required for the embryo stars to reach those stages of contraction. Massive stars go through all evolutionary stages faster than do low-mass stars. Stars more massive than the Sun reach the main sequence in a few thousand to a million years. The Sun took a few million years; tens of millions of years are required for stars to evolve to the lower main sequence.

Objects of extremely low mass never achieve a high enough central temperature to ignite nuclear reactions. The lower end of the main sequence terminates where stars have a mass just barely great enough to sustain nuclear reactions at a sufficient rate to stop gravitational contraction; this critical mass is calculated to be near $\frac{1}{12}$ that of the Sun.

(b)   The Range of Stellar Masses

Since objects with masses less than $\frac{1}{12}$ the mass of the Sun never become hot enough to ignite nuclear reactions, they cannot be considered true stars. Objects with masses between $\frac{1}{100}$ and $\frac{1}{12}$ times the mass of the Sun may produce energy for a brief time by means of nuclear reactions involving deuterium, but do not become hot enough to fuse protons to form helium. Such objects are called brown dwarfs or infrared dwarfs. Still smaller objects with masses less than $\frac{1}{100}$ times the mass of the Sun are true planets; remember that Jupiter has a mass of about 0.001 times the mass of the Sun. They may radiate energy produced by the radioactive elements that they contain. They may also radiate heat generated by slow gravitational contraction, but their interiors will never reach temperatures high enough for nuclear reactions to take place.

At the other extreme, the upper end of the main sequence terminates at the point where the mass of a star would be so high and the internal temperature so great that *radiation pressure* would dominate. The radiation produced from nuclear reactions would be so extreme that when absorbed by the stellar material it would impart to it a force greater than that produced by gravitation. Such a star could not be stable. The upper limit to stellar mass is thought to be about 100 solar masses.

In general, the pre-main-sequence evolution of a star slows down as the star moves along its evolutionary track toward the main sequence. For all stars, however, we can distinguish three evolutionary time scales:

1. The initial gravitational collapse from interstellar matter is relatively quick. Once the condensation is, say, 1000 AU in diameter, the time for it to reach hydrostatic equilibrium is measured in thousands of years.

2. Pre-main-sequence gravitational contraction is much more gradual. From the onset of hydrostatic equilibrium to the main sequence requires a few million years for stars with masses similar to that of the Sun.

3. Subsequent evolution on the main sequence is very slow because a star changes only as thermonuclear reactions alter its chemical composition. For a star of 1 solar mass, this gradual process requires billions of years.

(c)   Unanswered Questions

Star formation is one of the most active areas of astronomical research, and progress during the past 5 years has been nothing less than astounding. Despite the advances, many problems remain unsolved. We do not yet know what process (there may even be more than one) initiates star formation in a molecular cloud. We do not know what determines the masses of the stars that form. We know from the luminosity function that low-luminosity, and hence low-mass, stars are by far the most common, but we do not know why. There are even some star-forming regions where no high-mass stars are formed at all. We do not know what determines whether a collapsing dense core becomes a single or multiple star. We do not know what generates the strong stellar winds that apparently halt the infall of material onto the surface of a newly forming protostar. And, of course, the search for planets outside the solar system continues. These questions will challenge astronomers for years to come.

S U M M A R Y

28.1 It is possible to estimate the lifetime of a star by determining its total mass and by then calculating how rapidly its hydrogen must be converted to helium in order to account for its luminosity. The most massive stars live only a few million years. A star like the Sun will spend about 10 billion years on the main sequence. The changes in temperature and luminosity as a star evolves can be visualized by plotting these quantities on an H–R diagram.

28.2 The low temperature and relatively high densities of gas and dust within molecular clouds provide conditions favorable to the formation of stars. The best-studied molecular cloud is that of Orion, where star formation began about 12 million years ago and is moving progressively through the cloud. The formation of a star begins with a dense core of

material of stellar mass within a molecular cloud. The core accretes matter and collapses owing to gravity. The accumulation of material halts when the **protostar** develops a strong stellar wind. It is likely that nearly all protostars are surrounded by a disk containing an amount of mass that may be as large as 10 percent of the mass of the Sun—more than enough to form a planetary system.

28.3 Direct searches are under way for planets around nearby stars. Preliminary results suggest that the radial velocities of many solar-type stars vary by a few tens of meters per second. Continued measurements should establish whether these variations are due to the gravitational influence of orbiting planets.

28.4   The Drake equation is used to estimate the probability that there is life elsewhere in our Galaxy. Unfortunately, the uncertainties in the calculation are so large that we cannot predict whether life is likely to be very common or extremely rare.

28.5   The search for extraterrestrial intelligence is called SETI. The best way to conduct the search is by listening for radio signals broadcast by other civilizations.

28.6   Stars range in mass from about $1/12$ to 100 times the mass of the Sun. Objects with masses between that of true planets

($1/100$ the mass of the Sun) and a star are called **brown dwarfs** or infrared dwarfs. Protostars generate energy through gravitational contraction. The initial gravitational collapse takes several thousand years. After that, a slow contraction continues for, typically, millions of years until the star reaches the main sequence and nuclear reactions begin. The higher the mass of a star, the shorter the time it spends in each stage of evolution.

E X E R C I S E S

THOUGHT QUESTIONS

1. How do we know that the lifetime of an individual star cannot be infinitely long?

2. Why is star formation more likely to occur in cold molecular clouds than in regions where the temperature of the interstellar medium is several hundred thousand degrees?

3. What is the main factor that determines where a star falls along the main sequence?

4. Is it likely that there are planetary systems around other stars? Describe the evidence.

5. Describe what happens when a star forms. Begin with a dense core of material in a molecular cloud and trace the evolution up to the point at which the newly formed star reaches the main sequence.

6. Redo the calculation of the Drake equation, putting in values that you think are reasonable. What do you think are the extreme limits for the number of intelligent civilizations of any kind in the Galaxy?

PROBLEMS

7. The star Rigel has an absolute magnitude of -6.8 (about $4 \times 10^4 L_s$). Its mass is uncertain but is probably no more than 50 times the mass of the Sun. Estimate the lifetime of

Rigel. (Assume that the lifetime of the Sun is 10 billion years.)

8. Problem 6 in Chapter 24 asked you to construct a kind of H–R diagram for humans by plotting height against weight. Use that same diagram to show how people change as they age. In the terms used for stars, show how a person "moves" in the height–weight diagram as he or she evolves from infancy to old age.

9. Verify the statement in the text that Jupiter would be about 22 magnitudes fainter than the Sun as seen in visible light from a distant star.

10. Determine the angular distance of the Earth from the Sun as seen from distances of **(a)** 1 pc; **(b)** 10 pc; **(c)** 100 pc.

11. Calculate the total surface area if the Earth were broken up into particles with diameters of **(a)** 1 km; **(b)** 1 m; **(c)** 1 cm.

*12. Estimate the total brightness in visible light (relative to the brightness of the star) of a belt of asteroids circling a star at a distance of 10 AU and consisting of 10,000 objects with an average diameter of 10 km and an average reflectivity of 50 percent.

13. Suppose we could carry on a two-way radio communication with another civilization. How long would be the minimum time to receive an answer to a transmitted question if that civilization is **(a)** on the Moon? **(b)** on Jupiter? **(c)** on a planet in the Alpha Centauri system? **(d)** on a planet in the Tau Ceti system (see Appendix 13)? **(e)** at the Galactic center?

One of the first images taken with the repaired Hubble Space Telescope shows (inset at right) R136, a cluster of young stars at the center of the Tarantula Nebula, also known as the 30 Doradus Nebula (larger image at left). The nebula is an enormous region of gas and dust, located in one of our nearest neighbor galaxies, the Large Magellanic Cloud. The Tarantula is so large and bright that if it were located in our own galaxy—at the distance of the more familiar Orion Nebula—it would fill over half the sky and produce enough light for you to read this book at night. How many stars R136 contained was a matter of great debate among astronomers; some even thought for a while that it contained just one or a few incredibly massive superstars. Now the Hubble has settled the issue: preliminary analysis indicates the presence of more than 3000 closely packed stars in the dense cluster, the most massive of which are about 100 times the mass of the Sun.

Our models of how stars evolve—which we continue developing in this chapter—require rigorous testing, just like any other scientific model. The problem is that the scales of time and space in the Galaxy, which we need for those models, are not so easy to establish. Strictly speaking, for example, we know the age of the Sun only from having measured the age of the Earth, and from realizing that the Sun must be at least as old as our planet. And, as we saw in earlier chapters, measuring distances significantly beyond our immediate cosmic neighborhood—distances we need to establish the absolute luminosity of the stars—is a dauntingly difficult task.

But nature has been kind to us in some ways. There is a type of variable star, called a cepheid, whose changes in light output, when properly calibrated, allow us to measure distances to stars far beyond the range for parallax. And there are, scattered throughout the Galaxy, clusters of stars—with dozens to hundreds of thousands of members—that can be used to establish a time scale for stellar evolution. For astronomers, the stars in each cluster, having formed in the same place and at the same time, provide an unmatched laboratory for developing and then testing their ideas on the stages through which stars evolve and the time they take in each stage.

STAR CLUSTERS: TESTS OF STELLAR EVOLUTION

Henrietta Swan Leavitt (1868–1921) joined the staff of the Harvard College Observatory in 1892. While studying variable stars in the Magellanic Clouds, she discovered the period-luminosity relation for cepheid variable stars, which later made it possible for Edwin Hubble to demonstrate that our Galaxy is but one among billions in the universe. (Harvard College Observatory)

As soon as a newly formed star has reached the main sequence, it derives its energy almost entirely from the thermonuclear conversion of hydrogen to helium. It remains on the main sequence for most of its "life." The main-sequence phase of evolution was the first to be satisfactorily explained by theoretical calculations. Since changes in main-sequence stars occur very slowly, they are easier to model on a computer. The stars themselves have emerged from the clouds within which they were born and can be easily observed. There are even clusters that contain hundreds or thousands of main-sequence stars, all of a common age and composition, and testing of the theory is particularly straightforward for these groups of stars.

29.1 EVOLUTION FROM THE MAIN SEQUENCE TO RED GIANTS

When four hydrogen atoms fuse to form helium, only 0.7 percent of the mass is converted to energy. During the main-sequence phase of evolution, therefore, stars do not change mass appreciably. As hydrogen fusion continues, however, the chemical composition in the central regions does change as hydrogen is depleted and helium accumulates. This change of composition forces stars to change in structure, luminosity, temperature,

and size. Eventually, the positions where stars are plotted on the H–R diagram evolve away from the main sequence. The original main sequence, corresponding to stars of homogeneous chemical composition, is called the **zero-age main sequence.**

(a) Evolution on the Main Sequence

As helium accumulates in the center of a star, calculations show that the temperature and density in that region must increase. The rate at which fusion reactions occur also increases with temperature (as T^4 for the proton-proton cycle). Consequently, the rate of nuclear energy generation increases, despite the depletion of hydrogen "fuel," and the luminosity of the star slowly rises. Initially, these changes are small, and stars remain close to the zero-age main sequence for most of their lifetime. Astronomers often use the term "burning" to describe the depletion of hydrogen or other elements by nuclear reactions. Bear in mind that this "nuclear burning" is not burning in the chemical sense.

When the hydrogen has been depleted completely in the central part of a star, a core develops that contains only helium, "contaminated" by whatever small percentage of heavier elements the star had to begin with. The energy source from hydrogen burning is now used up, and with nothing more to supply heat, the helium core begins again to contract gravitationally. Once more, the star's energy is partially supplied by gravitational potential energy released from the contracting core. The rest of its energy comes from hydrogen burning that now begins in the region immediately surrounding the core. These changes result in a substantial and rather rapid readjustment of the star's entire structure, so that the star leaves the vicinity of the main sequence altogether.

About 10 percent of a star's mass must be depleted of hydrogen before the star evolves away from the main sequence.

The lifetime of a star depends on two things—how much fuel it has, and how fast it uses up that fuel. Since hydrogen is the main fuel source and most of the mass of a star consists of hydrogen, the amount of fuel is proportional to the total mass of the star (M). The rate at which the fuel is used depends on luminosity (L). The lifetime of a star is therefore proportional to M/L. According to the mass-luminosity relation (Section 23.2), L is proportional to M^4, so that the lifetime of stars varies as $1/M^3$. *The most massive stars have the shortest lifetimes.* Although massive stars have more fuel, they burn it so prodigously that their lifetimes are very short. The most massive stars spend only a few million years on the main sequence. A star of 1 solar mass remains there for about 10^{10} years, and a spectral-type M0 V star of about 0.4 solar mass has a main-sequence life of some 2×10^{11} years, a value much longer than the age of the universe.

The lifetimes of main-sequence stars are summarized in Table 29.1.

(b) Evolution to Red Giants

As the central core contracts, it releases gravitational potential energy, some of which heats up the hydrogen surrounding the core. In this hot hydrogen shell, the conversion of hydrogen to helium accelerates, causing most stars to increase in luminosity. The stars also expand to enormous proportions. The expansion of the outer layers causes the temperature at the stellar surface to decrease. After leaving the main sequence, then, stars

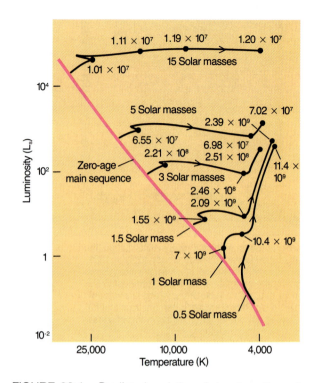

FIGURE 29.1   Predicted evolution of stars from the main sequence to red giants. See text for explanation. (Based on calculations by I. Iben)

move to the high-luminosity and low-temperature (upper right) portion of the H–R diagram. They become *red giants.* Red giants have a small core composed of helium, which is surrounded by a thin shell in which hydrogen is being converted to helium. No nuclear reactions are occurring in the outermost layers.

Figure 29.1, based on theoretical calculations by University of Illinois astronomer Icko Iben, shows the tracks of evolution on the H–R diagram from the main sequence to red giants for stars of several representative masses and with chemical composition similar to that of the Sun. The broad band is the zero-age main sequence. The numbers along the tracks indicate the times, in years, required for the stars to reach those points in their evolution after leaving the main sequence.

This description of stellar evolution is based entirely on calculations of stellar models. No star completes its main-sequence lifetime or its evolution to a red giant quickly enough for us to observe these structural changes. Fortunately, nature has provided us a way to test the calculations. Instead of observing the evolution of a single star, we can look at a group or cluster of stars that all formed at the same time but that have different masses. Since stars with higher masses evolve more quickly, we can hope to find clusters in which massive stars have already completed their main-sequence phase of evolution and have become red giants or supergiants, while stars of lower-mass in the same cluster are still

TABLE 29.1	LIFETIMES OF MAIN-SEQUENCE STARS	
Spectral Type	Mass (Mass of Sun = 1)	Lifetime on Main Sequence
O5	40	1 million years
B0	16	10 million years
A0	3.3	500 million years
F0	1.7	2.7 billion years
G0	1.1	9 billion years
K0	0.8	14 billion years
M0	0.4	200 billion years

on the main sequence or even undergoing pre–main-sequence gravitational contraction.

One main advantage of using clusters of stars rather than individual stars, like those in the solar neighborhood, to check the theory of stellar evolution is that the many stars in a cluster are at the same distance, so that their luminosities can be directly compared. If the stars in a group are very close together in space, it is also reasonable to assume that they all formed nearly at the same time, from the same cloud, and with the same composition.

In the next section, we will first describe star clusters—where they are found, how big they are, what kind of stars they contain. Then we shall show that clusters of various ages do indeed differ from one another in the way that theory predicts.

29.2 STAR CLUSTERS

In our Galaxy, there are two basic types of clusters. **Globular clusters** were formed about 13 to 15 billion years ago and contain only very old stars. **Open clusters** are much younger, some being only a few million years old or even still forming stars. Many open clusters are young enough to contain massive hot stars.

(a) Globular Clusters

About a hundred globular clusters are known in our Galaxy, most of them in a spherical halo surrounding the flat, wheel-like shape formed by the spiral arms. All are very far from the Sun, and some are found at distances of 20,000 pc or more from the galactic plane. One of the most famous globular clusters is M13 (Figure 29.2) in the constellation of Hercules, which passes

FIGURE 29.2   The globular cluster M13. (U.S. Naval Observatory)

nearly overhead on a summer evening at most places in the U.S.

A good picture of a typical globular cluster shows it to be a nearly circularly symmetrical system of stars, with the highest concentration of stars near its own center. In the central regions of the cluster, the stars are so closely packed that most of them cannot be resolved as individual points of light but appear as a fairly uniform and continuous glow. Images of a globular cluster made in light of two different colors, say red and blue, show that the brightest stars are red. From measurements of the distances to globular clusters, we know that the brighter red stars must be red giants.

Linear diameters of globular clusters range from 20 to 100 pc or more. In one of the nearer globular clusters, more than 30,000 stars have been counted. There are probably hundreds of thousands of additional member stars that are too faint to be observable. The combined light from all these stars gives a typical globular cluster an absolute magnitude in the range -5 to -10, or 10^4 to 10^6 times the luminosity of the Sun.

The average star density in a globular cluster is about 0.4 star per cubic parsec. In the dense center of the cluster, the star density may be as high as 100 or even 1000 per cubic parsec. There is plenty of space between the stars, however, even in the center of a cluster. The "solid" photographic appearance of the central regions of a globular cluster results from the finite resolution of the telescope and seeing effects in the Earth's atmosphere. (With the Hubble Space Telescope, which is above the blurring effects of the atmosphere, it is possible to resolve and analyze individual stars near the centers of globular clusters.) If the Earth revolved not about the Sun, but about a star in the densest part of a globular cluster, the nearest neighboring stars, light months away, would still appear as points of light. Thousands of stars, however, would be scattered uniformly over the sky. The Milky Way would be hard to see, and even on the darkest of nights the brightness of the sky would be comparable to faint moonlight.

Globular clusters revolve about the nucleus of the Galaxy on orbits of high eccentricity and high inclination to the galactic plane (rather like the orbits of comets in the solar system). Obeying Kepler's second law, a cluster spends most of its time far from the nucleus; a typical cluster probably has a period of revolution of the order of 10^8 years.

Because most globular clusters lie outside the plane of the Milky Way, nearly all of them have probably already been discovered. A few dozen, hidden by obscuring dust clouds, may remain undiscovered in the disk and nucleus of the Galaxy.

(b) Open Clusters

In contrast to the rich, partially unresolved globular clusters, open clusters appear comparatively loose and

FIGURE 29.3 The Jewel Box (NGC 4755). This famous cluster of young, bright stars is an open cluster some 2500 pc from the Sun. It was named the Jewel Box from its description by Sir John Herschel as "a casket of variously colored precious stones." (Anglo-Australian Telescope Board)

"open" (hence their name). They contain far fewer stars than globular clusters and show little or no strong concentration of stars toward their own centers (Figure 29.3). Although open clusters are usually more or less round in appearance, they lack the high degree of spherical symmetry that characterizes a globular cluster. The stars in these clusters are usually fully resolved, even in the central regions.

Open clusters are found in the disk of the Galaxy, often associated with interstellar matter. Because of their locations, they are sometimes called **galactic clusters** rather than open clusters. They are low-velocity objects and are presumed to originate in or near spiral arms. Over 1000 open clusters have been catalogued, but many more are identifiable on good search photographs, such as those of the Palomar Sky Survey. Yet only the nearest open clusters can be observed, because of interstellar obscuration in the Milky Way plane. We conclude, therefore, that we see only a small fraction of the open clusters that actually exist in the Galaxy; possibly tens or even hundreds of thousands of them escape detection.

Several open clusters are visible to the unaided eye. Most famous among them is the Pleiades, which appears as a tiny group of six stars (some people see more than six) arranged like a tiny dipper in the constellation of Taurus (Figure 25.10). A good pair of binoculars shows dozens of stars in the cluster, and a telescope reveals hundreds. (The Pleiades is not the Little Dipper;

the latter is part of the constellation of Ursa Minor, which also contains the North Star.) The Hyades is another famous open cluster in Taurus. To the naked eye, the cluster appears as a V-shaped group of faint stars, marking the face of the bull. Telescopes show that the Hyades actually contains more than 200 stars. The naked-eye appearance of the Praesepe cluster, in Cancer, is that of a barely distinguishable patch of light. This group is often called the "Beehive" cluster, because its many stars, when viewed through a telescope, look like a swarm of tiny specks of light, which have fancifully been compared to a swarm of bees.

Typical open clusters contain several dozen to several hundred member stars. A few, such as M67, contain more than a thousand. Compared with globular clusters, open clusters are small, usually having diameters of less than 10 pc.

(c) Associations

For more than 50 years it has been known that the most luminous main-sequence stars—those of spectral types O and B—are not distributed at random in the sky but tend to be grouped into what are now called **associations,** lying along the spiral arms of our Galaxy. Because the stars of an association lie in the galactic plane and are spread over tens of parsecs, each revolves about the galactic center with a slightly different orbital speed. The different orbital speeds of the different members of an association will completely disrupt the group after a few million years. All associations, therefore, must be very young objects (astronomically speaking, of course!).

Associations are always linked with interstellar matter and consist of very young stars that have just formed from it. We distinguish between two kinds of associations: Those containing luminous, massive O and B stars are called O-associations; the others contain only low-mass stars and are called T-associations because they contain numerous T Tauri stars. The T Tauri stars are low-mass stars (masses less than three times that of the Sun) that have formed so recently that they are still contracting and have not yet reached the main sequence. Sometimes, as in Orion, T-associations and O-associations coexist.

An O-association appears as a group of several (say, 5 to 50) O stars and B stars scattered over a region of space some 30 to 200 pc in diameter. Because these stars are rare, it would be very unlikely for so many of them to exist by chance in so relatively small a volume of space. It is assumed, therefore, that the stars in an association are physically associated and have a common origin. Stars of other spectral types may also belong to associations, but these more common stars are not conspicuous against the general star field.

About 70 associations are now catalogued. Like ordinary open clusters, however, they lie in regions occupied by interstellar matter, and many others must be obscured. There are probably several thousand undiscovered associations in our Galaxy. Analyses of the statistics of associations and O and B stars have led some investigators to conclude that all such massive stars have originated in associations.

(d) Summary of Clusters

The properties of the different types of star clusters are summarized in Table 29.2. The sizes, absolute magnitudes, and numbers of stars listed for each type of cluster are approximate only and are intended as representative values.

(e) Stability of Star Clusters

A condition for the stability of an isolated cluster is that the average speed of its individual stars must not exceed the escape velocity from the cluster. Otherwise, the stars will escape into space, and the cluster will dissipate. If the stellar velocities are low enough to meet this condition, then we say that the cluster is *gravitationally bound*—that is, the force of gravity is strong enough to keep the member stars from escaping.

Clusters, however, are not completely isolated but move in various orbits in the Galaxy. Thus, an added condition for the stability or permanence of a cluster is that it be bound together with gravitational forces that are stronger than the disrupting tidal forces exerted on it by the Galaxy or other nearby stars. The more compact a cluster, the greater is its own gravitational *binding force* compared with the disrupting forces, and the better chance it has to survive to old age.

Globular clusters are highly compact systems and are, consequently, very stable. Most globular clusters can probably maintain their identity almost indefinitely. Even these clusters lose some stars, however—especially those of relatively small mass. A few stars in a cluster are always moving substantially faster than average. Every now and then one of them, through an encounter, will be given enough speed to escape the cluster. Some of the stars in the galactic halo must be stars that have, in the past, escaped in this way from globular clusters.

When a star escapes, it carries off energy, leaving less energy than before for the stars remaining in the cluster. The result is that, over time, the cluster develops a tightly bound core surrounded by a rarefied halo of stars. In the dense core, stars occasionally collide, and some of the debris eventually coalesces. It is predicted that this dynamical evolution will lead to the development of a massive black hole at the cluster center (Chapter 31). Meanwhile, a few stars in the outer parts of the cluster continue to escape.

The escape rate and dynamical evolution for the rich globular clusters, however, are so slow that the clusters can survive for many billions of years. The situation is analogous to the evaporation of molecules from a more or less permanent planetary atmosphere.

Matters are very different for most open clusters. Those that we have discovered are relatively close to the Sun. At

TABLE 29.2	CHARACTERISTICS OF STAR CLUSTERS		
	Globular Clusters	Open Clusters	Associations
Number known in Galaxy	125	~1000	70
Location in Galaxy	Halo and nuclear bulge	Disk (and spiral arms)	Spiral arms
Diameter (pc)	20–100	<10	30–200
Mass (solar masses)	10^4–10^5	10^2–10^3	10^2–10^3?
Number of stars	10^4–10^5	50–10^3	10–100?
Color of brightest stars ·	Red	Red or blue	Blue
Integrated luminosity of cluster (L_s)	10^4–10^6	10^2–10^6	10^4–10^7
Density of stars (M_s/pc^3)	0.5–1000	0.1–10	<0.01
Examples	Hercules Cluster (M13)	Hyades, Pleiades	Zeta Persei, Orion

mine. Its apparent brightness can also be measured. From the period-luminosity relation we can then determine the star's intrinsic luminosity. The apparent brightness and luminosity can then be used to derive the cepheid's distance by using the inverse square law for the propagation of light (Section 7.1). This technique gave the first accurate estimates of the size of our own Galaxy and of the distances to other galaxies nearby (Section 33.2). Before the period-luminosity relationship, we had no way to estimate the size of the universe.

(d) RR Lyrae Stars

Cepheids of the kind just described are massive stars with short lifetimes. Some are found in open clusters. None are found in a globular clusters, which contain only stars that are billions of years old.

Luckily, other types of pulsating variable stars are found among old stars. Among the most common types of pulsating variable stars are the **RR Lyrae stars,** named for RR Lyrae, best-known member of the group. These are old low-mass stars, and thousands of them are known in our Galaxy. Nearly all globular clusters contain at least a few RR Lyrae variables, and some contain hundreds.

The periods of RR Lyrae stars are less than 1 day; most periods fall in the range of 0.3 to 0.7 day. The amplitudes of their visual light curves never exceed two magnitudes, and most RR Lyrae stars have amplitudes less than one magnitude.

Measurements show that the RR Lyrae stars occurring in any particular globular cluster all have about the same median apparent magnitude. Since members of a cluster are all at approximately the same distance, it follows that all RR Lyrae stars within a cluster must also have nearly the same absolute magnitude. Because the RR Lyrae stars in different clusters are all similar to one another in observable characteristics, it is reasonable to assume that *all* RR Lyrae stars have about the same absolute magnitude. If we could learn what that absolute magnitude is, we could immediately calculate the distances to all globular clusters that contain these stars.

Unfortunately, as is true for cepheids, not a single RR Lyrae star is near enough to measure its parallax by direct triangulation. Like distances to the cepheids, distances to RR Lyrae stars have also had to be determined by statistical means, that is, by analyzing their proper motions and radial velocities. Figure 29.5 shows where the RR Lyrae variables are located in the period-luminosity diagram.

29.4 CAUSE OF PULSATION

What causes some stars to vary? And why are most stars constant in luminosity to an accuracy of 1 percent or less?

(a) What Varies?

A hint of the physical nature of the variability of cepheids comes from their spectra. At maximum light, their spectral classes correspond to higher surface temperatures than at minimum light. The light variations of a cepheid, therefore, are due in part to variations in the temperature of its outer layers. Further spectroscopic evidence reveals that the temperature fluctuations are accompanied by actual pulsations in the sizes of these stars. The lines in the spectrum of a cepheid show Doppler shifts that vary in exactly the same period as that of the star's light fluctuations. Evidently, the changes in light are associated with a periodic rise and fall of the cepheid's atmosphere.

A graph that displays changes in the Doppler shifts of the spectral lines of a cepheid with the lapse of time is called a *radial-velocity curve* (Figure 29.4). It is like the radial-velocity curve of a spectroscopic binary star, except that in a cepheid the Doppler shifts are due to the periodic rising and falling of its outer layers rather than to the orbital motion of the star as a whole. The mean value of the apparent radial velocity corresponds to the line-of-sight motion of the star itself; the photospheric pulsations cause variations about this mean value. When the photosphere expands, it is approaching us with respect to the rest of the star, and each spectral line is shifted to slightly shorter wavelengths than that of its mean position. When the photosphere contracts, the lines are shifted to slightly longer wavelengths. When the photosphere reaches its highest or lowest point—that is, when the star is at its largest or smallest size—the position of the spectral lines corresponds to the radial velocity of the star itself.

We can calculate the total distance through which the cepheid's photosphere rises or falls by multiplying the velocity of its rise at each point in the pulsation cycle by the time it spends at that velocity, and then adding the products (or, technically, by integrating the velocity curve over the time of rise or fall). For Delta Cephei, for example, we find by such a calculation that the photosphere pulsates up and down over a distance of somewhat under 3 million km. The mean diameter of Delta Cephei, as calculated from Stefan's law (Section 23.3), is about 40 million km. During a pulsation cycle, therefore, the radius of the star changes by almost 8 percent.

We might expect a pulsating star to be hottest when it is smallest and most compressed. Delta Cephei, however, like other cepheid variables, is hottest and brightest at about the time when its radiating surface is rushing outward at its maximum speed. Evidently, the greatest compression of the star as a whole does not correspond to the maximum temperature at its surface; the explanation is related to the mechanism by which energy is transferred outward through the outer layers of the star.

(b) Why Does It Vary?

In a stable star the weights of the various layers, pulled toward the center of the star by gravity, are just balanced by the pressure of the hot gases within (Section 27.3). A pulsating star, on the other hand, is something like a spring: As the star contracts, its internal pressures build up until they surpass the weights of its outer layers. Eventually, these pressures start the star pulsing outward, but because of their inertia, the outward-moving layers overshoot the equilibrium point where their weights will just balance the internal pressures. As the star expands further, the weights of the overlying layers decrease, but the internal pressure decreases faster. Hence, the overlying layers are not supported adequately, and the star begins to contract. As it does so, it overshoots again, and this time it becomes too highly contracted. Once more the inner pressures cause the star to expand—and so the pulsation continues.

Stars would pulsate indefinitely if there were no dissipation of energy. Most stars do not pulsate in this way, however, because with each pulsation some of the energy is radiated away, converted to convection of stellar gases or otherwise lost. Thus if a star were to start oscillating, its pulsations would quickly die out owing to these damping forces unless it experienced some kind of driving force.

What is the driving force that causes some stars to pulsate? A highly simplified explanation for the cepheids and stars with similar temperature follows. It is important, in reading this explanation, to understand that the pulsation is confined to the outer layers of the star. The deep interior does not pulsate.

During the phase when the star is smallest, that is, when it is most compressed, the temperature becomes hot enough to strip both electrons away from the helium nuclei in the star's atmosphere. Energy absorbed by the helium atoms in excess of that required for ionization is converted to kinetic energy of the freed electrons, which in turn heat the gas in the stellar atmosphere through collisions. Energy absorbed by the ionization process is thus dammed up in the atmosphere instead of being radiated into space. The temperature increases in the atmosphere where the ionization is occurring, pressure builds, and the outer layers of the star are forced to expand. Once most of the helium is ionized, however, then radiation can once again flow freely through the atmosphere, the atmosphere cools, the pressure drops, and the weight of the outer layers of the atmosphere compresses the star. As the pressure increases, electrons recombine with helium nuclei, and the process begins all over again as these helium nuclei are re-ionized and in the process again block the flow of energy. In some stars, the ionization of hydrogen can also drive pulsation.

All stars have helium. Why do only some of them pulsate? Again, in very simplified terms, in cool stars, the temperature becomes high enough to ionize helium so deep in the star that the pressure in this region is not great enough to lift all of the mass in the layers lying above it. In very hot stars, the ionization of helium occurs in a layer of the atmosphere so close to the outer boundary of the star that the weight of the layer above it is not great enough to compress the star.

This mechanism will, therefore, cause stars of intermediate temperature to pulsate. Figure 29.6 shows the *instability strip*, which is that portion of the H–R diagram where the temperature is right for helium ionization to cause pulsation. Cepheids and RR Lyrae stars lie within the instability strip. Even white dwarfs that fall within the instability strip are variable.

Although most stars do not pulsate, it is still interesting to ask what the pulsation period would be for an ordinary star if it were unstable. It turns out that the period is greater for a giant star of low mean density than for a smaller compact star of higher density—just as a long piano string vibrates more slowly than a short one. It can be shown that for pulsating stars of any one type the period of a particular star is inversely proportional to the square root of its mean density.

We can derive the relation between the period and the mean density of a pulsating star from a consideration of Kepler's third law. When a star is at its maximum size and there is no longer enough pressure to support its outer layers, they simply fall inward under the influence of the star's gravitation. Thus the matter at the surface of such a star is temporarily in free fall toward the star's center, as if it were in an elliptical orbit about the center of the star (but a degenerate ellipse that is a straight line). The period of revolution of such material in this hypothetical orbit is approximately equal to the period with which the star would pulsate. By Kepler's third law, this period, P, is related to the mass, M, of the star and the semimajor axis of the orbit—which is proportional to the radius, R, of the star—by the equation (Section 3.4)

$$MP^2 \propto R^3.$$

Upon solving for the period, we find

$$P \propto \frac{R^{3/2}}{M^{1/2}}.$$

Now the mass of the star is the product of its volume, $4\pi R^3/3$, and its mean density, ρ; therefore,

$$P \propto \frac{R^{3/2}}{R^{3/2}\rho^{1/2}} \propto \frac{1}{\rho^{1/2}}$$

or

$$P\rho^{1/2} = \text{constant}.$$

The surface of the star does not, of course, complete an actual orbit about its center. We have seen that in a cepheid, for example, the radius changes by less than 10 percent. On the other hand, if a point on the stellar surface could complete such a full orbit, it would still spend most of its time in the outer part of that orbit (in accord with Kepler's second law). Thus the representation is not as bad as it might seem. Even if the actual pulsation period differed by five or ten times from that derived by the above procedure, it would still be the case that the ratios of pulsation periods of different

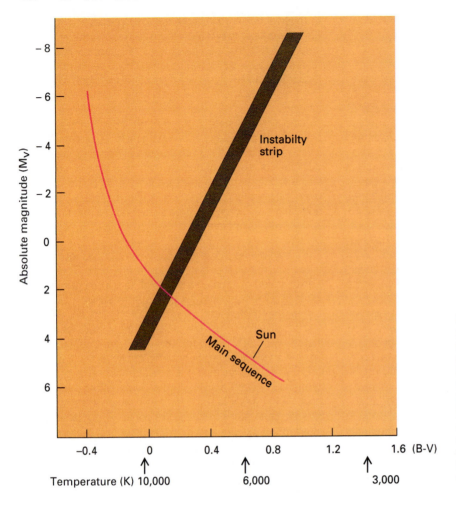

FIGURE 29.6   H–R diagram showing the instability strip where cepheids, RR Lyrae stars, and other pulsating variables are found. Cepheids with high luminosity lie at the top of the instability strip and have periods of about 50 days. RR Lyrae stars have luminosities of about 50 L_s and have periods of only about half a day. Pulsating white dwarfs also lie within the instability strip, which extends below the main sequence.

stars would be roughly inversely proportional to the square roots of their mean densities.

The shortest period with which a star could rotate is, approximately, the same period with which it would pulsate. This relation is easily seen; the speed at the equator of a rotating star cannot exceed the speed of a body at that point on a circular orbit, and that circular orbit has only twice the semimajor axis of the straight-line orbit of major axis equal to the star's radius. (For this straight-line orbit the minor axis is equal to zero.)

In sum, the shorter the period of pulsation of a star, the higher the mean density of the star. We can estimate the periods of pulsation of other kinds of stars by comparing their mean densities. Table 29.3 shows the results of such a comparison for stars all of 1 solar mass but of different radii. Because most stars have masses that differ from the Sun's by less than a factor of 10, their pulsation periods would differ from the tabulated values for 1-solar-mass stars of the same radius by only at most a factor of 3.

A note of caution is in order. While Table 29.3 indicates what the period of these various stars would be if they did pulsate, additional calculations are required to determine whether or not these stars are unstable and, if so, whether the amplitudes of pulsation are large enough to be detected. In the case of the Sun, we know that pulsation does occur, but the amplitude is very small—so small, in fact, that similar

pulsations could not be detected except in a few of the brightest (in terms of apparent magnitude) stars in the sky.

29.5 POPULATIONS OF STAR CLUSTERS

A form of the H–R diagram known as a color-magnitude diagram is extremely useful in the study of stellar evolution. A color-magnitude diagram is a plot of the

TABLE 29.3	PULSATION PERIODS FOR VARIOUS STARS OF ONE SOLAR MASS	
Radius (Solar Radii)	Period	Examples
1	1 hr	Sun
1000	4 yr	Red supergiants
100	1 mo	Cepheid
10	1 day	RR Lyrae star
0.1	2 min	
0.01	4 s	White dwarf
10^{-5}	10^{-4} s	Neutron star

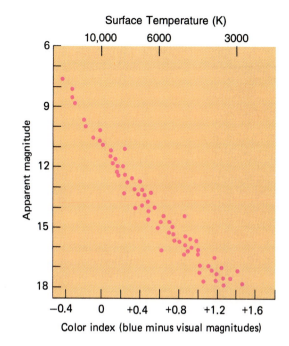

FIGURE 29.7 Color-magnitude diagram for a hypothetical open star cluster.

apparent magnitudes of the stars in a cluster versus their color indices, as shown in Figure 29.7. Recall that the color index of a star depends on the star's temperature (Section 22.2) and is thus related directly to its spectral class. Colors can be measured much more quickly than spectral types and so are frequently used in studies of large numbers of faint stars.

A plot of apparent magnitude versus color index is, with one exception, like a H–R diagram for the cluster. The exception is that a normal H–R diagram is a plot of

absolute magnitudes (rather than apparent magnitudes) of stars against their spectral classes (or color indices). All the stars in the cluster, however, are at very nearly the same distance from the Sun, and the difference between the apparent and absolute magnitudes is the same for every star and is equal to the distance modulus of the cluster (Section 22.1).

In this section we shall describe some of the properties of the color-magnitude diagrams for different kinds of clusters. The key point to remember is that the differences in the color-magnitude diagrams of open and globular clusters are a result of differences in *age*. Stars in globular clusters formed much longer ago than stars in open clusters.

(a) Color–Magnitude Diagrams of Globular Clusters

Globular clusters nearly all have very similar color-magnitude diagrams. Figure 29.8 shows the appearance of the color-magnitude diagram for a typical globular cluster of known distance, for which the apparent magnitudes have been converted to absolute magnitudes. The region from *a* to *b* is the main sequence. Presumably, the main sequence would extend farther down than *a* if the cluster were near enough for us to observe its fainter stars. Above point *b*, however, the main sequence seems to terminate; in most globular clusters, this point occurs at about absolute magnitude $M_v = +3.5$. From *b* to *c* there extends a sequence of stars that are yellow and red giants. The brightest and reddest of them (at *c*) at $M_v = -3$ are brighter than typical red giants in the solar neighborhood. A third

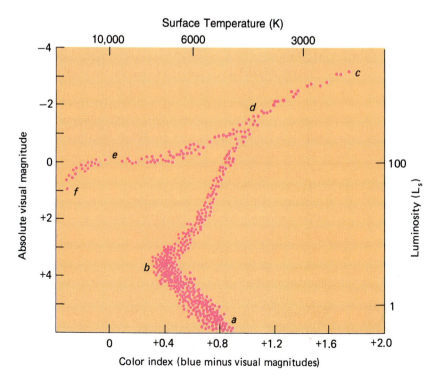

FIGURE 29.8 H–R diagram for a hypothetical globular star cluster.

FIGURE 29.9   The globular cluster 47 Tucanae (NGC 104) is one of the nearest globular clusters and is at a distance of 5000 pc. This group of old stars has a diameter of about 60 pc. (National Optical Astronomy Observatories)

sequence of stars extends from d to f; it is called the **horizontal branch** of the H–R diagram for a globular cluster. There is a gap in the horizontal branch near $M_v = 0$ (point e), where no star of constant light output is found. The stars observed in this gap are the RR Lyrae variables. The stars on the horizontal branch have already been through the red giant phase of evolution once (Chapter 30).

Figure 29.9 shows a picture of 47 Tucanae, which is one of the brightest and most thoroughly studied of the globular clusters. Figure 29.10 shows the color-magnitude diagram for this cluster.

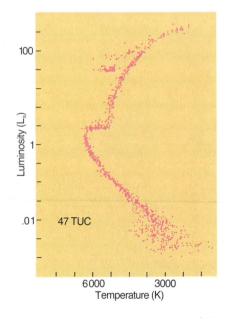

FIGURE 29.10   Color–magnitude diagram of the globular cluster 47 Tucanae. (Data by J. Hesser and collaborators)

The color-magnitude and H–R diagrams of any system of old stars are similar to those in Figures 29.8 and 29.10. The most massive stars in a cluster are the first ones to use up the hydrogen in their cores and evolve off the main sequence to become red giants. As more time goes on, stars of successively lower mass leave the main sequence, making it seem to burn down, like a candle. The globular clusters are 13 to 15 billion years old. The only stars still remaining on the main sequence have masses comparable to that of the Sun or less. All of the more massive stars have evolved away from the main sequence. Some are now red giants. Still others have completed their evolution and become white dwarfs or neutron stars (Chapter 30).

In the globular clusters, the giants are brighter than the brightest main-sequence stars and must, therefore, have increased in luminosity during their evolution from the main sequence. The red giants in globular clusters are even more luminous than are those in the oldest open clusters, such as M67. Calculations show these differences to be due to differences in chemical composition; stars in globular clusters have, on the average, lower abundances of heavier elements than do stars in open clusters.

Counts of stars of various kinds in old clusters show that the number of giants is very small compared with the number of main-sequence stars. All the giants now in these clusters must have evolved from a very short segment of the original main sequence, just above its present termination point. (In Figure 29.8, to avoid crowding, points are plotted for only about one in ten of the main-sequence stars that lie below the giant-branch turnoff.) In other words, all the giants in a single cluster are expected to have nearly the same mass and also to have had almost the same luminosity when they were on the main sequence. Those stars at the top of the giant branch on a cluster H–R diagram are, to be sure, further evolved than those, say, only halfway to the top, but they started from only very slightly greater luminosities on the main sequence, and so had only a slight "head start." We can conclude, therefore, that the sequence of stars forming the giant branch in the H–R diagram of an old cluster lies very nearly along the evolutionary tracks of individual stars.

The red giant stage is a relatively brief part of a star's life. In this stage of evolution, a star's nuclear fuel is consumed quickly, and further evolutionary changes soon follow.

(b) Color–Magnitude Diagrams of Young Clusters: Checking Out the Theory

What should the H–R diagram be like for a cluster whose stars have recently condensed from an interstellar cloud? After a few million years, the most massive stars should have completed their contraction phase and be on the main sequence, while the less massive ones should be off to the right, still on their way to the main sequence. Figure 29.11 shows the H–R diagram calculated by R. Kippenhahn and his associates at Munich for a hypothetical cluster at an age of 3 million years.

There are real star clusters that fit this description,

FIGURE 29.12   The cluster NGC 2264, which is at a distance of 800 pc. This region of newly formed stars is a complex mixture of red hydrogen gas, ionized by hot, embedded stars, and dark, obscuring dust lanes. (Anglo-Australian Telescope Board)

too. The first to be studied (about 1950) was NGC 2264, a cluster still associated with nebulosity (Figure 29.12). Figure 29.13 shows its H–R diagram. Among the several other star clusters in such an early stage is the one in the middle of the Orion Nebula (Figures 25.3 and 28.3).

After a short time—less than a million years after reaching the main sequence—the most massive stars use up the hydrogen in their cores and evolve off the main sequence to become red giants. As more time goes on, stars of successively lower mass leave the main sequence.

Figure 29.14 shows the H–R diagram of the real cluster NGC 3293 (Figure 29.15), which we judge to be a little less than 10 million years old. Note the gap that appears in the H–R diagram for NGC 3293 between the stars near the main sequence and the one red giant in the cluster. In the snapshot of stellar evolution represented by the H–R diagram, a gap does not necessarily represent a locus of temperatures and luminosities that stars avoid. In this particular case, it just represents a domain of temperature and luminosity through which a star

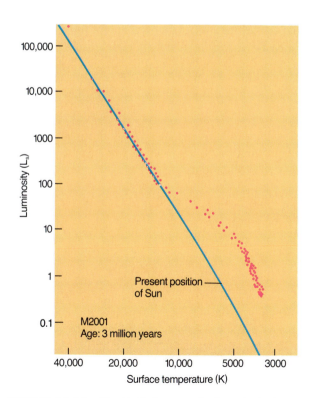

FIGURE 29.11   The H–R diagram of a hypothetical cluster at an age of 3 million years.

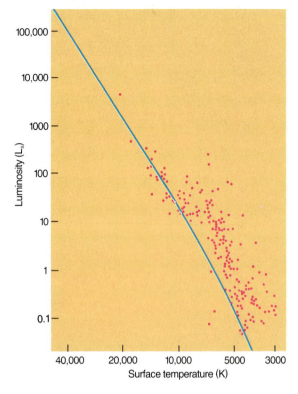

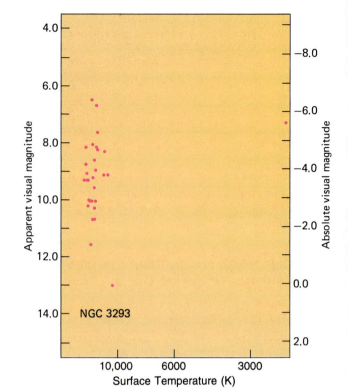

FIGURE 29.13   The H–R diagram of NGC 2264. (Data by M. Walker)

moves very quickly as it evolves. We see a gap because at this particular moment we have not caught a star in the process of scurrying across this part of the diagram. This gap, called the *Hertzsprung gap*, is broadest in the color-magnitude diagrams of clusters whose main sequences extend to high luminosities, and narrows for clusters whose main sequences terminate at successively lower luminosities. The gap disappears in globular clusters and in open clusters that are several billion years old.

At 4 billion years, stars only a few times as luminous as the Sun begin to leave the main sequence. Figure 29.16 shows a theoretical H–R diagram for such a cluster. (Since the calculations were terminated when the model stars became red giants, this diagram does not have horizontal-branch stars or RR Lyrae variables.) For comparison, Figure 29.17 shows schematically the observed H–R diagrams for several open clusters and for the globular cluster M3. Note that the H–R diagram of M67, the oldest of these open clusters, resembles that of the theoretical cluster in Figure 29.16.

The globular clusters have main sequences that terminate at a luminosity only slightly greater than the luminosity of the Sun. Star formation in these systems evidently ceased billions of years ago. Open clusters, on the other hand, are often located in regions of interstellar matter, where star formation can still take place. Indeed, we find open clusters of many ages, from less than 1 million to several billion years.

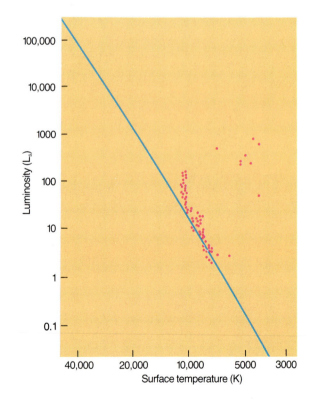

FIGURE 29.14   The color–magnitude diagram of NGC 3293 (*top*). Note the one red giant at the far right of the diagram. Only the brightest stars were observed, so only a portion of the main sequence is shown. A more complete H–R diagram for the slightly older cluster M41 (*bottom*) shows more of the main sequence and several red giants.

FIGURE 29.15   The open cluster of stars NGC 3293. All of the stars in this cluster formed at about the same time. The most massive stars, however, exhaust their nuclear fuel more rapidly and hence evolve more quickly than stars of low mass. As stars evolve, they become redder. The bright orange star in NGC 3293 is the member of the cluster that has evolved most rapidly. (Anglo-Australian Telescope Board)

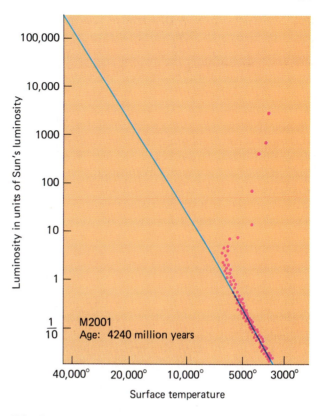

FIGURE 29.16   The H–R diagram of a hypothetical cluster at an age of 4.24 billion years.

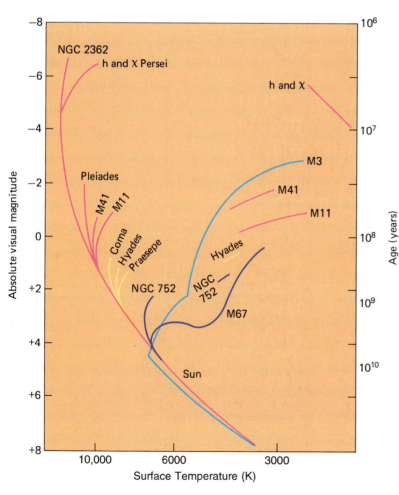

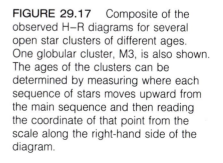

FIGURE 29.17   Composite of the observed H–R diagrams for several open star clusters of different ages. One globular cluster, M3, is also shown. The ages of the clusters can be determined by measuring where each sequence of stars moves upward from the main sequence and then reading the coordinate of that point from the scale along the right-hand side of the diagram.

(c)   Differences in Chemical Composition of Stars in Different Clusters

Hydrogen and helium, the most abundant elements in stars in the solar neighborhood, are also the most abundant constituents of the stars in all kinds of clusters. The exact abundances of the elements heavier than helium, however, vary from cluster to cluster. In the Sun and most of its neighboring stars, the combined abundance (by mass) of the heavy elements seems to be between 1 and 4 percent of the mass of the star. The strengths of the lines of heavy elements in the spectra of stars in most open clusters show that they, too, have 1 to 4 percent of their matter in the form of heavy elements.

Globular clusters, however, are a different story. Spectra of their brightest stars often show extremely weak lines of the heavy elements. The abundance of heavy elements in stars in typical globular clusters is found to range from only 0.1 to 0.01 percent of that of the Sun, or even less. Other very old stars outside globular clusters also have spectra that often indicate low abundances of heavy elements, although the difference between them and the Sun is not usually as extreme as for some of the stars in globular clusters. Differences in chemical composition are related to where and when stars were formed. The probable explanation of these phenomena is discussed in Chapters 30 and 32.

(d)   Color–Magnitude Diagrams and the Distances to Clusters

Distances to globular clusters can be determined from the period-luminosity relationship for RR Lyrae stars.

For open clusters and for globular clusters that do not contain RR Lyrae variables, there is another useful technique that makes use of the color-magnitude diagram such as the one shown in Figure 29.7. Most of the stars in a cluster lie along a main sequence similar to that defined by stars in the neighborhood of the Sun. We find a main sequence, therefore, in the color-magnitude diagram of a cluster, with the bluer, brighter-appearing stars (which are really the more luminous main-sequence stars in the cluster) farther up on the diagram than the redder, fainter-appearing stars. From the ordinary H–R diagram, we know what absolute magnitudes correspond to various color indices along the main sequence. The difference, at any given color index, between the apparent magnitude of the cluster stars and the absolute magnitude of known main-sequence stars of the same color is the distance modulus of the cluster.

As a numerical example, consider the hypothetical cluster whose color-magnitude diagram is shown in Figure 29.7. A main-sequence cluster star of color index $+0.6$ is seen to have an apparent magnitude of $+15$. But this is a star like the Sun, whose absolute magnitude is $+5$. At a distance of 10 pc, therefore, this star would appear 10 magnitudes brighter than it does at the actual distance of the cluster. Since 10 magnitudes corresponds to a factor of 10,000 in light, then the cluster must be 100 times more distant than 10 pc, or must be 1000 pc away.

Clusters, then, allow us to measure the ages and distances of stars and to learn what happens as they grow old. In the next chapter, we will see how stars die.

SUMMARY

29.1   The fusion of hydrogen to form helium changes the interior composition of stars, which in turn results in changes in temperature, luminosity, and radius. As stars age, they evolve away from the **zero-age main sequence** and become red giants. Calculations that show what happens as stars age can be checked by measuring the properties of stars in clusters. The members of a given cluster all formed at about the same time, that is, in a time short compared with the lifetimes of stars, and have the same composition, so that comparison of observations and theory is fairly straightforward.

29.2   **Globular clusters** have diameters of 20 to 100 pc, contain hundreds of thousands of stars, and are distributed in a halo around the galaxy. **Open** or **galactic clusters** typically contain hundreds of stars, are located in the plane of the Galaxy, and have diameters of less than 10 pc. **Associations** are found in regions of gas and dust and contain extremely young stars.

29.3   **Cepheids** and **RR Lyrae** variables are two types of **pulsating** stars. Both obey the **period-luminosity** relation, and so their distances can be derived from measurements of their periods. A few cepheids are found in open clusters. Most globular clusters contain at least a few RR Lyrae variables. Since all RR Lyrae stars have about the same average luminosity, they are very useful for measuring distances to globular clusters.

29.4   A pulsating variable changes size by expanding and contracting, and its temperature also changes. Both variations result in changes in brightness. Measurements of radial velocities show that the radius of a cepheid variable changes by a few percent during the pulsation cycle.

29.5   The H–R diagram of the stars in a cluster changes systematically as a cluster evolves. The most massive stars evolve the most rapidly. In the youngest clusters and associations, highly luminous blue stars are on the main sequence; the stars with the lowest masses lie to the right of the main sequence and are still contracting toward it. With passing time, stars of progressively lower mass evolve away from the main sequence. In globular clusters, which have ages typically of 13 to 15 billion years, there are no luminous blue stars at all. The composition of the young stars in open clusters is similar to that of the Sun. The old stars in globular clusters have abundances of elements heavier than helium that are only 0.1 to 0.01 percent, or even less, of those found in the Sun. Since the main sequence follows the same relationship between luminosity and temperature in all clusters, measurements of the apparent magnitudes and temperatures of main-sequence stars are enough to determine the distance to a cluster.

E X E R C I S E S

THOUGHT QUESTIONS

1. Use star charts to identify at least one open cluster that is visible at this time of year. The Pleiades and Hyades are good fall objects, and Praesepe is a good springtime object. Go out and look at these clusters with binoculars and describe what you see. How would you expect the appearance of a globular cluster, as seen through binoculars, to differ from that of an open cluster?

2. Table 29.2 indicates that stellar associations can emit even more light than a globular cluster. How is this possible if the associations have so few stars?

3. What color would a globular cluster appear? Why?

4. Explain how an H–R diagram can be used to determine the age of a cluster of stars.

5. The H–R diagram for field stars (that is, stars all around us in the sky) shows very luminous main-sequence stars and also various kinds of red giants and supergiants. Explain these features, and interpret the H–R diagram for field stars.

6. In the H–R diagrams for some young clusters, stars of very low and very high luminosity are off to the right of the main sequence, whereas those of intermediate luminosity are on the main sequence. Can you offer an explanation? Sketch an H–R diagram for such a cluster.

7. If the Sun were a member of the cluster NGC 2264, would it be on the main sequence yet? Why?

8. Explain how you could decide whether red giants seen in a star cluster probably had evolved away from the main sequence or were still evolving toward the main sequence.

9. If all the stars in a cluster have the *same age*, how can clusters be useful in studying evolutionary effects?

10. Suppose a star cluster were at such a large distance that it appeared as an unresolved spot of light through the telescope. What would you expect the color of the spot to be if it were the image of the cluster immediately after it was formed? How would the color differ after 10^{10} years. Why?

PROBLEMS

11. Suppose globular clusters have orbits about the galactic center with very high eccentricities—near unity. When a particular globular cluster is at its farthest from the center of the Galaxy, its distance from the center is 10^4 pc. What is its period of galactic revolution? (*Hint:* 1 pc $= 2 \times 10^5$ AU. Assume that the mass of the Galaxy is 10^{12} solar masses.)

12. From the data of Table 29.2, estimate the average mass of the stars in each of the three different cluster types. Is the average mass of stars in an open cluster likely to be larger or smaller than the average mass in a globular cluster?

13. What is the density in solar masses per cubic parsec of the following clusters: **(a)** a globular cluster 50 pc in diameter containing 10^5 stars? **(b)** a stellar association of 100 solar masses and 20 pc in radius?
Answer: (a) 1.5 solar masses per cubic parsec; **(b)** 0.003 solar masses per cubic parsec

14. The RR Lyrae stars in a particular globular cluster appear at apparent magnitude $+16$. How distant is the cluster? (Ignore interstellar absorption and assume that the RR Lyrae stars have absolute magnitudes of $+1$.)

15. A main-sequence star of color index 0 has an absolute magnitude of about $+1$. In the color-magnitude diagram of a certain cluster, it is noted that stars of 0 color index have apparent magnitudes of about $+6$. How distant is the cluster? (Ignore interstellar absorption.)

16. Suppose a star spends 10×10^9 years on the main sequence and uses up 10 percent of its hydrogen. Then it quickly becomes a red giant with a luminosity 100 times as great as the luminosity it had on the main sequence and remains a red giant until it exhausts the rest of its hydrogen. How long a time would it be a red giant? Ignore helium fusion and other nuclear reactions and assume that the star brightens from main sequence to red giant almost instantaneously.

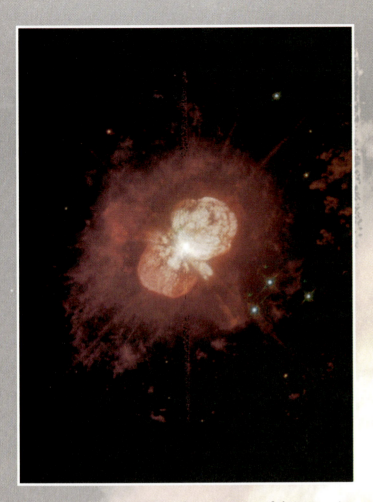

As stars evolve, they often eject some of their mass into space. This image, taken with the repaired Hubble Space Telescope, shows the material surrounding the star Eta Carinae. With a mass of at least 100 times the mass of the Sun, it is one of the most massive stars known. The red outer material, which was ejected in 1841 and is moving away from the star with a velocity of about 1000 km/s, is rich in nitrogen and other elements formed in the interior of the massive star. The bright blue-white nebulosity consists of material that was ejected at lower velocities and is therefore seen closer to the star. It appears blue-white because it contains dust and reflects the light of Eta Carinae, which is a hot supergiant.

(J. Hester/Arizona State University and NASA)

On July 4, 1054 AD—long before July 4 had its present-day meaning—Chinese astronomers saw some celestial fireworks that astonished them. A star had appeared in the sky that was so bright, it was visible during the daytime. They called it a "guest star" because, for a time, it was an honored guest of the Sun, where few other stars ever appeared.

When astronomers look at the position of that guest star today, our radio telescopes reveal an object that pulses with remarkable regularity, some thirty times each second. The observations indicate that the pulsing object is the spinning, collapsed remnant of that once brilliant star. This remnant is made of material so dense, that to duplicate it on Earth, we would have to take all the people in the world, and squeeze them into a single raindrop.

We now understand that the star of 1054 AD was a *supernova* explosion, which marked the death of a massive star. As much as 90% of a star's material can be expelled by such an explosion, whose temperatures can reach hundreds of billions of degrees. It is in the crucible of these exploding stars that many of our own atoms were once forged and it is through the action of such supernovae that the heavier elements made in stars make their way into the reservoirs of the Galaxy.

In this chapter we explore not only the death of stars, but the vast recycling scheme of the cosmos, where the death of one generation of stars enriches the generations that are to come, and the primordial hydrogen and helium are slowly transformed into the more interesting atoms required to produce our readers.

Subrahmanyan Chandrasekhar (b. 1910) was born in India and educated at Madras and Cambridge Universities. He has spent most of his career at the University of Chicago, where he has made fundamental contributions in almost every area of astrophysics, including the physical theory of white dwarfs. Chandrasekhar received the Nobel Prize in 1983.

EVOLUTION AND DEATH OF OLD STARS

Before dawn on February 24, 1987, Ian Shelton, a Canadian astronomer working at an observatory in Chile, pulled a photographic plate from the developer. Two nights earlier he had begun a survey of the Large Magellanic Cloud. Shelton planned to study variable stars in the Cloud, but the plates from the first two nights did not have sharp images. It was difficult to guide long exposures with the telescope that Shelton was using, which was more than 50 years old. The plate from the third night was a success. The images were sharp. Shelton examined the Tarantula Nebula, a region of bright glowing gas where active star formation is occurring. Nearby, where there should have been only faint stars, he saw a large spot on his plate. Concerned at first that his photograph was flawed, Shelton went outside to look at the Large Magellanic Cloud—and saw that a very bright new object had actually appeared in the sky (Figure 30.1).

What Shelton had discovered was an exploding star, a **supernova.** Now known as SN 1987A, since it was the first supernova discovered in 1987, this brilliant newcomer to the southern sky gave astronomers for the first time an opportunity to study in detail the death of a star.

In the chapter that follows, we trace the late stages of evolution first of stars like the Sun and then of more massive stars. By the late stages of evolution, we refer to what happens after a star becomes a red giant for the first time.

30.1 RED GIANTS

(a) Evolution Beyond the Main Sequence

Sooner or later all stars must die. Chapter 29 followed the evolution of stars up to the point where they leave the main sequence and become red giants. Already it has become apparent that the evolution of a star is shaped by the balance between two forces—pressure and gravity. When these two forces are in balance, a star is stable. If pressure forces exceed the force of gravity, a star will expand. If pressure, for any reason, becomes inadequate to resist gravity, a star will contract.

As a star consumes its supply of hydrogen and evolves away from the main sequence, these two forces do get out of balance. The star changes the way it generates energy, and related changes in its size and internal structure are required to achieve a new equilibrium. Each new stage of equilibrium is only temporary. At each phase of its evolution a star will ultimately exhaust the supply of whatever fuel it is using to generate energy, the balance between pressure and gravity will be destroyed, and the structure of the star will change in such a way as to restore equilibrium.

Sometimes the changes in structure are accompanied by loss of mass. This mass loss may occur gently, in the form of a stellar wind, but often it is explosive and can destroy most or all of the star. Any major loss of mass

FIGURE 30.1   Before and after pictures of the field around Supernova 1987A in the Large Magellanic Cloud. The difference in image quality between these pictures is an effect of the Earth's atmosphere, which was steadier when the plates used to make the presupernova picture were taken. (Anglo-Australian Telescope Board, 1987)

will influence the star's evolution, of course, and the final state of the object after collapse may be very different from what we would have expected from the initial mass of the star on the main sequence. We thus must be careful to distinguish between the initial, main-sequence mass of a star and the mass of the final collapsed object.

In the end, all sources of energy will be completely consumed, and the star will cease to shine. Its ultimate fate will depend on its mass. Stars that start their lives with masses similar to that of the Sun will end their lives as white dwarfs. Massive stars will become neutron stars or possibly black holes or may even be completely disrupted.

(b)   Degenerate Matter

To understand the final stages in the life of a star, we will first discuss a new kind of pressure—the pressure exerted by an electron-degenerate gas—that becomes important when the star begins to exhaust its store of nuclear energy.

The electrons in a neutral atom occupy certain allowed *states*. We saw in Section 7.4 how an electron can change states when the atom absorbs or emits energy. When the atom is ionized, at least one of its electrons is free, and we speak of them as occupying a *continuum* of states. Even free electrons, however, do not have an infinite number of states available to them.

Electrons, protons, and neutrons belong to a class of particles that must obey a rule of quantum mechanics that insists that no two of them occupy exactly the same state. What this means is that no two electrons can be in the same place at the same time doing the same thing. The rule is known as the *Pauli exclusion principle*, after the Austrian physicist Wolfgang Pauli, who enunciated it. We specify the *place* of the electron by its precise position in space, and we specify what it is doing by its momentum and the way it is spinning.

Quantum mechanics also shows (Section 7.4) that the simultaneous position and momentum of an electron (or anything else) cannot be known any more precisely than is allowed by the Heisenberg uncertainty principle. Specifically, this means that the amount by which the x-coordinate of its location is uncertain, Δx, and the precision of the x-coordinate of its momentum, Δp_x, are governed by the condition

$$\Delta x \Delta p_x \geq \frac{h}{2\pi},$$

where h is Planck's constant (6.626×10^{-34} joule·s). Since the state of an electron cannot be specified more precisely, it is fuzzy. The Pauli principle permits only one electron in each of these fuzzy states (actually, two electrons are permitted if they are spinning in opposite directions).

Imagine the free electrons in an ionized gas with a certain temperature. That temperature determines the distribution of velocities, and hence the range of momenta, for the electrons. If the temperature is high enough, the momentum range is large, and the electrons have plenty of possible momentum-position states to occupy without violating the Pauli principle. On the other hand, what if the temperature is *not* high enough? (How high is "high enough" depends on how many electrons are crowded into a given volume—that is, their density.) When all the available states (of position and momentum) are occupied, the electrons will resist further crowding with overwhelming pressure. Such electrons are said to be **degenerate,** and the gas is an electron-degenerate gas.

The electrons in a degenerate gas move about, as do particles in any gas, but not with freedom. A particular electron cannot change position or momentum until another electron in an adjacent state gets out of the way. It is as if all the particles were geared together. The situation is much like that in the parking lot at the end of a big football game. Cars are closely packed, and one cannot move until another next to it moves, leaving an empty space to be filled by the car behind. If one car leaves the lot, others can move over, and still others follow behind them, producing a flow of cars toward the exit. In a similar way, it turns out that heat can flow through an electron-degenerate gas with great ease.

Now what has all this to do with stars? Simply that if part or all of a star contracts enough and increases sufficiently in density, it will become electron-degenerate. The pressure of the degenerate electrons can halt the contraction.

In a typical stellar environment even an electron-degenerate gas is mostly empty space. The electrons are by no means packed into contact. Their densities are typically a million to a billion times lower than the density of the atomic nucleus. The nuclei themselves still move about freely among the electrons, obey the usual perfect gas law, and exert the normal pressure of particles of their masses and temperature. The pressure exerted by the degenerate electrons, however, generally swamps that of the nuclei, so it is the electron pressure that dominates and controls the structure of that region of the star.

Other particles become degenerate as well. In particular, there are also neutron stars formed of a degenerate-neutron gas. But neutrons are nearly 2000 times more massive than electrons and at the same temperature have very much greater momentum. Neutrons, therefore, must be crowded to enormously higher densities before filling their available momentum states and becoming degenerate.

(c)   From Main-Sequence Star to Red Giant

Now back to our story of how stars evolve. The way in which a star evolves depends on its mass. We will first consider stars with masses less than about twice the mass of the Sun. Like all main-sequence stars, these low-mass stars initially burn hydrogen in the central regions and in the process build up a small core of helium. When the hydrogen in the core is used up, the burning of hydrogen stops. Heat begins to leak out of the core. But if the center cools down, its pressure drops. A lower pressure cannot support the weight of the layers of the star that lie outside the core. The overlying layers crush the core and force it to shrink. As the core contracts, it releases gravitational potential energy, which heats up both the core and the region immediately surrounding it. Eventually this region becomes hot enough for hydrogen to begin to fuse in a shell around the core.

While the interior of the star is getting hotter, the outer portions of the star are expanding and cooling, and the star becomes a red giant (Chapter 29). The outer layers of the star transport energy by convection, and as time passes, this convective region becomes deeper. Eventually the convective currents reach deep enough into the star to dredge up material that has undergone fusion, and this nuclear-processed material is carried up to the surface of the star.

The core itself still has no source of thermonuclear energy and continues to shrink in diameter and increase in density. Ultimately, it becomes so dense that matter in the innermost part is electron-degenerate. Meanwhile, the surrounding shell soon exhausts its hydrogen and adds to the helium core. With no energy source, the core continues to shrink and grow hotter because of the release of additional gravitational energy. Thus the degenerate core continues to contract and heat. By the time the star reaches its maximum luminosity at the tip of the red giant branch on the H–R diagram, its central temperature exceeds 100 million K.

Ultimately the temperature becomes high enough to cause helium atoms to fuse together. When two helium atoms combine, they do not form a stable nucleus. The combination does sometimes, however, survive long enough to be joined by a third helium atom to form carbon. This process has been named the *triple-alpha process*, and it is expected to begin abruptly in the central core of a red giant. As soon as the temperature becomes high enough to start the triple-alpha process, the extra energy released is transmitted quickly through the entire degenerate core, producing a rapid heating of all the helium there. Because the gas in the core is degenerate, the red giant lacks a safety valve that was available to the star when it was still on the main sequence. In a main-sequence star, if the core of the star becomes hotter for any reason, then the core can expand slightly, the temperature will drop, nuclear reactions will slow, and the star will become stable again.

In a degenerate core, the pressure is determined primarily by degeneracy effects. Raising the temperature slightly does not increase the pressure, but it will speed nuclear reactions, which will in turn raise the temperature still further. Helium burning accelerates, and we have a runaway generation of energy. Once helium burning begins, the entire core is reignited in a flash.

The energy released in this *helium flash* is so great that it removes the degeneracy and expands the core. The expansion of the outer parts of the red giant reverses, and the outer layers of the star shrink rapidly. Calculations show that the star's surface temperature increases and its luminosity decreases. At this time, the core of the star is nondegenerate and is stable, fusing helium to form carbon. Often a newly formed carbon nucleus is joined by another helium nucleus to produce a nucleus of oxygen. Surrounding this helium-burning core is a shell in which hydrogen is fusing to form helium. Such a star is called a *horizontal-branch star* because of its location on the H–R diagram (Figure 30.2).

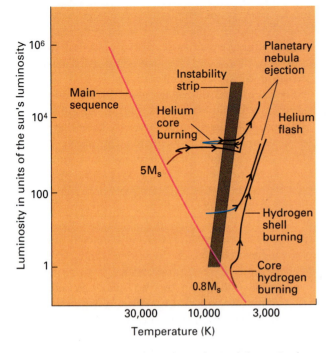

FIGURE 30.2 This diagram shows the evolutionary tracks for stars of two different masses. After exhausting hydrogen in its core, the 0.8-solar-mass star evolves to the tip of the red giant branch, where the helium flash occurs. The star then becomes a horizontal branch star and evolves again up the giant branch, where it ejects a planetary nebula, The 5-solar-mass star goes through similar stages of evolution except that it does not undergo a helium flash.

After a while, the helium in the central region will be used up, and energy is again no longer produced in the core. Now we have a core of carbon and oxygen surrounded by a shell where helium is still burning; farther out in the star is another shell where hydrogen is burning. On the H–R diagram the star moves back to the red giant domain. Calculations indicate that all of these evolutionary stages occur in tens or hundreds of millions of years—a brief time compared with the main-sequence lifetime.

Low-mass stars do not have enough mass to compress the carbon-oxygen core to initiate yet another stage of nuclear burning. At the tip of the red giant branch, the star ejects its envelope (Section 30.3) and leaves behind a white dwarf. The star then begins the last stages of its life.

30.2 WHITE DWARFS: ONE FINAL STAGE OF EVOLUTION

When a star with a mass similar to that of the Sun exhausts its store of nuclear energy, it can only contract and release more of its potential energy. Eventually, the shrinking star will attain an enormous density. We observe such stars—the extremely compact **white dwarf** stars (Section 24.3c), whose mean densities

range up to over 1 million times that of water. In a white dwarf the electrons are completely degenerate throughout the star, save for a very thin layer at the surface.

(a) Structure of White Dwarfs

White dwarfs are simpler than most stars because the pressure that supports a white dwarf in hydrostatic equilibrium is supplied almost entirely by the degenerate electrons. The internal pressure in a white dwarf, therefore, does not depend on temperature, but only on the density. The volume to which a star can be compressed before the electrons become degenerate depends on the amount of gravitational potential energy that can be released by the collapsing star, which in turn depends on its mass. The size of a white dwarf, therefore, depends on its mass—the more massive the white dwarf, the smaller its size. A white dwarf of one solar mass has a radius about the size of the Earth!

In the more massive white dwarfs, some of the electrons have speeds that are an appreciable fraction of that of light, and a rigorous treatment must include the effects of special relativity (Chapter 8). The first such rigorous models of white dwarfs were constructed by the Nobel prize–winning astrophysicist S. Chandrasekhar, who showed that it is possible to calculate the radius of a white dwarf if its mass is known. For a mass larger than 1.4 times the mass of the Sun, hydrostatic equilibrium cannot be achieved (Figure 30.3). Thus, 1.4 solar masses is the upper limit to the mass of a white dwarf. Stars that are more massive than 1.4 solar masses at the time they run out of sources of nuclear energy must continue to collapse to a size that is far smaller than that of a white dwarf. Precisely what these stars become will be discussed in Section 30.7.

White dwarfs have hot interiors—tens of millions of degrees Kelvin. At those temperatures and at the high

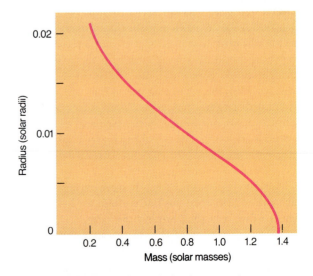

FIGURE 30.3 Theoretical relation between the masses and radii of white dwarf stars.

densities of these stars, any remaining hydrogen would undergo violent fusion into helium, giving the stars luminosities many times higher than observed. Consequently, white dwarfs can have no hydrogen in their interiors. The interiors of most white dwarfs are composed of carbon and oxygen, the principal products of helium burning. White dwarfs formed by very low-mass stars may be nearly pure helium. White dwarfs formed by stars with initial masses in the range of 8 to 12 solar masses may consist predominantly of oxygen, neon, and magnesium (Section 30.5).

(b)   Evolution of White Dwarfs

A white dwarf is presumed to have exhausted its available nuclear energy sources. It cannot contract and release gravitational potential energy because of the great pressure of the electron-degenerate gas. Thus its only source of energy is the thermal energy (that is, kinetic energy) of the nondegenerate *nuclei* of atoms, behaving like ordinary gas particles, scattered throughout the degenerate electrons. As these nuclei slow down (cool), the electron gas conducts their thermal energy to the surface. At the boundary of the star, the very thin, skin-like layer of nondegenerate gas radiates this energy into space. Only the opacity of this outer layer keeps the nuclei in the interior of the star from cooling off at once.

Gradually, however, a white dwarf does cool off, much like a hot pan when it is removed from a stove. The cooling is relatively rapid at first, but as the star's internal temperature drops, so does its cooling rate. Calculations indicate that its luminosity should drop to about 1 percent of the Sun's in the first few hundred million years of its existence as a white dwarf. Since the radius of a white dwarf is constant, its luminosity (by Stefan's law) is proportional to the fourth power of its effective temperature. Therefore, as a white dwarf cools, its track on the H–R diagram is along a diagonal

line toward the lower right, that is, toward low temperature and low luminosity.

Eventually, a white dwarf will cease to shine at all. It will then be a *black dwarf,* a cold mass of degenerate gas floating through space. (Do not confuse black dwarf with black hole; see Chapter 31.) A long time may be required, however, for a star to cool off to the black dwarf stage. It may be that the Galaxy is not old enough for any star to have yet had time to become a black dwarf.

(c)   Novae—Last Bursts of Glory

While isolated white dwarfs die in the unspectacular fashion we have just described, those white dwarfs that have a nearby stellar companion may call attention to their demise by becoming **novae.**

Nova literally means "new." An astronomical nova is an existing star that suddenly emits an outburst of light. In ancient times, when such an outburst brought a star's luminosity up to naked-eye visibility, it seemed like a new star. Novae remain bright for only a few days or weeks and then gradually fade. They seldom remain visible to the unaided eye for more than a few months. The Chinese, whose annals record novae from centuries before Christ, called them "guest stars." Only occasionally are novae visible to the naked eye, but on the average, two or three are found telescopically each year. Many must escape detection. Altogether there may be as many as two or three dozen nova outbursts per year in our Galaxy. The light curve of a typical nova is shown in Figure 30.4.

Novae occur in close binary-star systems in which one member is a white dwarf. The outburst of light occurs when mass is transferred from the normal star to its white dwarf companion. This mass transfer is a consequence of gravity's acting on the outer atmosphere of a star that is expanding as it evolves. Specifi-

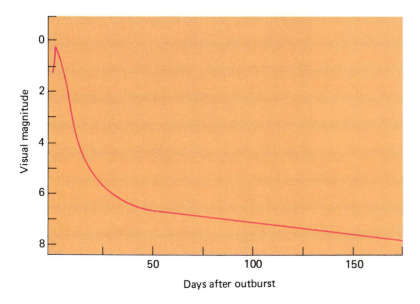

Figure 30.4 Light curve of Nova Puppis, 1942.

cally, on the line between the centers of the two stars of a binary system there is a point where the gravitational attraction between a small bit of matter and one of the stars would be equal to that between the matter and the other star. If the two stars were of equal mass, for example, that point would lie halfway between them.

Now suppose that one of the stars of a binary system, in becoming a red giant, increases in size, so that its outer surface extends through the point where the gravitational forces of the two stars are equal. The stellar material no longer knows which star it belongs to and is no longer gravitationally bound to the original star. An exchange of material from one star to the other may result.

If the star receiving the mass is a white dwarf, for a while (years to centuries) hydrogen-rich material from the outer layers of the non-white dwarf star just piles up on the surface of the white dwarf. Gradually, however, the weight of this matter increases, and hence so does the temperature, until it approaches that of the degenerate interior of the dwarf. Then, explosively, hydrogen burning ignites through the CNO cycle, and like a nuclear bomb it blows off the outer layer of matter that had accumulated on the white dwarf. Velocities of ejection up to 1000 km/s or more are observed. The shell of gas blown off in a nova explosion typically contains from 10^{-5} to 10^{-4} solar mass. Some months or years after the outburst, the expanding envelope may become visible on telescopic photographs.

After the outburst, the white dwarf settles down, but since mass is continually flowing onto it from its companion, the process eventually repeats. Some novae have long been known to be recurrent. Generally, the more intense the outburst, the longer the period of quiescence between nova flareups. According to the model, all novae recur on some time scale. The most violent novae, which reach absolute visual magnitudes of -6 to -9 (about $10^5 L_s$), may wait thousands of years between outbursts.

30.3   MASS LOSS

What kinds of stars eventually end their lives as white dwarfs? The critical determining factor is mass. White dwarfs can be no more massive than about 1.4 times the mass of the Sun. Yet somewhat surprisingly, observations indicate that stars that have masses larger than 1.4 times the mass of the Sun *at the time they are on the main sequence* also complete their evolution by becoming white dwarfs. One argument that this must be so comes from the sheer number of white dwarfs that have been identified. There are far too many to be accounted for if only stars with main-sequence masses less than 1.4 solar masses become white dwarfs.

A second argument is even more compelling. White dwarfs have been found in young open clusters—clusters so young that only stars with masses in the range six to eight times the mass of the Sun have had time within the lifetime of the cluster to exhaust their supply of nuclear energy and complete their evolution to the white dwarf stage. In the Pleiades, for example, stars with masses five times the mass of the Sun are still on the main sequence. This cluster also has at least one white dwarf, and its main-sequence mass must, therefore, have exceeded five solar masses.

(a)   Planetary Nebulae

If stars that initially have masses in the range six to eight times the mass of the Sun are to become white dwarfs, then somehow they must get rid of enough matter so that their total mass at the time when nuclear energy generation ceases is less than 1.4 masses. How do they do it?

FIGURE 30.5   The Helix Nebula, NGC 7293, is at a distance of about 125 pc and is the planetary nebula nearest the Sun. On photographs the Helix has a diameter about the same as that of the full moon. The greenish color is produced by emission lines of ionized oxygen; the red color is due to nitrogen and hydrogen. (Anglo-Australian Telescope Board)

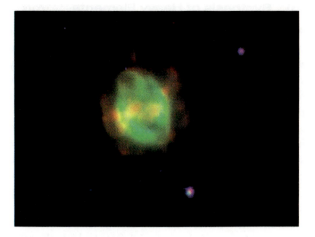 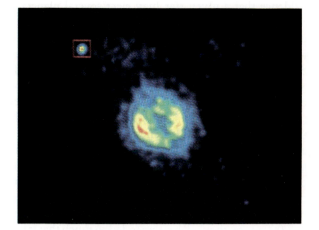

FIGURE 30.6 Optical *(left)* and radio *(right)* images of the planetary nebula NGC 7354. The optical image is color-coded in such a way that red corresponds to forbidden emission from singly ionized nitrogen, green corresponds to forbidden emission from doubly ionized oxygen, and blue corresponds to emission from ionized helium. The radio emission is produced by hydrogen gas glowing at 10,000 K. (Optical image courtesy of Bruce Balick, University of Washington; radio image courtesy of National Radio Astronomy Observatory/AUI)

For stars of no more than a few solar masses—including red giants in globular clusters—one of the most important mass-ejection mechanisms may be the planetary nebula phenomenon. **Planetary nebulae** are shells of gas ejected from red giant stars. The red giants then quickly evolve to become low-luminosity, hot stars, which illuminate the expanding gas shells (see Figure 30.2). Planetary nebulae derive their name from the fact that a few bear a superficial telescopic resemblance to planets. Actually they are thousands of times larger than the entire solar system and have nothing whatever to do with planets. Examples of planetary nebulae are shown in Figures 30.5 and 30.6.

The most famous planetary nebula is the Ring Nebula in the constellation Lyra (Figure 30.7). It is typical of many planetaries in that it is actually a hollow shell of material, even though it appears as a ring. The explanation is that when we observe near the center of the nebula, we are viewing through the thin front and back parts of the shell. When we observe near the edge of the nebula, our line of sight encounters a long path through the glowing material. Similarly, a soap bubble often appears to be a thin ring.

Altogether, about 1000 planetary nebulae have been catalogued in our own Galaxy. Doubtless there are many distant ones that have escaped detection, so there must be some tens of thousands in the Milky Way system. Nevertheless, among the tens of billions of stars in the Galaxy, planetary nebulae must be classed as rare objects.

An appreciable amount of material is ejected in the shell of a planetary nebula. From the light emitted by the shells, we calculate that they must have masses of 10 to 20 percent that of the Sun. The shells expand away from their parent stars at speeds of 20 to 30 km/s. There is also

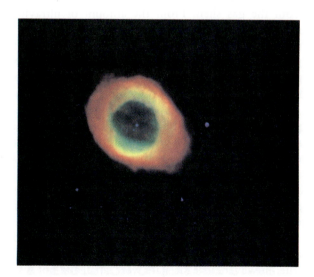

FIGURE 30.7 The Ring Nebula. The image is color-coded in the same way as the optical image in Figure 30.6. Note that the helium line, which requires the most energy to produce, is formed in the central regions of the nebula. The hydrogen line, which requires the least energy, is found near the outer boundary of the nebula, far from the central star, which provides the photons necessary to produce the spectral lines. The central star, which has no emission lines at these wavelengths, is invisible in this photograph but is clearly seen in broadband optical images. (Bruce Balick, University of Washington)

fusion of four hydrogen atoms to form helium. After the hydrogen in the core of the star is exhausted, the core shrinks under the weight of the star's outer layers. As the core contracts, it releases gravitational potential energy and heats the region surrounding it. Hydrogen then begins to burn in the region that surrounds the core.

The nuclei in the interiors of stars are all completely ionized and are positively charged. Since positive charges repel each other, nuclei must be moving at very high velocities in order to overcome this repulsion and undergo fusion. Velocity depends on temperature, and that is the reason why high temperatures are required for fusion. Furthermore, more massive nuclei have higher positive charge and require higher temperatures for fusion than do protons.

In stars with masses in the range of 2 to 8 solar masses, helium ignition begins when the temperature of the core is about 10^8 K. For these stars, the core is not degenerate, and so there is no helium flash. The intermediate-mass stars spend about 20 to 30 percent of their total lifetime converting helium to carbon and oxygen. When they exhaust their core helium, the stars again move up the red giant branch (Figure 30.2). Such red giants contain an inert core of carbon and oxygen, surrounded by a shell in which helium burning is taking place. Still farther out in the star there is a hydrogen-burning shell. What happens next is uncertain. Some stars may initiate explosive carbon burning and destroy themselves. Most of the stars probably eject material that forms a planetary nebula and become white dwarfs, just as the low-mass stars do.

In stars with masses greater than about 8 solar masses, the weight of the outer layers is sufficient to force the core of carbon and oxygen to contract. The carbon-oxygen core continues to shrink until it becomes hot enough to ignite carbon nonexplosively. The carbon can then form neon, still more oxygen, and magnesium. Stars that had masses between about 8 and 10 solar masses when they were on the main sequence probably end their lives as white dwarfs made of oxygen, neon, and magnesium.

In stars with masses greater than about 10 solar masses, the core can be crushed still further until it reaches a temperature high enough to lead to the fusion of even heavier nuclei. Each cycle lasts for a shorter time than the one that preceded it. When the temperature in the core finally becomes hot enough to fuse silicon to form iron, then energy generation must cease. Up to this point, each fusion reaction has released energy. Iron nuclei, however, are so tightly bound that energy is *required* to fuse iron with any other atomic nucleus.

At this stage of its evolution, a massive star resembles an onion in the sense that it has several distinct layers. There is an iron core, and at progressively larger distances from the center, layers of decreasing temperature are burning successively silicon, oxygen, neon, carbon, helium, and finally hydrogen (Figure 30.9).

FIGURE 30.9   Just before its final gravitational collapse, a massive star is layered like an onion. The iron core is surrounded by spherical shells of silicon, sulfur, oxygen, neon, carbon mixed with some oxygen, helium, and finally hydrogen.

(b)   Supernovae—The Final Act

What happens next to a star with an iron core? The computations become very complicated, but we can trace the events in a schematic way. In effect, a massive star builds a white dwarf in its center, where no nuclear reactions are taking place. For stars that begin their evolution with masses of at least 10 to 12 times the mass of the Sun, this white dwarf is made of iron. For stars with initial masses in the range eight to ten times the mass of the Sun, the white dwarf that forms the core is made of oxygen, neon, and magnesium because the star never gets hot enough to form elements as heavy as iron. Whatever its composition, the white dwarf embedded in the center of the star is supported against further gravitational collapse by degenerate electrons.

While no energy is being generated within the white dwarf core of the star, fusion does still occur in shells surrounding the core. As the core accretes the ashes of the fusion reactions going on in the surrounding shell, the mass of the core grows. Ultimately, it is pushed over the **Chandrasekhar limit** of 1.4 times the mass of the Sun. That is, it becomes so massive that the force exerted by degenerate electrons is no longer great enough to resist gravity. The electrons merge with protons inside the nuclei of iron and other atoms to produce neutrons. The removal of electrons removes the main source of support for the core, and it collapses.

This collapse occurs very rapidly. In less than a second, the core, which originally was approximately the same diameter as the Earth, collapses to a diameter that is less than 100 km. The speed with which material

TABLE 30.1	FATE OF STARS OF DIFFERENT MASSES
Initial Mass (Mass of Sun = 1)	Final Evolutionary State
< 0.01	Planet
$0.01 < M < 0.08$	Brown dwarf
$0.08 < M < 0.25$	Helium white dwarf
$0.25 < M < 8$	Carbon-oxygen white dwarf
$8 < M < 10$	Oxygen-neon-magnesium white dwarf
$10 < M < 40$	Supernova; neutron star
$40 < M$	Supernova; black hole

falls inward reaches one-fourth the speed of light. The collapse halts only when the density of the core reaches the density of an atomic nucleus. In effect, the matter in the core has merged to form a single gigantic nucleus.

This nuclear material strongly resists further compression, abruptly halting the collapse. The shock of the abrupt jolt generates waves throughout the outer layers of the star and causes the star to blow off those outer layers in a violent supernova explosion.

Table 30.1 summarizes the discussion so far about what happens to stars of different initial masses. The mass limits that correspond to various outcomes may change somewhat as models improve. It is also not certain whether or not stars with masses in the range four to eight times the mass of the Sun form white dwarfs. The nuclear reactions in these stars may go one step beyond helium fusion to reactions involving carbon, which may occur explosively and lead to a supernova explosion. If the uncertainties are kept in mind, Table 30.1 does provide a useful guide to what theorists think is the ultimate result of stellar evolution.

30.6 SUPERNOVAE

Five supernovae have been observed in our own Galaxy during the past 1000 years. The Chinese reported the temporary appearance of "guest stars" in the years 1006, 1054, and 1181. The first of these was reported to be nearly as bright as the quarter moon. The two remaining galactic supernovae occured in 1572 and 1604 and were observed in considerable detail by Tycho Brahe and Johannes Kepler, respectively. From analyses of supernovae in other galaxies, we estimate that one supernova explosion occurs somewhere in the Milky Way Galaxy every 25 to 100 years. Unfortunately, however, no supernova explosion has been detected in our Galaxy since the invention of the telescope.

The changes in brightness of a supernova are similar to those of a nova, except that supernovae are more luminous and stay bright longer. In contrast to novae, which increase in luminosity a paltry few thousands or at most tens of thousands of times, a supernova can flare up to hundreds of millions of times its former brightness. At maximum light, supernovae reach absolute magnitude -14 to -18, or possibly even -20, or about 10^{10} L_s. For a brief time, a supernova may outshine the entire galaxy in which it appears. (Supernovae appear faint only because they are at large distances from the Sun.) After maximum light, supernovae gradually fade in brightness. Within a few months to years, they disappear from telescopic visibility.

Figure 30.10 shows the light curve of SN 1987A, which is at a distance of about 160,000 light years and soared in brightness in a single day by a factor of about 1000, from an apparent magnitude of about 12 to a magnitude of 5. The star then continued to increase slowly in brightness until it was about the same apparent magnitude as the stars in the Little Dipper.

When a supernova explodes, the force of the blast hurls the material surrounding the iron core into space at velocities up to 10,000 km/s. This material is rich in heavy elements, which confirms the idea that heavy elements are manufactured inside stars. Most of the elements less massive than iron are built up by fusion reactions before the star explodes. The explosion produces a flood of energetic neutrons that can be absorbed by iron and other nuclei and so build up the elements that are more massive than iron, including such terrestrial favorites as gold and silver. When supernovae explode, these elements are returned to the interstellar medium and mixed with it, and so become available to be incorporated into succeeding generations of stars. Along with planetary nebulae, supernovae play a major role in building up the supply of chemical elements.

One of the elements newly formed in a supernova explosion is radioactive nickel, with an atomic mass of 56. Nickel-56 is unstable and changes spontaneously, with a half-life of about 6 days, to cobalt-56, which in turn decays with a half-life of about 77 days to iron-56, which is stable. Gamma rays are emitted when these radioactive nuclei decay, and those gamma rays, which are absorbed in the overlying gas and re-emitted at visible wavelengths, are responsible for virtually all of the radiation detected from SN 1987A after day 40. From the total amount of radiation seen, astronomers can estimate how much nickel was produced. The total mass of nickel-56 turns out to be about 7 or 8 percent of the mass of the Sun, and this mass is in agreement with theoretical predictions for a supernova explosion like 1987A.

The radioactive nuclei produced in the explosion will ultimately decay into stable elements and will cease to produce gamma rays. With no new source of energy, SN 1987A will slowly fade away. By the beginning of 1991, it was observable only with very large telescopes.

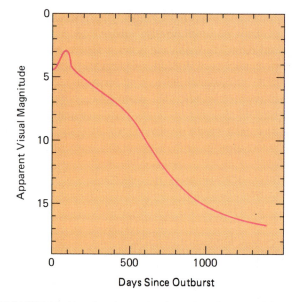

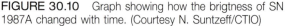

Days Since Outburst

FIGURE 30.10 Graph showing how the brigtness of SN 1987A changed with time. (Courtesy N. Suntzeff/CTIO)

(a) Types of Supernovae

The available evidence suggests that there are two distinct types of supernovae. One type, called Type II, marks the death of a massive star. SN 1987A was a Type II supernova. The second type is thought to occur in a binary system that initially contains a white dwarf and a nearby companion. The intense gravitational force exerted by the white dwarf attracts matter from the companion star (see Section 30.2c). The mass of the white dwarf, which initially must have been less than 1.4 solar masses, begins to build up, and eventually it may exceed the Chandrasekhar limit. At this point the white dwarf must begin to collapse. As it does so, it heats up, new nuclear reactions begin, and the energy released is so great that it completely disrupts the star. Gases are blown out into space at velocities of several thousand kilometers per second, and the temperature of the gas is typically several million degrees. No central star remains behind. The explosion completely destroys the white dwarf. Such supernovae are called Type Ia.

Observations are consistent with this model, in that Type Ia supernovae occur primarily in types of galaxies (ellipticals) or regions of galaxies (away from spiral arms) where there are large numbers of old, low-mass stars. In contrast, Type II supernovae appear in regions where

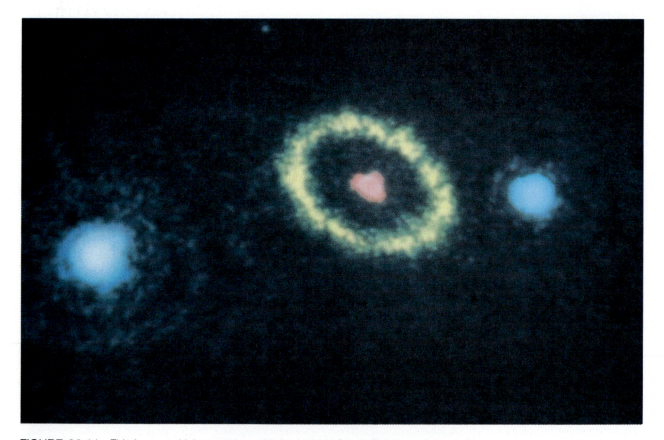

FIGURE 30.11 This image, which was taken with the Hubble Space Telescope, shows (in yellow) a ring of stellar material ejected by the star that ultimately became SN 1987A. This material was ejected long before the supernova explosion. Within 100 more years, the expanding debris from the supernova, which appears in red in this image, will plow into the ring and tear it apart. The blue stars to the left and right of the ring are not associated with the supernova. (NASA)

there are large numbers of young, massive stars. Type II supernovae are found in spiral arms and in regions where there is a great deal of dust and gas. They are seldom seen in elliptical galaxies.

The two types of supernovae can be distinguished by their light curves. Type Ia supernovae all have very similar light curves. The rise to maximum brightness takes a few weeks, and the star fades over a period of about 6 months. Type II supernovae light curves show considerable variety, but usually they fade more slowly than the light curves of Type Ia supernovae. Type II supernovae are typically about five times fainter at maximum light than Type Ia supernovae.

The observations by Tycho Brahe of the supernova of 1572 and by Kepler of the supernova of 1604, even though they were made without telescopes, are so accurate that we can construct light curves that show how the brightnesses of these two objects varied. Both were Type I supernovae (Figure 30.10). The supernova of 1006 was probably also a Type I because it was extremely bright. The remnants of these explosions have been identified, and there is no evidence of a central star in any of them. The light curves of the supernova of 1054 are imprecise but suggest that it was probably a Type II.

(b)   Supernova 1987A

The 1987 explosion of a Type II supernova in the Large Magellanic Cloud at long last provided astronomers with an opportunity to test their calculations of how stars die (Figure 30.11). Also, for the first time, we know what the star was like before it exploded.

By combining theory and observation, astronomers have reconstructed the life story of the star that became SN 1987A. Formed about 10 million years ago, the star originally had a mass of about 20 times the mass of the Sun. For 90 percent of its life it lived quietly on the main sequence, converting hydrogen to helium. At this time its luminosity was about 60,000 L_s, and it was spectral-type O. The temperature in its core was about 40 million K, and the central density was about 5 g/cm^3, or about the same as that of the Earth.

By the time hydrogen was exhausted at the center of the star, a helium core of about six times the mass of the Sun had developed, and hydrogen fusion was proceeding in a shell surrounding this core. The core contracted and grew hotter, until it reached a temperature of 170 million K and a density of 900 g/cm^3, at which time helium began to fuse to form carbon and oxygen. The surface of the star expanded to a radius of about 10^8 km, or about the distance from the Earth to the Sun. The star's luminosity nearly doubled to 100,000 L_s, and the star became a red supergiant. While it was a red supergiant, the star lost some mass in the form of a stellar wind (Figure 30.12).

Helium fusion lasted for only about 1 million years, forming a core of carbon and oxygen with a mass of

FIGURE 30.12   Red giant stars lose substantial quantities of mass in the form of a stellar wind, which is similar to the solar wind but contains much more material. The nebulosity is due to light reflected by dust surrounding the star. The dust is thought to be mainly silicate particles condensed from gas that the star is losing at a fairly steady rate. (Anglo-Australian Telescope Board, 1980)

about four times the mass of the Sun. When helium was exhausted at the center of the star, the core contracted again, the surface of the star also decreased in radius, and the star became a blue supergiant with a luminosity still about equal to 100,000 times the luminosity of the Sun, which was what it was when it exploded. When the contracting core reached a temperature of 700 million K and a density of 150,000 g/cm^3, nuclear reactions began to convert carbon to neon, sodium, and magnesium. This phase lasted only about 1000 years.

The core, having exhausted carbon as a fuel, again contracted and heated—this time to a temperature of 1.5 billion K and a density of 10^7 g/cm^3—at which point the conversion of neon to oxygen and magnesium, and then oxygen to silicon and sulfur, lasted for several years. At temperatures of 3.5 billion K and densities of 10^8 g/cm^3, the silicon began to melt into a sea of helium, neutrons, and protons, which then combined with some of the remaining silicon and sulfur nuclei to form iron.

When the silicon is converted to iron, and the mass of the core exceeds 1.4 times the mass of the Sun, the collapse begins. In the case of SN 1987A the collapse lasted only a few tenths of a second, and the velocity of the collapse in the outer portion of the iron core reached 70,000 km/s. The outer shells of neon, helium, and hydrogen, however, do not know about the collapse. Information travels through the star at the speed of sound, and cannot reach the surface in the few tenths of a second that is required for the collapse to occur. The

FIGURE 30.13 About 120 centuries ago, an inconspicuous star in the constellation Vela became a supernova. A portion of its expanding shell, which now covers 6° of the sky, is shown here. Ultimately, the shell expanding around SN 1987A will thin out and look like this. The matter ejected in supernova explosions is enriched in heavy elements. (ROE/Anglo-Australian Telescope Board, 1979)

surface layers hang suspended, much like a cartoon character that dashes off the edge of a cliff and hangs momentarily in space before he realizes that he no longer is held up by anything.

The collapse of the core continues until the densities rise to several times that of an atomic nucleus. The resistance to further collapse becomes very great, and the core rebounds. Infalling material runs into the brick wall of the rebounding core and is blown outward. The material that is ejected in the ensuing explosion is rich in heavy elements, and studies confirm that the composition is what would be predicted from the models of the composition of the stellar interior immediately prior to the explosion (Figure 30.13).

Table 30.2 summarizes the steps that led inexorably to the explosion that was SN 1987A. The data are taken from detailed calculations by S. Woosley (University of California at Santa Cruz) and his collaborators.

(c) Neutrinos from SN 1987A

Brilliant as SN 1987A was in the southern skies, less than $1/_{10}$ of 1 percent of the energy of the explosion appeared as optical radiation. One of the predictions of models of supernovae is that a large number of neutrinos should be ejected from the core of the star at the time of the collapse. When the collapse occurs, the electrons merge with protons to form neutrons, and this reaction also releases neutrinos, which escape from the star at the speed of light. The energy carried away by these neutrinos is truly astounding. In the first second, the total luminosity of the neutrinos is 10^{46} watts, which exceeds the luminosity of all the stars in all the galaxies in the part of the universe that we can observe. And the supernova generated this energy in a volume less than 50 km in diameter! Supernovae are by far the most violent events in the universe.

One of the most exciting results from observations of SN 1987A is that astronomers detected the neutrinos at the right time and in the expected quantity, thereby obtaining confirmation of the validity of the theoretical calculations. The neutrinos were detected by two instruments, which might be called "neutrino telescopes," about 3 hr before the brightening of the star was first detected in the optical region of the spectrum. Both "telescopes," one in Japan and the other under

TABLE 30.2	EVOLUTIONARY PHASES THAT PRECEDED SN 1987A		
Phase	Central Temperature (K)	Central Density (g/cm³)	Duration of Phase
Hydrogen fusion	40×10^6	5	9×10^6 years
Helium fusion	170×10^6	900	10^6 years
Carbon fusion	700×10^6	150,000	10^3 years
Neon fusion	1.5×10^9	10^7	Several years
Oxygen fusion	2.1×10^9	10^7	Several years
Silicon fusion	3.5×10^9	10^8	Days
Core collapse	200×10^9	2×10^{14}	Tenths of a second

Lake Erie, consist of several thousand tons of purified water surrounded by several hundred detectors that are sensitive to light. Incoming neutrinos interact with the water to produce positrons and electrons, which move rapidly through the water and emit deep blue light.

The Japanese system detected 11 neutrino events over an interval of 13 s, and the instrument beneath Lake Erie measured 8 events in the same time. Since the neutrino telescopes are located in the Northern Hemisphere, and the supernova occurred in the Southern Hemisphere, the neutrinos detected had already passed through the Earth and were on their way back out into space when they were captured!

Only a few neutrinos were detected because the probability that they will interact with matter is very, very low. It is estimated that the supernova actually released 10^{58} neutrinos. About 5×10^{14} of these neutrinos passed through every square meter on the Earth, and about a million people experienced a neutrino interaction within their bodies. Of course, this interaction had absolutely no biological effect and went completely unnoticed.

Since the neutrinos come directly from the heart of the supernova, their energies provide a measure of how hot the star was at the time of the explosion. The central temperature was about 2×10^{11} K.

Scientists came very close to missing the neutrinos altogether. The scientists in Japan had finished a regular set of calibration measurements only minutes before the supernova neutrinos arrived—had the calibration still been in progress the supernova neutrinos would have gone unnoticed—and experienced a power failure a couple of days later. A few hours before the neutrino burst, the Lake Erie instrument lost a power supply, and one-quarter of its detectors were not in operation at the crucial time.

What is left now that the explosion is over? According to theory, there should be an object with a density approximately equal to that of an atomic nucleus composed essentially entirely of neutrons at the center of the gaseous cloud that has been blown out into space. Unfortunately, the cloud is still so dense that we cannot yet see inside. As the cloud expands and disperses, astronomers will search for the evidence of a **neutron star.** Fortunately, from observations of the remnant of the much older Crab supernova, where the cloud has thinned enough so that we can observe to its center, we know exactly what to look for.

(d)   The Crab Nebula

The Crab Nebula in Taurus is a chaotic, expanding mass of gas (Figure 30.14). The outer parts of this gas cloud are observed to be moving away from the center at rates roughly proportional to their distances from it. The Doppler shifts of light from the center of the nebula show the gases there to be moving toward us at speeds up to 1450 km/s. If we assume that the nebula has always expanded at the same rate, we can calculate when the filaments of gas must have started their outward motion to reach their present size. It turns out that both the location and the computed time of formation of the Crab Nebula are strongly in agreement with the occurrence of the supernova of 1054. The Crab Nebula, therefore, must be the material ejected during that stellar explosion.

Observations of the center of the Crab Nebula demonstrate conclusively that a supernova explosion can leave behind a neutron star. The first neutron star to be discovered, however, did not lie in the Crab, and initially it presented astronomers with a mystery.

30.7   PULSARS

(a)   The Discovery of Neutron Stars

In 1967, Jocelyn Bell, a graduate research student at Cambridge University, was studying distant radio sources with one of the Cambridge radio telescopes. In the course of her investigation, Bell made a remarkable discovery—one that won her advisor, Antony Hewish, the Nobel Prize in physics, because his analysis of the object (and other similar ones) revealed the first evidence for neutron stars.

FIGURE 30.14   The Crab Nebula in the constellation Taurus. This nebula is the remnant of a supernova explosion, which was seen on Earth in the year A.D. 1054. It is located some 2000 pc away and is approximately 2 pc in diameter, still expanding outward. In this true color picture, the red filaments are tendrils of hot gas, which emits strongly at the wavelength corresponding to the red Balmer line of hydrogen. The pulsar is clearly visible just below the center. (W. Schoening and N. Sharp/National Optical Astronomy Observatories)

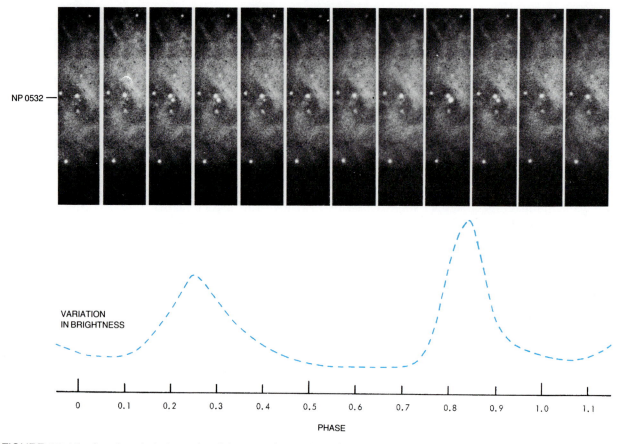

FIGURE 30.15   A series of photographs of the central part of the Crab Nebula taken by S. P. Maran at Kitt Peak National Observatory. Note the star that seems to blink on and off; it is the pulsar, which has a period of 1/30 s. (Kitt Peak National Observatory and National Optical Astronomy Observatories)

What Bell had found, in the constellation of Vulpecula, was a source of rapid, sharp, intense, and extremely regular pulses of radio radiation, the pulses arriving exactly every 1.33728 s. For a time there was speculation that they might be signals from an intelligent civilization. Radio astronomers half-jokingly dubbed the source "LGM," for "little green men," and withheld announcement, pending more careful study. Soon, however, three additional similar sources were discovered in widely separated directions in the sky; when it became apparent that this type of source was fairly common, astronomers concluded that it was highly unlikely that they were signals from other civilizations. By now more than 400 such sources have been discovered. They are called **pulsars,** for pulsating radio sources.

The pulse periods of different pulsars range from a little longer than $^1/_{1000}$ s to nearly 10 s. One pulsar is in the middle of the Crab Nebula, and it has a pulse period of 0.033 s. The period is observed to be very slowly increasing, showing that pulsars evolve, pulsing gradually more slowly as they age. The source of the pulses is what appears to be a 16th-magnitude star at the center of the nebula. In addition to pulses of radio energy, there are also pulses of optical and x-ray radiation every 0.033 s from the Crab (Figure 30.15). About 10 to 15 percent of the x-ray radiation from the Crab Nebula comes from the pulsar.

In addition to these pulsating radio sources, dozens of x-ray sources pulse in short, regular periods. Of these, only the Crab pulsar is seen also in visible light and radio waves. At least some x-ray pulsars are members of binary-star systems, and for four of these, enough information is available to calculate masses by techniques described in Chapter 23. These four x-ray pulsars have masses in the range 1.4 to 1.8 times that of the Sun.

(b)   Theory of Pulsars—Neutron Stars

The energy emitted by pulsars is not small. The Crab pulsar pours forth energy at a higher rate than the Sun. Thus pulsars are like stars in their output of radiation. Yet they emit this energy in pulses of up to nearly 1000 per second. What kind of object can vary so rapidly?

Theoretical calculations carried out long before the discovery of pulsars proved to hold the answer to this question. After the discovery of the neutron in 1932, theoreticians speculated that if the matter in a star could be subjected to such high pressure that the free electrons were forced into the atomic nuclei, the star could become a body composed entirely of neutrons. A few years later, Mt. Wilson astronomers Walter Baade and Fritz Zwicky suggested that supernova explosions might form neutron stars.

Neutrons, like electrons, obey the Pauli exclusion principle and can become degenerate if crowded into a sufficiently small volume for a given momentum range. Suppose a star has collapsed into degenerate neutrons. The neutrons, in such a condition, cannot decay into protons and electrons, for by the time the star is that collapsed, the allowable states for electrons would be filled. The structure of a neutron star is analogous to that of a white dwarf, except that neutron stars are much smaller. A neutron star of one solar mass would have a radius of only about 10 km. Such a star would have a density of 10^{14} to 10^{15} g/cm^3—comparable to that of the atomic nucleus itself. Table 30.3 compares the properties of neutron stars and white dwarfs.

There exists a mass-radius relation for neutron stars, and an upper mass limit as well. Although the exact theory is not yet certain, the mass limit for a neutron star is believed to be from two to three solar masses. The measured masses of the x-ray pulsars are all less than this limit.

Any magnetic field that existed in the original star is highly compressed if the core of the star collapses to a neutron star. Thus a moderate field of the order of 1 gauss in a star the size of the Sun increases to the order of 10^{10} to 10^{12} gauss around the neutron star. At the very surface of the collapsed star, neutrons decay into protons and electrons. Many of these charged particles should leave the stellar surface and move out in the vicinity of the magnetic poles into the circumstellar magnetic field. With such intense fields and high densities, many of the particles, especially the electrons, move at speeds close to the speed of light and emit electromagnetic energy by the synchrotron mechanism (Section 33.6). The radiation is very directional, however, and if the magnetic poles happen not to coincide with the poles of the axis of rotation (the rotation rate also increases greatly during the collapse, to conserve angular momentum), the rotation carries first one and then the other magnetic pole into our view. The consequence is that the radiation from the rotating magnetic field can be directed toward us twice each time the star turns completely around on its axis, much as a lighthouse directs its light toward a ship at sea briefly during each revolution.

(c)   Planets Around Pulsars

Recently, astronomers have reported that one pulsar is accompanied by at least two planets. The pulsar itself spins hundreds of times each second, and we can measure the rate of spin by measuring the time interval between successive pulses. If there are planets present, then the pulsar will also revolve in a small orbit around the center of mass of the system defined by the positions and masses of the pulsar and planets. As the pulsar moves in its orbit, the intervals between successive pulses will vary, being slightly longer when the pulsar is moving away from us and shorter when the pulsar is moving toward us. By analyzing the intervals between pulses, astronomers have concluded that the pulsar PSR1257 + 12 is being orbited by two objects with masses of at least 2.8 and 3.4 times the mass of the Earth. The orbital periods are 98 and 67 days, and the sizes of the orbits are similar to the size of Mercury's orbit. This system may also contain a third planet with an orbital period of about one year.

These planets would certainly not be hospitable abodes for life. The pulsar emits most of its energy in the form of an intense stellar wind, which would blast the planet with high energy particles moving at nearly the speed of light. Planets this close to the pulsar could certainly not have survived the supernova explosion that preceded the formation of the pulsar. Rather they must have been formed since the explosion, either from the debris blown off in the explosion or, more likely, from material blasted away from a nearby companion star, which has by now been completely vaporized by radiation from the pulsar. Theoretical calculations show that such planets could form on a time scale of 10^5–10^6 years.

The significance of this result is that it suggests that the formation of planets may be relatively easy. If

TABLE 30.3	PROPERTIES OF TYPICAL WHITE DWARFS AND NEUTRON STARS	
	White Dwarf	Neutron Star
Mass	1.0	1.5
(mass of Sun = 1)	(always <1.4)	(always <3)
Radius	5000 km	10 km
Density	5×10^5 g/cm^3	10^{14} g/cm^3

planets can form so quickly in a disk of material around a pulsar, it may be equally easy to form planets in the disks that surround protostars. If that is true, then potentially habitable planets may be very common. The next question astronomers want to answer is whether most pulsars that rotate hundreds of times per second have planets or whether the particular pulsar that has been studied is unique. Detailed studies of additional pulsars are now being made, but since the orbital periods are months to years and multiple cycles must be observed to make sure the results are right, it will be several years before we know whether most pulsars, or only a few, have planets around them.

(d)   Evolution of Pulsars

From observations of the several hundred pulsars discovered so far, astronomers have concluded that one new pulsar is born somewhere in the Galaxy every 30 to 100 years. Pulsars typically have velocities of about 100 km/s and are presumably accelerated to these high velocities at the time of the supernova explosion. Most pulsars lie fairly close to the plane of the Galaxy. Since no pulsars have managed to get more than a few hundred parsecs above or below the plane despite their high velocities, astronomers conclude that they must stop emitting pulses of radiation before they have time to

travel large distances. Calculations suggest that the typical lifetime of a pulsar is about 10 million years.

The energy radiated by a pulsar, evidently from matter ejected from the star interacting with the stellar magnetic field, robs the star of rotational energy. Thus, theory predicts that the rotating neutron stars should gradually slow down, and that the pulsars should slowly increase their periods. Indeed, as we have already mentioned, the Crab pulsar is actually observed to be increasing the interval between its pulses. According to present ideas, the Crab pulsar is rather young and of short period (we know it is only about 900 years old); the other, older pulsars have already slowed to longer periods. Pulsars thousands of years old have lost too much energy to emit pulses appreciably in the visible and x-ray wavelengths and are observed only as radio pulsars. Their periods are a second or more.

Only 3 of more than 400 pulsars discovered so far are embedded in visible nebulae. The lifetime of a pulsar is about 100 times longer than the length of time required for the expanding gas to disperse into interstellar space. This fact explains why we find only about 1 percent of the pulsars to be still surrounded by the gas ejected during the supernova explosion.

One other possible end state for a star is that of a black hole, which involves one of the more bizarre predictions of general relativity theory. We shall discuss black holes in the next chapter.

S U M M A R Y

30.1 According to the Pauli exclusion principle, no two electrons can be in the same spatial position and have the same momentum. A gas in which all possible momentum and positional states available to electrons are occupied is said to be **electron-degenerate.** When a contracting star becomes electron-degenerate, the pressure exerted by the electrons will halt the contraction. When hydrogen in the core of a star is exhausted, a star will evolve rapidly and become a red giant. Energy is generated by the fusion of hydrogen in a shell surrounding the helium core. The core contracts, heats, and becomes electron-degenerate. When the temperature becomes high enough, the core is re-ignited. Helium begins to fuse to form carbon and some oxygen, and the star quickly becomes a horizontal-branch star. For a star with a mass similar to that of the Sun, formation of the carbon-oxygen core marks the end of fusion in the center of the star.

30.2 When energy generation in the core stops, some stars become **white dwarfs,** which are supported against further collapse by degenerate electrons. **Novae** are outbursts that occur when matter is transferred to the surface of a white dwarf from a nearby companion, and hydrogen fusion begins explosively. The interiors of most white dwarfs are composed of carbon and oxygen.

30.3 Stars that begin their lives with masses up to 6 to 8 times the mass of the Sun shed enough mass during their evolution

to become white dwarfs with masses less than 1.4 times the mass of the Sun. **Planetary nebulae** are shells of gas ejected by stars during the red giant phase of evolution. Gas ejected during the late stages of evolution is enriched in elements heavier than helium. These heavy elements are produced by nucleosynthesis in the stellar interior.

30.4 The Sun required a few tens of millions of years to contract to the main sequence. It has been generating energy by hydrogen fusion for about 5 billion years and will continue to do so for another 5 billion years before becoming a red giant. The Sun's photosphere will then expand until it reaches nearly to the orbit of Mars; the Sun will engulf and vaporize the Earth.

30.5 In a massive star, hydrogen fusion in the core is followed by several other fusion reactions involving heavier elements. Just before it exhausts all sources of energy, a massive star has an iron core surrounded by shells of (in order of increasing distance from the center and decreasing temperature) silicon, oxygen, neon, carbon, helium, and hydrogen. When the mass of the iron core of a star exceeds the **Chandrasekhar limit** (1.4 solar masses), the core collapses until its density exceeds that of an atomic nucleus. The core rebounds and transfers energy outward, blowing off the outer layers of the star in a **supernova** explosion.

30.6   Type I supernovae occur in binary systems in which a white dwarf accretes enough matter to push its mass over the Chandrasekhar limit. Type II supernovae are the explosions of massive stars. Studies of SN 1987A, including the detection of neutrinos, have confirmed theoretical calculations of what happens during the explosion of massive stars. At least some supernovae leave behind a rotating **neutron star,** which is called a **pulsar.**

30.7   Pulsars emit pulses of radiation at regular intervals. Their periods are approximately in the range of 0.001 to 10 s and are related to the rotation period of the neutron star. Pulsars lose energy as they age, the rotation slows, and their periods increase.

E X E R C I S E S

THOUGHT QUESTIONS

1. Describe the evolution of a star with a mass of two times the mass of the Sun from the main-sequence phase of its evolution until it becomes a white dwarf.

2. Describe the evolution of a massive star up to the point at which it becomes a supernova. How does the evolution of a massive star differ from that of the Sun? Why?

3. You observe an expanding shell of gas through a telescope. What measurements would you make to determine whether you have discovered a planetary nebula or the remnant of a supernova explosion?

4. Arrange the following stars in order of age:
   a. A star with no nuclear reactions going on in the core, which is made primarily of carbon and oxygen
   b. A star of uniform composition from center to surface; the star contains hydrogen but has no nuclear reactions going on in the core
   c. A star that is fusing hydrogen to form helium in its core
   d. A star that is fusing helium to form carbon in the core and hydrogen to form helium in a shell around the core
   e. A star that has no nuclear reactions going on in the core but is fusing hydrogen to form helium in a shell around the core

5. Would you expect to find any white dwarfs in the Orion Nebula?

6. Suppose no stars had ever formed that were more massive than about two times the mass of the Sun. Would life as we know it have been able to develop?

7. Would you be more likely to observe a Type II supernova (the explosion of a massive star) in a globular cluster or in an open cluster? Why?

8. One group of astronomers has reported the detection of a pulsar formed in the 1987A supernova explosion. Other observations have, as of this writing, failed to confirm the detection. Use contemporary sources to determine whether or not the existence of a neutron star has been verified by observations.

9. Look elsewhere in this book for the necessary data and indicate what the final stage of evolution—white dwarf, neutron star, or black hole—will be for each of the following kinds of stars:
   a. Spectral-type O main-sequence star
   b. B main-sequence star
   c. A main-sequence star
   d. G main-sequence star
   e. M main-sequence star

10. Show the evolutionary track in the H–R diagram for the star that exploded as SN 1987A. Begin when the star was on the main sequence and end just before the explosion.

PROBLEMS

11. The gas shell of a particular planetary nebula is expanding at the rate of 20 km/s. Its diameter is 1 LY. Find its age. For this calculation, assume that there are 3×10^7 s/yr and 10^{13} km/LY.

12. Prepare a chart or diagram that exhibits the relative sizes of a typical red giant, the Sun, a typical white dwarf, and a neutron star of mass equal to the Sun's. You may have to be clever to devise such a diagram.

13. Suppose the central star of a planetary nebula is 16 times as luminous as the Sun, and 20 times as hot (about 110,000 K). Find its radius, in terms of the Sun's. Compare this radius with the radius of a typical white dwarf.

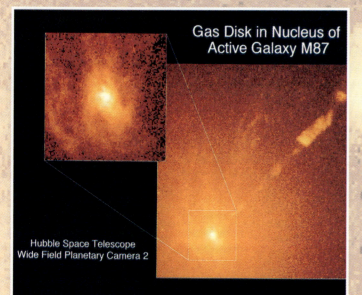

Gas Disk in Nucleus of
Active Galaxy M87

Hubble Space Telescope
Wide Field Planetary Camera 2

*The repaired Hubble Space Telescope discovered this
spiral-shaped disk of hot gas in the core of the giant
elliptical galaxy M87. Spectroscopic measurements of
the Doppler shifts of the spectral lines formed in the
disk show that it is rotating at a speed of about
550 km/s. Such a high orbital velocity indicates that
there must be about three billion solar masses
concentrated at the center of the disk. This mass is
almost certainly in the form of a black hole, a type of
collapsed object described by the theory of general
relativity. The image also shows a jet of high
speed electrons that is emitted from the vicinity
of the black hole.*

(Holland Ford, et al./Space Telescope
Science Institute and NASA)

"Apollo to Mission Control:
We're almost in reach of our goal;
But our readings of g
Seem excessive to me,
So we may be inside a black ho. . ."

[Limerick from an astronomy student newsletter]

Few objects in astrophysics are as mind-bog-
gling as black holes. Dozens of science fiction
stories have used their strange properties to
good advantage, the Walt Disney studios made
(a rather awful) film about them, and *Time*
magazine even used them to illustrate a cover
on the national deficit. Astronomers too find
them fascinating, especially as modern instruments reveal the strong possibility of black holes
at the centers of many galaxies of stars. If our
theories are correct, these collapsed remnants of
the most massive stars are places where gravity
has carved out a separate little structure in space
and time, where any path that would seem to
lead *out* instead becomes a path going *in*, and
where light itself is trapped inside forever.

Yet the ideas that led to black holes began
with thoughts that had nothing to do with the
death of stars. Einstein devised his general theory
of relativity to incorporate his results from the
special theory (see Chapter 8) into a theory of
gravitation. But when the general theory was
formulated, it quickly became clear that it had
profound implications in a wide range of
astronomical situations—from the ultimate
fate of stars to the birth and geometry of
the entire universe.

GENERAL RELATIVITY: CURVED SPACETIME AND BLACK HOLES

Albert Einstein (1879–1955) received the Nobel Prize in 1921, not for his theory of relativity but for the photoelectric effect. At that time his ideas on relativity were still at the frontier. Einstein believed that such seemingly diverse areas of physics as mechanics and electromagnetic phenomena— and even gravitation—were guided in the same way by underlying principles.

The theory of general relativity is one of the major intellectual achievements of the 20th century. Until recently, however, this theory had little impact either on our daily lives or on scientific research. For a long time there were only three tests of general relativity, none of them precise enough to provide compelling support for the theory. In the past two decades, however, general relativity has become essential in understanding pulsars, quasars, and other astronomical objects and events.

31.1 PRINCIPLE OF EQUIVALENCE

The fundamental insight that led to the formulation of general relativity is deceptively simple. Galileo noted that all bodies, despite their different masses, if dropped together, fall to the ground at the same rate. According to Newton's law of gravitation, the Earth pulls on a more massive object with a greater force than it does on a less massive one. The two objects fall together, however, because according to Newton's second law, a proportionately greater force is required to impart the same acceleration to the heavier object. Any two freely falling bodies, independent of their composition and internal structure, will follow identical paths. This similarity of behavior of all types of objects is called an **equivalence principle.**

Einstein broadened this equivalence principle and as a consequence reached sweeping conclusions about the very fabric of space and time. The basic postulate of Einstein's equivalence principle is that life in a freely falling laboratory is indistinguishable from, and hence equivalent to, life with no gravity. Similarly, life in a stationary laboratory in a gravitational field is indistinguishable from life in an accelerating laboratory far from any gravitational force.

(a)   Gravity or Acceleration?

To explore the implications of this idea, let's consider as an example the case of a foolhardy boy and girl who simultaneously jump into a bottomless chasm from opposite sides of its banks. If we ignore air friction, while they fall they accelerate downward at the same rate and feel no external force acting on them. They can throw a ball back and forth between them, aiming always in a straight line, as if there were no gravitation. The ball, falling along with them, moves directly between them.

It's very different on the surface of the Earth. Everyone knows that a ball, once thrown, falls to the ground. Thus in order to reach its target (the catcher), the ball must be aimed upward so that it follows a parabolic arc—falling as it moves forward—until it is caught at the other end.

Because our freely falling boy, girl, and ball are all falling at the same rate and in the same direction, we could enclose them in a large box falling with them. Inside that box, no one can be aware of any gravitational force. Nothing falls to the ground, or anywhere else, but moves in a straight line in the most simple natural way, obeying Newton's laws. By having our box fall with the boy and girl, we have removed the force of gravitation. We have selected a *coordinate system* that is accelerating at just the right rate to compensate for gravitation. Here is the principle of equivalence—a force of gravitation is equivalent to an acceleration of the coordinate system of the observer.

Einstein himself pointed out how our weight seems to be reduced in an elevator when it accelerates from a stop to a rapid downward velocity. Similarly, our weight seems to increase in an elevator that is increasing its upward velocity. Stand on a scale in an elevator and see what happens! In a *freely falling* elevator, with no air friction, we would lose our weight altogether.

This idea is not hypothetical. Astronauts in the Space Shuttle orbiting the Earth live in just such an environment (Figure 31.1). The Shuttle in orbit is, of course,

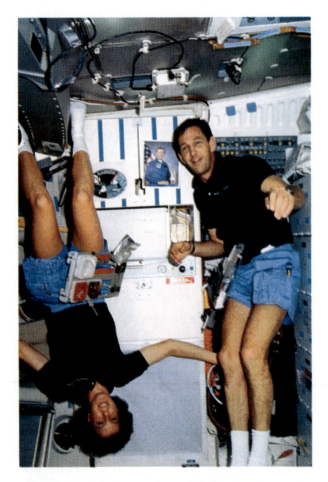

FIGURE 31.1 When the Space Shuttle is in orbit, everything stays put or moves uniformly because there is no apparent gravitation acting inside the spacecraft. (NASA)

falling freely around the Earth. While in free fall the astronauts live in a magical world where there seem to be no gravitational forces. One can give a wrench a shove, and it moves at constant speed across the orbiting laboratory. One can lay a pencil in midair and it remains there, as if no force is acting on it.

Appearances are misleading. There *is* a force. Neither the Shuttle nor the astronauts are *really* weightless, for they continually fall around the Earth, pulled by its gravity. But since all fall together—Shuttle, astronauts, wrench, and pencil—within the Shuttle all gravitational forces appear to be absent.

Thus the Shuttle provides an excellent example of the principle of equivalence—how local effects of gravitation can be removed by a suitable acceleration of the coordinate system. To the astronauts it is as if they are far off in space, remote from all gravitating objects. Suppose the astronauts *actually were* in remote space and activated the engines of their ship, producing acceleration. The ship would then push up against their feet, giving the impression of a gravitational tug. If one were to drop a small coin and a hammer, the floor of the ship would move up to meet both objects at the same time. To the astronauts, though, it would seem that the hammer and coin fell to the floor together. In other words, an acceleration of one's local environment produces exactly the same effect as a gravitational attraction. The two are indistinguishable—again, the principle of equivalence.

(b) Trajectories of Light and Matter

Einstein postulated that the principle of equivalence is a fundamental fact of nature. If so, then there must be *no* way in which an astronaut, at least by experiments within his local environment, can distinguish between his weightlessness in remote interstellar space and his free fall in a gravitational field about a planet like the Earth. Experiments must have the same result whatever the objects involved and whether or not the test involves falling objects, interactions of atoms, or light.

But how about light? If the equivalence principle really applies to light, then the consequences are profound. Suppose that astronauts shine a beam of light along the length of their ship, which is in a free-fall orbit about a planet. The ship is falling away from a straight-line path. If the beam of light follows a straight line, which from everyday experience we believe it does, then won't the light strike above its target (Figure 31.2)?

Not so, according to Einstein. If the principle of equivalence is correct, there must be *no* way of knowing whether one is accelerated. Hence the light beam *must fall with the ship* if that ship is in orbit about a gravitating body (Figure 31.2). Is light actually bent from its straight-line path by the force of gravity? Einstein preferred to think that light always follows the

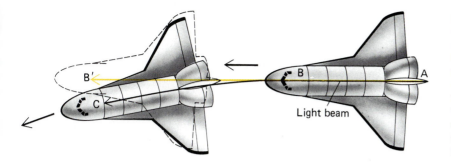

FIGURE 31.2 In a spaceship moving to the left (in this figure) in its orbit about a planet, light is beamed from the rear, *A,* toward the front, *B*. We might expect the light to strike at *B'*, above the target in the ship, which has fallen out of its straight path in its orbit about the planet. Instead, the light, bent by gravity, follows the curved path and strikes at *C*.

shortest path—but that path may not be straight. On the Earth, we know that the shortest distance between two points is not a straight line but follows the arc of a **great circle** (the equator is a great circle). Suppose we define **spacetime** to be a system of coordinates within which we can specify the time and place of any event. If spacetime is curved, then the shortest path between two points may be curved.

By examining the consequences of the equivalence principle, Einstein reached conclusions about the fundamental nature of spacetime. He proposed that we live in a *curved* spacetime and that gravitation is equivalent to, and indistinguishable from, curvature of spacetime. A massive body distorts space and time, and the path of a second body passing near such a mass is controlled by these distortions.

To understand this concept, we will explore in more detail what is meant by spacetime and by curvature of spacetime.

31.2 SPACETIME

(a) Events in Spacetime

There is nothing mysterious about four-dimensional spacetime. Imagine yourself in the rear seats at an outdoor concert at the Hollywood Bowl. The sound from the orchestra in the shell, hundreds of feet away, takes a goodly fraction of a second to reach you, and the players seem to be behind the beat of the conductor. When a piece is finished, you first hear the applause from people near you, and slightly later from the front of the amphitheater. Because of the finite travel time of sound, all people do not hear the same note of music at the same time; nor do events that appear simultaneous *visually* seem to be audibly simultaneous.

Light also has a finite speed, so we never see an instantaneous snapshot of events around us (as we saw in Chapter 8). The speed of light is so great that within a single room we obtain *effectively* an instantaneous snapshot, but this is certainly not the case astronomically. We see the Moon as it was just over a second ago,

and the Sun as it was about 8 min ago. At the same time, we see the stars by light that left them years ago, and the other galaxies as they were millions of years in the past. We do not observe the world about us at an instant in time, but rather we see different things about us as different *events* in spacetime.

Relatively moving observers do not even agree on the order of events. As an example, suppose that I am riding exactly in the middle of a boxcar on a train moving at uniform speed. Suppose you are standing on the ground beside the tracks and that, just at the instant that we are abreast of each other, two lightning bolts strike the ends of my boxcar and the ground at points *A* and *B* exactly below the ends of the boxcar (Figure 31.3).

You will receive the light signals from the lightning bolts that struck at *A* and *B* at the same time. Since these signals will have traveled identical distances at the speed of light to reach you, you will conclude that the strikes at *A* and *B* occurred simultaneously. That is not what I see. When you receive the signals from *A* and *B*, I have already moved forward (Figure 31.3). The signal from the lightning strike at the front of the boxcar will already have swept past me on its way to you; the signal from the strike at the back of the boxcar will not yet have reached me. Therefore, I must conclude that the front of the boxcar was struck first. Thus, we are forced to conclude that two events that are simultaneous to one observer will not appear to be simultaneous to a second observer who is moving with respect to the first.

At this point, you may wish to ask, "Well, who is really right?" According to relativity, both are correct—in their own worlds or their own *reference frames*. Furthermore, no one reference frame is to be preferred over any other. Space and time are inextricably connected. We need to describe the universe not just in terms of three-dimensional space, but in terms of four-dimensional spacetime.

(b) Visualizing Spacetime

We can easily represent the spatial positions of objects in two dimensions on a flat sheet of paper (for example, the

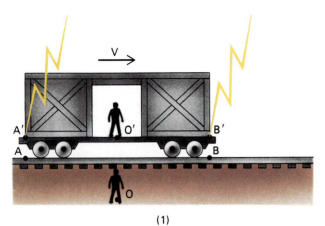

(1)

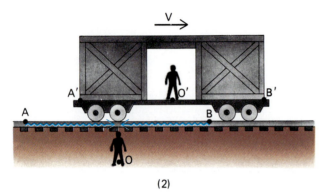

(2)

FIGURE 31.3 An experiment demonstrating the nonabsoluteness of simultaneity. When the boxcar passes the observer beside the tracks, lightning bolts strike the ends of the boxcar and points A and B on the ground. The stationary observer will see these events simultaneously. The observer on the train will think that the front of the boxcar was struck first.

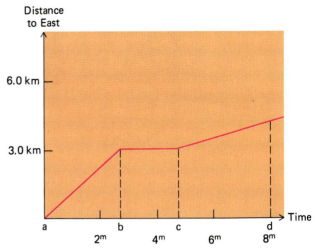

FIGURE 31.4 The progress of a motorist traveling east across town.

plan of a city). To plot three dimensions on a page, the draftsperson uses projections. Architectural drawings of a home generally show three projections: floor plan and two different elevations—say, the house as seen from the east and from the north—to give all necessary information. By the use of perspectives, we can also give an impression of a three-dimensional view. There is no easy way, though, to draw a four-dimensional perspective to include time.

There is no problem, however, in showing a two-dimensional projection of four-dimensional spacetime. Figure 31.4, for example, shows the progress of a motorist driving to the east across town. How much time has elapsed since he left home is shown on the horizontal axis, and how far he has traveled eastward is shown on the vertical axis. From a to b he drove at a uniform speed. From b to c he stopped for a traffic light and made no progress, and from c to d he drove more slowly because of increased traffic.

Figure 31.5 shows a rather conventional two-dimensional representation of spacetime. Time increases

upward in the figure, and one of the three spatial dimensions is shown horizontally. If we measure time in years and distance in light years, light goes one unit of distance in one unit of time, and so flows along diagonal lines as shown. "Here and now" is at the origin of the diagram. At this instant we can receive information of a past event along such a line as AO. In this case the messenger was going slower than light, so she covered less distance than light would in the same time. Because nothing can go faster than light, we cannot, right now, know of something happening at point B in spacetime, for the message along BO would have to travel faster than light. We will have to wait until we are at C in the future, before a light or radio beam can get us the word along path BC.

(c) Curvature of Spacetime

To understand what is meant by curvature of spacetime, let's consider a familiar analogy. A simple Mercator-type map, with lines of constant latitude running horizontally and lines of constant longitude running vertically, is fine for showing a small area of the Earth—say, a single city—without noticeable distortion. But such a map cannot show a large area of the curved Earth without distortion; everyone knows how distorted and enlarged countries near the poles appear on the usual flat world maps. We cannot map the Earth with ordinary (Euclidean) plane geometry.

Indeed, if we travel far enough in a straight line on the surface of the Earth, we end back at our starting point; our path is a great circle (the equator is such a great circle). More generally, if we take into account the slight polar flattening of the Earth, as well as the effects of such irregularities as mountains, our "straight-line" path is called a **geodesic,** which means "Earth divider."

Einstein showed how to find spacetime coordinates within which all objects move as they would if there were no forces. In a small local region, where a gravitational field is uniform, those coordinates are the ones used in plane geometry. A city map, for example, makes use of plane geometry. But to describe paths of objects over a large region, where the gravitational field varies, the geometry used to describe spacetime must be curved, just as we must use curved geometry to describe a large area of the spherical Earth.

The distribution of matter determines the nature of a gravitational field, so it is the distribution of matter that determines the geometry of curved spacetime. Within this curved spacetime, everything moves in the simplest possible way. In analogy with Earth geometry, the paths of light and material objects in spacetime are called geodesics.

In its journey from one location to another, light follows the shortest path. In spacetime that path happens to be curved, just as the shortest route for a plane traveling halfway around the Earth is a great circle route, which is also curved. If we want a simple mathematical description of the motion of light in spacetime (or of planes around the Earth), then we must choose a system of curved coordinates. In this curved coordinate system, motions can be described as "straight" in the same sense that great circles (or geodesics) are "straight" on the curved surface of the Earth.

31.3   TESTS OF GENERAL RELATIVITY

Relativity is different from Newtonian theory in several ways. First, the signals that govern gravitational interactions are not instantaneous, but travel with the speed of light. Matter and energy are equivalent, so that not only matter itself but also energy contributes to gravitation—that is, to the geometry of spacetime. Energy (light, for example) and mass are both affected by that geometry. Where speeds are low compared with that of light and where the gravitational field is relatively weak—and both of these conditions are met throughout most of the solar system—the predictions of general relativity must agree with those of Newton's theory, which has served us so admirably in our technology and in guiding space probes to the other planets. In familiar territory, therefore, the differences between predictions of the two theories are subtle and difficult to detect.

Einstein himself proposed three observational tests of general relativity. One, the *gravitational redshift*, is actually a test of the equivalence principle as it applies to light. The other two measure how much spacetime is curved and so test quantitatively the predictions of general relativity. One of these tests involves observation of the *deflection of starlight* that passes close to the

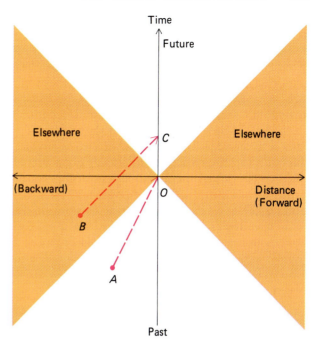

FIGURE 31.5   A spacetime diagram.

Sun, and the second depends on a subtle effect in the *motion of the planet Mercury*. With the advent of spacecraft, another and more precise test of general relativity has become possible.

(a)   The Gravitational Redshift

Let us consider an experiment with light in a freely falling laboratory (the Einstein elevator). Suppose we shine a light beam—say, a laser beam of a precise frequency—upward from floor to ceiling. Now the laboratory accelerates downward, gaining speed. By the time the light beam travels up to the ceiling, that ceiling is moving downward faster than the source on the floor was when the light left it. In other words, the receiver at the ceiling is approaching the place where the source was when the light left it. Therefore, wouldn't we expect to find the light at the ceiling blueshifted slightly because of the Doppler effect (Section 7.5)? But this would violate the principle of equivalence, for the blueshift would reveal our downward acceleration and show us we could not be weightless in free space. Therefore, Einstein postulated, there must be a redshift, due to the light's moving upward against gravity, that exactly compensates for the Doppler shift that would otherwise be observed. If so, that gravitational redshift should be observed in radiation climbing upward in a gravitational field—even at the surface of the Earth.

This idea was verified experimentally in the 1960s, at the Jefferson Physical Laboratory at Harvard University. Radioactive cobalt, which is a source of gamma rays, was placed in the basement of the building. The detector was a layer of cobalt placed at the top of the building, 20 m above the source. If there were no gravitational redshift, the gamma rays traveling upward from the basement should have been absorbed by the cobalt at the top of the building. In fact, they were not absorbed. In traveling upward against the Earth's gravitation, they suffered a gravitational redshift, which changed their wavelengths. To absorb them, the detector had to be moved slowly downward to produce a blueshift to compensate for the Earth's gravitational redshift. The actual motion of the detecting cobalt needed to make it absorb the gamma rays from the emitting cobalt in the basement was so slow that it would have required a full year to close the 20-m gap between emitter and detector. That speed produced a Doppler shift that agreed with the value needed to compensate for Einstein's predicted redshift to within 1 percent.

We can calculate the amount of gravitational redshift with elementary algebra. Let an elevator of height h be falling freely in the Earth's gravitational field. For definiteness, we will assume that the elevator is near the Earth's surface, where the uniform gravitational acceleration is g. Light, at speed c, takes a time $t = h/c$ to travel from the elevator floor to the ceiling. During this time, the elevator, accelerating uniformly, has increased its downward speed by

$$v = gt = \frac{gh}{c}.$$

The value gh/c is, then, the speed of the detector (when light reaches it) relative to the source (when light left it). The Doppler shift would then be (Section 7.5)

$$\frac{\Delta\lambda}{\lambda} = z = \frac{v}{c} = \frac{gh}{c^2}.$$

So long as the acceleration is uniform and $v \ll c$, so that the simple formula for the Doppler shift can be used, gh/c^2 is the gravitational redshift.

In the Jefferson Laboratory experiment, $h = 20$ m, so the gravitational redshift is

$$z = \frac{gh}{c^2} = \frac{9.80 \times 20}{(3 \times 10^8)^2} = 2.18 \times 10^{-15}.$$

The speed required to produce a Doppler shift of that amount is

$$v = 2.18 \times 10^{-15} c = 6.53 \times 10^{-7} \text{m/s}.$$

At that speed, the time for the detector to travel 20 m is

$$t = \frac{20}{6.53 \times 10^{-7}} = 3.06 \times 10^7 \text{s} \approx 1 \text{ year}.$$

It is also possible to calculate the gravitational redshift of light leaving a star. Within a freely falling spaceship (like the Space Shuttle) or in the Jefferson Laboratory, the gravitational field is essentially uniform. Such is not the case, however, for the light we observe leaving a star, because that light has to pass from the strong field near the star's surface on out through the continually weakening one as it gets farther and farther from the star. However, Einstein showed that we need only add up the tiny effects as the light passes through each small region within which gravity can be regarded as effectively constant to calculate the total gravitational redshift of light leaving the star. It works out that the wavelengths of light from the Sun should be increased by about two parts in a million—and recent measurements have verified this prediction.

The acceleration of gravity at a distance r from the center of a star of mass M is GM/r^2 (Section 3.2), where G is the gravitational constant. Over a small distance, dr, the small contribution, dz, to the gravitational redshift is

$$dz = \frac{1}{c^2} \frac{GM}{r^2} dr.$$

Readers familiar with calculus will recognize that a simple integration—adding up of contributions from the surface, R, of the star to a point very far away—gives the total gravitational redshift of light from the star:

$$z = \frac{1}{c^2} \frac{GM}{R}.$$

White dwarf stars, being very dense, have a much stronger surface gravity than does the Sun, and Einstein suggested that the gravitational redshift of the light from white dwarfs might be detectable. It can be observed, however, only for white dwarfs whose radial velocities are known from independent methods so that the gravitational redshift can be separated from the Doppler shift due to the stars' motions. Fortunately, several white dwarfs are members of binary-star systems, and their radial velocities can be deduced from those of their non–white dwarf companions, for which the gravitational redshift is negligible. The first reliable measurements of this effect were made in 1954 and verified Einstein's predictions to within about 20 percent.

Far higher accuracy has been attained recently in the near-Earth environment with space-age technology. In

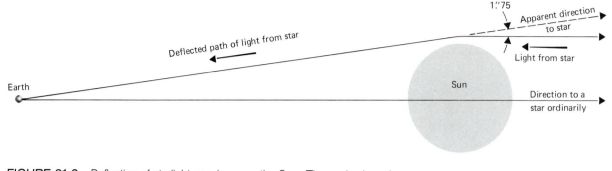

FIGURE 31.6   Deflection of starlight passing near the Sun. (The angles have been exaggerated.)

the mid–1970s, a hydrogen maser, which is a device that produces a microwave radio signal at a particular wavelength, was carried by a rocket to an altitude of 10,000 km. This maser was used to detect the radiation from a similar maser on the ground. That radiation showed a gravitational redshift due to the Earth's field that confirmed the relativity predictions to within a few parts in 10,000.

(b)   Deflection of Starlight

The strength of the gravitational acceleration at the surface of the Sun is 28 times its value at the surface of the Earth. Since spacetime is curved in regions where the gravitational field is strong, we would expect to find that light passing near the Sun would appear to follow a curved path (Figure 31.6). Einstein calculated from general relativity theory that starlight just grazing the Sun's surface should be deflected by an angle of 1.75 arcsec.

Stars cannot be seen or photographed near the Sun in bright daylight, but with difficulty they can be photographed close to the Sun at times of total solar eclipses. Einstein suggested an eclipse observation to test the light deflection in a paper he published during World War I. A single copy of that paper, passed through neutral Holland, reached the British astronomer Arthur S. Eddington. The next suitable eclipse was on May 29, 1919. The British organized two expeditions to observe it, one on the island of Principe, off the coast of West Africa, and the other in Sobral, in North Brazil. Despite some problems with the weather, both expeditions obtained successful photographs of stars near the Sun. Measures of their positions were then compared with measurements on photographs of the same stars taken at other times of the year, when the Sun was elsewhere in the sky. The stars seen near the Sun were indeed displaced, and, to the extent of accuracy of the measurements, which was about 10 percent, the shifts were

consistent with the predictions of relativity. It was a triumph that made Einstein a world celebrity.

The measurements made in 1919 were good enough to distinguish between Newton's theory of gravity, which predicts no deflection of starlight, and Einstein's theory of gravity, which does predict a deflection. Nevertheless, 10 percent accuracy is hardly a convincing demonstration that a theory is completely correct. Far higher accuracy has been obtained recently at radio wavelengths. Simultaneous observations of a source of radio waves with two telescopes far apart (a radio interferometer—see Chapter 10) can pinpoint the direction of the source very precisely. Observations of remote astronomical radio sources show that the difference in the directions to any two of them depends on their positions relative to the Sun. Measurements of the apparent shifts in direction as the Sun moves through the sky confirm the predictions made by Einstein to within 1 percent.

(c)   Advance of Perihelion of Mercury

According to relativity, the energy and momentum associated with the motion of a body, and even its gravitational energy, all contribute to its effective mass, and hence to the force of gravitation on it. Mercury has a fairly eccentric orbit, so that it is only about two-thirds as far from the Sun at perihelion as it is at aphelion. As required by Kepler's second law, Mercury moves fastest when nearest the Sun, which is also the time when the gravitational force exerted by the Sun is greatest. According to relativity, these effects combine to produce a very tiny additional push on Mercury, over and above that predicted by Newtonian theory, at each perihelion. The result of this effect is to make the major axis of Mercury's orbit rotate slowly in space. Each successive perihelion occurs in a slightly different direc-

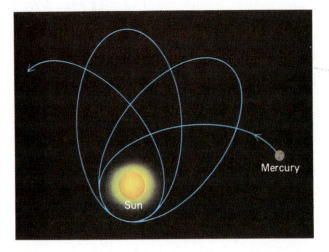

FIGURE 31.7 Rotation of the major axis of the orbit of a planet, such as Mercury, because of various perturbations.

tion as seen from the Sun (Figure 31.7). The prediction of relativity is that the direction of perihelion should change by 43 arcsec per century. It would thus take about 30,000 years for the major axis of the orbit to make a complete rotation.

The gravitational effects (perturbations) of the other planets on Mercury also produce an advance of its perihelion. According to Newtonian theory, the gravitational force exerted by Venus will cause Mercury's perihelion to advance by 277 arcsec per century, Jupiter contributes another 153 arcsec, the Earth accounts for 90 arcsec, and Mars and the remaining planets are responsi-

ble for an additional 10 arcsec. The total of these contributions amounts to about 531 arcsec per century. In the last century, however, it was observed that the actual advance is 574 arcsec per century. The discrepancy was first pointed out by Leverrier, codiscoverer of Neptune. In analogy with Neptune, it was assumed that an intramercurial planet was responsible. The hypothetical planet was even named for the god Vulcan. Vulcan, of course, never materialized, and that 43-arcsec anomaly was entirely explained by relativity. The relativistic advance of perihelion can also be observed in the orbits of several minor planets that come close to the Sun.

(d)   Time Delay of Light

According to general relativity, if light from a distant source passes very near the edge of the Sun, it will take longer for that light to reach the Earth than one would expect on the basis of Newton's law of gravity. The smaller the distance between the ray of light and the edge of the Sun at closest approach, the longer the delay in the arrival time. Half of the delay is a result of the gravitational redshift that occurs when light climbs out of a strong gravitational field. The other half of the delay is a consequence of the curvature of spacetime.

To see why curvature of spacetime causes a time delay, imagine a spacetime formed of a rubber sheet (Figure 31.8). Imagine the Sun as a heavy ball that causes the rubber sheet to sag in the middle. In this spacetime, light can travel only along the surface of the rubber sheet. When a light ray travels near the Sun, it must pass into the depression and therefore cover a longer distance

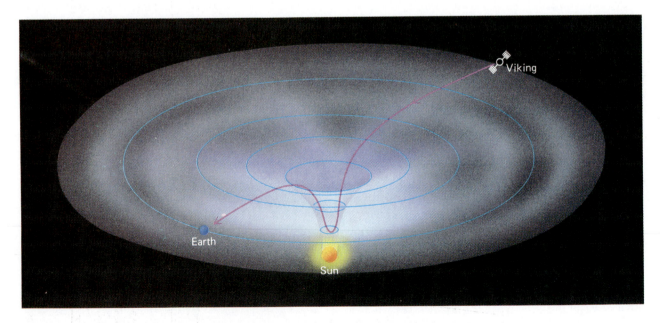

FIGURE 31.8 Radio signals from the Viking lander on Mars are delayed when they pass near the Sun, where spacetime is curved relatively strongly. In this picture, spacetime is pictured as a two-dimensional rubber sheet.

than it would if the Sun were not present and there were no curvature of spacetime in this region.

The time delay in light that passes near the Sun is very small. For example, when Mars is at superior conjunction, the delay in the arrival time of a radar signal sent from the Earth to Mars and reflected back again amounts to only 250-millionths of a second according to general relativity. The total time required for the round trip is 42 min. In 250-millionths of a second, light travels 75 km, so the delay can be estimated by determining the difference between the true distance to Mars, which is known from its orbit, and its apparent distance as measured from the time required for a signal to reach the Earth from Mars.

The most accurate measurements of time delay were made by using signals sent from the Viking spacecraft that landed on Mars. The predictions of general relativity turned out to be correct within the accuracy of the observations, which was 1 part in 1000.

Additional tests for relativity theory are in the planning or experimental stages—some at the frontier of modern technology. In a satellite experiment expected to fly by the end of the century, the behavior of a gyroscope will be carefully monitored. Relativity predicts an angular change in the orientation of the axis of the gyroscope due to its motion through the Earth's gravitational field of only 0.05 arcsec per year, but it is believed that even that small a change can be measured accurately enough to check the theory.

31.4 GRAVITATIONAL WAVES

According to special relativity, no information can travel from one place to another at a rate that exceeds the speed of light. This limitation applies to gravity as well. Suppose that two objects interact gravitationally. Suppose that one of them changes its shape from a sphere, say, to a sausage. The force of gravity that it exerts on the second object will change, but the second object will sense that change only after a time equal to the amount of time required for light to traverse the distance between the two objects. If the first object returns to its original shape, then information about the change in the gravitational force that it exerts will again propagate out into space at the speed of light.

The effect is much the same as what happens when you drop a stone into a pond. A wave is generated that moves outward, and a twig floating in the pond will move up and down only when the wave reaches it.

In general relativity, the geometry of spacetime depends on the distribution of matter. Any rearrangement of matter—say, from a sphere to a sausage shape—must result in an alteration of spacetime—that is, it creates a disturbance. This disturbance, which propagates

through space with the speed of light, is called a **gravitational wave.** Gravitational waves carry away energy from the object that produces them. This loss of energy turns out to hold the key to proving that general relativity is again correct—that gravitational waves actually do exist.

(a)   The Binary Pulsar PSR1913 + 16

In 1974, R. A. Hulse and J. H. Taylor, of the University of Massachusetts, observing with the 1000-ft radio telescope at Arecibo, Puerto Rico, discovered a remarkable pulsar, now designated PSR1913 + 16 (the numbers give the coordinates of its location in the sky). The unique thing about PSR1913 + 16 is that the period of the pulses shows cyclic variations over a short time interval of 7^h45^m. These period changes are due to the Doppler effect caused by the pulsar's revolution about another object. When the pulsar is approaching us in its orbit, each successive pulse has less far to travel to reach us on Earth, and we receive the pulses slightly closer together than average. Conversely, when the pulsar is moving away from us, we receive the pulses slightly spread out in time. Thus we can analyze the orbital motion of the pulsar just as we do that of a spectroscopic binary star (Section 23.2).

Such an analysis, combined with our knowledge of the expected properties of neutron stars, indicates that the pulsar is in mutual revolution in that 8-hr period around an invisible companion of comparable mass that is probably either a white dwarf or another neutron star.

There are two important points about this binary pulsar. First, at its orbital speed of approximately 0.1 percent that of light, it should radiate gravitational waves at a rate great enough to carry appreciable energy away from the system, causing the pulsar and its companion to spiral slowly closer together. Second, pulsars are superb clocks, remaining stable to 0.001 s over several years.

As the pulsar and its companion spiral together, their period of revolution shortens. The shortening is only about one-ten-millionth of a second per orbit, but the effect accumulates like a clock that runs a little faster each day. During the first seven years this system was observed, the time of periastron (when the two objects are closest together) shifted by more than a full second of time relative to when it would have occurred if the period had remained constant.

The shift in periastron time is just what general relativity predicts it should be as a result of the emission of gravitational radiation by two stars, each of about 1.5 solar masses, with an orbit like that of the binary pulsar. PSR1913 + 16 is now regarded as providing strong evidence that gravitational waves do exist, as Einstein predicted, and Hulse and Taylor received the 1993 Nobel prize in physics for this work.

(b) Gravitational Wave Astronomy (?)

In the case of the binary pulsar, we infer the presence of gravitational radiation from the observations of its effects—that it removes energy from the binary system and causes the pulsar and its companion to spiral ever closer together. Scientists are now trying to devise ways to go one step further and detect gravitational waves *directly* here on Earth. To judge whether or not such an experiment is feasible, we must know first how gravitational radiation affects matter so that we know how to build a gravitational wave detector. Second, we must calculate how much gravitational radiation is emitted by various possible sources so that we know how sensitive the detectors must be.

Suppose we arrange four masses in the pattern shown in Figure 31.9 and suppose these masses are free to move. If a gravitational wave passes by these four masses, then it will cause them to move relative to one another in the way shown in Figure 31.9. (In fact, each gravitational wave has two independent states of polarization, and the second polarization will produce a pattern of motion that is rotated with respect to the pattern shown in Figure 31.9 by 45°.)

To build a gravitational wave detector, we need only build a system that is sensitive to these motions. This sounds like a simple task, but in fact it is very difficult because the motions turn out to be very small indeed. Potentially, one of the strongest sources of gravitational radiation would be a rotating star collapsing to form a black hole (Section 31.5). If such an event were to occur in our own Galaxy, it would cause two masses on Earth, separated by 1 m, to move apart by 0.01 times the diameter of an atomic nucleus. One of the first gravitational wave detectors was a solid cylinder of aluminum several meters long. At room temperature, the thermal motions of the aluminum atoms produce random changes in the apparent length of the cylinder that are much larger than those expected from passing gravitational waves.

Newer gravitational wave detectors are cooled to very low temperatures to reduce thermal motions. Other gravitational wave detectors do not consist of solid cylinders of material, but resemble more the diagram in Figure 31.9. Mirrors are suspended in such a way that they can move in response to passing gravitational waves. The distance between the mirrors is measured by sending laser light to them and measuring how long it takes for the signal to return. Scientists in the U.S. have designed a gravitational wave detector in which mirrors would be mounted in two perpendicular directions separated from the central laboratory by 4 km. With such an arrangement it should be possible to measure a change in the length of one 4-km arm with respect to the other to an accuracy of 1 part in 10^{20}. Since the changes are so small, scientists would believe that a variation in the distance between the mirrors was caused

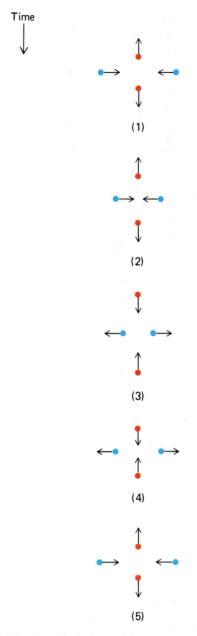

FIGURE 31.9 The simplest possible gravitational wave will cause masses to move in the pattern shown.

by gravitational waves only if a second and entirely independent gravitational wave detector saw the same thing at the same time. Accordingly, U.S. plans call for two detector systems, one on each side of the country.

Is a sensitivity of 1 part in 10^{20} enough to detect gravitational waves? Every planet, star, or galaxy is potentially a source of gravitational radiation as it moves in its orbit or changes its shape, but under normal conditions ordinary objects do not emit enough gravitational radiation to be detected. Even the binary pulsar is so weak a source of gravitational waves that no detector that we can now imagine could sense those waves.

The strongest emitters are likely to be supernova explosions, two binary stars that have finally spiraled so close together that they merge, and matter that is collapsing to form a black hole. If optimistic estimates of the amount of gravitational radiation produced by a supernova explosion are correct, then existing detectors are adequate to sense a supernova in our own Galaxy. Unfortunately, no supernova has been observed in our own Galaxy in nearly 400 years. To detect a supernova in the nearest large cluster of galaxies, detectors would have to measure changes in length to an accuracy of 1 part in 10^{22}. Greater sensitivity is required, since the intensity of the gravitational radiation decreases with the square of the distance from the source.

Despite the technical challenges of building detectors of the required sensitivity and the uncertainties in the calculations of the amount of gravitational radiation that might be emitted by astronomical sources, scientists in several countries are now seeking funds to build gravitational wave "telescopes." If these experiments are successful, and it will probably be well into the next century before we know the outcome, then gravitational waves could offer an important new window through which we might study the most violent events in the universe.

31.5 BLACK HOLES

(a) Definition of a Black Hole

Let's do a thought experiment—that is, an experiment that is just in our heads, since it is impossible to do this experiment in a laboratory. We already know that a rocket must be launched from the surface of the Earth at a very high velocity if it is to escape the pull of the Earth's gravity. In fact, any object—rocket, bullet, ball—that is thrown into the air with a velocity that is less than 11 km/s will fall back to the Earth's surface. Only those objects launched with a velocity greater than 11 km/s can escape into space.

For the Sun, the escape velocity is higher yet—618 km/s. Now imagine that we begin to compress the Sun and force it to shrink in diameter. When the Sun reaches the diameter of a neutron star (less than 100 km), the velocity required to escape the gravitational pull of the shrunken Sun is about half the speed of light. Suppose we continue to compress the Sun to a smaller and smaller diameter. Ultimately, the escape velocity will exceed the speed of light. If light is composed of particles, one might expect it, too, just like rockets, to feel the influence of gravity and so be unable to escape. If no light can escape, the object is invisible.

In modern terminology, we call such an object a **black hole,** a name suggested by the American scientist John Wheeler in 1969. The idea that such objects might exist is, however, not a new one. Cambridge professor and amateur astronomer John Michell wrote a paper in 1783 about the possibility that stars might exist for which the escape velocity exceeds that of light. In 1796, the French mathematician Pierre Simon, Marquis de Laplace, wrote about similar objects, which he termed "dark bodies."

The theory of general relativity is required to calculate what actually happens when the gravitational force becomes so large. We have already described one extremely dense object with a very strong gravitational field—the neutron star. We recall that light grazing the surface of the Sun is deflected by about 1.75 arcsec. Light grazing the surface of a white dwarf would be deflected by about 1 arcmin, and that grazing a typical neutron star by about 30°.

Consider now the light radiated from the surface of a neutron star. Light that emerges normal (perpendicular) to the surface flows out radially from the star. Light emitted in a nearly horizontal direction will follow a slightly curved path because of the gravitational deflection. Similarly, a rocket ship launched in a nearly horizontal direction will be deflected by the Earth's gravity. Now imagine a more massive star that shrinks to an even smaller size and higher density than a neutron star. The path followed by light emitted from such an object is even more sharply curved. Eventually the star reaches a size at which a horizontal beam of light enters a circular orbit. A surface of that radius is called the **photon sphere.**

Now suppose the star shrinks to a size even smaller than that of the photon sphere. To escape from the star, light must flow into a cone about the normal to the surface of half-angle θ (Figure 31.10). Light at a greater angle falls back on the star. The angle θ becomes smaller and smaller as the star collapses. When the radius of the star is two-thirds that of the photon sphere, θ becomes zero, and no light at all can escape. At this point the velocity of escape from the star equals the speed of light. As the star contracts still more, light as well as everything else is trapped inside, unable to escape through that surface where the escape velocity is the speed of light. That surface is called the **event horizon,** and its radius is the *Schwarzschild radius*, named for Karl Schwarzschild, who first described the situation within two months after Einstein published general relativity in its final form. This surface is the boundary of the black hole. All that is inside is hidden forever from us.

The size of the event horizon is proportional to the mass of the collapsed object. For a black hole of one solar mass, the event horizon is about 3 km in radius; thus the entire black hole is about one-third the size of a neutron star of that same mass. For a star of three solar masses, the radius of the event horizon is about 9 km, the same as the radius of a neutron star of three solar masses. If a collapsing star is much more massive than three solar masses, it reaches its event horizon before it can shrink to neutron star dimensions. This is another

only very near the surface of a black hole that its gravitation is so strong that Newton's laws break down. For a black hole of the mass of the Sun, light would have to come within 4.5 km of its center to be trapped. A solar-mass black hole, remember, is only 3 km in radius—a very tiny target. A star would be far, far safer to us as an interloping black hole than it would have been in its former stellar dimensions.

(e) A Trip into a Black Hole

Still, it is interesting to contemplate a trip into a black hole. Suppose that the invisible companion of the star associated with Cygnus X-1 is a black hole of ten solar masses. What would you see if a daring astronaut bravely flies into it in a spaceship?

At first he darts away from you as though he were approaching any massive star. However, when he nears the event horizon of Cygnus X-1—some 30 km in radius and presumably near the center of the accretion disk—things change. The strong gravitational field around the black hole makes his clocks run more slowly as seen by you. Signals from him reach you at greatly increased wavelengths because of the gravitational redshift. As he approaches the event horizon, his time slows to a stop—as seen by you—and his signals are redshifted through radio waves to infinite wavelength. He fades from view as he seems to you to come to a stop, frozen at the event horizon.

All matter falling into a black hole appears to an outside observer to stop and fade at the event horizon, frozen in place, and taking an infinite time to fall through it—including the matter of a star itself that is collapsing into a black hole. For this reason, black holes are sometimes called *frozen stars.*

This, however, is only as you, well outside the black hole, see things. To the astronaut, time goes at its normal rate, and he moves right on through the event horizon. In principle, he could even survive a trip into a very massive black hole. If the black hole has a mass of thousands of millions of solar masses, the event horizon is far away from the black hole and is very large. The astronaut notices nothing special as he passes through it. But if the black hole is small—on the order of 10 solar masses, for example—the astronaut could not survive. Suppose he is falling feet first toward the black hole. The gravitational force on his feet is greater than on his head. If the black hole is small, so is the event horizon, and the astronaut must get very close to the black hole in order to reach the event horizon. As he approaches the event horizon, the force on his feet will become so much stronger than the force on his head that he will be ripped apart.

Once inside, everything, including astronaut and light, is doomed to remain hidden forever from the universe outside. We can never know what goes on inside the event horizon. Theory predicts that after entering the black hole, the astronaut races irreversibly to the center, to a point of zero volume and infinite density. Mathematicians call such a point a **singularity.**

(f) Other Ways to Make Black Holes

While the collapse of a massive star seems to be the mechanism most likely to produce a black hole, calculations suggest that other processes may lead to the formation of black holes of all sizes, with masses ranging from 10^{-8} kg to 10^8 times the mass of the Sun.

As we shall see in Chapter 33, there is evidence that supermassive black holes (up to 10^8 times the mass of the Sun) can be found at the centers of many galaxies. Such massive black holes might be formed from the collapse of supermassive stars, through the collision and coalescence of stars in a very dense star cluster, through the coalescence of smaller black holes, or through the accretion and accumulation of matter into an originally much smaller black hole.

Most intriguing of all is the possibility that very small black holes—as small as 10^{-8} kg—may have formed during the first fraction of a second after the universe began. The density of material as it crosses the event horizon to form a black hole varies inversely as the mass of the black hole. An object with a mass of 10^9 times the mass of the Sun would have a density about equal to the density of water as it fell inside its event horizon. For an object with a mass equal to that of the Sun, the corresponding density is a thousand times the density of an atomic nucleus. Objects much less massive than the Sun would have to be compressed to incredible densities to form a black hole. We can think of no circumstances at the present time that seem likely to produce such an enormous compression. The necessary densities may, however, have been achieved during the first fraction of a second after the universe was formed, when all of the matter of the universe was confined to a very small volume (see Chapter 35).

There is no observational evidence that such primordial black holes actually exist, but they are of great interest to theoreticians because their properties are determined by quantum mechanics as well as by relativity. This possibility has been explored by the brilliant British theoretical astrophysicist Stephen Hawking.

We have seen (Chapter 27) that all fundamental particles have their antiparticles—for example, electrons and positrons, protons and antiprotons, and so on. Whenever a particle and its antiparticle come into contact, they annihilate each other, transforming completely into energy. Similarly, pure energy can be converted into pairs of particles—an electron and a

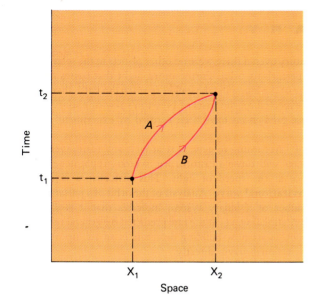

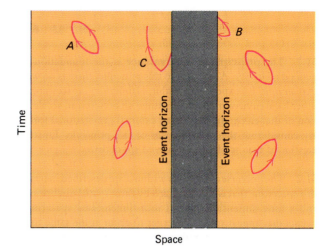

FIGURE 31.12 A spacetime diagram showing the creation of a particle and an antiparticle and their subsequent annihilation. At time t_1 and position x_1, the two particles are created. They may then move at different velocities and so separate in space. For example, the particle may move along path A in spacetime, and the antiparticle may move along trajectory B. Ultimately their mutual attraction will bring them back together to the same point in space at the same time (point t_2 and x_2), and they will destroy each other. Particles created in this way do not separate far enough for long enough to be observable.

FIGURE 31.13 The shaded area is a schematic representation of a black hole in spacetime (schematic because we do not show the curvature of spacetime that is present near a black hole). Far from the black hole (at A, for example), particles and antiparticles appear and annihilate each other. If a particle and an antiparticle are created near a black hole, however, the possibility exists that both particles will fall into the black hole (as at B) or that only one particle will fall into the black hole (as at C), leaving its partner free to escape.

positron, for example. This process is known as *pair production* and is observed regularly in the nuclear physics laboratory (Figure 31.12).

Now all this is possible because mass and energy are equivalent. According to quantum theory, it is even possible for matter (or energy) to be created from *nothing* for an exceedingly brief time. This possibility comes about because of the innate uncertainty in nature, at the microscopic level, of the measures of physical quantities such as mass and energy. This uncertainty does not violate conservation laws because any matter that comes into being almost immediately disappears again spontaneously, so on the average mass and energy (combined) are conserved.

But, Hawking points out, what if a positron and an electron (say) come into existence momentarily in the vicinity of a black hole (Figure 31.13). There is a chance that one or the other will fall into the hole and hence will not be able to be annihilated with its antiparticle, returning the energy it "borrowed" from nature. Its antiparticle, therefore, can escape unscathed. However, many such positrons and electrons so created near black holes and escaping from them do annihilate each other, creating energy. That energy cannot come from nothing;

according to Hawking's theory it must come from the black hole itself. Robbing the black hole of energy in this way robs it of mass (for they are equivalent), so the black hole must slowly evaporate through this process of pair production.

As esoteric as this idea may seem, it is generally thought to be possible by theoretical physicists. The process is only important, however, near very tiny black holes. Solar-mass black holes would evaporate in this way at an absolutely negligible rate. In fact, the only black holes that would have had time to evaporate in the age of the universe would be those of an original mass less than about 10^{12} kg (like a minor planet). Smaller ones would already be gone; those of about 10^{12} kg should be finishing off about now if they were formed when the universe began. Because the evaporation rate increases as the mass of the black hole goes down, at the end one would go off explosively, emitting a final burst of gamma radiation.

Nobody knows whether mini-black holes were created in the early universe. If they were, and if they really do evaporate, we would expect bursts of gamma rays to be emitted when mini-black holes explode. So far our instruments are not sensitive enough to detect these gamma rays, but the speculation remains an interesting possibility.

We have touched on many ideas in this chapter. Some, near the end, are highly speculative, but the main thrust of general relativity theory now appears to be firmly established. General relativity has come of age and is an important part of modern astronomy.

An image of the southern Milky Way taken from the Cerro Tololo Inter-American Observatory on April 14, 1986. Halley's Comet is seen in the center of the picture. Note the dark clouds of dust silhouetted against the stars in the plane of the Milky Way.

(Gabriel Martin/NOAO)

Our journey of exploration now takes us into a different and larger realm. Instead of considering single stars or star clusters, we expand our vision to consider the properties and behavior of entire *galaxies*—vast systems containing billions of stars. Appropriately, we begin with the Galaxy in which we live; but getting to know it has not been as simple as its proximity may lead you to believe. After all, getting a good picture of our Galaxy from our location within is like trying to get a good view of yourself from inside your knee-cap—not an easy task. Nevertheless, especially with the explosive growth of better astronomical instruments during the 20th century, we have made much progress in uncovering the structure and history of our Galaxy.

We now know, for example, that since the Sun and the Earth formed, some five billion years ago, we have circled the center of our rotating Galaxy almost 20 times. Ours is not the central place in the system; in fact, our location does not seem to be particularly distinguished in any other way. At our "suburban" site, the center of the Galaxy is quite hard to discern; and what strange "beast" lurks at the center is still a subject of great debate among astronomers.

Indeed, the size of our Galaxy staggers the imagination and dwarfs all measures of human existence—at the fastest speed any of our planetary probes have flown through space, they would take roughly 1.3 billion years to cross our Galaxy! Yet, as you will see, it is now becoming increasingly clear that all the parts of our Galaxy that we have come to know are only the tip of the cosmic iceberg. Surrounding the parts we can detect, there appears to be an even larger halo of dark, unknown matter which may compose the bulk of the Galaxy. Our journey to explore the complex system in which we find ourselves may, in many ways, still be at its beginning.

THE MILKY WAY GALAXY

Harlow Shapley (1885–1972) began his career as a newspaper reporter. In his twenties he enrolled in the University of Missouri, where he searched through the catalogue for a suitable major and found Astronomy under the As. After earning his bachelor's degree, he went to Princeton for his Ph.D. Subsequently, at Mt. Wilson, his study of globular clusters revealed the true extent of our Galaxy. (Yerkes Observatory)

One of the most striking features in a truly dark sky is a band of faint white light that stretches from one horizon to the other (Figure 32.1). Because of its appearance, this band of light is called the Milky Way. In the Northern Hemisphere, the band is brightest in the region of the constellation Cygnus and is best viewed in the summer. In the Southern Hemisphere, the Milky Way is even brighter—so bright, in fact, that Indians in South America gave names to various portions of it just as northern astronomers gave constellation names to conspicuous groupings of stars. Unfortunately, the Milky Way is not bright enough to be seen from urban areas with their artificial lighting, and so many city-dwellers have not seen the Milky Way.

In 1610, Galileo made the first telescopic observations of the Milky Way and discovered that it is composed of a multitude of individual stars. We now know that the Sun is located within a disk-shaped system of stars. The Milky Way is the light from nearby stars that lie more or less in the plane of the disk. We call this great stellar system, which includes all of the individual stars that we can see except with the largest telescopes, the **Milky Way Galaxy** or, more simply, just the **Galaxy.**

32.1 THE ARCHITECTURE OF THE GALAXY

(a) Overview of the Galaxy

The disk shape of the stellar system to which the Sun belongs was demonstrated quantitatively in 1785 by William Herschel, who used a telescope to count the numbers of stars that he saw in various directions (Section 24.4). Herschel found that the numbers of stars were about the same in any direction around the Milky Way, a result that seemed to show that the Sun was near the center of the Galaxy. We now know, however, that this conclusion is wrong. Interstellar dust absorbs the light from stars and restricts optical observations in the plane of the Milky Way to stars within about 2000 pc. Herschel was able to observe only a tiny fraction of the system of stars that surrounds us.

With modern instrumentation, astronomers can make observations at radio and infrared wavelengths, and electromagnetic radiation at these long wavelengths penetrates the dust easily. On the basis of these new observations, astronomers now picture the Galaxy as a thin, circular **disk** of luminous matter distributed across a region about 30,000 pc in diameter, with a thickness of about 300 pc. Dust and gas, the raw material from which stars form, are fairly closely

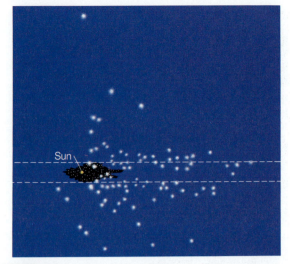

FIGURE 32.4   A copy of a diagram by Shapley, showing the distribution of globular clusters in a plane perpendicular to the Milky Way and containing the Sun and the center of the Galaxy. Herschel's diagram (Figure 24.8) is shown centered on the Sun, approximately to scale.

volume exceeds that of the main disk of the Galaxy by many times.

Individual RR Lyrae stars have been found in significant numbers as far away as 10,000 to 15,000 pc on either side of the galactic plane, which shows that the halo must have an overall thickness of at least 30,000 pc. One RR Lyrae star has recently been found at a distance of about 50,000 pc from both the Sun and the center of the Galaxy. A few globular clusters have distances from the galactic center of almost 80,000 pc. If further observations show that these objects are gravitationally bound to the Galaxy and are not intergalactic interlopers, then the halo extends far beyond the visible "rim" of the main disk of the Galaxy. Halos of some other galaxies have been traced to similar distances. We shall see in Section 32.4 that modern data on the rotation of our Galaxy give good reason to think that the halo contains a large fraction of the Galaxy's mass.

There is also gas in the halo, and some of this gas emits x-ray radiation. To emit x rays, this gas must be very hot—about 10^6 K—too hot to produce the emission lines seen in ordinary H II regions. The gas, probably heated by supernova shells and/or stellar winds (Section 25.3), extends only into the inner part of the halo and defines a region now usually referred to as the **galactic corona.**

(d)   Interstellar Matter

The main body of stars in the Milky Way forms a flat disk. In addition to stars, the disk of the Galaxy contains interstellar gas and dust.

Hydrogen makes up about three-quarters of the interstellar gas, and hydrogen and helium together compose 96 to 99 percent of it by mass. Most of the gas is cold and nonluminous. Gas clouds at temperatures of a few degrees Kelvin to about 30 K consist primarily of hydrogen molecules (H_2). Clouds at temperatures of a few tens to hundreds of degrees Kelvin contain primarily atomic hydrogen. Near very hot stars, hydrogen is ionized by the ultraviolet radiation from those stars (Figure 25.2).

Observations at 21 cm show that the neutral atomic hydrogen in the Galaxy is confined to an extremely flat layer. This hydrogen extends well beyond the Sun to a distance of about 25,000 pc from the center of the Galaxy. At the position of the Sun, the layer of atomic hydrogen is only about 125 pc thick.

In the inner regions of the Galaxy, dust is typically found in the same places as neutral hydrogen gas, and therefore the dust, too, is confined to the disk, with the highest concentrations occurring in the spiral arms. The thickness of the dust layer is also about 125 pc. There is very little emission from dust lying outside the Sun's orbit around the center of the Galaxy.

The coldest hydrogen gas in the Galaxy has temperatures less than about 30 K and is found mostly in the form of hydrogen molecules (H_2). These molecules are clumped into giant molecular clouds (Section 25.6), which are found in greatest numbers in a broad ring at distances between 3000 and 8500 pc from the center of the Galaxy. The molecular clouds are normally not found more than 100 pc above or below the plane of the Galaxy. Molecular hydrogen is also observed in the center of the Galaxy.

The most massive molecular clouds are found in the spiral arms. In many cases, individual clouds have gathered into large complexes containing a dozen or more discrete clumps. Since the large molecular clouds and molecular cloud complexes are the sites where star formation occurs, most young stars are also to be found in spiral arms. That is why the spiral arms shine so brilliantly in photographs of spiral galaxies (Figure 32.5).

32.2   SPIRAL STRUCTURE OF THE GALAXY

(a)   Spiral Arms

Studies of the 21-cm line have played a key role in determining the nature of the spiral structure in our own Galaxy. While the interpretation of the measurements is somewhat controversial, it appears likely that the Galaxy has four major spiral arms, with some smaller spurs (see Figure 32.2). The Sun appears to be near the inner

FIGURE 32.5   The spiral galaxy NGC 2997. The two spiral arms, which appear to originate in the yellow nucleus, are peppered with bright red blobs of ionized hydrogen that are similar to regions of star formation in our own Milky Way Galaxy. The hot blue stars that generate most of the light in the arms of the galaxy form within these hydrogen clouds. (Anglo-Australian Telescope Board)

edge of a short arm or spur called the *Orion arm*, which is about 5000 pc long and contains such conspicuous features as the North America Nebula, the Coalsack (near the Southern Cross), the Cygnus Rift (great dark nebula in the summer Milky Way), and the Orion Nebula. More distant, and therefore less conspicuous, are the *Sagittarius-Carina* and *Perseus arms*, located, respectively, about 2000 pc inside and outside the Sun's position with respect to the galactic center. Both of these arms, and the Cygnus arm, are about 25,000 pc long. The fourth arm is unnamed and is difficult to detect because emission from it is confused with strong emission from the central regions of the Galaxy, which lie between it and us.

(b)   Differential Galactic Rotation

At the Sun's distance from its center, the Galaxy does not rotate like a solid wheel. Stars in larger orbits do not keep abreast of those in smaller ones, but trail behind. This effect produces a shearing motion in the plane of the Galaxy, called *differential galactic rotation*. We can detect the differential galactic rotation from observations of proper motions and radial velocities of stars around us.

As an illustration, let us consider the effect of differential rotation on the radial velocities of stars in different directions in the plane of the Galaxy. Let us assume that all stars have circular orbits (only approximately true). If we look in the direction of the galactic center or in the opposite direction,

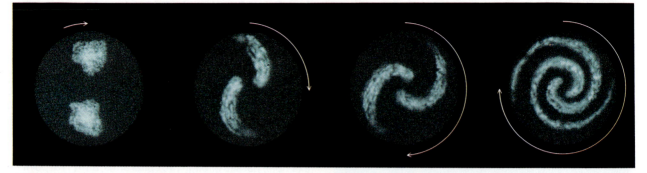

FIGURE 32.10 Hypothetical formation of two spiral arms from irregular clouds of interstellar material based simply on the rotation of the Galaxy, which is more rapid for objects closer to the center.

spirals. Figure 32.10 shows the development of spiral arms from two irregular blobs of interstellar matter, as the portions of the blobs closest to the galactic center move fastest, while those farther away trail behind.

In order to understand quantitatively how spiral arms form, astronomers have calculated how stars and gas clouds move as they orbit on circular paths about the galactic center under the influence of both the gravitational fields produced by the Galaxy as a whole and by the matter forming the spiral arms themselves. The calculations show that objects slow down in the regions of the spiral arms and linger there longer than elsewhere in their orbits. Thus a wave of higher density builds up where the spiral arms are. Stars, gas, and dust pass slowly through the spiral arms. This theory for the formation of spiral arms is referred to as the **spiral density wave** model.

As a good analogy for what happens, suppose you are driving on a freeway with three lanes. Suppose that there are cars that are moving unusually slowly in all three lanes ahead of you. Traffic behind these three cars will be forced to slow down, and the density of cars will increase. Some individual cars may manage to get past the three slowly moving cars, but others will take their place in the traffic jam. Viewed from high above the freeway, the point of maximum density of cars would appear to move along the freeway at the same speed as that of the three slowly moving cars. The place of maximum density would thus be moving more slowly than the cars either in front of the traffic jam or well behind it. In just the same way, stars and interstellar clouds slow down when they pass through the spiral arms, which are the places of maximum density.

As gas and dust clouds approach the inner boundaries of an arm and encounter the higher density of slower-moving matter, they collide with it. It is here, where the shock of the collision occurs, that theory predicts star formation is most likely to take place. We know from our own Galaxy that the youngest stars are found in the spiral arms. In some other galaxies, where we can view the spiral arms face on, we see young stars, along with the densest dust clouds, near the inner boundaries of spiral arms, just as theory predicts (Figure 32.11). Spiral density waves have also been observed directly in the rings of Saturn.

There is at least one other way to produce an elongated structure that looks something like a spiral arm. Star formation sometimes moves progressively through molecular clouds and produces extended regions of young stars that mimic spiral arms (Section 28.2).

32.3 STELLAR POPULATIONS IN THE GALAXY

(a) High- and Low-Velocity Stars

The majority of the stars near the Sun move nearly parallel to the Sun's path about the galactic nucleus, and their speeds with respect to the Sun are generally less than 40 or 50 km/s. These are said to be **low-velocity** stars. The radial velocities of nearby gas clouds are also low. The gas clouds, like the Sun, move in roughly circular orbits about the galactic nucleus.

Some stars, on the other hand, have speeds relative to the Sun in excess of 80 km/s and are called **high-velocity** stars. They move along orbits of rather high eccentricity that cross the Sun's orbit in the plane of the Galaxy at rather large angles (Figure 32.12). Nearby stars moving on such orbits are passing through the solar neighborhood and are only temporarily near us. Globular clusters and stars in the galactic halo also have orbits very different from the Sun's and are high-velocity objects.

The term "high velocity" or "low velocity" refers to the speed of an object *with respect to the Sun* and has nothing to do with its motion in the Galaxy. Most high-velocity stars lag behind the Sun in its motion about the

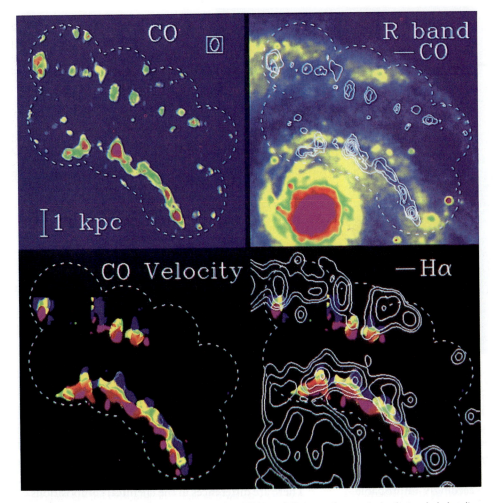

FIGURE 32.11 Measurements of M51, the Whirlpool Galaxy, demonstrate that spiral density waves lead to star formation. All four panels show the northwest part of M51. The upper left picture shows the locations of molecular clouds, which are detected by measurements of CO. Two segments of spiral arms are seen. In this picture, the molecular gas is moving counterclockwise. In the upper right picture, contour lines show the locations of the same molecular clouds relative to the visible spiral arms (colored yellow). Note that the clouds lie in dusty regions along the inner part of each spiral arm. In the lower pictures, the molecular clouds of highest velocity have been color-coded red; those of lower velocity are blue. As the gas moves from red to blue, it slows by 20 to 30 km/s. The contours in the right hand plot show H II regions (ionized hydrogen gas); these are regions containing massive hot stars, and so star formation has occurred recently here. Note that star formation occurs downstream from the molecular clouds. This spatial relationship shows that star formation occurs *after* the molecular clouds have been slowed down and compressed by their passage through regions of higher density. (Courtesy, S.N. Vogel, University of Maryland)

galactic center and hence are actually revolving about the Galaxy with speeds *less* than those of the low-velocity stars near the Sun.

The component of velocity of an individual star in a direction perpendicular to the plane of the Galaxy (sometimes called the z *component of its velocity*) cannot, in general, be determined from its radial velocity alone. However, the average of the z velocities for stars of a given class or group can be found from a statistical analysis of their radial velocities. Observations show that low-velocity stars usually have lower velocity components perpendicular to the galactic plane than do

high-velocity stars. Consequently, high-velocity stars tend to be less strongly concentrated in the plane of the Galaxy than do low-velocity stars. Stars in the halo are extreme examples and usually have very high z velocities.

It is, of course, the high velocities of halo objects perpendicular to the galactic plane that carry them to large distances above and below the disk and account for their spheroidal distribution in the Galaxy. Globular clusters, in particular, are believed to revolve about the nucleus of the Galaxy in orbits of high eccentricity and inclination to the galactic plane, rather like the comets

forms an *accretion disk* of material around it. As the material spirals ever closer to the black hole, it accelerates and heats through compression to millions of degrees. This hot matter then would be the source of the radio emission from the galactic center.

Suppose there is a black hole of several million solar masses in the center of the Galaxy. Where did the mass come from? At the present time, matter from such sources as colliding gas clouds in the ring is falling into the galactic center at the rate of about 1 solar mass per 1000 years. If matter had been falling in at the same rate for about 5 billion years—and we do not know whether or not this is the case—then it would have been possible to accumulate the matter needed to form a black hole with a mass of several million solar masses.

The density of stars near the galactic center is such that we would expect a star to pass near the black hole and be disrupted by it every few thousand years. As this material falls into the black hole there should be a brilliant outburst. Perhaps such an outburst drove gas and dust out of the galactic center a few tens of thousands of years ago, leaving behind the cavity and the clumpy ring of molecular clouds that we now see. Between outbursts, according to this idea, the black hole should be fairly quiescent, as it is now.

So scientists find a great deal of circumstantial evidence that there is a black hole in the center of the Galaxy. But a black hole may not be the only possibility. Gamma rays and radio emission can, for example, also be produced in an accretion disk around a neutron star. Over the next few years, astronomers will turn their new observational tools—gamma-ray, x-ray, infrared, and radio telescopes—to the galactic center to try to prove or disprove the existence of a black hole at the heart of our own Galaxy. As we shall see in later chapters, there is evidence that black holes exist in other galaxies and in quasars as well.

32.6 THE FORMATION OF THE GALAXY

During the past decade there has been a revolution in our ideas about how the Galaxy formed. The flattened shape of the Galaxy suggests that the basic process was similar to the way in which the Sun and solar system formed (Section 20.3), but now we know that this is only part of the story.

Initially, we believe that the Galaxy formed from a rotating cloud. Since the oldest stars that we know—stars in the halo and in globular clusters—are distributed in a sphere centered on the nucleus of the Galaxy, we assume that the protogalactic cloud was spherical. The oldest stars have ages of 13 to 15 billion years, and so we assume that this is about when the Galaxy began to form. The cloud slowly collapsed, and as it contracted it

began to spin faster in order to conserve angular momentum. After a few billion years, it formed a thin disk, just as the collapsing solar nebula formed a disk around the protosun. The stars that were formed far from the disk before the cloud collapsed did not participate in the collapse but continue to orbit in the halo (Figure 32.19).

The oldest stars in the disk are not much younger than the stars in the halo and in globular clusters. The collapse must therefore have been fairly rapid, and it may have taken only a few hundred million years. Gravitational forces within the disk then caused the gas in the disk to fragment into clouds or clumps with masses like those of star clusters. These individual clouds then fragmented further to form individual stars.

Star formation in the disk appears to occur in bursts that last a few billion years. In between bursts, the disk is quiescent for periods of about 1 to 3 billion years. The rate of star formation is high at the present time. Star formation may be stopped when many supernovae explode, injecting energy and momentum into the interstellar matter and making it harder for the gas and dust to coalesce to form stars. Over time, the interstellar matter cools off, recontracts, and a new round of star formation can begin.

Supernovae also eject material enriched in heavy elements, so that each generation of stars contains a larger fraction of heavy elements than the generations that preceded it. Some low-mass members of the earliest generations of stars are still shining. The abundance of elements heavier than helium in these very old stars is 1 percent or less of the abundance of heavy elements in the Sun. Most of the stars in the disk are several billion years younger than the halo stars and have high abundances of elements heavier than hydrogen and helium.

The bulk of the material in the Galaxy came from the initial rotating cloud. Recently, however, astronomers have discovered that globular clusters differ from one another in age by as much as 3 billion years. Astronomers have also found that the younger globular clusters orbit the center of the Galaxy in a direction opposite to that of the majority of the globulars and of the galactic disk. These "backward" globular clusters also tend to be found at large distances from the center of the Galaxy.

Based on this evidence, some astronomers have suggested that the Galaxy formed in two stages. The first stage proceeded as shown in Figure 32.19. About 2 billion years after the formation of the majority of globular clusters, which rotate in the forward direction, the Galaxy acquired additional stars and globular clusters from one or more satellite galaxies that ventured too close to the Milky Way Galaxy and were captured by it. (Remember that some of the satellites in our own solar system have retrograde orbits and are thought to be captured asteroids or comets.)

It appears that the Milky Way is on the brink of

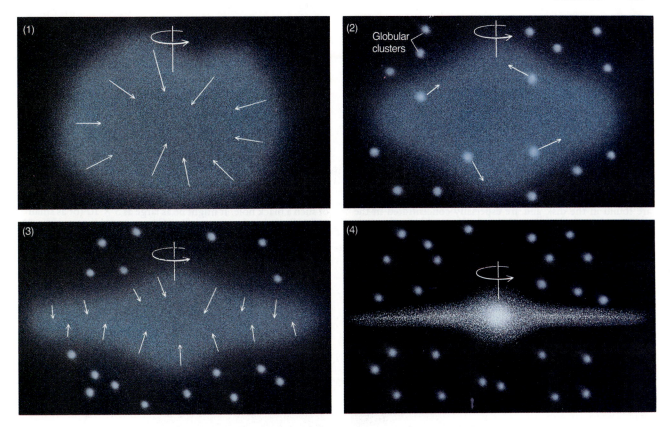

FIGURE 32.19 The Galaxy probably formed from an isolated, rotating cloud of gas that collapsed owing to gravity. Halo stars and globular clusters either formed prior to the collapse or were formed elsewhere and attracted to the Galaxy early in its history by gravity. Stars in the disk formed late, and the gas from which they were made was contaminated with heavy elements produced in early generations of stars.

another collision. The Large and Small Magellanic Clouds, our two nearest satellite galaxies, are spiraling ever closer to the Milky Way and will be captured by it some time in the future. The destruction of the Clouds has already begun. There is a stream of neutral hydrogen gas trailing behind the Magellanic Cloud galaxies, pulled away from the Clouds by the gravitational force of the Galaxy. Future astronomers may well find that new globular clusters and other stars have been added to our own Galaxy, while the Magellanic Clouds will have disappeared.

S U M M A R Y

32.1 The Sun is located in the outskirts of the **Milky Way Galaxy.** The **Galaxy** consists of a **disk,** which contains dust, gas, and young stars; a **nuclear bulge,** which contains old stars; and a spherical **halo,** which contains very old stars, including the members of globular clusters and RR Lyrae variables. Analysis of the distribution of globular clusters gave the first indication that the Sun is not located at the center of the Galaxy. Hot (10^6 K) gas is found in the inner galactic halo, which is called the **galactic corona;** this gas was ejected from the disk and is probably heated by supernova explosions. Radio observations at 21 cm show that atomic hydrogen is confined to a flat disk, which has a thickness of only 125 pc

near the Sun. Atomic hydrogen extends beyond the Sun to a distance of at least 25,000 pc from the galactic center. Dust is found in the same locations as atomic hydrogen inside the Sun's orbit; there is very little dust outside the Sun's orbit. Molecular clouds are concentrated at distances between 3000 and 8500 pc from the galactic center. The most massive molecular clouds, where star formation is active, are concentrated in the spiral arms.

32.2 Studies of atomic hydrogen show that the Galaxy has four main **spiral arms.** The Galaxy does not rotate like a solid body. The stars within it follow orbits that obey Kepler's

These two images of the central region of spiral galaxy M100 were taken before (top) and after (bottom) the repair of the Hubble Space Telescope. Before the repair, the mirror's spherical aberration blurred the image and required extensive image processing to bring out some additional detail. Following the repair, it is now possible to resolve details as small as 30 light years across in this galaxy, which is tens of millions of light years away. The repaired HST can be used to study the stars and gas in much more distant galaxies, to determine accurate distances to a wider range of galaxies, and to analyze how galaxies change over the lifetime of the universe.

(NASA and the Space Telescope Science Institute)

This chapter includes two simple but profound ideas about the universe, which astronomers realized during the first third of this century. One is the existence of other galaxies beyond the Milky Way. The other is that the galaxies participate in a general expansion of the entire universe, a result so unexpected that, as we shall see, even Einstein resisted it.

The confirmation of these ideas came about, like so much progress in astronomy, through the use of better instruments—in this case, the 100-inch reflector atop Mount Wilson in Southern California. When it went into operation, it was the largest optical telescope in the world. This was the instrument that Edwin Hubble used to make the pioneering observations that would establish his reputation and earn him the honor of having his name on our large telescope in space.

In reading about such work in a textbook decades later, it is easy to lose track of how difficult the work of these early observers really was. The faint light of the most distant objects each new telescope could detect had to be collected sometimes for several days and focused precisely onto glass plates coated with light-sensitive chemicals. A cautionary tale told to astronomy graduate students in the 1960's concerned Rudolph Minkowski, a legendary observer at Mount Palomar, known for his heroic efforts to take long exposures of the faintest galaxies known. Once he took an exposure that lasted four, or depending on who is telling the story, five nights, taking things up each evening carefully where he had left off. When he was finally finished, he rushed to the observatory dark room to develop the plate. But in his excitement, he put the plate in the wrong chemical—and ruined it irretrievably.

Today, electronic light detectors have replaced photographic plates, but the quest to wring every last photon out of each observation and to glean what information we can about the most distant objects still continues.

E. Margaret Burbidge (b. 1919) has held a variety of posts in British and American astronomy, including the Directorship of the Royal Observatory of Greenwich and the Presidency of the American Association for the Advancement of Science. Her research has concentrated on stellar nucleosynthesis, galaxies, and especially the nature of quasars. (AIP Niels Bohr Library)

GALAXIES

The "analogy [of the nebulae] with the system of stars in which we find ourselves . . . is in perfect agreement with the concept that these elliptical objects are just [island] universes—in other words, Milky Ways . . ."

So wrote Immanuel Kant (1724–1804) concerning the faint patches of light that telescopes revealed in large numbers. Despite Kant's speculation that these patches of light are giant systems of stars like our own Milky Way Galaxy, their true nature remained a subject of controversy until 1924.

33.1 GALACTIC OR EXTRAGALACTIC?

(a) Catalogues of Nebulae

Faint star clusters, glowing gas clouds, dust clouds reflecting starlight, and galaxies all appear as faint, unresolved luminous patches when viewed visually with telescopes of moderate size. Since the true natures of these various objects were not known to early observers, all of them were called "nebulae." *Nebula* (plural *nebulae*) literally means "cloud."

One of the earliest catalogues of nebulous-appearing objects was prepared in 1781 by the French astronomer Charles Messier (1730–1817). Messier was a comet hunter, and he made a list of 103 fuzzy-looking objects that might be mistaken for comets. Messier's list contains some of the most conspicuous star clusters, nebulae, and galaxies in the sky, and these objects are often referred to by their numbers in his catalogue—for example, M31, the great galaxy in Andromeda.

By 1908 nearly 15,000 nebulae had been catalogued and described. Some had been correctly identified as star clusters and others as gaseous nebulae (such as the Orion Nebula). The nature of most of them, however, still remained unexplained. Were they luminous clouds of gas in our own Galaxy? Or were they remote unresolved systems of thousands of millions of stars—galaxies in their own right?

(b) The Resolution of the Controversy

The controversy was finally resolved by the discovery of variable stars in some of the nearer nebulae in 1923 and 1924. Edwin Hubble, working with the 100 in. (2.5 m) telescope at the Mt. Wilson Observatory, analyzed the light curves of variables he had discovered in M31, M33, and NGC 6822 and found that they were cepheids. Although cepheid variables are supergiant stars, the ones studied by Hubble were very faint—near apparent magnitude 18. Those stars, therefore, and the systems in which they were found, must be very far away. The "nebulae" had been established as galaxies (Figure 33.1).

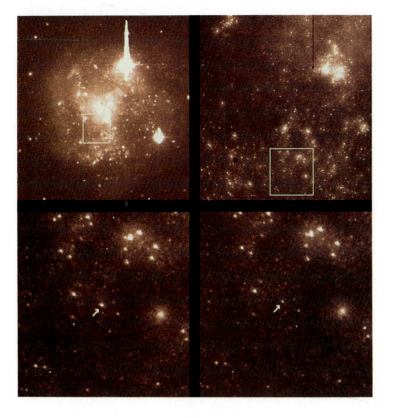

FIGURE 33.1 A series of images showing a cepheid in the spiral galaxy IC 4182, which is at a distance of 16 million LY. (Top left) A ground-based image obtained with the 200-in. Hale Telescope on Palomar. The inset box shows the field of view of the Hubble Space Telescope. (Top right) An HST view of a small region of IC 4182. The inset box marks the position of the enlarged region in the two lower frames. (Bottom left) The arrow points to a cepheid variable star. (Bottom right) The same region of the galaxy imaged by HST 5 days earlier. The arrow points to the same cepheid, which is clearly fainter. These images were taken before the HST was repaired. (NASA)

33.2 THE EXTRAGALACTIC DISTANCE SCALE

One of the most important, difficult, and controversial problems in modern observational astronomy is that of measuring the distances to galaxies. Galaxies are much too far away to display parallaxes or proper motions. Until very recently, the only way to determine the distance to a galaxy was to identify objects within it whose intrinsic luminosities were already known. Hubble, for example, used the cepheids in M31. Measurements of the apparent luminosities of these objects then made it possible to calculate the distance. Remember that the apparent brightness of an object decreases inversely as the square of its distance (Section 7.1).

In the past few years, astronomers have devised three new techniques for measuring distances to galaxies. These new techniques all give the same answer for the distances to galaxies with an accuracy of about ten percent, and so it appears at long last that we really do have a fairly good estimate for the size of the visible universe.

(a)   Distances from Standard Candles

To determine distances to galaxies, we must resort to a multistep process. The traditional approach goes roughly as follows. First, we derive distances to individual nearby stars in our own Galaxy by measuring parallaxes and proper motions. With knowledge of the intrinsic luminosities of these nearby stars, we can then determine distances to clusters, which contain stars similar to those with known luminosities. Once we measure the distance to a cluster, we know the luminosity of every star within the cluster. Fortunately, clusters contain some stars, including cepheid variables, that are much more luminous than any of the nearby stars for which we can obtain parallaxes by direct measurement. We then search for similar highly luminous stars in other galaxies. Since we can measure the apparent brightness of these stars and already know their true luminosity from studies of stars in clusters in our own Galaxy, we can use the inverse-square law for the propagation of light to determine the distances to the galaxies to which they belong. Any object whose intrinsic luminosity is known is referred to as a **standard candle**.

Several different types of stars have been used as standard candles. The most accurate distances for nearby galaxies are based on measurements of cepheid variable stars. Cepheids have the advantage of being relatively luminous (maximum luminosity about $2 \times 10^4 \, L_s$). Because they vary in brightness, cepheids are easily identified from multiple images of a galaxy. The luminosity of a cepheid can be determined from its period through the period-luminosity relation. Comparison of their known absolute magnitudes and observed apparent magnitudes enables us to find their distances and hence the distances to the galaxies in which they occur, with the inverse-square law of light.

With so many ways (
one might think that th
Unfortunately, it is not.
to make. Experts differ
interpretation of the ob
standard candles, and
reliable. That is why th
here represent such a
niques are all independe
same distances within a
astronomer has accepte
tests of these new meth
are very encouraging.
very close to having so
just how far away the
Hubble with his obser
now be nearing a co

33.3 THE EXPAI

Early in the present c
lect evidence that all
nearest ones) are mo
expansion of everyth
cations for the struct
we shall further disc
section, we will de
dence was obtained a
the distances to remo

(a) The Evider

The universe is *exp*
tion underlies all
Curiously, this fact
the search for dist

In 1894, Percival
Flagstaff, Arizona,
life in the universe.
ulae might be sola
tion—like the sola
asked one of the
Vesto M. Slipher,
the spiral nebulae
chemical composi
forming planets.

The Lowell Of
24-in. refracting
suited to observat
nology available i
to 40 hr long! Sli
the Andromeda
known to be a gal
planets but did di
lines that indicat

The amount of work involved in finding cepheids and measuring their periods can be enormous. Hubble, for example, obtained 350 photographs of M31 over a period of 18 years and identified only 40 cepheids. Even though the cepheids are fairly luminous stars, they can be detected in only about 30 of the nearest galaxies with the world's largest ground-based telescopes. One of the main goals of the Hubble Space Telescope is to measure cepheids in more distant galaxies (out to distances of at least 15 million pc) (Figure 33.1).

Other standard candles that are even more luminous than cepheids have been used to measure distances that are more than 15 million parsecs away. Supergiant stars can be as bright as $10^6 L_s$, and so can extend the distance scale to more than six times the distance to which cepheids can be seen. Whole systems of stars are also used as standard candles. Globular clusters, for example, are more luminous than any single star, with the exception of supernovae.

Exploding stars, both novae and supernovae, are also used as standard candles. Particularly useful are supernovae of Type Ia, which all reach the same luminosity (about $10^{10} L_s$) at maximum light and can be detected out to distances of 200 million pc. This type of supernova occurs when a white dwarf star, which already has a mass near the Chandrasekhar limit, accretes additional matter from a companion star. When the mass of the star becomes too great to be supported by the pressure of degenerate electrons, the star collapses. A thermonuclear explosion then occurs and disrupts the entire star, emitting copious amounts of energy. The major problem with using Type Ia supernovae as standard candles is that none have so far been observed in nearby galaxies whose distances are already known from some other technique. Therefore, while all such supernovae seem to have about the same intrinsic luminosity, we do not yet know very accurately what that luminosity actually is.

At still greater distances, we must use the total light emitted by an entire galaxy as a standard candle. As we shall see later in this chapter, galaxies span an enormous range in intrinsic luminosity. Furthermore, most galaxies do not have distinguishing characteristics that enable us to estimate their luminosities. We can only tell which galaxies are highly luminous and which ones are not if we see a collection of them of various brightnesses, side by side in a cluster of galaxies. Thus we can use the apparent brightness of the brightest members (say, the average of the brightest five) in a large cluster to estimate the distance to the cluster.

(b) New Techniques

Recently three new techniques have been devised for measuring distances to galaxies. All of these techniques

---

assume that the distances measured to nearby galaxies through use of the cepheids are correct. Characteristics of these nearby galaxies are then calibrated and used to measure distances to galaxies that lie beyond the limits where measurements of the cepheids are possible.

The first method makes use of the observed fact that the luminosity of a spiral galaxy like the Milky Way is related to its rotational velocity. This method is called the Tully-Fisher technique, named for the two astronomers who devised it in the late 1970s. We can see in principle why this technique works by rewriting Kepler's third law in the form

$$M = \frac{rv^2}{G},$$

where M is the mass of the galaxy inside a circular orbit of radius r and v is the orbital speed of the stars and gas at distance r (Section 32.4). If we measure v, we can calculate M. If the ratio of mass to luminosity, M/L, is the same in all spirals, then M can be used to estimate L.

In practice, it is somewhat surprising that this technique works, since as we have seen much of the matter associated with galaxies does not contribute at all to the luminosity. There is also no obvious reason why M/L should be the same for all spiral galaxies. Nevertheless, observations show that measuring the rotational velocity of a galaxy is enough to determine its intrinsic luminosity and hence its distance. An advantage of this technique relative to the use of cepheids is that it will work for any spiral galaxy for which it is possible to obtain a spectrum with which to measure the Doppler shift produced by the galaxy's rotation.

The second of the new techniques makes use of planetary nebulae. If we look at the location of the central stars of planetary nebulae on an H-R diagram (Figure 30.8), it does not seem as if the stars would be bright enough to be seen in other galaxies. However, the central stars are very bright in the ultraviolet part of the spectrum. The most luminous ones have total luminosities, including ultraviolet radiation, of 10^4–10^5 L_s. The ultraviolet light emitted by the central star is absorbed in the surrounding nebula, and much of it is re-emitted in the form of emission lines in the optically visible part of the spectrum, where it can be easily observed.

Planetary nebulae are easy to find in other galaxies. All one has to do is photograph the galaxy in the light of one of the emission lines. Ordinary stars will not emit strongly at this wavelength, and the planetary nebulae will therefore stand out from the background light of other stars.

Observations show that the number of planetaries in a galaxy depends on their intrinsic luminosity. Suppose you were to count all the planetaries in a galaxy whose

distance you alre
in it. If you then
within each int
there will be rel
will also disco
creases with de
you have done
planetary nebul
study planetari
this luminosity

All you have
whose distanc
number of plat
brightness. Sin
every galaxy,
luminosity of
sults with Fig

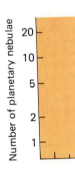

FIGURE 33.2
found in vario
the intrinsic lu
radiated in a
of 500.7 nm.
increases with
negative mag
relationship b
the same for

TABLE 33.

Method

Cepheids
Brightest s
Planetary
Novae
Globular c
Surface b
　fluctuati
Supernov
21-cm line
Total light
Brightest
Radial ve

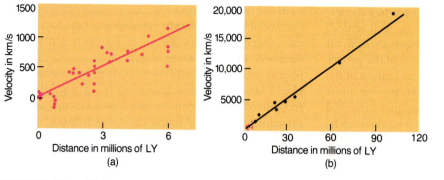

FIGURE 33.3　(a) Hubble's original velocity-distance relation, adapted from his 1929 paper in the *Proceedings of the National Academy of Sciences.* (b) Hubble and Humason's velocity-distance relation, adapted from their 1931 paper in *The Astrophysical Journal.* The red dots at the lower left are the points in the diagram in the 1929 paper (a). Comparison of the two graphs shows how rapidly the determination of distances and redshifts of galaxies progressed in the two years between these publications.

(c)　Expansion Requires a Velocity–Distance Relation

The fact that galaxies obey the Hubble law shows beyond doubt that the universe is expanding uniformly. A uniformly expanding universe—that is, one that is expanding at the same rate everywhere—*requires* that we and all other observers within it, no matter where they are located, *must* observe a proportionality between the velocities and distances of remote galaxies.

To see why, imagine a ruler made of flexible rubber, with the usual lines marked off at each centimeter. Now suppose someone with strong arms grabs each end of the ruler and slowly stretches it, so that, say, it doubles in length in 1 min (Figure 33.4). Consider an intelligent ant sitting on the mark at 2 cm—intentionally not at either end or in the middle. This ant measures how fast other ants, sitting at the 4-, 7-, and 12-cm marks, move away from him as the ruler stretches. The one at 4 cm, originally 2 cm away, has doubled its distance; it has moved 2 cm/min. Similarly, the ones at 7 cm and 12 cm, which were originally 5 and 10 cm distant, have had to move away at 5 and 10 cm/min, respectively, to reach their current distances of 10 cm and 20 cm. All ants move at speeds proportional to their distance.

Now repeat the analysis, but put the intelligent ant on some other mark, say, on 7 or 12, and you'll find that in all cases, as long as the ruler stretches uniformly, this ant finds that every other ant moves away at a speed proportional to its distance.

For a three-dimensional analogy, look at the raisin bread in Figure 33.5. The cook has put too much yeast in the dough, and when he sets the bread out to rise, it doubles in size during the next hour and all the raisins move farther apart. Some representative distances from one of the raisins (chosen arbitrarily, but not at the center) to several others are shown in the figure. Since each distance doubles during the hour, each raisin must move away from the one selected as origin at a speed proportional to its distance. The same is true, of course, no matter which raisin you start with. But the analogy must not be carried too far; in the bread it is the expanding dough that carries the raisins apart, but in the universe no pervading medium is presumed to separate the galaxies.

From the foregoing, it should be clear that if the universe is uniformly expanding, all observers everywhere, including us, must see all other objects moving away from them at speeds that are greater in proportion to their distances. As Hubble and Humason showed, that is precisely what observers on Earth do see.

As telescopes have grown larger and detectors have become more sensitive, it has become possible to observe more and more remote galaxies with greater and greater speeds of recession (Figure 33.6). The velocities

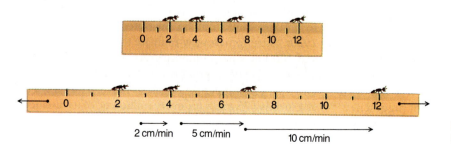

FIGURE 33.4　Stretching a ruler. See text for explanation.

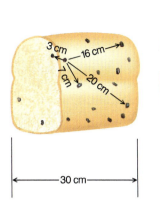
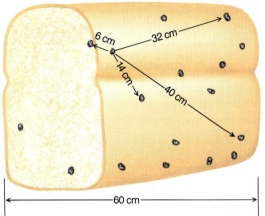
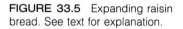

FIGURE 33.5 Expanding raisin bread. See text for explanation.

of recession or redshifts are often expressed in terms of the quantity z, which is given by the following expression:

$$z = \frac{(\lambda - \lambda_0)}{\lambda_0} = \frac{\Delta\lambda}{\lambda_0}.$$

In words, z is the ratio of the amount by which the wavelength of a line in the spectrum of a galaxy is shifted ($\Delta\lambda$) to the laboratory or rest wavelength (λ_0) of that line.

The current (April 1992) record holder as the most distant galaxy so far discovered is 4C 41.17, which has a redshift of z = 3.800. What this means is that the line of Lyman α, which has a rest wavelength of 121.5 nm and is so far into the ultraviolet region that it cannot be observed from the ground, is redshifted to the yellow part of the spectrum at 583 nm and is easily observed. If we use the equation for the relativistic Doppler effect (Section 7.5), we find that this galaxy is moving away from us at 92 percent of the speed of light. There are

FIGURE 33.6   A distant cluster of galaxies in Hydra, with a radial velocity of about 20 percent that of light. (National Optical Astronomy Observatories)

objects called quasars that have even higher velocities, provided that the redshift of their spectral lines is indeed produced by the cosmological Doppler effect.

The expansion of the universe does not imply that the galaxies and clusters of galaxies themselves are expanding. The raisins in our raisin bread analogy do not grow in size as the loaf expands. Similarly, gravitation holds galaxies and clusters together, and they simply separate as the universe expands, just as do the raisins in the bread. Galaxies in clusters do, of course, have individual motions of their own superimposed on the general expansion. Galaxies in pairs, for example, revolve about each other, and those in clusters move about within the clusters. In fact, a few galaxies in nearby groups and clusters move fast enough within those systems so that they are actually approaching us, even though the clusters of which they are a part are, as a group, moving away.

33.4 TYPES OF GALAXIES

Galaxies differ a great deal among themselves, but the majority of optically bright galaxies fall into two general classes: spirals and ellipticals. A minority of them are classed as irregular.

(a) Spiral Galaxies

Our own Galaxy and the Andromeda Galaxy (Figure 33.7), which is believed to be much like it, are typical large **spiral galaxies.** Like our Galaxy (Chapter 32), a spiral consists of a nucleus, a disk, a halo, and spiral arms. Interstellar material is usually spread throughout the disks of spiral galaxies. Bright emission nebulae are present, and absorption of light by dust is also often apparent, especially in those systems turned almost edge on to our line of sight (Figure 33.8). The spiral arms contain the young stars, which include luminous supergiants. These bright stars and the emission nebulae make the arms of spirals stand out like the arms of a Fourth-of-July pinwheel (see Figure 33.1). Open star clusters can be seen in the arms of nearer spirals, and globular clusters are often visible in their halos; in M31, for example, more than 200 globular clusters have been identified. Spiral galaxies contain both young and old stars.

FIGURE 33.7   The nearby spiral galaxy in Andromeda, which is similar in size and structure to the Milky Way Galaxy. Its distance is about 725,000 pc, and its diameter is about 30,000 pc. This galaxy contains over 300 billion stars. (National Optical Astronomy Observatories)

FIGURE 33.8 NGC 253, one of the dustiest of the spiral galaxies. Much of the interior structure of this galaxy is obscured by dust, but two spiral arms and many blueish clusters of stars can be seen around the outer edge. This galaxy is only 3 million pc distant. It appears elongated because we view it nearly edge on. (Anglo-Australian Telescope Board)

FIGURE 33.9 NGC 4650, a barred spiral of type SBa in the constellation Centaurus (National Optical Astronomy Observatories)

Perhaps a third or more of spiral galaxies display conspicuous bars running through their nuclei. The spiral arms of such a system usually begin from the ends of the bar, rather than winding out directly from the nucleus. These are called barred spirals (Figure 33.9). Some astronomers believe that almost all spirals, including probably the Milky Way Galaxy, contain at least a weak bar. Studies of the rotations of some barred spirals show that their inner parts (out to the ends of the bars) are rotating approximately like solid bodies. In the absence of differential shearing rotation, the straight bar can persist, rather than winding up. The detailed structures and dynamics of barred spirals, however, are just beginning to be understood.

In both normal and barred spirals we observe a gradual transition of morphological types. At one extreme, the nuclear bulge is large and luminous, the arms are faint and tightly coiled, and bright emission nebulae and supergiant stars are inconspicuous. At the other extreme are spirals in which the nuclear bulges are small—

almost lacking—and the arms are loosely wound, or even wide open. In these latter galaxies, there is a high degree of resolution of the arms into luminous stars, star clusters, and emission nebulae. Our Galaxy and M31 are both intermediate between these two extremes. Photographs of spiral galaxies, illustrating this transition of types, are shown in Figures 33.10 and 33.11. All spirals and barred spirals rotate in the sense that their arms trail, as does our own Galaxy.

Spiral galaxies range in diameter from about 6000 to more than 30,000 pc, and the atomic hydrogen in the

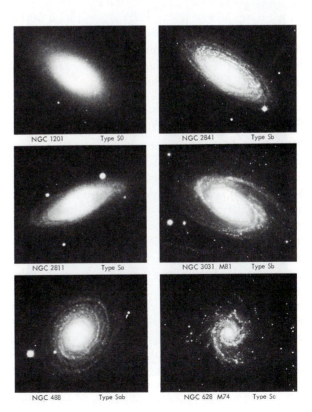

Figure 33.10 Types of spiral galaxies. (Palomar Observatory, California Institute of Technology)

FIGURE 33.16 The Small Magellanic Cloud. This dwarf irregular galaxy is a satellite of our own Milky Way Galaxy. (National Optical Astronomy Observatories)

that took place some 200 million years ago. It is now being pulled apart by the gravitation of the Milky Way.

(d) Classification of Galaxies

Of the several classification schemes that have been suggested for galaxies, one of the earliest and simplest, and the one most used today, was invented by Hubble during his study of galaxies in the 1920s. Hubble's scheme consists of three principal classification sequences: ellipticals, spirals, and barred spirals. The ir-

regular galaxies form a fourth class of objects in Hubble's classification.

The ellipticals are classified according to their degree of flattening or ellipticity. Hubble denoted the spherical galaxies by E0, and the most highly flattened by E7. The classes E1 through E6 are used for galaxies of intermediate ellipticity. (Each of the numbers 0 through 7 that describe the flattening of a galaxy is defined in terms of the major and minor axes of the image of the galaxy, a and b, respectively, by $10(a - b)/a$.) Hubble's classification of elliptical galaxies is based on the appearance of

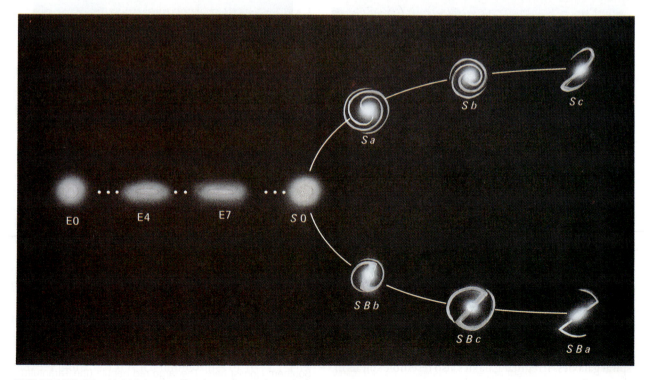

FIGURE 33.17 Hubble's classification scheme for galaxies.

their images, not upon their true shapes. An E7 galaxy, for example, must really be a relatively flat elliptical galaxy seen nearly edge on, but an E0 galaxy could be one of any degree of ellipticity, seen face on. Analyses indicate that some elliptical galaxies are oblate (like a pumpkin), others are prolate (like a football), and still others are triaxial. A triaxial galaxy is one in which the three perpendicular axes through the center to the edge are unequal in length.

Hubble classed the normal spirals as S and the barred spirals as SB. Lowercase letters a, b, and c are added to denote the extent of the nucleus and the tightness with which the spiral arms are coiled. For example, Sa and SBa galaxies are spirals and barred spirals, respectively, in which the nuclei are large and the arms tightly wound. Sc and SBc are spirals of the opposite extreme. Our Galaxy and M31 are classed as Sb.

In rich clusters, galaxies are observed that have the disk shape of spirals but no trace of spiral arms. Hubble regarded these as galaxies of a type intermediate between spirals and ellipticals and classed them S0.

Hubble's classification scheme for all but irregular galaxies is illustrated in Figure 33.17, in which the morphological forms are sketched and labeled, with the three principal sequences joined at S0. The diagram is based on one by Hubble himself. Figure 33.17 provides an easy way to remember the shapes of galaxies, but it does not represent an evolutionary sequence. As spirals consume their gas, star formation will stop, and the spiral arms will gradually become less conspicuous. Over long periods, spirals will therefore begin to look more like S0 galaxies. Collisions of spiral galaxies in the centers of dense clusters of galaxies may play a role in forming the most massive elliptical galaxies. It is likely, however, that most of today's elliptical galaxies have always been ellipticals.

The most important characteristics of the different kinds of galaxies are summarized in Table 33.2. Many of the figures given, especially for mass, luminosity, and diameter, are very rough and are intended to illustrate only orders of magnitude.

33.5 PROPERTIES OF GALAXIES

The linear size of that part of a galaxy that corresponds to its observed angular size can be calculated once the distance to the galaxy is known, just as we calculate the diameter of the Sun or of a planet. If we know the distance to a galaxy, we can apply the inverse-square law of light and calculate its total luminosity, or the absolute visual magnitude, from the amount of light flux we receive from it. Thus our knowledge of the radii and luminosities of galaxies is dependent on the accuracy of our estimates of the distances to galaxies. It also follows that an error in the extragalactic distance scale will lead to systematic errors in the derived sizes and luminosities.

The determination of masses of galaxies, however, is more difficult and, in fact, is possible (by present techniques) for only a small fraction of them. There are several techniques for measuring the masses of galaxies.

(a) Mass of Galaxies from Internal Motions

We determine the masses of galaxies, like those of other astronomical bodies, by measuring their gravitational influences on other objects or on the stars within them. We must assume, of course, that Newton's law of gravitation is valid over extragalactic distances. Some theorists have challenged this assumption, but most astronomers agree that observations to date are consistent with it.

Internal motions in galaxies provide the most reliable methods of measuring their masses. The procedure for spiral galaxies is to observe the rotation of a galaxy from the Doppler shifts of either features in the optical spectrum or the 21-cm line of neutral hydrogen, and then to compute its mass with the help of Kepler's third law.

As an illustration we shall consider the rotation of M31, the Andromeda Galaxy (Figure 33.7). The galaxy is inclined at an angle of only about 15° to our line of sight, so we see it highly foreshortened. There is evi-

TABLE 33.2 CHARACTERISTICS OF GALAXIES OF DIFFERENT TYPES			
	Spirals	Ellipticals	Irregulars
Mass (solar masses)	10^9 to 10^{12}	10^5 to 10^{13}	10^8 to 10^{11}
Diameter (thousands of pc)	5–50	1–200	1–10
Luminosity (solar units)	10^8 to 10^{11}	10^6 to 10^{11}	10^7 to 2×10^9
Absolute visual magnitude	−15 to −22.5	−9 to −23	−13 to −20
Population content of stars	Old and young	Old	Old and young
Composite spectral type	A to K	G to K	A to F
Interstellar matter	Both gas and dust	Almost no dust; little gas	Much gas; some are deficient in dust; others contain large quantities of dust
Mass-light ratio	2–20	100	1

which have very high mass-light ratios. Rotation curves for spiral galaxies, for example, are usually more or less like that for M31 (Figure 33.18), which rises to a maximum velocity and then flattens. Furthermore, from observations of the 21-cm line of atomic hydrogen, we know that the rotation curves remain flat even beyond the point where the visible radiation begins to drop off. The only way that the rotational velocity can remain high is if the visible matter is supplemented by invisible matter, that is, by matter that does not emit detectable radiation. This dark matter still does, of course, exert gravitational force, and this force is what keeps the material in the outer portion of the galaxy, with its high rotational velocity and correspondingly large centripetal force, from flying off into space (see Section 32.4 for a more detailed explanation).

The mass-light ratios measured for various types of galaxies are given in Table 33.2.

The outer parts of at least some galaxies, therefore, and perhaps clusters of galaxies as well, contain matter that has not yet been identified by its light or by means other than its gravitational influence. The anomaly is sometimes called the missing mass problem. In fact, the mass is there; it is the light that is missing.

What is this dark matter? Extensive searches have been made for both hot and cold gas, and neither is present in sufficient quantity to account for the dark matter. The most likely explanation seems to be that the dark matter in galaxies is composed of massive collapsed objects (for example, black holes), stellar-like objects that are not massive enough to burn hydrogen (brown dwarfs), or massive neutrinos or exotic subnuclear particles (see Chapter 35).

Whatever the composition of the dark matter, these measurements of other galaxies support the conclusion reached already from studies of the rotation of our own Galaxy—namely, that probably 90 percent or more of all the material in the universe cannot be observed directly in any part of the electromagnetic spectrum. The light that we see from galaxies does not trace the bulk of material that is present in space, and so may give us a very misleading picture of the large-scale structure of the universe. An understanding of the properties and distribution of this invisible matter is therefore crucial. Through the gravitational force that it exerts, dark matter probably plays a dominant role in the formation of galaxies. As we shall see in Chapter 35, it may also determine the ultimate fate of the universe.

33.6 QUASARS

(a) Discovery

The Sun is very faint at radio wavelengths. Since the Sun is a typical star, we would not expect to see strong radio emission from other stars. Astronomers were, therefore, surprised when in 1960 two radio sources were identified with what appeared to be stars. There seemed to be no chance that the identifications were in error, because the precise positions of the radio sources were pinned down by noting the exact instants they were occulted by the Moon. By 1963 the number of such "radio stars" had increased to four (Figure 33.19). These radio stars were especially perplexing because their optical spectra showed emission lines that at first could not be identified with known chemical elements.

The breakthrough came in 1963 when Maarten Schmidt, at Caltech's Palomar Observatory, recognized the emission lines in one of the objects to be the Balmer lines of hydrogen (Section 7.4) shifted far to the red from their normal wavelengths. If the redshift is caused by the Doppler effect, the object is receding from us at about 15 percent the speed of light! With this hint, the emission lines in the other objects were re-examined to see if they too might be well-known lines with large redshifts. Such proved to be the case, but the other objects were found to be receding from us at even greater speeds. Evidently, they could not be neighboring stars; their stellar appearance must be due to the fact that they are very distant. They are called, therefore, quasi-stellar radio sources, or simply quasi-stellar

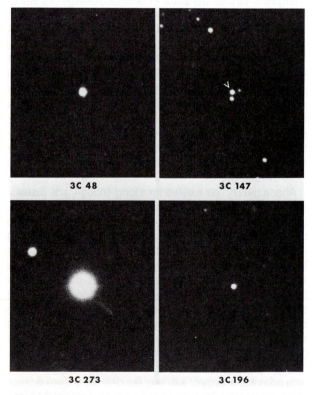

FIGURE 33.19 Quasi-stellar radio sources photographed with the 5-m telescope. (Palomar Observatory, California Institute of Technology)

objects (abbreviated QSO). Later, similar objects were found that are not sources of strong radio emission. Today they are all designated by the term **quasar**, and only about one percent of all quasars are radio sources. Some astronomers think that radio–emitting quasars are a temporary phase in the evolution of quasars.

Thousands of quasars have now been discovered. All the spectra show large to very large redshifts. The largest redshift measured to date (April 1992) corresponds to a relative shift in wavelength of $\Delta\lambda/\lambda = 4.9$. The Lyman α line of hydrogen, which has a laboratory wavelength of 121.5 nm in the ultraviolet portion of the spectrum, is shifted all the way through the visible range to 700 nm! If we apply the exact formula for the Doppler shift (Section 7.5), we find that this redshift corresponds to a velocity of more than 94 percent the speed of light.

What are the quasars? They are at the distances of galaxies, but they are certainly not normal galaxies. Galaxies contain stars, so the spectra of normal galaxies have absorption lines, just as do the spectra of stars. Quasar spectra are dominated by emission lines. Accounting for the energy emitted by quasars presents another fundamental problem. The difficulty is that if the quasars obey the Hubble law and are at the distances that correspond to their redshifts, then they are more luminous than the brightest galaxies. Furthermore, as we shall see, they must generate this enormous amount of energy in a volume of space that is no larger in diameter, and is sometimes much smaller, than the distance from the Sun to the nearest star.

(b) Luminosities of Quasars

We can determine the distance to a galaxy if we know its redshift. Let us assume for the moment that quasars also obey the Hubble law, and that we can estimate their distances accurately by measuring their velocities. If this assumption is true, then quasars are *extremely* luminous compared with ordinary galaxies. In visible light most are far more energetic than the brightest elliptical galax-

ies. Quasars also emit energy at x-ray wavelengths, and many are radio sources as well. Some quasars have total luminosities as large as 10^{14} L_s, or 10 to 100 times the brightness of the brighter elliptical galaxies.

Finding a mechanism to produce this much energy would be difficult under any circumstance. But the quasars present an additional problem. Quasars vary in luminosity on time scales of months, weeks, or even in some cases, days. This variation is irregular, evidently at random, and can amount to a few tens of percent. Since quasars are highly luminous, a change in brightness by, for example, a factor of two means an extremely large amount of energy is released rather suddenly—equivalent to 10^{14} L_s or to the total conversion of about 10 Earth-masses per minute from mass into energy. Moreover, because the fluctuations occur in such short times, the part of a quasar responsible for the light (and radio) variations must be smaller than the distance light travels in a month or so.

To see why this must be so, consider a cluster of stars 10 LY in diameter (Figure 33.20) at a very large distance from Earth. Suppose every star in this cluster brightens simultaneously and remains bright. We would first see the light from stars on the near side; five years later we would see light from stars at the center. It would be ten years before we detected light from stars on the far side. Even though this cluster brightened instantaneously, from Earth it would appear that 10 years elapsed before maximum brightness was reached. In other words, if an object brightens suddenly, it will seem to us to brighten over a period of time that is equal to the time it takes light to travel across the object from its far side.

In general, the time scale for significant changes in brightness sets an upper limit on the size of the region that brightened. Since quasars vary on time scales of months, the region where the energy is generated can be no larger than a few light months. Some quasars vary on even shorter time scales, and for them the energy must be generated in a region that is even smaller. The challenge then, is to devise a power source that can generate

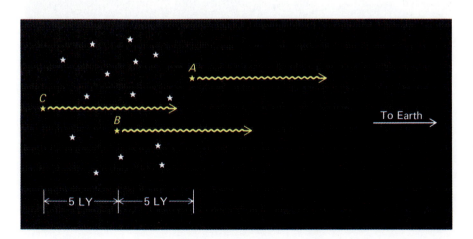

FIGURE 33.20 A diagram showing why light variations from a large region in space appear to last for an extended period as viewed from Earth. Suppose all the stars in this cluster brighten simultaneously and instantaneously. In this example, star *A* will appear to the observer to brighten five years before star *B*, which in turn appears to brighten five years earlier than star *C*.

TABLE 33.3	NUMBERS OF GALAXIES OF DIFFERENT TYPES
Type	Number/10^{15} Cubic Parsecs
Luminous spirals	10^7
Seyfert galaxies	10^5
Radio galaxies	10^3
QSOs (without radio emission)	10^2
Quasars (with radio emission)	1

FIGURE 33.23 An image of M87 obtained with the Hubble Space Telescope. The strong concentration at the center of the galaxy probably indicates that a 2.6 billion solar mass black hole is located in the nucleus of the galaxy. Both the nucleus and the jet ejected from it are sources of radio emission. (NASA)

(f) Black Holes—The Power Behind the Quasars?

The observations of quasars and of all the various types of galaxies that are unusually active emitters of optical, x-ray, and radio radiation suggest that what these objects have in common is a compact source of enormous energy, evidently buried in the nucleus of a galaxy. Many models have been offered to account for this energy source, including stellar collisions in dense galactic cores, supermassive stars, extraordinarily powerful supernovae, and others.

The most widely accepted model at the present time is that quasars, and presumably other types of active galaxies as well, derive their energy output from an enormous black hole at the center of what would otherwise be a normal galaxy. The black hole must be very large—perhaps a billion solar masses. Given such a massive black hole, relatively modest amounts of additional material—only about 10 solar masses per year—falling into the black hole would be adequate to produce as much energy as a thousand normal galaxies and could account for the total energy of a quasar.

A black hole itself can, of course, radiate no energy. The energy comes from material very close to the black hole. The black hole attracts matter—stars, dust, and gas—which is orbiting around in the dense nuclear regions of the galaxy. This material then spirals in toward the black hole and forms an accretion disk of material around it. As the material spirals ever closer to the black hole, it accelerates and heats through compression to millions of degrees. This hot matter can radiate prodigious amounts of energy as it falls into the black hole.

One of the strongest pieces of observational evidence that massive black holes do indeed exist in the centers of galaxies has been obtained with the Hubble Space Telescope. Figure 33.23 shows that the stars at the center of the giant elliptical M87 become densely concentrated toward the center, forming a bright sharp core. The central density of stars in M87 is at least 300 times greater than expected for a normal giant elliptical galaxy, and over a thousand times denser than the distribution of stars in the neighborhood of our own Sun. HST

measurements of the orbital velocity of gas show that there must be at least 2×10^9 solar masses of material within 20 pc of the center of M87. So much mass in such a small volume is almost certainly a black hole. A massive black hole can attract stars and produce the high stellar density in the core of M87.

This black hole may have formed from the merger of small black holes created by the explosion of massive stars when M87 was a young galaxy. Once formed, the black hole would grow by feeding on gas and stars that passed too close to it. Stars orbiting near the nucleus of the galaxy would be pulled into orbits around the black hole, thereby producing the concentration of light we now see. Some of these stars may ultimately fall into the black hole, thereby increasing its mass still more.

A number of the phenomena that we observe can be explained naturally in terms of this model. First and foremost, it can produce the amount of energy that is actually observed to be emitted by quasars and active galactic nuclei. Detailed calculations show that about 10 percent of the rest mass of matter falling into a black hole is converted to energy. Remember that during the entire course of the evolution of a star like the Sun, only a tiny fraction of its rest mass will be converted to energy by nuclear fusion. Infall into a black hole is a very efficient way to produce energy.

Since the black hole is also fairly compact in terms of its circumference, the emission produced by infalling matter comes from a small volume of space. As we recall, this condition is required to explain the fact that quasars vary on a time scale of weeks to months.

All galaxies emit radio waves. The supernova remnants in the Milky Way Galaxy, for example, are

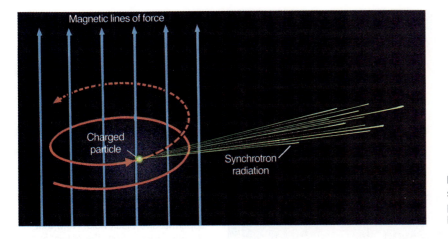

FIGURE 33.24 The emission of synchrotron radiation by a charged particle moving at nearly the speed of light in a magnetic field.

sources of radio emission. Quasars and other active galactic nuclei are, however, much stronger sources of radio waves, emitting 10^8 times (or even more) as much radio energy as do normal galaxies such as our own. The radio radiation from quasars is in the form of *synchrotron radiation*. When a charged particle enters a magnetic field, the field compels it to move in a circular or spiral path around the lines of force. The particle is thus accelerated and radiates energy (Figure 33.24). If the speed of the particle is nearly the speed of light, the energy it radiates is called synchrotron radiation. This terminology was chosen because particles radiate in the same way when they are accelerated to these speeds in a laboratory synchrotron. The compact radio source in the nucleus of the Galaxy (Section 32.5) and the Crab nebula (Section 30.7) are also sources of synchrotron radiation.

In the case of quasars, the synchrotron radiation produces emission not only at radio wavelengths but also in the visible and x-ray regions of the spectrum.

The quasar emission lines must originate from ionized gas at not too high a temperature. The emission lines certainly do not come from the same region as the x-ray and synchrotron radiation. At the temperatures required to produce x-ray emission, the gas would be completely ionized and no atomic emission lines could be produced. The strengths of the emission lines vary on time scales of months, however, so they cannot be spread throughout the galaxy in which the quasar is found. Models suggest that the broad emission lines are formed in relatively dense clouds within about 0.5 LY of the black hole. The broadening of the lines is produced by the Doppler effect, but it is not known whether the motions of the gas are caused by turbulence, rotation around the black hole, expansion, or contraction. It is also not known whether these clouds may play a role in providing fuel for the black hole.

As we have seen, quasars and other active galaxies emit jets that extend far beyond the limits of the parent

galaxy. Observations have traced these jets to within 3–30 LY of the parent quasar or galactic nucleus. The black hole and accretion disk are, of course, much smaller than 1 LY, but it is presumed that the jets originate from the vicinity of the black hole.

Why are energetic particles ejected into jets, or often into two oppositely directed jets, rather than in all directions? It may be that the accretion disk around the black hole is dense enough to prevent radiation from escaping in all but the two directions perpendicular to the disk (Figure 33.25). The basic idea behind the formation of jets is that matter in the accretion disk will move inward toward the black hole. Some of this matter will not actually fall into the black hole but will feed the jets. That is, some infalling matter will be accelerated by the intense radiation pressure in the vicinity of the black hole and will be blown out into space along the rotation axis of the black hole in a direction perpendicular to the plane of the accretion disk. The detailed mechanism for converting the energy associated with infall into an outward flowing jet remains a matter of controversy for theorists.

Figure 33.26 summarizes what we think we know about structure of quasars and of the regions surrounding them.

If matter in the accretion disk is continually being depleted by falling into the black hole or being blown out from the galaxy in the form of jets, then a quasar can continue to radiate only so long as there is gas available to replenish the accretion disk. Where does this matter come from? One possibility is that very dense star clusters form near the centers of galaxies. These stars might then supply the fuel, either through gas that is lost during the normal course of stellar evolution by means of stellar winds and supernovae explosions or because the tidal forces exerted by the black hole are strong enough to tear the stars apart. An alternate source of fuel may come from collisions of galaxies. That is, if two galaxies collide and merge, then gas and dust from one

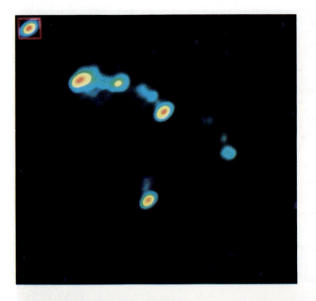

FIGURE 33.29 A radio image of the double quasar 0957+561. A massive galaxy acting as a gravitational lens forms multiple images of a quasar lying far beyond it. The two images of the quasar are the bright point-like objects just above and below the center of the picture. The weak blue image just above the lower image of the quasar is the galaxy that forms the gravitational lens. Several other radio sources are seen above and to the left of the double quasar. (National Radio Astronomy Observatory/AUI)

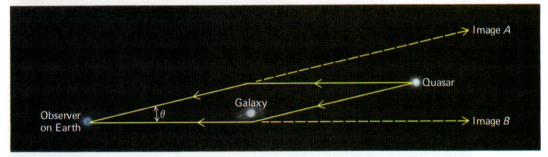

FIGURE 33.30 Gravitational lens associated with the double quasar 0957+561. Two light rays from the quasar are shown being bent in passing a foreground galaxy, and they arrive together at the Earth. We thus see two images of the quasar, in directions A and B. This simple schematic does not reflect the subtle complexities produced by the finite size of the galaxy or by the cluster of which it is a member, but it does illustrate how we can observe multiple images of an object. The angular separation, θ, of the two images at Earth is greatly exaggerated. The actual separation is only 6 arcsec.

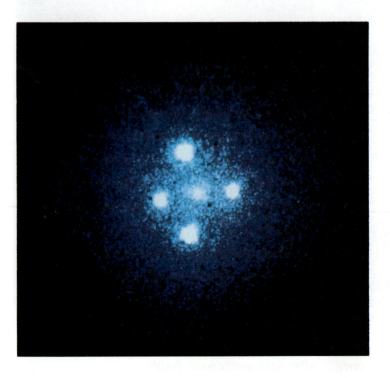

FIGURE 33.31 This image of a gravitational lens, which is referred to as Einstein's cross, was taken with the HST. The four concentrations of light at the ends of the cross bars are four images of a single distant quasar. The quasar is at a distance of approximately 8 billion LY. A galaxy at a distance of 400 million LY serves as the gravitational lens. The diffuse central object is the bright core of this intervening galaxy. (NASA)

In 1979, astronomers D. Walsh, R. F. Carswell, and R. J. Weymann of the University of Arizona noticed that a pair of quasars, separated by only 6 arcsec and known collectively as 0957 + 561 (the numbers give their coordinates in the sky), are remarkably similar in appearance and spectra (Figure 33.29). They are both about 17th magnitude, and both have a redshift ($\Delta\lambda/\lambda$) of 1.4. The astronomers suggested that the two quasars might actually be only one, and that we are seeing two images produced by an intervening object, acting as a gravitational lens (Section 31.3).

We now know that there is an 18th-magnitude galaxy that lies in the same direction as one of the quasars. In fact, the galaxy turns out to be a member of a cluster of galaxies, which has a redshift of 0.39 and thus is much closer than the quasar. The geometry and estimated mass of the galaxy are correct to produce the gravitational lens effect. A schematic of the lens is shown in Figure 33.30.

There is fairly convincing evidence for several other gravitational lenses, and searches are under way to discover still more. The search is difficult. If theoretical calculations are correct, the light from only about one quasar in a thousand will pass close enough to a galaxy so that the galaxy can act as a gravitational lens and produce a double image of the background quasar.

Gravitational lenses can produce not only double images but also multiple images (Figure 33.31) and even arcs and rings. Images produced by point-like gravitational lenses can appear much brighter than the actual source would appear to be in the absence of lensing. Some of the brightest quasars may owe their apparently high luminosities to enhancement by gravitational lensing.

Galaxies may not be the only gravitational lenses. As we have already seen, a large fraction of the material in the universe is not luminous. This dark matter may also act as a gravitational lens. Recent searches for the lensing effects produced by dark matter have produced additional evidence that only a small fraction of the matter in clusters of galaxies can actually be seen in the visible part of the spectrum (Section 34.6).

SUMMARY

33.1 Faint star clusters, clouds of glowing gas, dust clouds reflecting starlight, and galaxies all appear as faint, unresolved patches of light in telescopes of the quality available at the beginning of the 20th century. It was only when the discovery of cepheid variables in the Andromeda Galaxy was announced in 1924 that it became firmly established that there are other galaxies similar to the Milky Way in size and content.

33.2 Astronomers determine the distances to galaxies by measuring the apparent magnitudes of objects whose absolute magnitudes are known and then using the inverse-square law for light. Such objects are known as **standard candles.** Some useful standard candles are cepheids, planetary nebulae, novae, supernovae, and globular clusters. The rotational velocities of spiral galaxies are correlated with their absolute magnitudes and are very useful indicators of distance.

33.3 The universe is expanding. Observations show that the lines in the spectra of distant galaxies are **redshifted,** and their velocities of recession are proportional to their distances from us. The relationship between velocity and distance is known as the **Hubble law.** The rate of recession, which is known as the **Hubble constant,** is approximately 75 km/s per million parsecs. We are not at the center of this expansion. An observer in any other galaxy would see the same expansion that we do.

33.4 The majority of bright galaxies are either **spirals** or **ellipticals.** Spiral galaxies contain both old and young stars, as well as interstellar matter. Typical masses are in the range 10^9 to 10^{12} solar masses. Our own Galaxy is a large spiral. Ellipticals are spheroidal or elliptical systems that consist

almost entirely of old stars, with very little interstellar matter. Elliptical galaxies range in size from giant ellipticals, which are more massive than any spiral, down to dwarf ellipticals, which have masses of only about 10^6 solar masses. A small percentage of galaxies are classified as **irregular.** The Milky Way's nearest neighbors in space are the Large and Small Magellanic Clouds, which are both irregular galaxies.

33.5 The masses of spiral galaxies are determined from measurements of their rates of rotation. The masses of elliptical galaxies are estimated from statistical analyses of the random motions of the stars within them. Galaxies are characterized by their **mass-light ratios.** The luminous parts of galaxies with active star formation have mass-light ratios typically in the range 1 to 10; the luminous parts of elliptical galaxies, which contain only old stars, have mass-light ratios of typically 10 to 20. The mass-light ratios of whole galaxies, including their halos, are much higher, indicating that a great deal of dark matter is present.

33.6 Most astronomers now view **quasars** as the most extreme example of a class of peculiar galaxies that generate large amounts of energy in a small region at their centers. These **active galactic nuclei** are thought to derive their energy from material falling toward, and forming a hot accretion disk around, a massive (up to 10^9 solar masses) black hole. Quasars are strong sources of radio synchrotron emission. Some quasars have small, discrete sources of radio emission that appear to be separating from one another at speeds that exceed the speed of light. The apparent speeds are illusory and can be explained if the motion of the emitting objects is directed almost exactly toward us. Observations of

NGC 4038/9, two galaxies in the process of merging. This recent image, the first ever taken of this pair in true color, shows many bright, young star clusters (surrounded by reddish regions of ionized hydrogen) formed as a result of the interaction of the two galaxies. The nuclei of the two galaxies can be seen as two diffuse yellow "blobs" of light. Note the massive cloud of interstellar matter located in the region between the galaxies. Interactions of galaxies result in bursts of star formation that dramatically affect the subsequent evolution of both systems.

(Image by David Malin, copyright by the Anglo-Australian Telescope Board)

The final part of our journey of exploration takes us into the realms where the galaxies themselves become mere building blocks in the larger puzzle. The ultimate questions astronomers can ask concern the beginning, evolution, and eventual fate of the entire universe. For millenia of human thought, such issues were mostly the province of religious thinkers or philosophers; but in our century, enough information has been accumulated to bring them into the purview of science. The ideas that have come out of this investigation will be the subject of our final chapter; here we first assemble as many observations as we can about galaxies, their arrangement in space, and their development through time.

In looking at the evolution of galaxies, astronomers have had to make a major change in their thinking in the last decade or two. The prevailing idea about galaxies used to be that their development depended mainly on the initial characteristics with which they were endowed at birth—how much material they contained, how quickly they rotated, what shape they took on as a result. In parallel with human growth, we might have said that each galaxy was viewed as developing as a result of its "genetic endowments". Today, we understand that an equally important role in the story of each galaxy is played by the environment into which the galaxy is born—just as in humans, we now know that "nature and nurture" are each important. In crowded groups, galaxies can interact with one another and become transformed or even consumed by the process.

Woven into the story of the galaxies (as into so much of astronomy) is the thread of the great extent of the universe—the development we describe in this chapter takes place on a playing field of enormous scope. The planets and stars we have studied in earlier chapters now seem to dwindle to microscopic size, and we see immense chains of galaxy groups, enormous voids of mostly empty space, and vast reservoirs of dark matter come to center stage. Many astronomers (and astronomy enthusiasts) enjoy contemplating these large-scale questions precisely because our everyday human concerns seem very small in comparison.

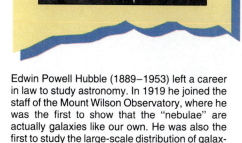

Edwin Powell Hubble (1889–1953) left a career in law to study astronomy. In 1919 he joined the staff of the Mount Wilson Observatory, where he was the first to show that the "nebulae" are actually galaxies like our own. He was also the first to study the large-scale distribution of galaxies. In 1929 he gave the first evidence for the expansion of the universe. (Caltech)

STRUCTURE AND EVOLUTION OF THE UNIVERSE

Celestial objects rarely journey through space alone. The Earth is but one of nine planets orbiting the Sun. More than half of all stars are at least double, and many are members of star clusters. All of the stars and clusters in the Milky Way Galaxy are gravitationally bound, orbit a common center, and will complete their evolution in proximity. Does this cosmic togetherness persist on still larger scales? Are most galaxies to be found in clusters of galaxies? What is the structure of the universe as a whole? How are galaxies distributed in space? Are there as many in one direction of the sky as in any other? And if we count fainter and fainter galaxies, presumably farther and farther away, do we find that their numbers increase in the way they should if galaxies are distributed uniformly in depth? Edwin Hubble began to try to answer these questions only a few years after he first showed that galaxies existed.

34.1 DISTRIBUTION OF GALAXIES IN SPACE

(a) Surveys of Faint Galaxies

Hubble had at his disposal what were then the world's largest telescopes—the 100-in. (2.5-m) and 60-in. (1.5-m) reflectors on Mt. Wilson. But although those telescopes can probe to great depths, they can do so only in small fields of view. To photograph the entire sky with the 100-in. telescope would take thousands of years. So, instead, Hubble sampled the sky in many regions, much as Herschel did with his star gauging (Section 24.4). In the 1930s Hubble photographed 1283 sample areas, and on each photograph he carefully counted the number of galaxy images.

The results of Hubble's survey are shown in Figure 34.1, which is a map of the sky shown in what are called *galactic coordinates* (Appendix 7). The Milky Way, across the middle of the plot, defines the galactic equator, and the top and bottom of the map—the galactic poles—are 90° away from the Milky Way. The empty sectors at the lower right and left are the parts of the sky too far south to observe from Mt. Wilson. Each symbol represents one of the regions of the sky surveyed by Hubble, and the size of the symbol indicates the relative number of galaxies he could observe in that area.

The first thing to notice is that we do not see galaxies in the direction of the Milky Way. The obscuring clouds of dust in our Galaxy hide what lies in those directions. Hubble called this part of the sky the *zone of avoidance*. Near the Milky Way, the counts of galaxies are below average and are denoted by open circles. The farther away we look from the plane of the Milky Way, the less obscuring foreground dust lies in our line of sight and the more galaxies we see. From the counts of galaxies in different directions, Hubble determined that light is dimmed by about 0.25 magnitude or about 25

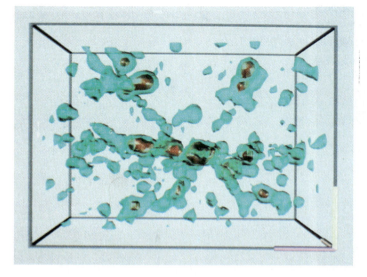

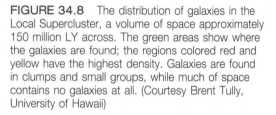

FIGURE 34.8   The distribution of galaxies in the Local Supercluster, a volume of space approximately 150 million LY across. The green areas show where the galaxies are found; the regions colored red and yellow have the highest density. Galaxies are found in clumps and small groups, while much of space contains no galaxies at all. (Courtesy Brent Tully, University of Hawaii)

FIGURE 34.9   A three-dimensional slice of the universe showing the distribution of galaxies in space as seen from our location, which is the point at the bottom of the figure where all of the slices come together. Note the concentration of galaxies in narrow bands or lanes with large voids between them. An analogous distribution would be obtained if we took a slice through a collection of bubbles of various sizes. The Great Wall is the band of galaxies stretching from left to right across the middle of the picture. (Courtesy Margaret Geller/Harvard-Smithsonian Center for Astrophysics)

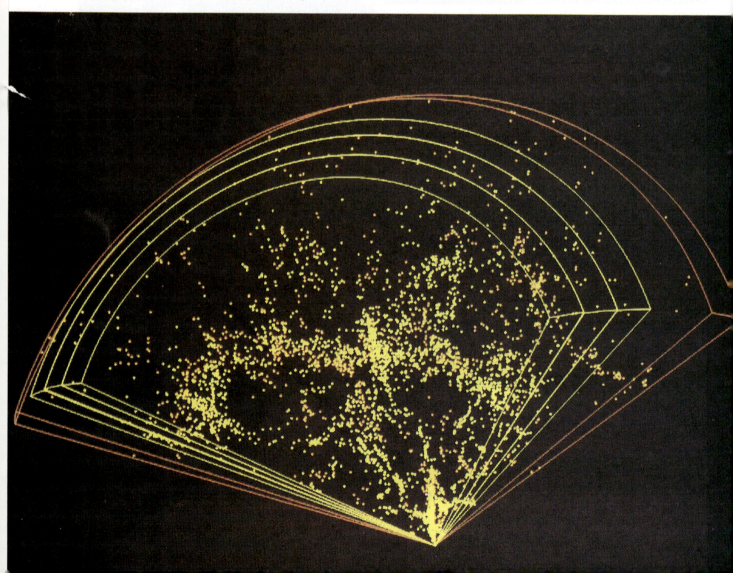

filamentary superclusters that extend over distances of at least a few hundred million light years. The largest structure seen so far in the universe is a sheet of galaxies that is at least 500 million LY long, 200 million LY high, and about 15 million LY thick. Referred to as the "Great Wall," this sheet of galaxies is about 250 million LY from our own Galaxy. The mass of the Great Wall is estimated to be 2×10^{16} solar masses, a factor of 10 greater than the mass of the Local Supercluster.

These filamentary structures are separated by voids, that is, by large holes where few galaxies can be found. You can imagine the voids as giant bubbles that have typical diameters of 150 million LY, with the clusters of galaxies concentrated along their boundaries.

One way to visualize these bubbles is by plotting the distribution of galaxies in a "slice" taken through space, as in the example shown in Figure 34.9. In such a picture, it is clear that most of the galaxies lie along the boundaries of the voids, which are roughly circular. The image is similar to what you would obtain by taking a slice through a sponge or a piece of Swiss cheese with large holes.

The discovery that most of space consists of voids has come as a surprise to astronomers. Most would probably have predicted that the regions between giant clusters of galaxies are filled with many small groups of galaxies or even with isolated individual galaxies. Careful searches within these voids have so far confirmed that they are not—few galaxies of any kind are found there. Apparently, 90 percent of the galaxies occupy less than 10 percent of the volume of space.

Knowledge of exactly *how* galaxies are distributed may provide clues as to *why* they are distributed that way. We know from observations of microwave radiation that the universe, only a few hundred thousand years after it was formed, was extremely smooth (Chapter 35). The challenge for the theoretician is to understand how that featureless universe changed into the complex one that we see today, with dense regions containing many galaxies and nearly empty regions containing almost none.

For example, based on data like that plotted in Figure 34.9, which show that galaxies are distributed on the walls of bubble-like structures, some astronomers have argued that matter used to be inside the bubbles and that it had somehow been cleaned out and pushed toward the walls. But how? One possibility is that there were giant explosions during the first few billion years of the universe that swept the gas into bubble-like shells, and that galaxies subsequently formed in these shells. Unfortunately, despite considerable ingenuity, astronomers have yet to devise a way to produce energetic enough explosions to account for the largest voids.

Even if we do not have a good explanation for why the voids exist, this discussion should serve to show how measurements of the distribution of galaxies in space can help generate ideas about what kinds of events must

have occurred when superclusters, clusters, and galaxies were just beginning to form.

If all this structure exists, you may want to ask whether the universe can really be described as homogeneous and isotropic. The answer is probably yes, provided we consider regions of the universe large enough to include a number of superclusters and voids. Consider the following analogy. All the residents of your neighborhood taken together may be fairly typical of the residents of the city as a whole in which you live. But the people within the particular house in which you live are probably not typical in number, age, or other characteristics. Just as we must take a fairly large sample of the population to obtain a representative group of people, so must we consider a fairly large volume of space to find within it the characteristics and kinds of objects that are typical of other large volumes of space.

34.2   EVOLUTION OF GALAXIES: THE OBSERVATIONS

It is only in the past decade that observational and theoretical studies of the evolution of galaxies have begun to make real progress in developing a picture of how galaxies change over the lifetime of the universe. Progress has been slow for several reasons. First, galaxies are made up of stars, and only after the evolution of individual stars was well understood could astronomers sensibly begin to explore how whole systems of stars change with time. Twenty years ago, any attempt to describe the evolution of galaxies would have been pointless—we simply did not know enough about the life histories of stars.

A second problem with trying to study galaxies is that they are very, very faint. Even with the biggest telescopes in the world, we cannot see individual stars or even determine the shapes of the most remote galaxies.

We do have one advantage, however, in studying galaxy evolution. The universe itself is a kind of time machine, which permits us to observe galaxies as they were when the universe was young. When we look at distant galaxies, we see them as they were when the light that we now measure left them. If we observe a galaxy that is 1 billion LY distant, we are seeing it as it was when the light left it 1 billion years ago. By observing more and more distant objects, we look ever further backward toward a time when both galaxies and the universe were young. Because of the relationship between distance and time, we will use light years rather than parsecs to measure distances for the remainder of this text.

Three types of observations have provided most of the clues to how galaxies evolve—spectroscopy, measurements of the colors of galaxies, and determination of their shapes.

(a)   Spectra, Colors, and Shapes

A spectrum of a galaxy provides a great deal of information. Measurement of the radial velocity can be used to estimate the distance. Studies of the rotation of galaxies can be used to estimate their masses. Detailed analysis of the spectral lines can determine what types of stars inhabit a galaxy, what their composition is, and whether a galaxy contains large amounts of interstellar matter.

Unfortunately, many galaxies are so faint that it is impossible to collect enough photons, even with the world's biggest telescopes, to produce a measurable spectrum. Astronomers thus have to use colors to estimate what kinds of stars inhabit the faintest galaxies (Figure 34.10). To understand how this works, remember that hot luminous blue stars have lifetimes of only a few million years. If we see a very blue galaxy, we know

that it must have many hot luminous blue stars, and that star formation must have occurred in the past few million years. If we see a reddish galaxy, however, it must contain mostly old stars, formed billions of years before the light that we now see was emitted by the galaxy.

Another important clue to the nature of a galaxy is its shape. As we have seen, spiral galaxies contain young stars and large amounts of interstellar matter. Elliptical galaxies have mostly old stars and very little interstellar matter. For whatever reason, elliptical galaxies turned most of their interstellar matter into stars many billions of years ago, while star formation has continued until the present day in spiral galaxies. As a result, spiral galaxies are bluer than elliptical galaxies. Unfortunately, very distant galaxies appear small on the sky. Often we cannot tell whether a distant galaxy is a spiral or an elliptical. One of the main goals of the Hubble Space Telescope is to obtain images of distant galaxies, unblurred by the Earth's atmosphere, in order to determine whether the same types of galaxies that we see nearby in the present-day universe existed billions of years ago.

(b)   The Ages of Galaxies

One starting point for all theories of galaxy formation is the observation that most galaxies are very old indeed. For example, there are stars in globular clusters in our own Galaxy that are 13 to 15 billion years old. Therefore, the Milky Way must be at least that old.

Astronomers have discovered that the universe itself is not significantly older than the globular cluster stars. The age of the universe is derived from the observed rate of expansion. As we have seen (see Section 33.3), galaxies are moving farther and farther apart. If we project this expansion back in time, we find that all of the galaxies were very close together sometime between 10 and 15 billion years ago. The major uncertainty in the age of the universe is caused by uncertainties in estimates of how far away galaxies are from us at the present time. These ideas are discussed in Chapter 35, but it does appear that the globular cluster stars in the Milky Way Galaxy must have formed during the first 2 billion years after the expansion of the universe began.

The most distant elliptical galaxies for which we have some information on composition emitted the light that we observe when the universe was only about half its present age. Yet some of these galaxies have about the same luminosity and colors, and hence about the same stellar content, as do galaxies that are only a few million light years distant, which are therefore about twice as old. The similarity of ellipticals that span half the age of the universe suggests that star formation in this type of galaxy has either been absent or nearly so for the last several billion years. Star formation probably began about 1 billion years after the universe began, and new stars continued to form for at most a few billion years.

FIGURE 34.10   Elliptical galaxy NGC 1199 and companions. NGC 1199 is the large E3 elliptical galaxy at bottom left. It is accompanied by a face-on barred spiral galaxy (NGC 1189) seen toward the upper left and an S0 galaxy (NGC 1190) of spindly appearance near the right edge. We can see the yellow-orange color characteristic of older population II stars in the elliptical galaxy and in the central bar and bulge of the barred spiral. Younger population I stars produce the bluish color of the disk of the barred spiral. This picture shows how colors can be used to determine the ages of stars found in distant galaxies. (N.A. Sharp/NOAO)

We can probe still farther back in time, and still closer to the beginning of the universe, by observing quasars, which are much brighter than normal galaxies and can be seen at larger distances. Quasars were most common when the universe was only about 20 percent its present age. Remember that quasars are found in the centers of galaxies, and so large numbers of galaxies must have formed very early.

Another clue to when galaxies formed comes from the study of the gas in quasars. Astronomers find that the gas in distant quasars contains not only hydrogen and helium but also heavier elements such as carbon, nitrogen, and oxygen. We think that these heavy elements were not present when the universe began but were manufactured in stars that evolved within newly formed galaxies. Since quasars contain large amounts of heavy elements, at least one generation of stars had already completed its evolution even before the light that we now see was emitted. Given the distance to quasars, this means that some galaxies must have formed when the universe was less than 20 percent as old as it is now.

(c) Star Formation in Galaxies

The evolution of spiral and elliptical galaxies differs in a fundamental way. In spirals, star formation is a continuous process that is still occurring today, although on average at a somewhat lower rate than several billion years ago. In elliptical galaxies, even the youngest stars are older than the Sun. Since there is very little dust or gas in ellipticals, star formation cannot take place in the present era.

Where did the gas and dust go? Much of it must have been consumed very rapidly in the formation of the first generations of stars. But star formation alone would not be efficient enough to consume all of the gas and dust originally present in elliptical galaxies. In any case, as stars evolve, they lose mass through the action of stellar winds, by forming planetary nebulae, or by exploding. In the process, they replenish the interstellar material.

It must be that gas and dust are somehow efficiently removed from elliptical galaxies. One possibility is that the gas is swept out. Ellipticals occur in clusters of galaxies, not in isolated and otherwise empty regions of space. In these clusters, gas is present between the galaxies, as we know from x-ray observations. As an elliptical galaxy orbits about within a cluster, it moves rapidly (typical velocities are 1000 km/s) through the gas that lies within the cluster but outside the galaxies. This intergalactic gas bombards whatever small amount of gas may lie within the boundaries of the elliptical galaxy and drives the gas from the galaxy (Figure 34.7). The pressure of the intergalactic gas is too small to affect the motions of the stars in the galaxy.

Observations show that about 80 to 90 percent of the galaxies in the high-density environments in the centers of clusters of galaxies are ellipticals and disk-shaped galaxies that have very little gas, no spiral arms, and no recent star formation. In the present era, relatively few spirals are found near the centers of rich clusters. Conversely, isolated galaxies found in regions outside of clusters or groups of galaxies, where the density of material is low, are mostly spirals. Spiral galaxies, like the Milky Way and the Andromeda Galaxy, have managed to retain their gas and dust until the present era because they lie isolated in regions of space where the density of intergalactic gas is too low to sweep them clean.

(d) Colliding Galaxies

There are other ways galaxies can influence each other's evolution. Over the past 20 years, astronomers have learned that galaxies in clusters and groups frequently collide with one another and that these collisions play an important role in determining how galaxies change with time.

Astronomers can calculate what happens when two galaxies collide by using the equations for gravitational tidal effects (Section 4.3). Three properties of tidal interactions determine what happens: (1) The tidal force is proportional to the inverse cube of the separation of the galaxies (see Section 4.3). (2) Tidal forces on an object tend to elongate it; thus there are tidal bulges on both the near and the far sides of each galaxy with respect to the other. (3) The perturbed galaxies are generally rotating before the tidal encounter, and the subsequent distributions of their material must therefore reflect the conservation of their angular momenta.

The calculations show that material is pulled out of two colliding galaxies and may form bridges between them. There are also "tails" of material that string out away from each galaxy in a direction opposite to that of the other (Figure 34.11). Because of the rotation of the galaxies, the tails and bridges can take on unusual shapes. The strange shapes calculated for the model galaxies resemble what is actually seen for real galaxies.

Stars in colliding galaxies are not much affected. Because the stars are very far apart, a direct collision of two stars is very unlikely, although their orbits may be altered. Interstellar matter is much more affected by galaxy interactions than are the stars. Interstellar gas clouds are large and likely to experience direct impacts with other clouds. These violent collisions compress the gas in the clouds, and the increased density can lead to star formation (Figure 34.12). In some interacting galaxies, star formation is so intense that all of the available gas will be exhausted in only a few million years, so that the burst of star formation is clearly only a temporary phenomenon. Bursts of star formation are very rare in isolated galaxies.

Additional evidence that galaxy collisions stimulate star formation comes from the Hubble Space Telescope (HST). Figure 34.13 shows an HST image of the peculiar galaxy NGC 1275. There are about 50 globular clusters in this picture. The surprising thing about these

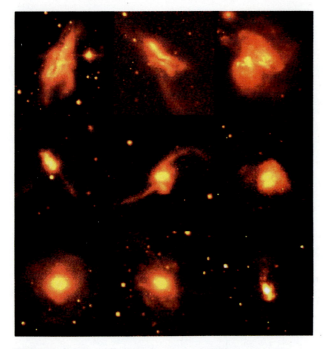

FIGURE 34.11 A series of images of different merging galaxies. These are false-color images, with color being used to indicate brightness. By comparison with theoretical calculations, we can arrange these galaxies in a sequence, from left to right and top to bottom, that corresponds to the time since the merger began. In the upper left we see two galaxies just beginning to merge. As time passes, the two galaxies lose their separate identities and the nuclei merge. The first galaxy in the second row is in this stage; infrared images show that there are still two nuclei behind the dust lane. Tidal tails form, and then wrap around and become part of the overall light distribution. Finally, as seen in the lower right-hand image, the galaxy takes on the form of an elliptical galaxy. True color images would show that the colors of the galaxies also change along this sequence from the blue colors of a galaxy with a high rate of star formation at the beginning of the merger to the older, yellowish color of an elliptical galaxy when the interaction is completed. The galaxies shown are, from top left to bottom right, NGC 3690, NGC 520, NGC 6240, Arp 220, NGC 2623, IC 883, NGC 4194, NGC 7252, and NGC 7585. (Courtesy, W. Keel)

clusters is that they contain young, hot, massive stars. Remember that all of the stars in globular clusters in the Milky Way are billions of years old. Astronomers think that NGC 1275 may actually be two galaxies—a giant elliptical galaxy and a smaller spiral galaxy—colliding with each other. The globular clusters may have formed recently as a result of this collision.

If two galaxies have low relative velocities when they collide, they may coalesce to form a single galaxy. The large elliptical galaxies found in the centers of clusters of galaxies are probably formed by the mergers of several smaller galaxies. Slow collisions and mergers can transform spiral galaxies into giant elliptical galaxies.

The term **merger** is used to refer to the interaction of two galaxies of comparable size. The swallowing of a small galaxy by one that is much larger is described as **galactic cannibalism.** The larger galaxy captures stars

through two mechanisms. The first is *tidal stripping.* If a small galaxy approaches a large one too closely, then its self-gravity may be inadequate to retain the stars and gas in its outer regions. The tidal forces of the larger galaxy will dominate and will rip stars away from the galaxy of lower mass. The physics is the same as that discussed in Section 17.4, which considered what happens to a small satellite in the vicinity of a large planet.

A large galaxy can swallow, or *cannibalize,* the dense core of a smaller galaxy through the second mechanism, which is referred to as *dynamical friction.* The basic idea is that if the core of the smaller galaxy is moving through the envelope of stars of the larger galaxy, it will lose energy and decelerate. This process causes the smaller galaxy to slow and spiral into the massive one.

Rich clusters of galaxies often have one or more supergiant galaxies near their centers. These "monster" galaxies, which are called cD galaxies, frequently have more than one nucleus and have probably acquired their unusually high luminosities by swallowing nearby galaxies. The multiple nuclei are the remnants of their victims.

(e)   How Galaxies Change with Time

The galaxies we see today appear to differ from the types of galaxies that were most common several billion years ago. Rich clusters at distances of about 5 billion years contain many more blue galaxies than do nearby rich clusters. These blue galaxies are mainly spiral galaxies (Figure 34.14). Since a blue galaxy must contain young stars, this difference in color indicates that more spiral galaxies were actively forming stars then than now. The rate of star formation has on average declined dramatically during the past 5 billion years or so.

Two processes have been suggested for producing a high rate of star formation. First, as we saw in the previous section, collisions of galaxies can compress the gas within them and stimulate the formation of stars. Another possibility is that rich clusters of galaxies may contain gas between the galaxies that has a high temperature and high pressure. A galaxy, as it moves on its orbit through the cluster, may suddenly run into some of this high-temperature gas. In the ensuing collision, the cold molecular clouds in the galaxy may be compressed, again accelerating the rate of star formation.

Why have these blue spirals vanished from the universe over the past 5 billion years? One possibility is that mergers of galaxies have reduced the number of galaxies and therefore reduced the likelihood of collisions and enhanced rates of star formation.

34.3   EVOLUTION OF GALAXIES: THE THEORIES

At the beginning, as we shall see in the next chapter, the universe was very smooth. In the present era, however,

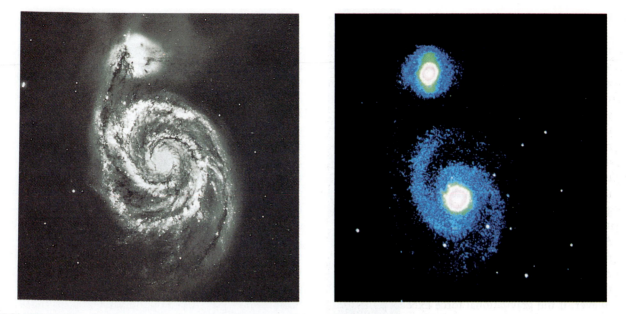

FIGURE 34.12 Optical (left) and infrared (right) images of M51, the Whirlpool Galaxy. The outlying arm, which reaches to the companion galaxy, is much brighter in the optical than in the infrared. This difference in color suggests that most of the stars in this arm are hot, young stars, which are blue in color, and that the formation of these stars was stimulated by the interaction of the two galaxies (NOAO).

the universe, with its clusters and superclusters of galaxies, is clumpy on many scales. The challenge for the theorist is to understand how an initially smooth (or nearly smooth) distribution of matter in the universe could give rise to the complex structure that we now see.

(a) Top-Down or Bottom-Up

There are many ideas about how this structure might have formed. Here we will look at the two possibilities that astronomers have explored in the most detail. Top-

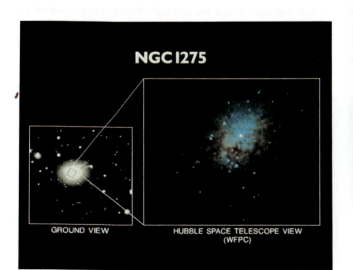

FIGURE 34.13 An image from the Hubble Space Telescope of the Galaxy NGC 1275. About 50 bluish globular clusters can be seen in this picture. (NASA)

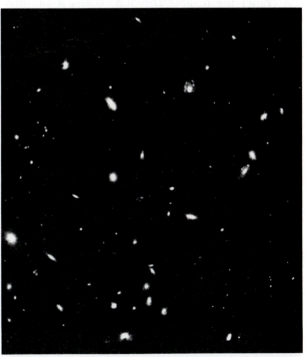

FIGURE 34.14 A Hubble Space Telescope image of a remote cluster of galaxies. We are seeing this cluster as it was when the universe was only two thirds as old as it is now (redshift = 0.4). This cluster contains many more spirals than do clusters in the present era. During the past few billion years, many spiral galaxies have either disappeared, probably through mergers with other galaxies, or grown much fainter as star formation has faded away. (A. Dressler/NASA)

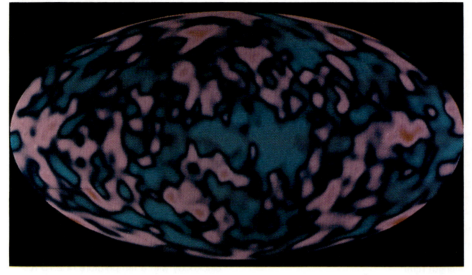

FIGURE 35.10 Microwave map of the whole sky made from one year's data obtained by the COBE satellite. Processing has removed all microwave radiation coming from sources other than the cosmic background radiation (CBR). The red (warmer) and blue (cooler) regions represent temperature differences 0.001 percent above or below the average sky temperature of 2.73 K. Regions with slightly different temperatures also have different densities.

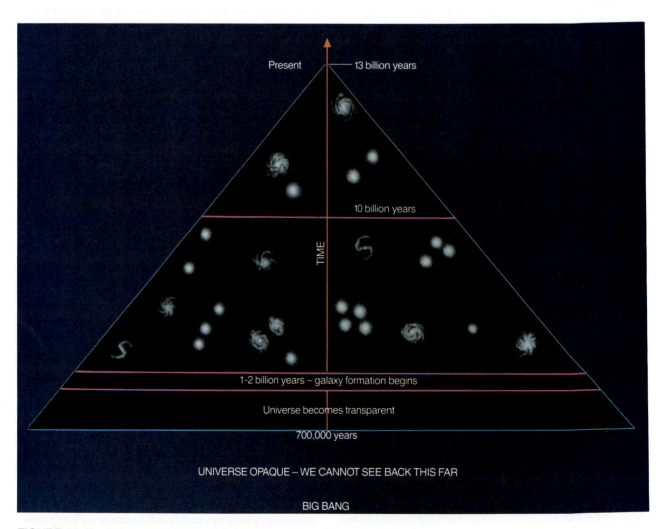

FIGURE 35.11 As we look to larger and larger distances and farther back into time, we see more and more galaxies. Ultimately, we see back to the point at which the universe became transparent, which occurred a few hundred thousand years after the expansion began. Because the universe was opaque before this time, we cannot directly observe earlier eras.

(d)   A Look Back to the Big Bang

Since the CBR comes from the time when the fireball first became transparent, it is at the farthest point in space and time to which we can now observe. If we could see that radiation visually, it would be as if it were coming from an opaque wall that completely surrounds us. No radiation from a more distant source can ever reach us, because such a source would lie farther back in time, behind that opaque wall.

As the time since the big bang becomes greater, we must look farther into the past to see the fireball, and hence farther away in space (Figure 35.11). At earlier times, we could have seen (had we been here) only relatively nearer objects, and there were fewer galaxies between us and the threshold provided by the fireball itself. As time passes, more and more galaxies come into view. Thus, not only does the universe expand with time, but the part of it accessible to observation becomes greater as well.

The concept that there is a limiting distance beyond which we cannot see, combined with a finite age for the universe, provides an answer to an apparently simple yet surprisingly profound question: Why is the sky dark at night? Suppose we lived in an infinite, nonexpanding, universe (with no curvature) that was uniformly filled with stars. Then in every direction in which we looked, we should see, at some distance great or small, a star. Since the total number of stars is infinite, their total contribution to the night sky is potentially very great. In fact, it is possible to show mathematically that if the universe were infinite, the entire night sky should be as bright as the Sun. The fact that the sky is dark rather than bright is referred to as Olbers' paradox, named after the German physician and amateur astronomer who is the best remembered of those who pointed out the problem.

The universe, however, is no more than about 15 billion years old. Light from stars more than 15 billion LY away simply has not had time to reach the Earth. The universe may indeed be infinite, but we can observe only that portion that lies within a sphere centered on the Earth with a radius of about 15 billion LY. Within this sphere, the density of stars is so low in most directions that our line of sight does not intercept a star. There is a boundary to the observable universe, but it is a boundary in time, not in space.

35.4   THE INFLATIONARY UNIVERSE

(a)   Problems with the Standard Big Bang Model

The hot big bang model is remarkably successful. It accounts for the expansion of the universe, and it correctly predicts the abundances of the light elements hydrogen, deuterium, helium, and lithium, which are formed in first few minutes of the expansion. As it turns out, this model also predicts that there should be exactly three types of neutrinos, and this prediction has been confirmed by experiments with high-energy accelerators. The theory also accounts for the observations of the CBR.

The theory is not complete, however. The standard big bang model does not explain why there is more matter than antimatter in the universe, nor does it account for the origin of the density fluctuations that ultimately grew into galaxies. It also does not explain the remarkable *uniformity* of the universe. The CBR is the same, no matter which direction we look, to an accuracy of about 1 part in 100,000. This sameness might be expected if all the parts of the visible universe were in contact at some point and had time to come to the same temperature—just as when we put ice in a glass of water and wait a while, we have no ice and all the water is at the same temperature. However, if we accept the standard big bang model, all parts of the visible universe were *not* in contact at any time. There is a maximum distance that light can have traveled since the time the universe began. This distance is called the *horizon distance*, because any two objects separated by this distance cannot ever have been in contact. No information, no physical process can propagate faster than the speed of light. One region of space separated by more than the horizon distance from another has been completely isolated from it through all eternity.

If we measure the CBR in two opposite directions in the sky, we are observing regions that were separated by more than 90 times the horizon distance at the time the CBR was emitted (Figure 35.12). We can see both, but they can never have seen each other. Why, then, are their temperatures so precisely the same? According to the standard big bang model, they have never been able to exchange information, and there is no reason why they should have identical temperatures. The only explanation is that the universe somehow started out being absolutely uniform. Scientists are always uncomfortable, however, when they must appeal to a special set of initial conditions to account for what they see.

Another problem with the standard big bang model is that it does not explain why the density of matter in the universe is so close to the critical density. As we have seen, observations are unable to tell us whether the expansion of the universe will continue forever or will ultimately slow and perhaps even come to a halt and reverse itself. The interesting fact, however, is that the universe is so nearly balanced between these two possibilities that we cannot yet determine which is correct. There could have been, after all, so little matter that it would be obvious that the universe is open and that the expansion will continue forever. Alternatively, there could have been so much matter that the universe would be clearly and unambiguously closed. Instead, the amount of matter present is within a factor of 5 or so

TABLE 35.5	AMOUNTS OF MASS PRESENT IN ASTRONOMICAL OBJECTS OF VARIOUS SIZES	
Object	Ratio of Measured Density to Critical Density
Stars	<0.01
Individual galaxies	0.01–0.03
Rich clusters of galaxies	0.2
Flow of galaxies toward the Great Attractor	0.5–1.0

surface of the balloon now looks much flatter than it did before it was blown up.

But is the universe actually flat now, as the inflation model predicts it must be? If it is, then the current density must be exactly equal to the critical density. Table 35.5 summarizes the amount of matter that is estimated to be present in various types of astronomical objects. Luminous matter in galaxies contributes less than 1 percent of the mass required to reach critical density. Even if we add to the stars and interstellar matter in galaxies the invisible dark matter that is detected through its gravitational influence on luminous objects, we are up to only 20 percent of the critical density. The only observational evidence that supports the idea that we live in a critical-density universe is the motion of nearby galaxies toward the Great Attractor (Section 34.4). Those observations are so new that some independent confirmation will be required before we can be confident about the conclusion.

(c) Dark Matter Again

In order for the ideas about the inflationary universe to be correct, most of the matter in the universe must be invisible. How, then, should astronomers go about looking for it? The techniques depend on what the dark matter is made of. We shall examine several possibilities in turn.

The first possibility is that the dark matter is made of ordinary matter—protons and neutrons. This type of matter is called *baryonic matter*, because protons and neutrons are called *baryons* in particle physics. This baryonic matter could take many forms. In particular, the protons and neutrons could be in the form of low-mass stars, brown dwarfs, or extremely massive black holes. Various experiments are in progress, such as the search for MACHOs described in the previous chapter, to try to identify such objects.

Even if MACHOs are found, we have not solved the problem. The abundance of deuterium gives us an estimate of the density of protons and neutrons in the universe. The best estimates are that protons and neutrons amount to no more than about 5 percent, and perhaps as little as 1 percent, of the critical density. This estimate is sufficiently uncertain that baryons might account for all of the dark matter in the halos of galaxies, but there are certainly not enough of them to produce a critical-density universe.

A second possibility is that neutrinos have sufficient mass to produce a critical-density universe. The mass of the neutrino is surely very small, but there is no experimental evidence that it is actually zero. Another argument in favor of neutrinos is that we know for a fact that they were actually produced in the big bang. Attempts are now being made with high-energy accelerators to try to measure the mass of the neutrino by detecting the change in momentum or energy of another particle that collides with it. Another experiment involves the study of neutrinos emitted by supernovae. If a supernova were to go off in our own Galaxy, then we could measure the times of arrival of the three different types of neutrinos. If they arrive at different times, then at least some of them must have mass. If neutrinos have zero mass, then all of the types will travel at the speed of light and will arrive at the same time.

A third possibility is that there is some type of elementary particle that we have not yet detected here on Earth that has mass and exists in sufficient abundance to close the universe. GUTs predict the existence of such particles. One class of particles has been given the name WIMPs, which stands for weakly interacting massive particles. Since these particles do not participate in nuclear reactions leading to the production of deuterium, the deuterium abundance puts no limits on how many WIMPs might be in the universe.

If WIMPs do exist, then some of them should be passing through our physics laboratories right now. The trick is to catch them. Because they are weakly interacting particles, the chances that they will have a measurable effect are small. Physicists are, however, now devising experiments to try to detect WIMPs. The basic idea is that a WIMP moving through such a detector might collide with an atomic nucleus and cause it to move. This motion would then be transferred to other particles in the detector causing a very (!) small change in temperature. Experiments based on this principle will be conducted in the next decade.

(d) Dark Matter and the Formation of Galaxies

Dark matter *must* be present if the inflationary models are correct. The reason is related to the formation of galaxies. Galaxies must have grown from density fluctuations in the early universe. The observations with COBE give us information on the size of those density fluctuations. It turns out that they are too small, at least according to our current theories, to have formed galaxies in the first billion years or so after the big bang. Yet observations of quasars suggest that galaxies were indeed formed that early.

The COBE data, however, give us information about density fluctuations only for the type of matter that interacts with radiation. What we measure with COBE is differences in intensity of radiation, which are related to differences in density. Suppose there is a type of matter that does not interact with light at all—namely, dark matter. This matter could have much stronger variations in density that we would not be able to detect. This matter might form a kind of gravitational trap that would attract ordinary matter and cause it to form high-density regions that could subsequently form galaxies quickly. Ordinary matter could not fall into these traps before the universe became transparent, that is, before the time the CBR observed by COBE was produced. When the universe was opaque to radiation, ordinary matter and radiation interacted so closely that any structure would have been quickly wiped out. Not so for dark matter that is unaffected by radiation.

The size of the gravitational traps depends on the nature of the dark matter. If the dark matter is moving near the speed of light—so-called *hot dark matter*—as neutrinos would, small-scale density fluctuations would be smoothed out by the rapidly streaming particles as they move from high- to low-density regions. In this case, large-scale structure would form first. If, on the other hand, the dark matter moves slowly—so-called *cold dark matter*—then the particles would not have time to move far enough to smooth out small-scale fluctuations in density. In this case, relatively small structures, the size of globular clusters or individual galaxies, would be likely to form first.

Neither hot nor cold dark matter is entirely successful in explaining the distribution of galaxies discussed in Chapter 34. Hot dark matter models predict that all galaxies should be found in large sheet-like structures, which is not seen. Cold dark matter cannot produce voids, walls, and long structures like the Great Wall. Now theories are being developed that incorporate both hot and cold dark matter. Even though current models are not adequate to explain how galaxies form, it is important to note that it is difficult to form galaxies at all unless a substantial amount of dark matter of some kind is present.

(e) Conclusion

From the beginning to the end of this book, we have traveled a long way. We began with a brief look at historical ideas about astronomy, which placed first the Earth and then the Sun at the center of the universe. Now we know that the Sun is a rather ordinary star, located at the outskirts of a rather ordinary galaxy, which occupies a completely ordinary part of the universe. It is this very ordinariness, however, that has allowed us to use observations from our own local neighborhood to develop a picture of the origin and evolution of the universe as a whole. That picture, and specifically the big bang model of cosmology, is extremely successful in explaining a large portion of what we see. But the picture is not complete. Attempts to explain what the big bang model does not—isotropy, flatness, the formation and distribution of galaxies—have now suggested radical new ideas about the universe, which must be tested by observations. If it is really true that most of the matter of the universe is made of some type of particle that we have not yet discovered, then we must accept the challenge of trying to detect it. The search is on—in huge accelerators, in university laboratories around the world, and deep in underground mines, where scientists are trying to trap elusive dark matter particles just as they once succeeded in capturing neutrinos. Astronomers are continuing to probe the depths of the universe with ever bigger telescopes to observe directly the formation of galaxies. And closer to home, the search is on for planets around other stars in an effort to understand whether our own planet Earth is, like the Sun and the Galaxy, rather ordinary or quite extraordinary.

Over the past decade the contents of this book have changed a remarkable amount. Dark matter earned barely a skeptical mention 10 years ago, and now we believe it may hold the secret to both the formation and ultimate fate of the universe. The next decade will undoubtedly yield many more surprising discoveries in astronomy. Whatever those discoveries may be, it is quite clear that we will be able to understand the beginnings of our universe much more completely than we would have thought possible only 10 years ago.

S U M M A R Y

35.1   The universe is expanding. From the rate of expansion of the universe, we can estimate that all of the matter in the universe was concentrated in an infinitesimally small volume 10 to 15 billion years ago. The beginning of the expansion is what astronomers mean when they talk about the beginning of the universe.

35.2   The most widely accepted model of the universe is the **big bang** model. The factor that controls the evolution of the universe is the density of matter (and energy). If the density is

high, then the rate of expansion will slow and possibly even reverse direction so that the galaxies all come together again (a **closed universe**). Observations suggest that the density is actually so low that the expansion will continue forever (an **open universe**).

35.3   The early universe was so hot that the collision of photons could produce material particles. As the universe expanded and cooled, protons and neutrons formed first, then electrons and positrons. Fusion reactions then produced helium nuclei. Finally, the universe became cool enough to form hydrogen atoms. At this time the universe became transparent to radiation. Scientists have detected the **cosmic background radiation (CBR)** from the hot early universe. The CBR now has the energy distribution of a perfect blackbody at a temperature of 2.73 K.

35.4 The big bang model does not explain why the CBR has the same temperature in all directions. Neither does it explain why there was originally more matter than antimatter, nor why the density of the universe is so close to the **critical density.** The new **inflationary universe,** which makes use of **grand unified theories (GUTs)** of the electromagnetic, **strong,** and **weak** forces, is one theory that has been developed to try to explain these observations. One prediction of this new theory is that the density of the universe should be exactly equal to the critical density. This prediction can be true only if 99 percent of the matter in the universe is invisible.

E X E R C I S E S

THOUGHT QUESTIONS

1. What is the most useful probe of the early evolution of the universe—a giant elliptical galaxy or an irregular galaxy like the Large Magellanic Cloud? Why?

2. What are the advantages and disadvantages of using quasars to probe the early history of the universe?

3. Consider the plot of radial velocities against the distances of remote clusters of galaxies. Are the measured distances and radial velocities that are plotted the *present* values of these quantities for the clusters? In each case, why? If not, try to describe how the diagram might differ if we did plot the present-day values of cluster distances and velocities.

4. Suppose the universe will expand forever. Describe what will become of the radiation from the primeval fireball. What will the future evolution of galaxies be like?

5. In this text we have discussed many motions of the Earth as it travels through space with the Sun. Describe as many of these motions as you can.

6. There are a variety of ways of estimating the ages of various objects in the universe. Describe some of these ways and indicate how well they agree with one another and with the age of the universe itself as estimated by its expansion.

7. In the 19th century, both geology and biology were based on the idea that evolution was a slow process. In the 20th century, we have come to the conclusion that violent events have played a significant role in shaping the evolution of the universe and everything in it, including the evolution of life on Earth. Discuss some of these violent events, starting with the big bang and including some events that have affected the Earth directly.

8. Since the time of Copernicus, each revolution in astronomy has moved humans farther from the center of the universe. Now it appears that we may not even be made of the most common form of matter. Trace the changes in scientific thought about the central nature of the Earth, the Sun, and our Galaxy on a cosmic scale.

9. Construct a time line for the universe and indicate when various significant events, from the beginning of the expansion to the formation of the Sun to the appearance of humans on Earth, occurred.

PROBLEMS

10. The Andromeda Galaxy is approaching the Sun at a velocity of about 300 km/s. Does this indicate that the universe is not expanding? Compare this velocity with the velocity of the Sun in its orbit around the center of the Galaxy. Suppose Andromeda is orbiting the Milky Way Galaxy with a period of 10 billion years. What velocity would it have?

11. Show that if $H = 75$ km/s per million parsecs, then the maximum age of the universe is 13 billion years.

12. Assume that the radial velocities of galaxies have always been the same and are given *at this instant of time* by $V = Hr$, where $H = 50$ km/s per million parsecs. Note, however, that we do not observe the *present* distances of galaxies, but the distances they had when light left them on its journey to us. Now plot the relation between velocity and distance that would be obtained directly for *observations* (that is, corresponding to *measured* distances, not present distance). Consider several distances, out to 2×10^9 pc. Discuss the shape of the curve. How would this curve differ if the expansion rate were decreasing (say, because of gravitational attraction between galaxies)? What if it were increasing?

At the end of our journey of exploration, like all weary travelers, we want to stop and recollect where we have been and what the journey has taught us. At the same time we look ahead and ask where else we might go if we continue our exploration in days to come. Let us therefore conclude by surveying the larger themes of astronomy that we have covered in this book, and a few of the important questions astronomers will be trying to answer in the next century of exploration.

CONCLUSIONS FROM THE STUDY OF ASTRONOMY

1. **The laws of nature are universal.** The same physical laws we have discovered on Earth apply on other worlds and in the most distant galaxies.

2. **The Earth is a planet.** It orbits the Sun with eight others (and assorted smaller bodies), and there seems to be nothing extraordinary about its position, its motion, or its composition.

3. **The Sun is a star.** It is a typical G-type main-sequence star, and only one of billions of stars in our Galaxy. Its location about 3/5 of the way out from the center of the Milky Way is also not particularly noteworthy.

4. **Stars are organized into galaxies and galaxies into groups.** Although we have much to learn, in the 20th century we have begun to glimpse the great levels of organization that permeate the universe, from galaxies, to clusters, to superclusters, to the filaments and voids our surveys are now beginning to reveal.

5. **It's a big universe out there.** The extent of the universe is enormous compared to human scales. The stars in galaxies are typically separated by light years; galaxies in groups are separated by hundreds of thousands of light years.

6. **The universe had a beginning.** The expanding universe can be traced back to a time of creation (the big bang) between 10 and 20 billion years ago. Our solar system is about 5 billion years old, roughly ⅓ of the age of the universe.

7. **The universe evolves.** Characteristics of the universe (such as density and composition) change with time. To quote Harlow Shapley, "In the beginning was the word; and the word was hydrogen." As time goes on, more and more hydrogen is changed into heavier elements by nuclear reactions in stars.

8. **Planets evolve.** Conditions on the surfaces and in the atmospheres of planets and satellites have changed over time by the action of geological and chemical processes. One possible outcome of such evolution is life.

9. **Most of the matter in the universe may be invisible.** There is good evidence that the bulk of the matter in the universe may be dark; the stars and other luminous matter could well be just the froth on the cosmic ocean.

10. **Progress in astronomy depends on new technology and new instruments.** Throughout the history of astronomy better observations and theories have come as a result of the development of more sophisticated tools for probing the universe. Future progress is likely to depend in the same way on advances and breakthroughs in instrumentation.

A FEW QUESTIONS FOR THE FUTURE

1. **Is the solar system unique?** Are there other planetary systems like our own, and, if so, how common are they?

2. **Is there life out there?** Did the chemical spark that led to life on Earth occur on other worlds, either in the solar system or around other stars?

3. **Is there intelligent life out there?** If life did develop elsewhere, has any example of it evolved to the stage where students are taking astronomy courses (an obvious sign of intelligence)?

4. **What happens inside a black hole?** Some scientists have speculated that rotating black holes may, at least in theory, provide paths to elsewhere or elsewhen in the universe. What is the exact nature of space and time inside a black hole and at a singularity?

Milky Way   The band of light encircling the sky, which is due to the many stars and diffuse nebulae lying near the plane of the Galaxy.

minerals   The solid compounds (often primarily silicon and oxygen) that form rocks.

minor axis (of ellipse)   The smallest or least diameter of an ellipse.

minor planet   See *asteroid*.

model atmosphere (or photosphere)   The result of a theoretical calculation of the run of temperature, pressure, density, and so on, through the outer layers of the Sun or a star.

molecule   A combination of two or more atoms bound together; the smallest particle of a chemical compound or substance that exhibits the chemical properties of that substance.

momentum   A measure of the inertia or state of motion of a body; the momentum of a body is the product of its mass and velocity. In the absence of a force, momentum is conserved.

monochromatic   Of one wavelength or color.

nadir   The point on the celestial sphere 180° from the zenith.

nanosecond   One thousand-millionth (10^{-9}) s.

nautical mile   The mean length of one minute of arc on the Earth's surface along a meridian.

navigation   The art of finding one's position and course at sea or in the air.

nebula   Cloud of interstellar gas or dust.

neutrino   A fundamental particle that has little or no rest mass and no charge but that does have spin and energy.

neutron   A subatomic particle with no charge and with mass approximately equal to that of the proton.

neutron star   A star of extremely high density composed almost entirely of neutrons.

new moon   Phase of the Moon when its longitude is the same as that of the Sun.

Newtonian focus   An optical arrangement in a reflecting telescope, in which a flat mirror intercepts the light from the primary before it reaches the focus and reflects it to a focus at the side of the telescope tube.

Newton's laws   The laws of mechanics and gravitation formulated by Isaac Newton.

node   The intersection of the orbit of a body with a fundamental plane—usually the plane of the celestial equator or of the ecliptic.

nongravitational force   The force that acts on comets to change their orbits, due not to the gravitational influence of the other members of the planetary system, but rather to the rocket effect of gases escaping from the cometary nucleus.

nonthermal radiation   See *synchrotron radiation*.

north point   That intersection of the celestial meridian and astronomical horizon lying nearest the north celestial pole.

nova   A star that experiences a sudden outburst of radiant energy, temporarily increasing its luminosity by hundreds to thousands of times.

nuclear   Referring to the nucleus of the atom.

nuclear bulge   Central part of our Galaxy.

nuclear transformation   Transformation of one atomic nucleus into another.

nuclear winter   A term coined in the early 1980s for the period of global environmental disturbance, including blockage of sunlight and rapid decline in surface temperature, that would occur following a major nuclear weapons exchange and the widespread fires that would result.

nucleosynthesis   The building up of heavy elements from lighter ones by nuclear fusion.

nucleus (of atom)   The heavy part of an atom, composed mostly of protons and neutrons, and about which the electrons revolve.

nucleus (of comet)   The solid chunk of ice and dust in the head of a comet.

nucleus (of galaxy)   Central concentration of matter at the center of a galaxy.

objective   The principal image-forming component of a telescope or other optical instrument.

oblate spheroid   A solid formed by rotating an ellipse about its minor axis.

oblateness   A measure of the "flattening" of an oblate spheroid; numerically, the ratio of the difference between the major and minor diameters (or axes) to the major diameter (or axis).

obliquity   The angle between the equator of a planet and the plane of its orbit; hence the tilt of the rotational axis of a planet.

obliquity of the ecliptic   Angle between the planes of the celestial equator and the ecliptic; about 23.5°.

occultation   The passage of an object of large angular size in front of a smaller object, such as the Moon in front of a distant star or the rings of Saturn in front of the Voyager spacecraft.

Oort comet cloud   The spherical region around the Sun from which most "new" comets come, representing objects with aphelia at about 50,000 AU, or extending about a third of the way to the nearest other stars.

opacity   Absorbing power; capacity to impede the passage of light.

open cluster   A comparatively loose or "open" cluster of stars, containing from a few dozen to a few thousand members, located in the spiral arms or disk of the Galaxy; sometimes referred to as a galactic cluster.

opposition   Configuration of a planet when its elongation is 180°.

optical double star   Two stars at different distances nearly lined up in projection so that they appear close together, but that are not really gravitationally associated.

optics   The branch of physics that deals with light and its properties.

orbit   The path of a body that is in revolution about another body or point.

oscillation   A periodic motion; in the case of the Sun, a periodic or quasi-periodic expansion and contraction of the whole Sun or some portion of it.

outgassing   The process by which the gases of a planetary atmosphere work their way out from the crust of the planet.

oxidizing   In chemistry, referring to conditions in which oxygen dominates over hydrogen, so that most other elements form compounds with oxygen. In very oxidizing conditions, such as are found in the atmosphere of the Earth, free oxygen gas (O_2) or even atomic oxygen (O) is present.

ozone A heavy molecule of oxygen that contains three atoms rather than the more normal two. Designated O_3.

Pangaea Name given to the hypothetical continent from which the present continents of the Earth separated.

parabola A conic section of eccentricity 1.0; the curve of the intersection between a circular cone and a plane parallel to a straight line in the surface of the cone.

parallactic ellipse A small ellipse that a comparatively nearby star appears to trace out in the sky, which results from the orbital motion of the Earth about the Sun.

parallax An apparent displacement of an object due to a motion of the observer.

parallax (stellar) An apparent displacement of a nearby star that results from the motion of the Earth around the Sun; numerically, the angle subtended by 1 AU at the distance of a particular star.

parent In referring to the process of radioactive decay, the name given to the radioactive isotope that is destroyed in the decay process to produce a new, or daughter, isotope.

parent body In planetary science, any larger original object that is the source of other objects, usually through breakup or ejection by impact cratering; for example, the asteroid 4 Vesta is thought to be the parent body of the eucrite meteorites.

parent molecules The original molecules (for example, H_2O, CO_2, CO, CH_4) that dissociate to form the radicals (C_2, CN, OH) actually observed in comets.

parsec The distance of an object that would have a stellar parallax of one second of arc; 1 parsec (pc) = 3.26 LY.

partial eclipse An eclipse of the Sun or Moon in which the eclipsed body does not appear completely obscured.

Pauli exclusion principle Quantum mechanical principle by which no two particles of the same kind can have the same position and momentum.

peculiar velocity The velocity of a star with respect to the local standard of rest; that is, its space motion, corrected for the motion of the Sun with respect to our neighboring stars.

penumbra The portion of a shadow from which only part of the light source is occulted by an opaque body.

penumbral eclipse A lunar eclipse in which the Moon passes through the penumbra, but not the umbra, of the Earth's shadow.

perfect cosmological principle The assumption that, on the large scale, the universe appears the same from every place and at all times.

perfect gas An "ideal" gas that obeys the perfect gas laws.

perfect gas laws Certain laws that describe the behavior of an ideal gas: Charles' law, Boyle's law, and the equation of state for a perfect gas.

perfect radiator Blackbody; a body that absorbs and subsequently re-emits all radiation incident upon it.

periastron The place in the orbit of a star in a binary-star system where it is closest to its companion star.

perigee The place in the orbit of an Earth satellite where it is closest to the center of the Earth.

perihelion The place in the orbit of an object revolving about the Sun where it is closest to the Sun's center.

period A time interval; for example, the time required for one complete revolution.

period-density relation Proportionality between the period and the inverse square root of the mean density for a pulsating star.

period-luminosity relation An empirical relation between the periods and luminosities of cepheid variable stars.

periodic comet A comet whose orbit has been determined to have an eccentricity of less than 1.0.

permafrost A region in the crust of a planet where the temperature is always below the freezing point of water; also, the frozen water mixed with soil that is usually found in such cold regions.

perturbation The disturbing effect, when small, on the motion of a body as predicted by a simple theory, produced by a third body or other external agent.

photometry The measurement of light intensities.

photon A discrete unit of electromagnetic energy.

photon sphere A surface surrounding a black hole, of radius about 1.4 times that of the event horizon, where a photon can have a closed circular orbit.

photosphere The region of the solar (or a stellar) atmosphere from which continuous radiation escapes into space.

pixel An individual picture element in a detector; for example, a particular silicon diode in a CCD or a grain in a photographic emulsion.

plage A bright region of the solar surface observed in the monochromatic light of some spectral line.

Planck's constant The constant of proportionality relating the energy of a photon to its frequency.

planet Any of the nine largest bodies revolving about the Sun, or any similar non–self-luminous bodies that may orbit other stars.

planetarium An optical device for projecting on a screen or domed ceiling the stars and planets and their apparent motions in the sky.

planetary nebula A shell of gas ejected from, and enlarging about, a certain kind of extremely hot star that is nearing the end of its life.

planetary system A term used in this text to refer to all of the solar system except the Sun: the planets, their satellites, rings, comets, asteroids, meteoroids, dust, and the solar wind.

planetesimals The hypothetical objects, from tens to hundreds of kilometers in diameter, that formed in the solar nebula as an intermediate step between tiny grains and the larger planetary objects we see today. The comets and some asteroids may be leftover planetesimals.

plasma A hot ionized gas.

plasma tail (of a comet) A cometary tail, usually narrow and bluish in color, extending straight away from the Sun and consisting of plasma streaming away from the head under the influence of the solar wind. Also called an ion tail.

plate tectonics The motion of segments or plates of the outer layer of the Earth over the underlying mantle.

polar axis The axis of rotation of the Earth; also, an axis in the mounting of a telescope that is parallel to the Earth's axis.

Population I and II Two classes of stars (and systems of stars), classified according to their spectral characteristics, chemical compositions, radial velocities, ages, and locations in the Galaxy.

position angle Direction in the sky of one celestial object from another; for example, the angle, measured to the east from the north, of the fainter component of a visual binary star in relation to the brighter component.

positron An electron with a positive rather than negative charge; an antielectron.

postulate An essential prerequisite to a hypothesis or theory.

potential energy Stored energy that can be converted into other forms; especially gravitational energy.

precession (of Earth) A slow, conical motion of the Earth's axis of rotation, caused principally by the gravitational torque of the Moon and Sun on the Earth's equatorial bulge. Lunisolar precession, precession caused by the Moon and Sun only; planetary precession, a slow change in the orientation of the plane of the Earth's orbit caused by planetary perturbations; general precession, the combination of these two effects on the motion of the Earth's axis with respect to the stars.

precession of the equinoxes Slow westward motion of the equinoxes along the ecliptic that results from precession.

pressure Force per unit area; expressed in units of atmospheres or pascals.

prime focus The point in a telescope where the objective focuses the light.

prime meridian The terrestrial meridian passing through the site of the old Royal Greenwich Observatory; longitude 0°.

primitive In planetary science and meteoritics, an object or rock that is little changed, chemically, since its formation, and hence representative of the conditions in the solar nebula at the time of formation of the solar system. Also used to refer to the chemical composition of an atmosphere that has not undergone extensive chemical evolution.

primitive meteorite A meteorite that has not been greatly altered chemically since its condensation from the solar nebula; called in meteoritics a chondrite (either ordinary chondrite or carbonaceous chondrite).

primitive rock Any rock that has not experienced great heat or pressure and therefore remains representative of the original condensates from the solar nebula—never found on any object large enough to have undergone melting and differentiation.

principle of equivalence Principle that a gravitational force and a suitable acceleration are indistinguishable within a sufficiently local environment.

principle of relativity Principle that all observers in uniform relative motion are equivalent; the laws of nature are the same for all, and no experiment can reveal an absolute motion or state of rest.

prism A wedge-shaped piece of glass that is used to disperse white light into a spectrum.

prolate spheroid The solid produced by the rotation of an ellipse about its major axis.

prominence A phenomenon in the solar corona that commonly appears like a flame above the limb of the Sun.

proper motion The angular change per year in the direction of a star as seen from the Sun.

proton A heavy subatomic particle that carries a positive charge; one of the two principal constituents of the atomic nucleus.

proton-proton cycle A series of thermonuclear reactions by which nuclei of hydrogen are built up into nuclei of helium.

protoplanet (or -star or -galaxy) The original material from which a planet (or a star or galaxy) condensed.

pulsar A variable radio source of small angular size that emits radio pulses in very regular periods that range from 0.03 to 5 s.

pulsating variable A variable star that pulsates in size and luminosity.

q_0 See *deceleration parameter* (q_0).

quadrature A superior planet is at quadrature when a line from the Earth to the Sun makes a right angle with the line from the Earth to the planet.

quantum efficiency The ratio of the number of photons incident on a detector to the number actually detected.

quantum mechanics The branch of physics that deals with the structure of atoms and their interactions with one another and with radiation.

quark A hypothetical subatomic particle. Quarks of six different kinds, in various combinations, are presumed to make up all other particles in the atomic nucleus.

quarter moon Either of the two phases of the Moon when its longitude differs by 90° from that of the Sun; the Moon appears half full at these phases.

quasar A stellar-appearing object of very high redshift, presumed to be extragalactic and highly luminous; an active galactic nucleus.

RR Lyrae variable One of a class of giant pulsating stars with periods less than one day.

radar The technique of transmitting radio waves to an object and then detecting the radiation that the object reflects back to the transmitter; used to measure the distance to, and motion of, a target object.

radial velocity The component of relative velocity that lies in the line of sight.

radial velocity curve A plot of the variation of radial velocity with time for a binary or variable star.

radiant (of meteor shower) The point in the sky from which the meteors belonging to a shower seem to radiate.

radiation A mode of energy transport whereby energy is transmitted through a vacuum; also the transmitted energy itself.

radiation pressure The transfer of momentum carried by electromagnetic radiation to a body that the radiation impinges upon.

radical A bond of two or more atoms that does not, in itself, constitute a molecule, but that has characteristics of its own and enters into chemical reactions as if it were a single atom.

radio astronomy The technique of making astronomical observations in radio wavelengths.

radio galaxy A galaxy that emits greater amounts of radio radiation than average.

radio telescope A telescope designed to make observations in radio wavelengths.

radioactive dating The technique of determining the ages of rocks or other specimens by the amount of radioactive decay of certain radioactive elements contained therein.

radioactivity (radioactive decay) The process by which certain kinds of atomic nuclei naturally decompose, with the spontaneous emission of subatomic particles and gamma rays.

Rayleigh scattering Scattering of light (photons) by molecules of a gas.

red giant A large, cool star of high luminosity; a star occupying the upper right portion of the Hertzsprung-Russell diagram.

reddening (interstellar) The reddening of starlight passing through interstellar dust, caused because dust scatters blue light more effectively than red.

redshift A shift to longer wavelengths of the light from remote galaxies; presumed to be produced by a Doppler shift.

reducing In chemistry, referring to conditions in which hydrogen dominates over oxygen, so that most other elements form compounds with hydrogen. In very reducing conditions free hydrogen (H_2) is present and free oxygen (O_2) cannot exist.

reflecting telescope A telescope in which the principal optical component (objective) is a concave mirror.

reflection The return of light rays by an optical surface.

reflection nebula A relatively dense dust cloud in interstellar space that is illuminated by starlight.

reflectivity Measure of the brightness of an object relative to a sphere of equal size made of a perfectly reflecting diffuse white material (also known as the geometric albedo).

refracting telescope A telescope in which the principal optical component (objective) is a lens or system of lenses.

refraction The bending of light rays passing from one transparent medium (or a vacuum) to another.

regolith The pulverized surface soil of the Moon (or any airless body) produced by meteorite impacts.

regular satellites Planetary satellites that have orbits of low or moderate eccentricity in approximately the plane of the planet's equator.

relative orbit The orbit of one of two mutually revolving bodies referred to the other body as origin.

relativistic particle (or electron) A particle (electron) moving at nearly the speed of light.

relativity A theory formulated by Einstein that describes the relations between measurements of physical phenomena by two different observers who are in relative motion at constant velocity (the special theory of relativity) or that describes how a gravitational field can be replaced by a curvature of spacetime (the general theory of relativity).

resolution The degree to which fine details in an image are separated, or resolved.

resolving power A measure of the ability of an optical system to resolve, or separate, fine details in the image it produces; in astronomy, the angle in the sky that can be resolved by a telescope.

resonance An orbital condition in which one object is subject to periodic gravitational perturbations by another, most commonly arising when two objects orbiting a third have periods of revolution that are simple multiples or fractions of each other.

rest mass The mass of an object or particle as measured when it is at rest in the laboratory.

retrograde (rotation or revolution) Backward with respect to the common direction of motion in the solar system; counterclockwise as viewed from the north, and going from east to west rather than from west to east.

retrograde motion An apparent westward motion of a planet on the celestial sphere or with respect to the stars.

revolution The motion of one body around another.

rift zone In geology, a place where the crust is being torn apart by internal forces, generally associated with the injection of new material from the mantle and with the slow separation of tectonic plates.

right ascension A coordinate for measuring the east-west positions of celestial bodies; the angle measured eastward along the celestial equator from the vernal equinox to the hour circle passing through a body.

Roche limit See *tidal stability limit*.

rotation Turning of a body about an axis running through it.

runaway greenhouse effect A process whereby the heating of a planet leads to an increase in its atmospheric greenhouse effect and thus to further heating, thereby quickly altering the composition of its atmosphere and the temperature of its surface.

Russell-Vogt theorem The theorem that the mass and chemical composition of a star determine its entire structure if it derives its energy entirely from thermonuclear reactions.

satellite A body that revolves about a planet.

scale (of telescope) The linear distance in the image corresponding to a particular angular distance in the sky; say, so many centimeters per degree.

Schwarzschild radius See *event horizon*.

science The attempt to find order in nature or to find laws that describe natural phenomena.

scientific method A specific procedure in science: (1) the observation of phenomena or the results of experiments; (2) the formulation of hypotheses that describe these phenomena and that are consistent with the body of knowledge available; (3) the testing of these hypotheses by noting whether or not they adequately predict and describe new phenomena or the results of new experiments.

secondary minimum (in an eclipsing binary light curve) The middle of the eclipse of the cooler star by the hotter, in which the light of the system diminishes less than during the eclipse of the hotter star by the cooler.

secular Not periodic.

secular parallax A mean parallax for a selection of stars, derived from the components of their proper motions that reflect the motion of the Sun.

sedimentary rock Any rock formed by the deposition and cementing of fine grains of material. On Earth, sedimentary rocks are usually the result of erosion and weathering, followed by deposition in lakes or oceans; however, breccias formed on the Moon by impact processes are also considered sedimentary rocks.

seeing The unsteadiness of the Earth's atmosphere, which blurs telescopic images.

seismic waves Vibrations traveling through the Earth's interior that result from earthquakes.

seismology (geology) The study of earthquakes and the conditions that produce them and of the internal structure of the Earth as deduced from analyses of seismic waves.

seismology (solar) The study of small changes in the radial velocity of the Sun as a whole or of small regions on the surface of the Sun. Analyses of these velocity changes can be used to infer the internal structure of the Sun.

semimajor axis   Half the major axis of a conic section.

separation (in a visual binary)   The angular separation of the two components of a visual binary star.

SETI   The search for extraterrestrial intelligence, usually applied to searches for radio signals from other civilizations.

Seyfert galaxy   A galaxy belonging to the class of those with active galactic nuclei; one whose nucleus shows bright emission lines; one of a class of galaxies first described by C. Seyfert.

shadow cone   The umbra of the shadow of a spherical body (such as the Earth) in sunlight.

shepherd satellite   Informal term for a satellite that is thought to maintain the structure of a planetary ring through its close gravitational influence—specifically, the two Saturn satellites, Prometheus and Pandora, that orbit just inside and outside the F ring.

shield volcano   A broad volcano built up through the repeated nonexplosive eruption of fluid basalts to form a low dome or shield shape, typically with slopes of only 4° to 6°, often with a large caldera at the summit. Examples include the Hawaiian volcanoes on Earth and the Tharsis volcanoes on Mars.

shower (meteor)   Many meteors, all seeming to radiate from a common point in the sky, caused by the encounter by the Earth of a swarm of meteoroids moving together through space.

sidereal day   The interval between two successive meridian passages of the vernal equinox.

sidereal month   The period of the Moon's revolution about the Earth with respect to the stars.

sidereal period   The period of revolution of one body about another with respect to the stars.

sidereal time   The local hour angle of the vernal equinox.

sidereal year   Period of the Earth's revolution about the Sun with respect to the stars.

sign (of zodiac)   Astrological term for any of 12 equal sections along the ecliptic, each of length 30°. Starting at the vernal equinox, and moving eastward, the signs are Aries, Taurus, Gemini, Cancer, Leo, Virgo, Libra, Scorpio, Sagittarius, Capricorn, Aquarius, and Pisces.

simultaneity   The occurrence of two events at the same time. In relativity, absolute simultaneity is seen not to have meaning, except for two simultaneous events occurring at the same place.

sine (of angle)   One of the trigonometric functions; the sine of an angle (in a right triangle) is the ratio of the length of the side opposite the angle to that of the hypotenuse.

sine curve   A graph of the sine of an angle plotted against the angle.

singularity   A theoretical point of zero volume and infinite density to which any object that becomes a black hole must collapse, according to the general theory of relativity.

small circle   Any circle on the surface of a sphere that is not a great circle.

SNC meteorite   One of a class of basaltic meteorites now believed by many planetary scientists to be impact-ejected fragments from Mars.

solar activity   Phenomena of the solar atmosphere: sunspots, plages, and related phenomena.

solar antapex   Direction away from which the Sun is moving with respect to the local standard of rest.

solar apex   The direction toward which the Sun is moving with respect to the local standard of rest.

solar constant   Mean amount of solar radiation received per unit time, by a unit area, just outside the Earth's atmosphere, and perpendicular to the direction of the Sun; the numerical value is 1370 watts/m².

solar eclipse   An eclipse of the Sun by the Moon, caused by the passage of the Moon in front of the Sun. Solar eclipses can occur only at the time of new moon.

solar motion   Motion of the Sun, or the velocity of the Sun, with respect to the local standard of rest.

solar nebula   The cloud of gas and dust from which the solar system formed.

solar parallax   Angle subtended by the equatorial radius of the Earth at a distance of 1 AU.

solar seismology   The study of pulsations or oscillations of the Sun in order to determine the characteristics of the solar interior, including composition and rotation rate.

solar system   The system of the Sun and the planets, their satellites, the minor planets, comets, meteoroids, and other objects revolving around the Sun.

solar time   A time based on the Sun; usually the hour angle of the Sun plus 12 hr.

solar wind   A radial flow of plasma leaving the Sun.

solidification age   The most common age determined by radioactive dating techniques; the time since the rock or mineral grain being tested solidified from the molten state, thus isolating itself from further chemical changes.

solstice   Either of two points on the celestial sphere where the Sun reaches its maximum distances north and south of the celestial equator.

space velocity or space motion   The velocity of a star with respect to the Sun.

spacetime   A system of one time and three spatial coordinates, with respect to which the time and place of an event can be specified; also called spacetime continuum.

spectral class (or type)   A classification of a star according to the characteristics of its spectrum.

spectral line   Radiation at a particular wavelength of light produced by the emission or absorption of energy by an atom.

spectral sequence   The sequence of spectral classes of stars arranged in order of decreasing temperatures of stars of those classes.

spectroheliogram   A picture of the Sun or a part of the Sun obtained in the monochromatic light of a single spectral line.

spectrometer   An instrument for obtaining a spectrum; in astronomy, usually attached to a telescope to record the spectrum of a star, galaxy, or other astronomical object.

spectrophotometry   The measurement of the intensity of light from a star or other source at different wavelengths.

spectroscopic binary star   A binary star in which the components are not resolved optically, but whose binary nature is indicated by periodic variations in radial velocity, indicating orbital motion.

spectroscopic parallax   A parallax (or distance) of a star that is derived by comparing the apparent magnitude of the star with its absolute magnitude as deduced from its spectral characteristics.

spectroscopy   The study of spectra.

spectrum   The array of colors or wavelengths obtained when light from a source is dispersed, as in passing it through a prism or grating.

spectrum analysis   The study and analysis of spectra, especially stellar spectra.

speed   The rate at which an object moves without regard to its direction of motion; the numerical or absolute value of velocity.

spicule   A jet of rising material in the solar chromosphere.

spiral arms   Arms of interstellar material and young stars that wind out in a plane from the central nucleus of a spiral galaxy.

spiral density wave   A mechanism for the generation of spiral structure in galaxies; a density wave interacts with interstellar matter and triggers the formation of stars. Spiral density waves are also seen in the rings of Saturn.

spiral galaxy   A flattened, rotating galaxy with pinwheel-like arms of interstellar material and young stars winding out from its nucleus.

spring tide   The highest tidal range of the month, produced when the Moon is near either the full or the new phase.

sputtering   The process by which energetic atomic particles striking a solid alter its chemistry and eject additional atoms or molecular fragments from the surface.

standard candle   An astronomical object of known luminosity; such an object can be used to determine distances.

standard time   The local mean solar time of a standard meridian, adopted over a large region to avoid the inconvenience of continuous time changes around the Earth.

star   A self-luminous sphere of gas.

star cluster   An assemblage of stars held together by their mutual gravitation.

statistical parallax   The mean parallax for a selection of stars, derived from the radial velocities of the stars and the components of their proper motions that cannot be affected by the solar motion.

steady state (theory of cosmology)   A theory of cosmology embracing the perfect cosmological principle and involving the continuous creation of matter.

Stefan-Boltzmann law   A formula from which the rate at which a blackbody radiates energy can be computed; the total rate of energy emission from a unit area of a blackbody is proportional to the fourth power of its absolute temperature.

stellar evolution   The changes that take place in the sizes, luminosities, structures, and so on, of stars as they age.

stellar model   The result of a theoretical calculation of the run of physical conditions in a stellar interior.

stellar parallax   The angle subtended by 1 AU at the distance of a star; usually measured in seconds of arc.

stellar wind   The outflow of gas, sometimes at speeds as high as hundreds of kilometers per second, from a star.

stony meteorite   A meteorite composed mostly of stony material.

stony-iron (meteorite)   A type of meteorite that is a blend of nickel-iron and silicate materials.

stratosphere   The layer of the Earth's atmosphere above the troposphere (where most weather takes place) and below the ionosphere.

Strömgren sphere   A region of ionized gas in interstellar space surrounding a hot star; an H II region.

strong nuclear force or strong interaction   The force that binds together the parts of the atomic nucleus.

subduction zone   In terrestrial geology, a region where one crustal plate is forced under another, generally associated with earthquakes, volcanic activity, and the formation of deep ocean trenches.

submillimeter (part of the electromagnetic spectrum)   That part of the electromagnetic spectrum with wavelengths of a few hundred micrometers.

subtend   To have or include a given angular size.

summer solstice   The point on the celestial sphere where the Sun reaches its greatest distance north of the celestial equator.

Sun   The star about which the Earth and other planets revolve.

sunspot   A temporary cool region in the solar photosphere that appears dark by contrast against the surrounding hotter photosphere.

sunspot cycle   The semiregular 11-year period with which the frequency of sunspots fluctuates.

supercluster   A large region of space (50 to 100 million pc across) where matter is concentrated into galaxies, groups of galaxies, and clusters of galaxies; a cluster of clusters of galaxies.

supergiant   A star of very high luminosity.

superior conjunction   The configuration of a planet in which it and the Sun have the same longitude, with the planet being more distant than the Sun.

superior planet   A planet more distant from the Sun than is the Earth.

supernova   An explosion that marks the final stage of evolution of a star. A Type I supernova is thought to occur when a white dwarf accretes enough matter to exceed the Chandrasekhar limit, collapses, and explodes. A Type II supernova is thought to mark the final collapse of a massive star.

surface gravity   The weight of a unit mass at the surface of a body.

surveying   The technique of measuring distances and relative positions of places over the surface of the Earth (or elsewhere); generally accomplished by triangulation.

synchrotron radiation   The radiation emitted by charged particles being accelerated in magnetic fields and moving at speeds near that of light.

synodic month   The period of revolution of the Moon with respect to the Sun, or its cycle of phases.

synodic period   The interval between successive occurrences of the same configuration of a planet; for example, between successive oppositions or successive superior conjunctions.

tail (of a comet)   See *dust tail (of a comet)* and *plasma tail (of a comet)*.

tangent (of angle)   One of the trigonometric functions; the tangent of an angle (in a right triangle) is the ratio of the length of the side opposite the angle to that of the shorter of the adjacent sides.

tangential (transverse) velocity The component of a star's space velocity that lies in the plane of the sky.

tau component (of proper motion) The component of a star's proper motion that lies perpendicular to a great circle passing through the star and the solar apex.

tectonic Activity and motion that result from expansion and contraction of the crust of a planet. See *plate tectonics.*

tektites Small, rounded glassy bodies found on the Earth, apparently ejecta from major impact craters.

telescope An optical instrument used to aid in viewing or measuring distant objects.

temperature (absolute or Kelvin) Temperature measured in centigrade (Celsius) degrees from absolute zero.

temperature (Celsius; formerly centigrade) Temperature measured on scale where water freezes at 0° and boils at 100°.

temperature (color) The temperature of a star as estimated from the intensity of the stellar radiation at two or more colors or wavelengths.

temperature (effective) The temperature of a blackbody that would radiate the same total amount of energy that a particular object, such as a star, does.

temperature (excitation) The temperature of a star as estimated from the relative strengths of lines in its spectrum that originate from atoms in different stages of excitation.

temperature (Fahrenheit) Temperature measured on a scale where water freezes at 32° and boils at 212°.

temperature (ionization) The temperature of a star as estimated from the relative strengths of lines in its spectrum that originate from atoms in different stages of ionization.

temperature (Kelvin) Absolute temperature measured in Celsius degrees.

temperature (kinetic) A measure of the mean energy of the molecules in a substance.

temperature (radiation) The temperature of a blackbody that radiates the same amount of energy in a given spectral region as does a particular body.

terrestrial planet Any of the planets Mercury, Venus, Earth, Mars, and sometimes Pluto.

theory A set of hypotheses and laws that have been well demonstrated to apply to a wide range of phenomena associated with a particular subject.

thermal energy Energy associated with the motions of the molecules in a substance.

thermal equilibrium A balance between the input and outflow of heat in a system.

thermal radiation The radiation emitted by any body or gas that is not at absolute zero.

thermodynamics The branch of physics that deals with heat and heat transfer among bodies.

thermonuclear energy Energy associated with thermonuclear reactions or that can be released through thermonuclear reactions.

thermonuclear reaction A nuclear reaction or transformation that results from encounters between nuclear particles that are given high velocities (by heating them).

thermosphere The region of the Earth's atmosphere lying between the mesosphere and the exosphere.

tidal force A differential gravitational force that tends to deform a body.

tidal heating Generation of heat in a planetary object through repeated tidal stresses from a larger nearby object, probably important for the Moon early in its history and responsible today for the high level of volcanic activity on Io.

tidal stability limit The distance—approximately 2.5 planetary radii from the center—within which differential gravitational forces (or tides) are stronger than the mutual gravitational attraction between two adjacent orbiting objects. Within this limit, fragments are not likely to accrete or assemble themselves into a larger object. Also called the Roche limit.

tide Deformation of a body by the differential gravitational force exerted on it by another body; in the Earth, the deformation of the ocean surface by the differential gravitational forces exerted by the Moon and Sun.

ton (metric) One million grams (2204.6 lb).

total eclipse An eclipse of the Sun in which the Sun's photosphere is entirely hidden by the Moon, or an eclipse of the Moon in which it passes completely into the umbra of the Earth's shadow.

transit An instrument for timing the exact instant a star or other object crosses the local meridian. Also, the passage of a celestial body across the meridian; or the passage of a small body (say, a planet) across the disk of a large one (say, the Sun).

transition region The region in the Sun's atmosphere where the temperature rises very rapidly from the relatively low temperatures that characterize the chromosphere to the high temperatures of the corona.

triangulation The operation of measuring some of the elements of a triangle so that other ones can be calculated by the methods of trigonometry, thus determining distances to remote places without having to span them directly.

trigonometry The branch of mathematics that deals with the analytical solutions of triangles.

triple-alpha process A series of two nuclear reactions by which three helium nuclei are built up into one carbon nucleus.

Trojan asteroid One of a large number of asteroids that share Jupiter's orbit about the Sun, but either preceding or following Jupiter by 60°.

Tropic of Cancer Parallel of latitude 23.5° N.

Tropic of Capricorn Parallel of latitude 23.5° S.

tropical year Period of revolution of the Earth about the Sun with respect to the vernal equinox.

troposphere Lowest level of the Earth's atmosphere, where most weather takes place.

turbulence Random motions of gas masses, as in the atmosphere of a star.

21-cm line A spectral line of neutral hydrogen at the radio wavelength of 21 cm.

ultraviolet radiation Electromagnetic radiation of wavelengths shorter than the shortest (violet) wavelengths to which the eye is sensitive; radiation of wavelengths in the approximate range 10 to 400 nm.

umbra The central, completely dark part of a shadow.

uncertainty principle See *Heisenberg uncertainty principle.*

uncompressed density The density that a planetary object would have if it were not subject to self-compression from its own gravity, hence the density that is characteristic of its bulk material independent of the size of the object.

universal time   The local mean time of the prime meridian.

universe   The totality of all matter and radiation and the space occupied by same.

upsilon component (of proper motion)   The component of a star's proper motion that lies along a great circle passing through the star and the solar apex.

variable star   A star that varies in luminosity.

vector   A quantity that has both magnitude and direction.

velocity   A vector that denotes both the speed and the direction a body is moving.

velocity of escape   The speed with which an object must move in order to enter a parabolic orbit about another body (such as the Earth), and hence move permanently away from the vicinity of that body.

vernal equinox   The point on the celestial sphere where the Sun crosses the celestial equator passing from south to north.

very-long-baseline interferometry (VLBI)   A technique of radio astronomy whereby signals from telescopes thousands of kilometers apart are combined to obtain very high resolution with interferometry.

virial theorem   A relation between the potential and kinetic energies of a system of mutually gravitating bodies in statistical equilibrium.

visual binary star   A binary star in which the two components are telescopically resolved.

volatile materials   Materials that are gaseous at fairly low temperatures. This is a relative term, usually applied to the gases in planetary atmospheres and to common ices (H_2O, CO_2, and so on), but it is also sometimes used for elements such as cadmium, zinc, lead, and rubidium that form gases at temperatures up to 1000 K. (These are called volatile elements, as opposed to refractory elements.)

volume   A measure of the total space occupied by a body.

wandering of the poles   A semiperiodic shift of the body of the Earth relative to its axis of rotation; responsible for variation of latitude.

watt   A unit of power.

wavelength   The spacing of the crests or troughs in a wave train.

weak nuclear force or weak interaction   The nuclear force involved in radioactive decay. The weak force is characterized by the slow rate of certain nuclear reactions—such as the decay of the neutron, which occurs with a half-life of 11 min.

weather   The state of a planetary atmosphere—its composition, temperature, pressure, motion, and so on—at a particular place and time.

weight   A measure of the force due to gravitational attraction.

white dwarf   A star that has exhausted most or all of its nuclear fuel and has collapsed to a very small size; such a star is near its final stage of evolution.

Wien's law   Formula that relates the temperature of a blackbody to the wavelength at which it emits the greatest intensity of radiation.

winter solstice   Point on the celestial sphere where the Sun reaches its greatest distance south of the celestial equator.

Wolf-Rayet star   One of a class of very hot stars that eject shells of gas at very high velocity.

x rays   Photons of wavelengths intermediate between those of ultraviolet radiation and gamma rays.

x-ray stars   Stars (other than the Sun) that emit observable amounts of radiation at x-ray frequencies.

year   The period of revolution of the Earth around the Sun.

Zeeman effect   A splitting or broadening of spectral lines due to magnetic fields.

zenith   The point on the celestial sphere opposite to the direction of gravity; or the direction opposite to that indicated by a plumb bob.

zenith distance   Arc distance of a point on the celestial sphere from the zenith; 90° minus the altitude of the object.

zero-age main sequence   Main sequence for a system of stars that have completed their contraction from interstellar matter and are now deriving all their energy from nuclear reactions, but whose chemical composition has not yet been altered by nuclear reactions.

zodiac   A belt around the sky 18° wide centered on the ecliptic.

zodiacal light   A faint illumination along the zodiac, which is due to sunlight that has been reflected and scattered by interplanetary dust.

zone of avoidance   A region near the Milky Way where obscuration by interstellar dust is so heavy that few or no exterior galaxies can be seen.

POWER OF TEN NOTATION

In astronomy and other sciences, it is often necessary to deal with very large or very small numbers. For example, the Earth is 150,000,000,000 m from the Sun, and the mass of the hydrogen atom is 0.00000000000000000000000000167 kg. Instead of writing and carrying so many zeros, the numbers are usually written as figures between 1 and 10 multiplied by the appropriate power of 10. For example, 150,000,000,000 is 1.5×10^{11}, and the mass of the hydrogen atom given above is written simply as 1.67×10^{-27} kg. Additional examples are given below:

one hundredth	=	0.01	$= 10^{-2}$
one tenth	=	0.1	$= 10^{-1}$
one	=	1	$= 10^{0}$
ten	=	10	$= 10^{1}$

one hundred	=	100	$= 10^{2}$
one thousand	=	1000	$= 10^{3}$
one million	=	1,000,000	$= 10^{6}$
one billion	=	1,000,000,000	$= 10^{9}$

The powers-of-ten notation is not only compact and convenient, it also simplifies arithmetic. To multiply two numbers expressed as powers of ten, you need only add the exponents. And to divide, you subtract the exponents. Following are several examples:

$$100 \times 100{,}000 = 10^{2} \times 10^{5} = 10^{2+5} = 10^{7}$$
$$0.01 \times 1{,}000{,}000 = 10^{-2} \times 10^{6} = 10^{6-2} = 10^{4}$$
$$1{,}000{,}000 \div 1000 = 10^{6} \div 10^{3} = 10^{6-3} = 10^{3}$$
$$100 \div 1{,}000{,}000 = 10^{2} \div 10^{6} = 10^{2-6} = 10^{-4}$$

4

UNITS

In the American system of measure (originally developed in England), the fundamental units of length, mass, and time are the yard, pound, and second, respectively. There are also, of course, larger and smaller units, which include the ton (2240 lb), the mile (1760 yd), the rod (16$\frac{1}{2}$ ft), the inch ($\frac{1}{36}$ yd), the ounce ($\frac{1}{16}$ lb), and so on. Such units are inconvenient for conversion and arithmetic computation.

In science, therefore, it is more usual to use the metric system, which has been adopted in virtually all countries except the United States. The fundamental units of the metric system are

length: 1 meter (m)
mass: 1 kilogram (kg)
time: 1 second (s)

A meter was originally intended to be 1 ten-millionth of the distance from the equator to the North Pole along the surface of the Earth. It is about 1.1 yd. A kilogram is about 2.2 lb. The second is the same in metric and American units.

The most commonly used quantities of length and mass of the metric system are the following:

LENGTH

1 km	= 1 kilometer	= 1000 meters	= 0.6214 mile
1 m	= 1 meter	= 1.094 yards	= 39.37 inches
1 cm	= 1 centimeter	= 0.01 meter	= 0.3937 inch
1 mm	= 1 millimeter	= 0.001 meter	= 0.1 cm
			= 0.03937 inch
1 μm	= 1 micrometer	= 0.000 001 meter	= 0.0001 cm
1 nm	= 1 nanometer	= 10^{-9} meter	= 10^{-7} cm

also: 1 mile = 1.6093 km
   1 inch = 2.5400 cm

MASS

1 metric ton = 10^6 grams = 1000 kg = 2.2046×10^3 lb
1 kg = 1000 grams = 2.2046 lb
1 g = 1 gram = 0.0022046 lb = 0.0353 oz
1 mg = 1 milligram = 0.001 g

also: 1 lb = 0.4536 kg
   1 oz = 28.3495 g

*Celsius is now the name used for centigrade temperature; it has a more modern standardization but differs from the old centigrade scale by less than 0.1°.

Three temperature scales are in general use:

1. Fahrenheit (F); water freezes at 32°F and boils at 212°F.

2. Celsius or centigrade★ (C); water freezes at 0°C and boils at 100°C.

3. Kelvin or absolute (K); water freezes at 273 K and boils at 373 K.

All molecular motion ceases at $-459°F = -273°C = 0$ K. Thus, Kelvin temperature is measured from this lowest possible temperature, called *absolute zero*. It is the temperature scale most often used in astronomy. Kelvins are degrees that have the same value as centigrade or Celsius degrees, since the difference between the freezing and boiling points of water is 100 degrees in each.

On the Fahrenheit scale, water boils at 212 degrees and freezes at 32 degrees; the difference is 180 degrees. Thus, to convert Celsius degrees or Kelvins to Fahrenheit, it is necessary to multiply by 180/100 = 9/5. To convert from Fahrenheit to Celsius degrees or Kelvins, it is necessary to multiply by 100/180 = 5/9.

Example 1: What is 68°F in Celsius and in Kelvins?

$$68°F - 32°F = 36°F \text{ above freezing.}$$

$$\frac{5}{9} \times 36° = 20°;$$

thus,

$$68°F = 20°C = 293 \text{ K.}$$

Example 2: What is 37°C in Fahrenheit and in Kelvins?

$$37°C = 273° + 37° = 310 \text{ K;}$$

$$\frac{9}{5} \times 37° = 66.6 \text{ Fahrenheit degrees;}$$

thus,

$$37°C \text{ is } 66.6°F \text{ above freezing}$$

or

$$37°C = 32° + 66.6° = 98.6°F.$$

SOME USEFUL CONSTANTS

PHYSICAL CONSTANTS

speed of light	c	$= 2.9979 \times 10^8$ m/s
constant of gravitation	G	$= 6.672 \times 10^{-11}$ N m^2/kg^2
Planck's constant	h	$= 6.626 \times 10^{-34}$ joules
mass of hydrogen atom	m_H	$= 1.673 \times 10^{-27}$ kg
mass of electron	m_e	$= 9.109 \times 10^{-31}$ kg
charge of electron	e	$= 4.803 \times 10^{-10}$ eu
Rydberg constant	R	$= 1.0974 \times 10^7$ per m
Stefan-Boltzmann constant	σ	$= 5.670 \times 10^{-8}$ joule/m^2·deg^4
constant in Wien's law	$\lambda_{max}T$	$= 2.898 \times 10^{-3}$ m·deg
electron volt (energy)	eV	$= 1.602 \times 10^{-19}$ joules
energy equivalent of 1 ton TNT	E	$= 4.2 \times 10^9$ joules

ASTRONOMICAL CONSTANTS

astronomical unit	AU	$= 1.496 \times 10^{11}$ m
light year	LY	$= 9.461 \times 10^{15}$ m
parsec	pc	$= 3.086 \times 10^{16}$ m
sidereal year	yr	$= 3.158 \times 10^7$ s
mass of Earth	M_E	$= 5.977 \times 10^{24}$ kg
equatorial radius of Earth	R_E	$= 6.378 \times 10^6$ m
obliquity of ecliptic	ϵ	$= 23°\ 27'$
surface gravity of Earth	g	$= 9.807$ m/s^2
escape velocity of Earth	v_E	$= 1.119 \times 10^4$ m/s
age of Earth	A_E	$= 4.55 \times 10^9$ yr
mass of Sun	M_S	$= 1.989 \times 10^{30}$ kg
equatorial radius of Sun	R_S	$= 6.960 \times 10^8$ m
luminosity of Sun	L_S	$= 3.83 \times 10^{26}$ watts
solar constant (at Earth)	S	$= 1.37 \times 10^3$ watts/m^2
Hubble constant	H	$= 75 \pm 15$ km/s/mpc
Age of "empty" universe	1/H	$= 1.3 \times 10^{10}$ yr
Critical density of universe	ρ	$= 10^{-29}$ g/cm^3 (approximately)

ASTRONOMICAL COORDINATE SYSTEMS

Several astronomical coordinate systems are in common use. In each of these systems the position of an object in the sky, or on the celestial sphere, is denoted by two angles. These angles are referred to as a *reference plane,* which contains the observer, and a *reference direction,* which is a direction from the observer to some arbitrary point lying in the reference plane. The intersection of the reference plane and the celestial sphere is a great circle, which defines the "equator" of the coordinate system. At two points, each 90° from this equator, are the "poles" of the coordinate system. Great circles passing through these poles intersect the equator of the system at right angles.

One of the two angular coordinates of each coordinate system is measured from the equator of the system to the object along the great circle passing through it and the poles. Angles on one side of the equator (or reference plane) are reckoned as positive; those on the opposite are negative. The other angular coordinate is measured along the equator from the reference direction to the intersection of the equator with the great circle passing through the object and the poles.

The system of terrestrial latitude and longitude provides an excellent analogue. Here the plane of the terrestrial equator is the fundamental plane, and the Earth's equator is the equator of the system; the North and South terrestrial Poles are the poles of the system. One coordinate, the *latitude* of a place, is reckoned north (positive) or south (negative) of the equator along a meridian passing through the place. The other coordinate, *longitude,* is measured along the equator to the intersection of the equator and the meridian of the place from the intersection of the equator and the Greenwich meridian. The direction (from the center of the Earth) to this latter intersection is the reference direction. Terrestrial longitude is either east or west (whichever is less), but the corresponding coordinate in celestial systems is generally reckoned in one direction from 0 to 360° (or, equivalently, from 0 to 24^h).

The following table lists the more important astronomical coordinate systems and defines how each of the angular coordinates is defined.

System	Reference Plane	Reference Direction	"Latitude" Coordinate	Range	"Longitude" Coordinate	Range
Horizon	Horizon plane	North point (formerly the south point was used by astronomers)	Altitude, h; toward the zenith (+) toward the nadir (−)	±90°	Azimuth, A; measured to the east along the horizon from the north point	0 to 360°
Equator	Plane of the celestial equator	Vernal equinox	Declination, δ; toward the north celestial pole (+) toward the south celestial pole (−)	±90°	Right ascension, α or R.A.; measured to the east along the celestial equator from the vernal equinox	0 to 24^h
Ecliptic	Plane of the Earth's orbit (ecliptic)	Vernal equinox	Celestial latitude, β; toward the north ecliptic pole (+) toward the south ecliptic pole (−)	±90°	Celestial longitude, λ; measured to the east along the ecliptic from the vernal equinox	0 to 360°
Galactic	Mean plane of the Milky Way	Direction to the galactic center	Galactic latitude, b; toward the north galactic pole (+) toward the south galactic pole (−)	±90°	Galactic longitude, l; measured along the galactic equator to the east from the galactic center	0 to 360°

ELEMENTS OF AN ORBIT

The elements of an orbit are those numbers needed to specify both the nature of the orbit and the location of an object in its orbit at any time. Two numbers (usually eccentricity and semimajor axis) are needed to describe the size and shape of the orbit. Three other numbers are needed to specify the orbit's orientation in space. The final two elements give the orbital period and specify where the object is at some particular time, so that its location at other times can be computed.

The table on this page lists the set of elements that is conventional for describing orbits of objects revolving about the Sun. However, other equivalent data can also specify an orbit and may be more convenient for calculation, depending on the exact problem addressed.

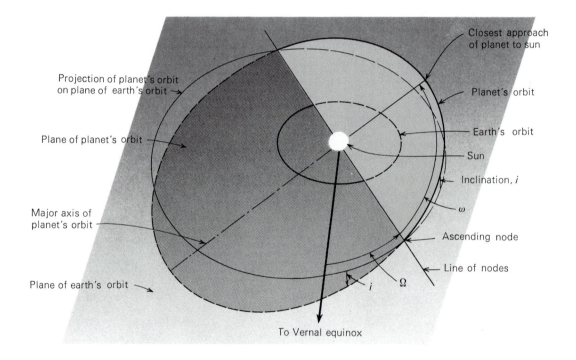

Name	Symbol	Definition
Semimajor axis	a	Half of the distance between the points nearest the foci on the conic that represents the orbit (usually measured in astronomical units).
Eccentricity	e	Distance between the foci of the conic divided by the major axis.
Inclination	i	Angle of intersection between the orbital planes of the object and of the Earth.
Longitude of the ascending node	Ω	Angle from the vernal equinox (where the ecliptic and celestial equator intersect with the Sun crossing the equator from south to north), measured to the east along the ecliptic plane, to the point where the object crosses the ecliptic traveling from south to north (the ascending node).
Argument of perihelion	ω	Angle from the ascending node, measured in the plane of the object's orbit and in the direction of its motion, to the perihelion point (its closest approach to the Sun).
Time of perihelion passage	T	One of the precise times that the object passed the perihelion point.
Period	P	The sidereal period of revolution of the object about the Sun.

SOME NUCLEAR REACTIONS OF IMPORTANCE IN ASTRONOMY

Given here are the series of thermonuclear reactions that are most important in stellar interiors. The subscript to the left of a nuclear symbol is the atomic number; the superscript to the left is the atomic mass number. The symbols for the positive electron (positron) and electron are e^+ and e^-, respectively, for the neutrino is ν, and for a photon (generally of gamma-ray energy) is γ.

1. THE PROTON-PROTON CHAINS

(Important below 15×10^6 K)

There are three ways the proton-proton chain can be completed. The first (a_1, b_1, c_1) is the most important, but depending on the physical conditions in the stellar interior, some energy is released by one or both of the following alternatives: a_1, b_1, c_2, d_2, e_2, and a_1, b_1, c_2, d_3, e_3, f_3.

(a_1) $^1_1\text{H} + ^1_1\text{H} \rightarrow ^2_1\text{H} + e^+ + \nu$
(b_1) $^2_1\text{H} + ^1_1\text{H} \rightarrow ^3_2\text{He} + \gamma$
(c_1) $^3_2\text{He} + ^3_2\text{He} \rightarrow ^4_2\text{He} + 2^1_1\text{H}$

or (c_2) $^3_2\text{He} + ^4_2\text{He} \rightarrow ^7_4\text{Be} + \gamma$
(d_2) $^7_4\text{Be} + e^- \rightarrow ^7_3\text{Li} + \nu$
(e_2) $^7_3\text{Li} + ^1_1\text{H} \rightarrow 2^4_2\text{He}$

or (d_3) $^7_4\text{Be} + ^1_1\text{H} \rightarrow ^8_5\text{B} + \gamma$
(e_3) $^8_5\text{B} \rightarrow ^8_4\text{Be} + e^+ + \nu$
(f_3) $^8_4\text{Be} \rightarrow 2^4_2\text{He}$

2. THE CARBON-NITROGEN CYCLE

(Important above 15×10^6 K)
(a) $^{12}_6\text{C} + ^1_1\text{H} \rightarrow ^{13}_7\text{N} + \gamma$
(b) $^{13}_7\text{N} \rightarrow ^{13}_6\text{C} + e^+ + \nu$
(c) $^{13}_6\text{C} + ^1_1\text{H} \rightarrow ^{14}_7\text{N} + \gamma$
(d) $^{14}_7\text{N} + ^1_1\text{H} \rightarrow ^{15}_8\text{O} + \gamma$
(e) $^{15}_8\text{O} \rightarrow ^{15}_7\text{N} + e^+ + \nu$
(f) $^{15}_7\text{N} + ^1_1\text{H} \rightarrow ^{12}_6\text{C} + ^4_2\text{He}$

3. THE TRIPLE-ALPHA PROCESS

(Important above 10^8 K)
(a) $^4_2\text{He} + ^4_2\text{He} \rightarrow ^8_4\text{Be} + \gamma$
(b) $^4_2\text{He} + ^8_4\text{Be} \rightarrow ^{12}_6\text{C} + \gamma$

ORBITAL DATA FOR THE PLANETS

Planet	Semimajor Axis		Sidereal Period		Mean Orbital Speed (km/s)	Orbital Eccentricity	Inclination of Orbit to Ecliptic (°)
	AU	10⁶ km	Tropical Years	Days			
Mercury	0.3871	57.9	0.24085	87.97	47.9	0.206	7.004
Venus	0.7233	108.2	0.61521	224.70	35.0	0.007	3.394
Earth	1.0000	149.6	1.000039	365.26	29.8	0.017	0.0
Mars	1.5237	227.9	1.88089	686.98	24.1	0.093	1.850
(Ceres)	2.7671	414	4.603		17.9	0.077	10.6
Jupiter	5.2028	778	11.86		13.1	0.048	1.308
Saturn	9.538	1427	29.46		9.6	0.056	2.488
Uranus	19.191	2871	84.07		6.8	0.046	0.774
Neptune	30.061	4497	164.82		5.4	0.010	1.774
Pluto	39.529	5913	248.6		4.7	0.248	17.15

Adapted from *The Astronomical Almanac* (U.S. Naval Observatory), 1981.

PHYSICAL DATA FOR THE PLANETS

Planet	Diameter		Mass (Earth = 1)	Mean Density (g/cm³)	Rotation Period (days)	Inclination of Equator to Orbit (°)	Surface Gravity (Earth = 1)	Velocity of Escape (km/s)
	(km)	(Earth = 1)						
Mercury	4878	0.38	0.055	5.43	58.6	0.0	0.38	4.3
Venus	12,104	0.95	0.82	5.24	−243.0	177.4	0.91	10.4
Earth	12,756	1.00	1.00	5.52	0.997	23.4	1.00	11.2
Mars	6794	0.53	0.107	3.9	1.026	25.2	0.38	5.0
Jupiter	142,796	11.2	317.8	1.3	0.41	3.1	2.53	60
Saturn	120,000	9.41	94.3	0.7	0.43	26.7	1.07	36
Uranus	52,400	4.11	14.6	1.3	−0.65	97.9	0.92	21
Neptune	50,450	3.81	17.2	1.5	0.72	29	1.18	24
Pluto	2200	0.17	0.0025	2.0	−6.387	118	0.09	1

SATELLITES OF THE PLANETS

Planet	Satellite Name	Discovery	Semimajor Axis (km × 1000)	Period (days)	Diameter (km)	Mass (10^{20} kg)	Density (g/cm³)
Earth	Moon	—	384	27.32	3476	735	3.3
Mars	Phobos	Hall (1877)	9.4	0.32	23	1×10^{-4}	2.0
	Deimos	Hall (1877)	23.5	1.26	13	2×10^{-5}	1.7
Jupiter	Metis	Voyager (1979)	128	0.29	20	—	—
	Adrastea	Voyager (1979)	129	0.30	40	—	—
	Amalthea	Barnard (1892)	181	0.50	200	—	—
	Thebe	Voyager (1979)	222	0.67	90	—	—
	Io	Galileo (1610)	422	1.77	3630	894	3.6
	Europa	Galileo (1610)	671	3.55	3138	480	3.0
	Ganymede	Galileo (1610)	1070	7.16	5262	1482	1.9
	Callisto	Galileo (1610)	1883	16.69	4800	1077	1.9
	Leda	Kowal (1974)	11,090	239	15	—	—
	Himalia	Perrine (1904)	11,480	251	180	—	—
	Lysithea	Nicholson (1938)	11,720	259	40	—	—
	Elara	Perrine (1905)	11,740	260	80	—	—
	Ananke	Nicholson (1951)	21,200	631 (R)	30	—	—
	Carme	Nicholson (1938)	22,600	692 (R)	40	—	—
	Pasiphae	Melotte (1908)	23,500	735 (R)	40	—	—
	Sinope	Nicholson (1914)	23,700	758 (R)	40	—	—
Saturn	Unnamed	Voyager (1985)	118.2	0.48	15?	3×10^{-5}	—
	Pan	Voyager (1985)	133.6	0.58	20	3×10^{-5}	—
	Atlas	Voyager (1980)	137.7	0.60	40	—	—
	Prometheus	Voyager (1980)	139.4	0.61	80	—	—
	Pandora	Voyager (1980)	141.7	0.63	100	—	—
	Janus	Dollfus (1966)	151.4	0.69	190	—	—
	Epimetheus	Fountain, Larson (1980)	151.4	0.69	120	—	—
	Mimas	Herschel (1789)	186	0.94	394	0.4	1.2
	Enceladus	Herschel (1789)	238	1.37	502	0.8	1.2
	Tethys	Cassini (1684)	295	1.89	1048	7.5	1.3
	Telesto	Reitsema et al. (1980)	295	1.89	25	—	—
	Calypso	Pascu et al. (1980)	295	1.89	25	—	—
	Dione	Cassini (1684)	377	2.74	1120	11	1.4
	Helene	Lecacheux, Laques (1980)	377	2.74	30	—	—
	Rhea	Cassini (1672)	527	4.52	1530	25	1.3
	Titan	Huygens (1655)	1222	15.95	5150	1346	1.9
	Hyperion	Bond, Lassell (1848)	1481	21.3	270	—	—
	Iapetus	Cassini (1671)	3561	79.3	1435	19	1.2
	Phoebe	Pickering (1898)	12,950	550 (R)	220	—	—
Uranus	Cordelia	Voyager (1986)	49.8	0.34	40?	—	—
	Ophelia	Voyager (1986)	53.8	0.38	50?	—	—
	Bianca	Voyager (1986)	59.2	0.44	50?	—	—
	Cressida	Voyager (1986)	61.8	0.46	60?	—	—
	Desdemona	Voyager (1986)	62.7	0.48	60?	—	—
	Juliet	Voyager (1986)	64.4	0.50	80?	—	—
	Portia	Voyager (1986)	66.1	0.51	80?	—	—
	Rosalind	Voyager (1986)	69.9	0.56	60?	—	—
	Belinda	Voyager (1986)	75.3	0.63	60?	—	—
	Puck	Voyager (1985)	86.0	0.76	170	—	—

SATELLITES OF THE PLANETS
(Continued)

Planet	Satellite Name	Discovery	Semimajor Axis (km × 1000)	Period (days)	Diameter (km)	Mass (10²⁰ kg)	Density (g/cm³)
	Ariel	Lassell (1851)	191	2.52	1160	13	1.6
	Umbriel	Lassell (1851)	266	4.14	1190	13	1.4
	Titania	Herschel (1787)	436	8.71	1610	35	1.6
	Oberon	Herschel (1787)	583	13.5	1550	29	1.5
Neptune	Naiad	Voyager (1989)	48	0.30	50	—	
	Thalassa	Voyager (1989)	50	0.31	90	—	
	Despina	Voyager (1989)	53	0.33	150	—	
	Galatea	Voyager (1989)	62	0.40	150	—	
	Larissa	Voyager (1989)	74	0.55	200	—	
	Proteus	Voyager (1989)	118	1.12	400	—	
	Triton	Lassell (1846)	355	5.88 (R)	2720	220	2.1
	Nereid	Kuiper (1949)	5511	360	340	—	—
Pluto	Charon	Christy (1978)	19.7	6.39	1200	—	—

TOTAL SOLAR ECLIPSES FROM 1972 THROUGH 2030

Date	Duration of Totality (*min*)	Where Visible
1972 July 10	2.7	Alaska, Northern Canada
1973 June 30	7.2	Atlantic Ocean, Africa
1974 June 20	5.3	Indian Ocean, Australia
1976 Oct. 23	4.9	Africa, Indian Ocean, Australia
1977 Oct. 12	2.8	Northern South America
1979 Feb. 26	2.7	Northwest U.S., Canada
1980 Feb. 16	4.3	Central Africa, India
1981 July 31	2.2	Siberia
1983 June 11	5.4	Indonesia
1984 Nov. 22	2.1	Indonesia, South America
1987 March 29	0.3	Central Africa
1988 March 18	4.0	Philippines, Indonesia
1990 July 22	2.6	Finland, Arctic Regions
1991 July 11	7.1	Hawaii, Mexico, Central America, Brazil
1992 June 30	5.4	South Atlantic
1994 Nov. 3	4.6	South America
1995 Oct. 24	2.4	South Asia
1997 March 9	2.8	Siberia, Arctic
1998 Feb. 26	4.4	Central America
1999 Aug. 11	2.6	Central Europe, Central Asia
2001 June 21	4.9	Southern Africa
2002 Dec. 4	2.1	South Africa, Australia
2003 Nov. 23	2.0	Antarctica
2005 April 8	0.7	South Pacific Ocean
2006 March 29	4.1	Africa, Asia Minor, U.S.S.R.
2008 Aug. 1	2.4	Arctic Ocean, Siberia, China
2009 July 22	6.6	India, China, South Pacific
2010 July 11	5.3	South Pacific Ocean
2012 Nov. 13	4.0	Northern Australia, South Pacific
2013 Nov. 3	1.7	Atlantic Ocean, Central Africa
2015 March 20	4.1	North Atlantic, Arctic Ocean
2016 March 9	4.5	Indonesia, Pacific Ocean
2017 Aug. 21	2.7	Pacific Ocean, U.S.A., Atlantic Ocean
2019 July 2	4.5	South Pacific, South America
2020 Dec. 14	2.2	South Pacific, South America, South Atlantic Ocean
2021 Dec. 4	1.9	Antarctica
2023 April 20	1.3	Indian Ocean, Indonesia
2024 April 8	4.5	South Pacific, Mexico, East U.S.A.
2026 Aug. 12	2.3	Arctic, Greenland, North Atlantic, Spain
2027 Aug. 2	6.4	North Africa, Arabia, Indian Ocean
2028 July 22	5.1	Indian Ocean, Australia, New Zealand
2030 Nov. 25	3.7	South Africa, Indian Ocean, Australia

THE NEAREST STARS

Star	Distance (LY)	Radial Velocity (km/s)	Spectra of Components			Visual Magnitudes of Components			Visual Luminosities of Components (L_S)		
			A	B	C	A	B	C	A	B	C
Sun			G2V			−26.8			1.0		
Proxima Centauri*	4.3	−16	M5V			+11.05			5.8×10^{-5}		
α Centauri	4.4	−22	G2V	K0V		−0.01	+1.33		1.4	0.44	
Barnard's Star	5.9	−108	M5V			+9.54			4.4×10^{-4}		
Wolf 359	7.7	+13	M8V			+13.53			1.7×10^{-5}		
Lalande 21185	8.2	−84	M2V			+7.50			5.2×10^{-3}		
Luyten 726-8	8.5	+30	M5.5V	M5.5V		+12.45	+12.95		6.3×10^{-5}	4.0×10^{-5}	
Sirius	8.6	−8	A1V	wd		−1.46	+8.68		2.3	1.9×10^{-3}	
Ross 154	9.5	−4	M4.5V			+10.6			4.0×10^{-4}		
Ross 248	10.2	−81	M6V			+12.29			$+1 \times 10^{-4}$		
ε Eridani	10.7	+16	K2V			+3.73			0.30		
Ross 128	10.8	−13	M5V			+11.10			3.3×10^{-4}		
Luyten 789-6	10.8	−60	M6V			−12.18			1.20×10^{-4}		
61 Cygni	11.1	−64	K5V	K7V		+5.22	+6.03		0.076	0.036	
ε Indi	11.2	−40	K5V			+4.68			0.13		
τ Ceti	11.3	−16	G8V			+3.50			0.44		
Procyon	11.4	−3	F51V-V	wd		+0.37	+10.7		7.6	5.2×10^{-4}	
BD + 59°1915	11.5	+5	M4V	M5V		+8.90	+9.69		2.8×10^{-3}	1.4×10^{-3}	
BD + 43°44	11.6	+17	M1V	M6V		+8.07	+11.04		6.3×10^{-3}	4.0×10^{-4}	
CD − 36°15693	11.7	+10	M2V			+7.36			0.012		
G51-15	11.9		MV			+14.8			1.3×10^{-5}		
Luyten 725-32	12.4		M5V			+11.5			3.0×10^{-4}		
BD + 5°1668	12.4	+26	M5V			+9.82			1.3×10^{-3}		
CD − 39°14192	12.6	+21	M0V			+6.67			0.025		
Kapteyn's Star	12.7	+245	M0V			+8.81			4.0×10^{-3}		
Kruger 60	12.8	−26	M3V	M4.5V		+9.85	+11.3		1.4×10^{-3}	4.0×10^{-4}	
Ross 614	13.4	+24	M7V	?		+11.07	+14.8		4.8×10^{-4}	1.6×10^{-5}	
BD − 12°4523	13.7	−13	M5V			+10.12			1.2×10^{-3}		
Wolf 424	13.9	−5	M5.5V	M6V		+13.16	+13.4		8.3×10^{-5}	6.9×10^{-5}	
v. Maanen's Star	14.1	+54	wd			+12.37			1.6×10^{-4}		
CD − 37°15492	14.5	+23	M3V			+8.63			5.8×10^{-3}		
Luyten 1159-16	14.7		M8V			+12.27			2.3×10^{-4}		
BD + 50°1725	15.0	−26	K7V			+6.59			0.040		
CD − 46°11540	15.1		M4V			+9.36			3.3×10^{-3}		
CD − 49°13515	15.2	+8	M3V			+8.67			6.3×10^{-3}		
CD − 44°11909	15.3		M5V			+11.2			6.3×10^{-4}		
BD + 68°946	15.3	−22	M3.5V			+9.15			4.0×10^{-3}		
G158 − 27	15.4		MV			+13.7			6.3×10^{-5}		
G208-44/45	15.5		MV	MV		+13.4	+14.0		8.3×10^{-5}	4.8×10^{-5}	
Ross 780	15.6	+9	M5V			+10.7			1.6×10^{-3}		
40 Eridani	15.7	−43	K0V	wd	M4.5V	+4.43	+9.53	+11.17	0.33	3.0×10^{-3}	6.9×10^{-4}
Luyten 145-141	15.8		wd			+11.44			5.2×10^{-4}		
BD + 20°2465	16.1	+11	M4.5V			+9.43			3.3×10^{-3}		
70 Ophiuchi	16.1	−7	K1V	K5V		+4.2	+6.0		0.44	0.83	
BD + 43°4305	16.3	−2	M4.5V			+10.2			1.7×10^{-3}		

*Proxima Centauri is sometimes considered an outlying member of the α Centauri system.

Adapted from data supplied by the U.S. Naval Observatory.

THE TWENTY BRIGHTEST STARS

Star	Right Ascension (1950) (h)	(m)	Declination (1950) °	'	Distance* (pc)	Proper Motion (arcsec/yr)	Spectra of Components A	B	C	Visual Magnitudes of Components A	B	C	Absolute Visual Magnitudes of Components A	B	C
Sirius	6	42.9	−16	39	2.7	1.33	A1V	wd		−1.46	+8.7		+1.4	+11.6	
Canopus	6	22.8	−52	40	30	0.02	F01b-II			−0.72			−3.1		
α Centauri	14	36.2	−60	38	1.3	3.68	G2V	K0V		−0.01	+1.3		+4.4	+5.7	
Arcturus	14	13.4	+19	27	11	2.28	K2IIIp			−0.06			−0.3		
Vega	18	35.2	+38	44	8.0	0.34	A0V			+0.04			+0.5		
Capella	5	13.0	+45	57	14	0.44	GIII	M1V	M5V	+0.05	+10.2	+13.7	−0.7	+9.5	+13
Rigel	5	12.1	−8	15	250	0.00	B8 Ia	B9		+0.14	+6.6		−6.8	−0.4	
Procyon	7	36.7	+5	21	3.5	1.25	F5IV-V	wd		+0.37	+10.7		+2.6	+13.0	
Betelgeuse	5	52.5	+7	24	150	0.03	M2Iab			+0.41v			−5.5		
Achernar	1	35.9	−57	29	20	0.10	B5V			+0.51			−1.0		
β Centauri	14	00.3	−60	08	90	0.04	B1III	?		+0.63	+4		−4.1	−0.8	
Altair	19	48.3	+8	44	5.1	0.66	A7IV-V			+0.77			+2.2		
α Crucis	12	23.8	−62	49	120	0.04	B1IV	B3		+1.39	+1.9		−4.0	−3.5	
Aldebaran	4	33.0	+16	25	16	0.20	K5III	M2V		+0.86	+13		−0.2	+12	
Spica	13	22.6	−10	54	80	0.05	B1V			+0.91v			−3.6		
Antares	16	26.3	−26	19	120	0.03	MIb	B4eV		+0.92v	+5.1		−4.5	−0.3	
Pollux	7	42.3	+28	09	12	0.62	KOIII			+1.16			+0.8		
Fomalhaut	22	54.9	−29	53	7.0	0.37	A3V	K4V		+1.19	+6.5		+2.0	+7.3	
Deneb	20	39.7	+45	06	430	0.00	A2Ia			+1.26			−6.9		
β Crucis	12	44.8	−59	24	150	0.05	B0.5IV			+1.28v			−4.6		

*Distances of the more remote stars have been estimated from their spectral types and apparent magnitudes, and are only approximate.

Note: Several of the components listed are themselves spectroscopic binaries. A "v" after a magnitude denotes that the star is variable, in which case the magnitude at median light is given. A "p" after a spectral type indicates that the spectrum is peculiar. An "e" after a spectral type indicates that emission lines are present. When the luminosity classification is rather uncertain, a range is given.

VARIABLE STARS

Type of Variable	Kind of Star	Peak Absolute Magnitude	Peak Luminosity (L_S)	Period *(days)*	Description	Example
Cepheids (type I)	F and G supergiants	−2 to −5	10^3 to 10^4	3 to 50	Regular pulsation as a stage in the late evolution of moderately massive stars of solar-type composition. Period-luminosity relation exists.	δ Cep
Cepheids (type II)	F and G supergiants	−3	about 10^3	5 to 30	Regular pulsation as a stage in the late evolution of moderately massive stars depleted in metals. Period-luminosity relation exists.	W Vir
RR Lyrae	A and F giants	0	about 50	0.5 to 1	Very regular, small-amplitude pulsations as a stage in the late evolution of stars depleted in metals.	RR Lyr
Long-period	M red giants	−1 to −5	10^2 to 10^4	80 to 600	Large-amplitude, semiperiodic variations in evolved, luminous red giants that are losing mass. Much of luminosity is in infrared.	o Ceti (Mira)
T Tauri	Young stars G to M	0 to +8	$\frac{1}{20}$ to 50	irregular	Rapid and irregular variations of young stars still embedded in gas and dust from their formation.	T Tau
Novae	O to A binaries	−5 to −8	10^4 to 10^5	—	Eruptive event with ejection of shell due to explosive hydrogen fusion in the atmosphere of one of two binaries exchanging mass. Star brightens in a few days by as much as 10,000 times.	GK Per
Supernovae (type I)	White dwarfs in binary systems	−15 to −20	10^8 to 10^{10}	—	Catastrophic disruption of white dwarf that accretes mass from its binary companion until it collapses. Star brightens in a few days by 10^{10} or more can outshine an entire galaxy at maximum.	—
Supernovae (type II)	Massive red supergiants	−15 to −18	10^8 to 10^9	—	Catastrophic ejection of most of the mass from the collapsing stellar core of a massive, evolved star. Star brightens in a few days by 10^6 or more, can outshine an entire galaxy at maximum.	SN1987A

THE BRIGHTEST MEMBERS OF THE LOCAL GROUP

Galaxy	Type	Right Ascension h m	Declination (Degrees)	Distance (1000 LY)	Absolute Magnitude	Apparent Magnitude	Diameter (1000 LY)
Milky Way	Sbc	17 42	−28		−20.6		100
Andromeda M31; NGC 224	Sb	00 40	+41	2200	−21.6	4.4	130
M33; NGC598	Sc	01 31	+30	2500	−19.1	6.3	60
Large Magellanic Cloud	Irr	05 24	−69	170	−18.4	0.6	30
Small Magellanic Cloud	Irr	00 51	−73	300	−17.0	2.8	25
IC 10	Irr	00 17	+59	4000	−16.2	11.7	
NGC 205	E5pec	00 37	+41	2200	−15.7	8.6	16
M32; NGC 221	E2	00 40	+40	2200	−15.5	9.0	8
NGC 6822	Irr	19 42	−14	1800	−15.1	9.3	9
WLM	Irr	23 59	−15	2000	−15.0	11.3	
IC 5152	Sd	21 59	−51	2000	−14.6	11.7	
NGC 185	E3pec	00 36	+48	2200	−14.6	10.1	8
IC 1613	Irr	01 02	+01	2500	−14.5	10.0	16
NGC 147	E5	00 30	+48	2200	−14.4	10.4	10
Leo A	Irr	09 56	+30	5000	−13.5	12.7	
Pegasus	Irr	23 26	+14	5000	−13.4	12.4	
Fornax	E3	02 37	−34	500	−12.9	8.5	15
GR8	Irr	12 56	+14	4000	−11.0	14.6	1.5
DDO 210	Irr	20 44	−13	3000	−11.0	15.3	
Sagittarius	Irr	19 27	−17	4000	−10.6	15.6	
Sculptor	E3	00 57	−33	230	−10.6	9.1	7
Andromeda I	E3	00 43	+37	2200	−10.6	14.0	1.6
Andromeda III	E5	00 32	+36	2200	−10.6	14.0	0.9
Andromeda II	E2	01 13	+33	2200	−10.6	14.0	2.3
Pisces, LGS 3	Irr	01 01	+21	3000	−9.7	15.5	
Leo I	Irr	10 05	+12	600	−9.6	11.8	5
Leo II	E0	11 10	+22	600	−9.2	12.3	5.2
Ursa Minor	E5	15 08	+67	300	−8.2	11.6	3
Draco	E3	17 19	+57	300	−8.0	12.0	4.5
Carina	E4	06 40	−50	300	>−5.5	>13.0	4.8
Sgr(Anon)	Epec			50			

APPENDIX
17

THE MESSIER CATALOGUE OF NEBULAE
AND STAR CLUSTERS

M	NGC or (IC)	Right Ascension (1980) h	m	Declination (1980) °	'	Apparent Visual Magnitude	Description
1	1952	5	33.3	+22	01	8.4	"Crab" nebula in Taurus; remains of SN 1054
2	7089	21	32.4	−0	54	6.4	Globular cluster in Aquarius
3	5272	13	41.2	+28	29	6.3	Globular cluster in Canes Venatici
4	6121	16	22.4	−26	28	6.5	Globular cluster in Scorpio
5	5904	15	17.5	+2	10	6.1	Globular cluster in Serpens
6	6405	17	38.8	−32	11	5.5	Open cluster in Scorpio
7	6475	17	52.7	−34	48	3.3	Open cluster in Scorpio
8	6523	18	02.4	−24	23	5.1	"Lagoon" nebula in Sagittarius
9	6333	17	18.1	−18	30	8.0	Globular cluster in Ophiuchus
10	6254	16	56.1	−4	05	6.7	Globular cluster in Ophiuchus
11	6705	18	50.0	−6	18	6.8	Open cluster in Scutum Sobieskii
12	6218	16	46.3	−1	55	6.6	Globular cluster in Ophiuchus
13	6205	16	41.0	+36	30	5.9	Globular cluster in Hercules
14	6402	17	36.6	−3	14	8.0	Globular cluster in Ophiuchus
15	7078	21	28.9	+12	05	6.4	Globular cluster in Pegasus
16	6611	18	17.8	−13	47	6.6	Open cluster with nebulosity in Serpens
17	6618	18	19.6	−16	11	7.5	"Swan" or "Omega" nebula in Sagittarius
18	6613	18	18.7	−17	08	7.2	Open cluster in Sagittarius
19	6273	17	01.4	−26	14	6.9	Globular cluster in Ophiuchus
20	6514	18	01.2	−23	02	8.5	"Trifid" nebula in Sagittarius
21	6531	18	03.4	−22	30	6.5	Open cluster in Sagittarius
22	6656	18	35.2	−23	56	5.6	Globular cluster in Sagittarius
23	6494	17	55.8	−19	00	5.9	Open cluster in Sagittarius
24	6603	18	17.3	−18	26	4.6	Open cluster in Sagittarius
25	(4725)	18	30.5	−19	16	6.2	Open cluster in Sagittarius
26	6694	18	44.1	−9	25	9.3	Open cluster in Scutum Sobieskii
27	6853	19	58.8	+22	40	8.2	"Dumbbell" planetary nebula in Vulpecula
28	6626	18	23.2	−24	52	7.6	Globular cluster in Sagittarius
29	6913	20	23.3	+38	27	8.0	Open cluster in Cygnus
30	7099	21	39.2	−23	16	7.7	Globular cluster in Capricornus
31	224	0	41.6	+41	10	3.5	Andromeda galaxy
32	221	0	41.6	+40	46	8.2	Elliptical galaxy; companion to M31
33	598	1	32.7	+30	33	5.8	Spiral galaxy in Triangulum
34	1039	2	40.7	+42	43	5.8	Open cluster in Perseus
35	2168	6	07.5	+24	21	5.6	Open cluster in Gemini
36	1960	5	35.0	+34	05	6.5	Open cluster in Auriga
37	2099	5	51.1	+32	33	6.2	Open cluster in Auriga
38	1912	5	27.3	+35	48	7.0	Open cluster in Auriga
39	7092	21	31.5	+48	21	5.3	Open cluster in Cygnus
40		12	21	+59			Close double star in Ursa Major
41	2287	6	46.2	−20	43	5.0	Loose open cluster in Canis Major
42	1976	5	34.4	−5	24	4	Orion nebula

M	NGC or (IC)	Right Ascension (1980) h	m	Declination (1980) °		Apparent Visual Magnitude	Description
43	1982	5	34.6	−5	18	9	Northeast portion of Orion nebula
44	2632	8	39	+20	04	3.9	Praesepe; open cluster in Cancer
45		3	46.3	+24	03	1.6	The Pleiades; open cluster in Taurus
46	2437	7	40.9	−14	46	6.6	Open cluster in Puppis
47	2422	7	35.7	−14	26	5	Loose group of stars in Puppis
48	2548	8	12.8	−5	44	6	"Cluster of very small stars"
49	4472	12	28.8	+8	06	8.5	Elliptical galaxy in Virgo
50	2323	7	02.0	−8	19	6.3	Loose open cluster in Monoceros
51	5194	13	29.1	+47	18	8.4	"Whirlpool" spiral galaxy in Canes Venatici
52	7654	23	23.3	+61	30	8.2	Loose open cluster in Cassiopeia
53	5024	13	12.0	+18	16	7.8	Globular cluster in Coma Berenices
54	6715	18	53.8	−30	30	7.8	Globular cluster in Sagittarius
55	6809	19	38.7	−30	59	6.2	Globular cluster in Sagittarius
56	6779	19	15.8	+30	08	8.7	Globular cluster in Lyra
57	6720	18	52.8	+33	00	9.0	"Ring" nebula; planetary nebula in Lyra
58	4579	12	36.7	+11	55	9.9	Spiral galaxy in Virgo
59	4621	12	41.0	+11	46	10.0	Spiral galaxy in Virgo
60	4649	12	42.6	+11	40	9.0	Elliptical galaxy in Virgo
61	4303	12	20.8	+4	35	9.6	Spiral galaxy in Virgo
62	6266	16	59.9	−30	05	6.6	Globular cluster in Scorpio
63	5055	13	14.8	+42	07	8.9	Spiral galaxy in Canes Venatici
64	4826	12	55.7	+21	39	8.5	Spiral galaxy in Coma Berenices
65	3623	11	17.9	+13	12	9.4	Spiral galaxy in Leo
66	3627	11	19.2	+13	06	9.0	Spiral galaxy in Leo; companion to M65
67	2682	8	50.0	+11	53	6.1	Open cluster in Cancer
68	4590	12	38.4	−26	39	8.2	Globular cluster in Hydra
69	6637	18	30.1	−32	23	8.0	Globular cluster in Sagittarius
70	6681	18	42.0	−32	18	8.1	Globular cluster in Sagittarius
71	6838	19	52.8	+18	44	7.6	Globular cluster in Sagittarius
72	6981	20	52.3	−12	38	9.3	Globular cluster in Aquarius
73	6994	20	57.8	−12	43	9.1	Open cluster in Aquarius
74	628	1	35.6	+15	41	9.3	Spiral galaxy in Pisces
75	6864	20	04.9	−21	59	8.6	Globular cluster in Sagittarius
76	650	1	41.0	+51	28	11.4	Planetary nebula in Perseus
77	1068	2	41.6	−0	04	8.9	Spiral galaxy in Cetus
78	2068	5	45.7	0	03	8.3	Small emission nebula in Orion
79	1904	5	23.3	−24	32	7.5	Globular cluster in Lepus
80	6093	16	15.8	−22	56	7.5	Globular cluster in Scorpio
81	3031	9	54.2	+69	09	7.0	Spiral galaxy in Ursa Major
82	3034	9	54.4	+69	47	8.4	Irregular galaxy in Ursa Major
83	5236	13	35.4	−29	31	7.6	Spiral galaxy in Hydra
84	4374	12	24.1	+13	00	9.4	Elliptical galaxy in Virgo
85	4382	12	24.3	+18	18	9.3	Elliptical galaxy in Coma Berenices
86	4406	12	25.1	+13	03	9.2	Elliptical galaxy in Virgo
87	4486	12	29.7	+12	30	8.7	Elliptical galaxy in Virgo
88	4501	12	30.9	+14	32	9.5	Spiral galaxy in Coma Berenices
89	4552	12	34.6	+12	40	10.3	Elliptical galaxy in Virgo
90	4569	12	35.8	+13	16	9.6	Spiral galaxy in Virgo
91	omitted						
92	6341	17	16.5	+43	10	6.4	Globular cluster in Hercules
93	2447	7	43.7	−23	49	6.5	Open cluster in Puppis
94	4736	12	50.0	+41	14	8.3	Spiral galaxy in Canes Venatici
95	3351	10	42.9	+11	49	9.8	Barred spiral galaxy in Leo
96	3368	10	45.7	+11	56	9.3	Spiral galaxy in Leo
97	3587	11	13.7	+55	07	11.1	"Owl" nebula; planetary nebula in Ursa Major
98	4192	12	12.7	+15	01	10.2	Spiral galaxy in Coma Berenices
99	4254	12	17.8	+14	32	9.9	Spiral galaxy in Coma Berenices
100	4321	12	21.9	+15	56	9.4	Spiral galaxy in Coma Berenices
101	5457	14	02.5	+54	27	7.9	Spiral galaxy in Ursa Major
102	5866(?)	15	05.9	+55	50	10.5	Spiral galaxy (identification as M102; in doubt)
103	581	1	31.9	+60	35	6.9	Open cluster in Cassiopeia

M	NGC or (IC)	Right Ascension (1980)		Declination (1980)		Apparent Visual Magnitude	Description
		h	m	°	'		
104*	4594	12	39.0	−11	31	8.3	Spiral galaxy in Virgo
105*	3379	10	46.8	+12	51	9.7	Elliptical galaxy in Leo
106*	4258	12	18.0	+47	25	8.4	Spiral galaxy in Canes Venatici
107*	6171	16	31.4	−13	01	9.2	Globular cluster in Ophiuchus
108*	3556	11	10.5	+55	47	10.5	Spiral galaxy in Ursa Major
109*	3992	11	56.6	+53	29	10.0	Spiral galaxy in Ursa Major
110*	205	0	39.2	+41	35	9.4	Elliptical galaxy (companion to M31)

*Not in Messier's original (1781) list; added later by others.

18

THE CHEMICAL ELEMENTS

Element	Symbol	Atomic Number	Atomic Weight* (Chemical Scale)	Number of Atoms per 10^{12} Hydrogen Atoms†
Hydrogen	H	1	1.0080	1×10^{12}
Helium	He	2	4.003	8×10^{10}
Lithium	Li	3	6.940	2×10^{3}
Beryllium	Be	4	9.013	3×10^{1}
Boron	B	5	10.82	9×10^{2}
Carbon	C	6	12.011	4.5×10^{8}
Nitrogen	N	7	14.008	9.2×10^{7}
Oxygen	O	8	16.0000	7.4×10^{8}
Fluorine	F	9	19.00	3.1×10^{4}
Neon	Ne	10	20.183	1.3×10^{8}
Sodium	Na	11	22.991	2.1×10^{6}
Magnesium	Mg	12	24.32	4.0×10^{7}
Aluminum	Al	13	26.98	3.1×10^{6}
Silicon	Si	14	28.09	3.7×10^{7}
Phosphorus	P	15	30.975	3.8×10^{5}
Sulfur	S	16	32.066	1.9×10^{7}
Chlorine	Cl	17	35.457	1.9×10^{5}
Argon	Ar(A)	18	39.944	3.8×10^{6}
Potassium	K	19	39.100	1.4×10^{5}
Calcium	Ca	20	40.08	2.2×10^{6}
Scandium	Sc	21	44.96	1.3×10^{3}
Titanium	Ti	22	47.90	8.9×10^{4}
Vanadium	V	23	50.95	1.0×10^{4}
Chromium	Cr	24	52.01	5.1×10^{5}
Manganese	Mn	25	54.94	3.5×10^{5}
Iron	Fe	26	55.85	3.2×10^{7}
Cobalt	Co	27	58.94	8.3×10^{4}
Nickel	Ni	28	58.71	1.9×10^{6}
Copper	Cu	29	63.54	1.9×10^{4}
Zinc	Zn	30	65.38	4.7×10^{4}
Gallium	Ga	31	69.72	1.4×10^{3}
Germanium	Ge	32	72.60	4.4×10^{3}
Arsenic	As	33	74.91	2.5×10^{2}
Selenium	Se	34	78.96	2.3×10^{3}
Bromine	Br	35	79.916	4.4×10^{2}
Krypton	Kr	36	83.80	1.7×10^{3}
Rubidium	Rb	37	85.48	2.6×10^{2}
Strontium	Sr	38	87.63	8.8×10^{2}
Yttrium	Y	39	88.92	2.5×10^{2}
Zirconium	Zr	40	91.22	4.0×10^{2}
Niobium (Columbium)	Nb(Cb)	41	92.91	2.6×10^{1}
Molybdenum	Mo	42	95.95	9.3×10^{1}
Technetium	Tc(Ma)	43	(99)	—
Ruthenium	Ru	44	101.1	68
Rhodium	Rh	45	102.91	13
Palladium	Pd	46	106.4	51
Silver	Ag	47	107.880	20
Cadmium	Cd	48	112.41	63

Element	Symbol	Atomic Number	Atomic Weight* (Chemical Scale)	Number of Atoms per 10^{12} Hydrogen Atoms†
Indium	In	49	114.82	75
Tin	Sn	50	118.70	1.4×10^2
Antimony	Sb	51	121.76	13
Tellurium	Te	52	127.61	1.8×10^2
Iodine	I(J)	53	126.91	33
Xenon	Xe(X)	54	131.30	1.6×10^2
Cesium	Cs	55	132.91	14
Barium	Ba	56	137.36	1.6×10^2
Lanthanum	La	57	138.92	17
Cerium	Ce	58	140.13	43
Praseodymium	Pr	59	140.92	6
Neodymium	Nd	60	144.27	31
Promethium	Pm	61	(147)	—
Samarium	Sm(Sa)	62	150.35	10
Europium	Eu	63	152.0	4
Gadolinium	Gd	64	157.26	13
Terbium	Tb	65	158.93	2
Dysprosium	Dy(Ds)	66	162.51	15
Holmium	Ho	67	164.94	3
Erbium	Er	68	167.27	9
Thulium	Tm(Tu)	69	168.94	2
Ytterbium	Yb	70	173.04	8
Lutecium	Lu(Cp)	71	174.99	2
Hafnium	Hf	72	178.50	6
Tantalum	Ta	73	180.95	1
Tungsten	W	74	183.86	5
Rhenium	Re	75	186.22	2
Osmium	Os	76	190.2	27
Iridium	Ir	77	192.2	24
Platinum	Pt	78	195.09	56
Gold	Au	79	197.0	6
Mercury	Hg	80	200.61	19
Thallium	Tl	81	204.39	8
Lead	Pb	82	207.21	1.2×10^2
Bismuth	Bi	83	209.00	5
Polonium	Po	84	(209)	—
Astatine	At	85	(210)	—
Radon	Rn	86	(222)	—
Francium	Fr(Fa)	87	(223)	—
Radium	Ra	88	226.05	—
Actinium	Ac	89	(227)	—
Thorium	Th	90	232.12	1
Protactinium	Pa	91	(231)	—
Uranium	U(Ur)	92	238.07	1
Neptunium	Np	93	(237)	—
Plutonium	Pu	94	(244)	—
Americium	Am	95	(243)	—
Curium	Cm	96	(248)	—
Berkelium	Bk	97	(247)	—
Californium	Cf	98	(251)	—
Einsteinium	E	99	(254)	—
Fermium	Fm	100	(253)	—
Mendeleevium	Mv	101	(256)	—
Nobelium	No	102	(253)	—

* Where mean atomic weights have not been well determined, the atomic mass numbers of the most stable isotopes are given in parentheses.

† Provided by L. H. Aller.

THE CONSTELLATIONS

Constellation (Latin name)	Genitive Case Ending	English Name or Description	Abbre-viation	Approximate Position α h	δ °
Andromeda	Andromedae	Princess of Ethiopia	And	1	+40
Antila	Antilae	Air pump	Ant	10	−35
Apus	Apodis	Bird of Paradise	Aps	16	−75
Aquarius	Aquarii	Water bearer	Aqr	23	−15
Aquila	Aquilae	Eagle	Aql	20	+5
Ara	Arae	Altar	Ara	17	−55
Aries	Arietis	Ram	Ari	3	+20
Auriga	Aurigae	Charioteer	Aur	6	+40
Boötes	Boötis	Herdsman	Boo	15	+30
Caelum	Caeli	Graving tool	Cae	5	−40
Camelopardus	Camelopardis	Giraffe	Cam	6	+70
Cancer	Cancri	Crab	Cnc	9	+20
Canes Venatici	Canum Venaticorum	Hunting dogs	CVn	13	+40
Canis Major	Canis Majoris	Big dog	CMa	7	−20
Canis Minor	Canis Minoris	Little dog	CMi	8	+5
Capricornus	Capricorni	Sea goat	Cap	21	−20
Carina*	Carinae	Keel of Argonauts' ship	Car	9	−60
Cassiopeia	Cassiopeiae	Queen of Ethiopia	Cas	1	+60
Centaurus	Centauri	Centaur	Cen	13	−50
Cepheus	Cephei	King of Ethiopia	Cep	22	+70
Cetus	Ceti	Sea monster (whale)	Cet	2	−10
Chamaeleon	Chamaeleontis	Chameleon	Cha	11	−80
Circinus	Circini	Compasses	Cir	15	−60
Columba	Columbae	Dove	Col	6	−35
Coma Berenices	Comae Berenices	Berenice's hair	Com	13	+20
Corona Australis	Coronae Australis	Southern crown	CrA	19	−40
Corona Borealis	Coronae Borealis	Northern crown	CrB	16	+30
Corvus	Corvi	Crow	Crv	12	−20
Crater	Crateris	Cup	Crt	11	−15
Crux	Crucis	Cross (southern)	Cru	12	−60
Cygnus	Cygni	Swan	Cyg	21	+40
Delphinus	Delphini	Porpoise	Del	21	+10
Dorado	Doradus	Swordfish	Dor	5	−65
Draco	Draconis	Dragon	Dra	17	+65
Equuleus	Equulei	Little horse	Equ	21	+10
Eridanus	Eridani	River	Eri	3	−20
Fornax	Fornacis	Furnace	For	3	−30
Gemini	Geminorum	Twins	Gem	7	+20
Grus	Gruis	Crane	Gru	22	−45
Hercules	Herculis	Hercules, son of Zeus	Her	17	+30
Horologium	Horologii	Clock	Hor	3	−60
Hydra	Hydrae	Sea serpent	Hya	10	−20
Hydrus	Hydri	Water snake	Hyi	2	−75
Indus	Indi	Indian	Ind	21	−55
Lacerta	Lacertae	Lizard	Lac	22	+45
Leo	Leonis	Lion	Leo	11	+15

Constellation (Latin name)	Genitive Case Ending	English Name or Description	Abbreviation	Approximate Position	
				α h	δ °
Leo Minor	Leonis Minoris	Little lion	LMi	10	+35
Lepus	Leporis	Hare	Lep	6	−20
Libra	Librae	Balance	Lib	15	−15
Lupus	Lupi	Wolf	Lup	15	−45
Lynx	Lyncis	Lynx	Lyn	8	+45
Lyra	Lyrae	Lyre or harp	Lyr	19	+40
Mensa	Mensae	Table Mountain	Men	5	−80
Microscopium	Microscopii	Microscope	Mic	21	−35
Monoceros	Monocerotis	Unicorn	Mon	7	−5
Musca	Muscae	Fly	Mus	12	−70
Norma	Normae	Carpenter's level	Nor	16	−50
Octans	Octantis	Octant	Oct	22	−85
Ophiuchus	Ophiuchi	Holder of serpent	Oph	17	0
Orion	Orionis	Orion, the hunter	Ori	5	+5
Pavo	Pavonis	Peacock	Pav	20	−65
Pegasus	Pegasi	Pegasus, the winged horse	Peg	22	+20
Perseus	Persei	Perseus, hero who saved Andromeda	Per	3	+45
Phoenix	Phoenicis	Phoenix	Phe	1	−50
Pictor	Pictoris	Easel	Pic	6	−55
Pisces	Piscium	Fishes	Psc	1	+15
Piscis Austrinus	Piscis Austrini	Southern fish	PsA	22	−30
Puppis*	Puppis	Stern of the Argonauts' ship	Pup	8	−40
Pyxis* (= Malus)	Pyxidus	Compass on the Argonauts' ship	Pyx	9	−30
Reticulum	Reticuli	Net	Ret	4	−60
Sagitta	Sagittae	Arrow	Sge	20	+10
Sagittarius	Sagittarii	Archer	Sgr	19	−25
Scorpius	Scorpii	Scorpion	Sco	17	−40
Sculptor	Sculptoris	Sculptor's tools	Scl	0	−30
Scutum	Scuti	Shield	Sct	19	−10
Serpens	Serpentis	Serpent	Ser	17	0
Sextans	Sextantis	Sextant	Sex	10	0
Taurus	Tauri	Bull	Tau	4	+15
Telescopium	Telescopii	Telescope	Tel	19	−50
Triangulum	Trianguli	Triangle	Tri	2	+30
Triangulum Australe	Trianguli Australis	Southern triangle	TrA	16	−65
Tucana	Tucanae	Toucan	Tuc	0	−65
Ursa Major	Ursae Majoris	Big bear	UMa	11	+50
Ursa Minor	Ursae Minoris	Little bear	VMi	15	+70
Vela*	Velorum	Sail of the Argonauts' ship	Vel	9	−50
Virgo	Virginis	Virgin	Vir	13	0
Volans	Volantis	Flying fish	Vol	8	−70
Vulpecula	Vulpeculae	Fox	Vul	20	+25

*The four constellations Carina, Puppis, Pyxis, and Vela originally formed the single constellation, Argo Navis.

INDEX

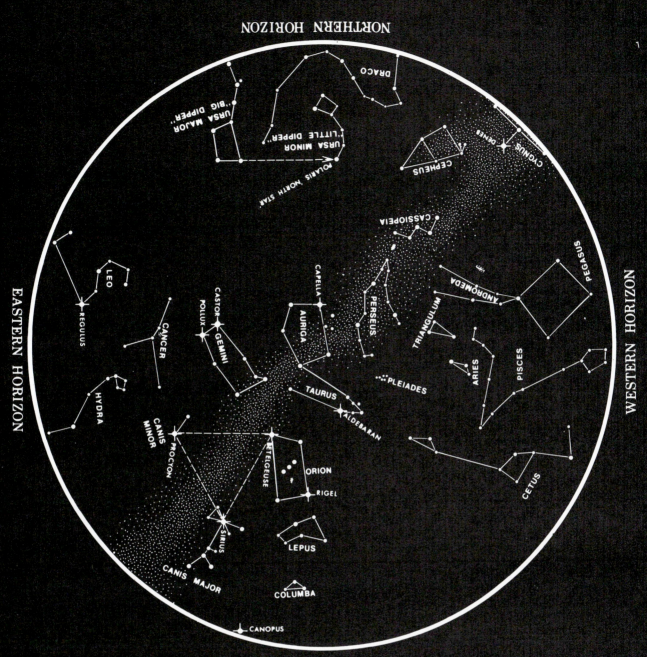

THE NIGHT SKY IN JANUARY

Latitude of chart is 34°N, but it is practical throughout the continental United States.

To use: Hold chart vertically and turn it so the direction you are facing shows at the bottom.

Chart time (Local Standard):

10 p.m. First of month

9 p.m. Middle of month

8 p.m. Last of month

Star Chart from GRIFFITH OBSERVER, Griffith Observatory, Los Angeles

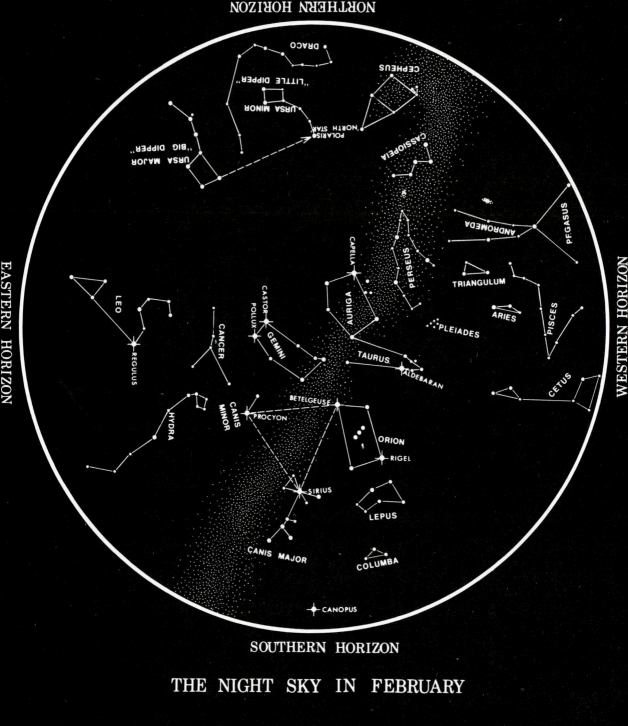

THE NIGHT SKY IN FEBRUARY

Latitude of chart is 34°N, but it is practical throughout the continental United States.

To use: Hold chart vertically and turn it so the direction you are facing shows at the bottom.

Chart time (Local Standard):

10 p.m. First of month

9 p.m. Middle of month

8 p.m. Last of month

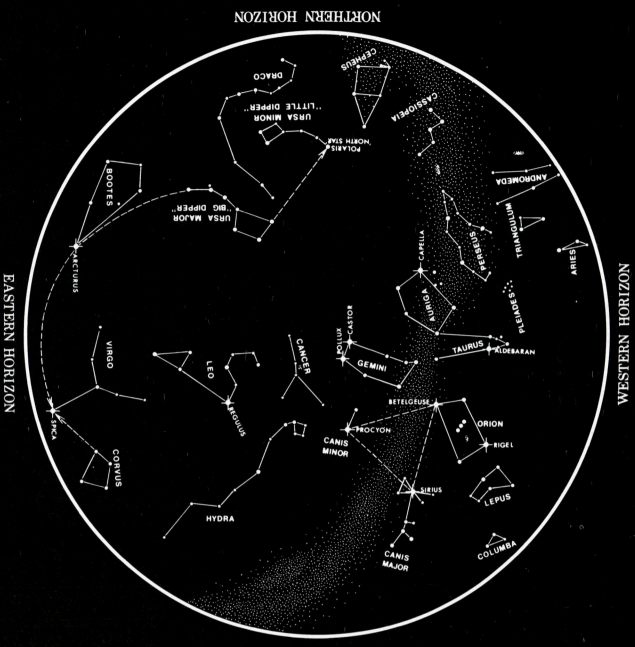

NORTHERN HORIZON

CEPHEUS
DRACO
CASSIOPEIA
URSA MINOR
"LITTLE DIPPER"
POLARIS "NORTH STAR"
ANDROMEDA
BOOTES
URSA MAJOR "BIG DIPPER"
TRIANGULUM
ARCTURUS
PERSEUS
ARIES
CAPELLA
AURIGA
PLEIADES
VIRGO
CASTOR
POLLUX
TAURUS
ALDEBARAN
LEO
CANCER
GEMINI
REGULUS
BETELGEUSE
ORION
SPICA
PROCYON
RIGEL
CORVUS
CANIS MINOR
LEPUS
HYDRA
SIRIUS
COLUMBA
CANIS MAJOR

EASTERN HORIZON

WESTERN HORIZON

SOUTHERN HORIZON

THE NIGHT SKY IN MARCH

Latitude of chart is 34° N, but it is
practical throughout the continental
United States.

To use: Hold chart vertically and turn
it so the direction you are facing
shows at the bottom.

Chart time (Local Standard):

10 p.m. First of month

9 p.m. Middle of month

8 p.m. Last of month

Star Chart from *GRIFFITH OBSERVER*, Griffith Observatory, Los Angeles

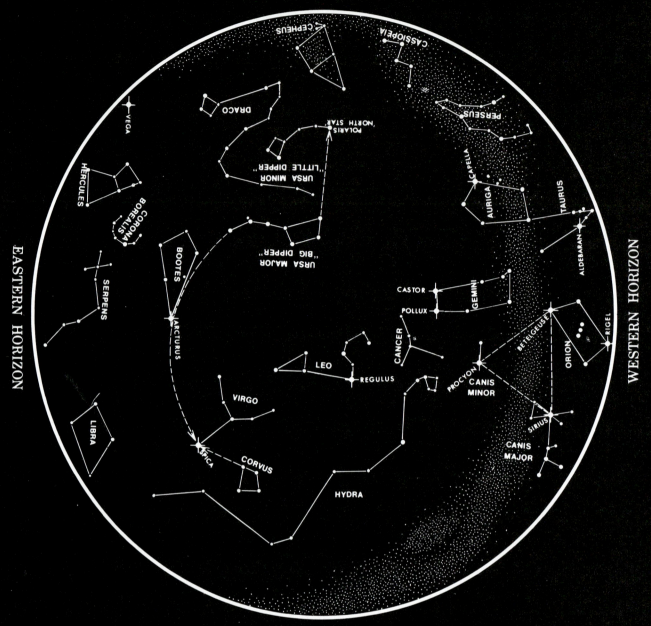

THE NIGHT SKY IN APRIL

Latitude of chart is 34°N, but it is
practical throughout the continental
United States.

To use: Hold chart vertically and turn
it so the direction you are facing
shows at the bottom.

Chart time (Local Standard):

10 p.m. First of month

9 p.m. Middle of month

8 p.m. Last of month

Star Chart from *GRIFFITH OBSERVER*, Griffith Observatory, Los Angeles

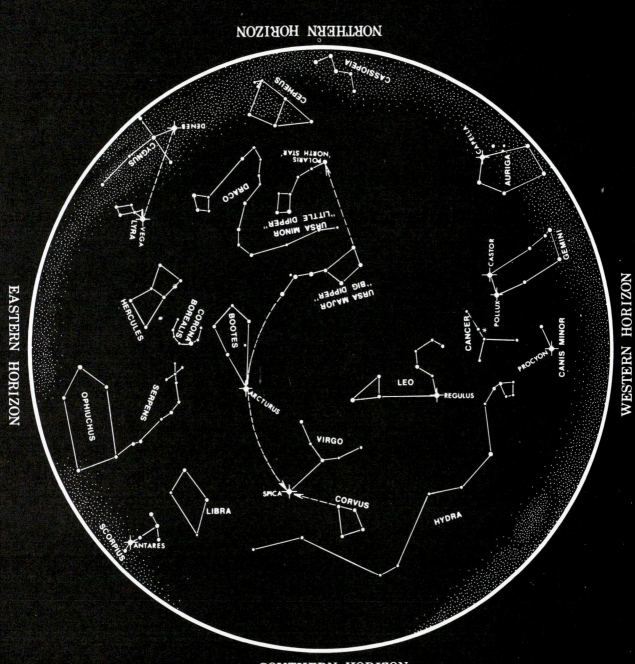

NORTHERN HORIZON

EASTERN HORIZON

WESTERN HORIZON

SOUTHERN HORIZON

THE NIGHT SKY IN MAY

Latitude of chart is 34°N, but it is
practical throughout the continental
United States.

To use: Hold chart vertically and turn
it so the direction you are facing
shows at the bottom.

Chart time (Local Standard):

10 p.m. First of month

9 p.m. Middle of month

8 p.m. Last of month

Star Chart from *GRIFFITH OBSERVER*, Griffith Observatory, Los Angeles

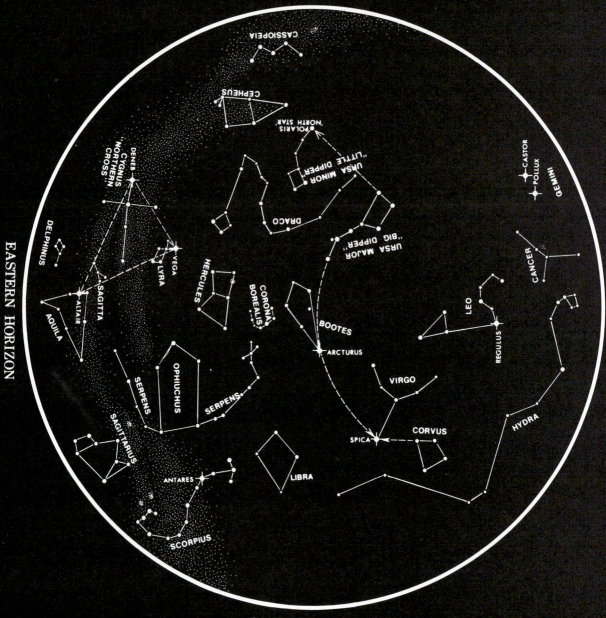

NORTHERN HORIZON

EASTERN HORIZON

WESTERN HORIZON

SOUTHERN HORIZON

THE NIGHT SKY IN JUNE

Latitude of chart is 34°N, but it is
practical throughout the continental
United States.

To use: Hold chart vertically and turn
it so the direction you are facing
shows at the bottom.

Chart time (Local Standard):

10 p.m. First of month

9 p.m. Middle of month

8 p.m. Last of month

Star Chart from *GRIFFITH OBSERVER*, Griffith Observatory, Los Angeles

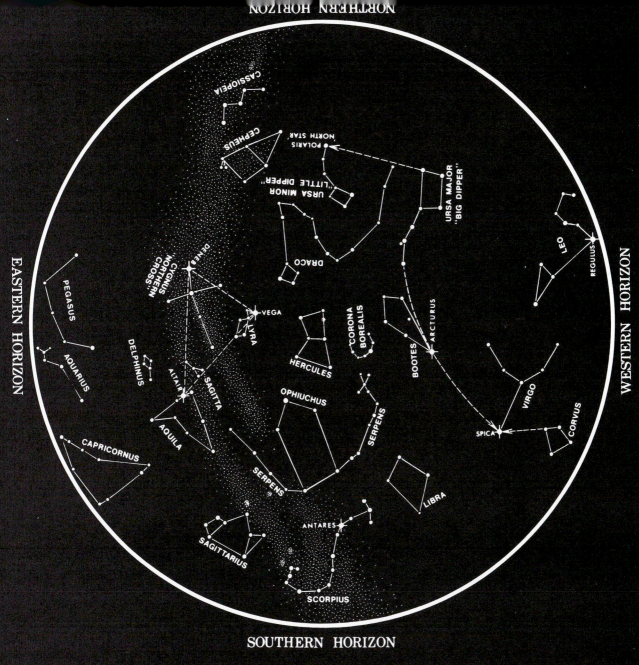

SOUTHERN HORIZON

THE NIGHT SKY IN JULY

Latitude of chart is 34°N, but it is
practical throughout the continental
United States.

To use: Hold chart vertically and turn
it so the direction you are facing
shows at the bottom.

Chart time (Local Standard):

10 p.m. First of month

9 p.m. Middle of month

8 p.m. Last of month

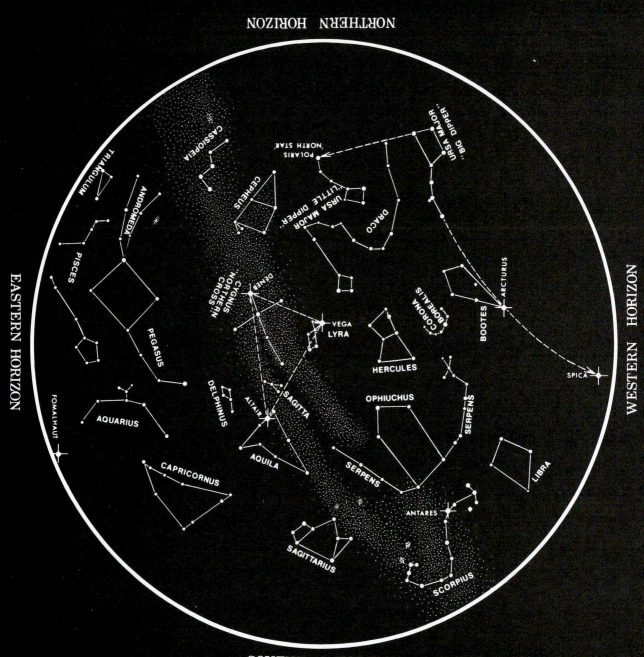

NORTHERN HORIZON

EASTERN HORIZON

WESTERN HORIZON

SOUTHERN HORIZON

THE NIGHT SKY IN AUGUST

Latitude of chart is 34°N, but it is practical throughout the continental United States.

To use: Hold chart vertically and turn it so the direction you are facing shows at the bottom.

Chart time (Local Standard):

10 p.m. First of month

9 p.m. Middle of month

8 p.m. Last of month

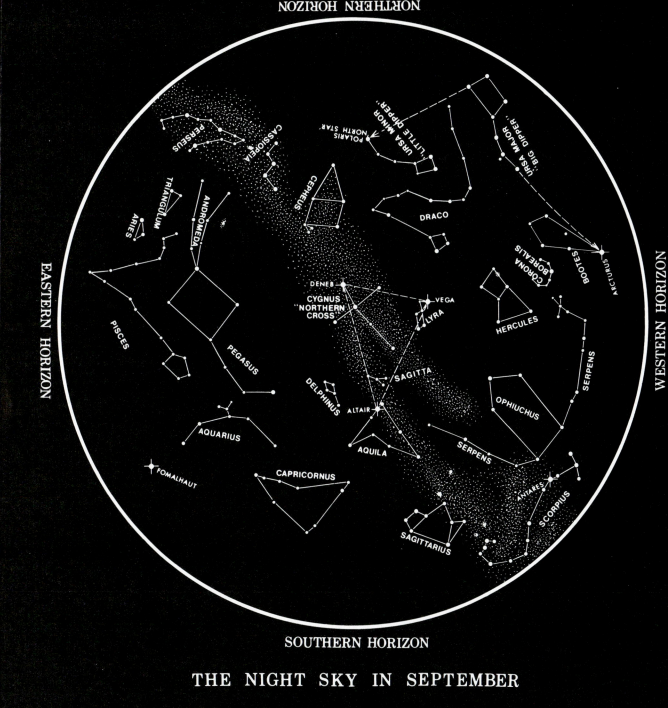

NORTHERN HORIZON

EASTERN HORIZON

WESTERN HORIZON

PERSEUS
CASSIOPEIA
TRIANGULUM
ARIES
ANDROMEDA
CEPHEUS
URSA MINOR "LITTLE DIPPER"
POLARIS "NORTH STAR"
URSA MAJOR "BIG DIPPER"
DRACO
BOÖTES
ARCTURUS
CORONA BOREALIS
PISCES
PEGASUS
DENEB
CYGNUS "NORTHERN CROSS"
VEGA
LYRA
HERCULES
SERPENS
AQUARIUS
DELPHINUS
SAGITTA
OPHIUCHUS
ALTAIR
AQUILA
SERPENS
FOMALHAUT
CAPRICORNUS
ANTARES
SCORPIUS
SAGITTARIUS

SOUTHERN HORIZON

THE NIGHT SKY IN SEPTEMBER

Latitude of chart is 34°N, but it is
practical throughout the continental
United States.

To use: Hold chart vertically and turn
it so the direction you are facing
shows at the bottom.

Chart time (Local Standard):

10 p.m. First of month

9 p.m. Middle of month

8 p.m. Last of month

Star Chart from *GRIFFITH OBSERVER*, Griffith Observatory, Los Angeles

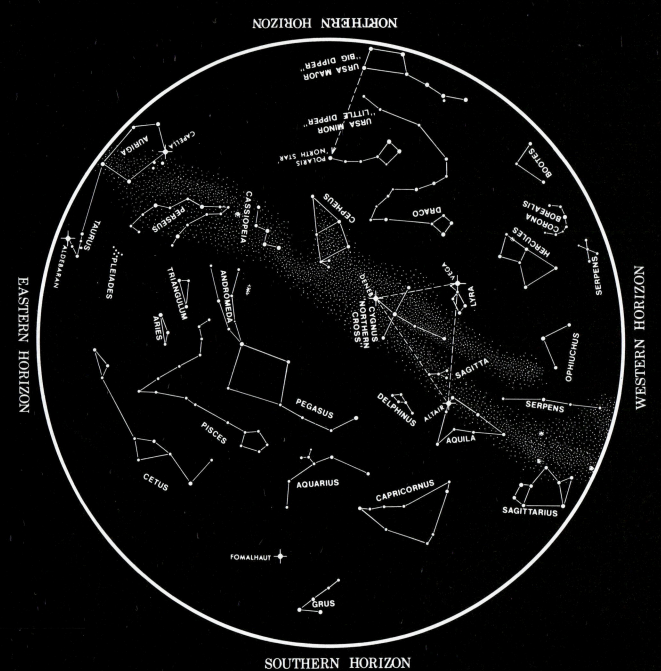

THE NIGHT SKY IN OCTOBER

Latitude of chart is 34°N, but it is
practical throughout the continental
United States.

To use: Hold chart vertically and turn
it so the direction you are facing
shows at the bottom.

Chart time (Local Standard):

10 p.m. First of month

9 p.m. Middle of month

8 p.m. Last of month

Star Chart from GRIFFITH OBSERVER, Griffith Observatory, Los Angeles

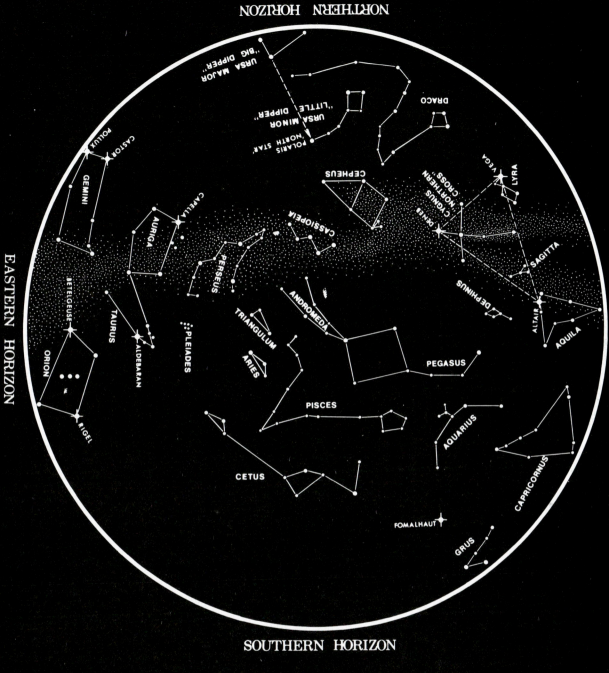

THE NIGHT SKY IN NOVEMBER

Latitude of chart is 34°N, but it is practical throughout the continental United States.

To use: Hold chart vertically and turn it so the direction you are facing shows at the bottom.

Chart time (Local Standard):

10 p.m. First of month

9 p.m. Middle of month

8 p.m. Last of month

Star Chart from GRIFFITH OBSERVER, Griffith Observatory, Los Angeles

NORTHERN HORIZON

EASTERN HORIZON

WESTERN HORIZON

SOUTHERN HORIZON

THE NIGHT SKY IN DECEMBER

Latitude of chart is 34°N, but it is
practical throughout the continental
United States.

To use: Hold chart vertically and turn
it so the direction you are facing
shows at the bottom.

Chart time (Local Standard):
10 p.m. First of month
9 p.m. Middle of month
8 p.m. Last of month

Star Chart from *GRIFFITH OBSERVER*, Griffith Observatory, Los Angeles

THE PLANETS AND LARGE SATELLITES

Name	Semimajor Axis (AU)	Diameter (Earth = 1)	Mass (Earth = 1)	Density (g/cm^3)
Mercury	0.4	0.38	0.06	5.4
Venus	0.7	0.95	0.82	5.3
Earth	1.0	1.00	1.00	5.5
Moon	—	0.27	0.01	3.3
Mars	1.5	0.53	0.11	3.9
Jupiter	5.2	11.2	318.	1.3
Callisto	—	0.38	0.02	1.9
Ganymede	—	0.41	0.02	1.9
Europa	—	0.25	0.01	3.0
Io	—	0.29	0.01	3.6
Saturn	9.5	9.5	94.	0.7
Titan	—	0.40	0.02	1.9
Uranus	19.2	4.0	15.	1.2
Neptune	30.1	3.9	17.	1.5
Triton	—	0.21	0.003	2.1
Pluto	39.4	0.17	0.002	2.0